Communications and Control Engineering

Springer
London
Berlin
Heidelberg
New York
Barcelona
Hong Kong
Milan
Paris
Santa Clara
Singapore
Tokyo

Published titles include:

Parametrizations in Control, Estimation and Filtering Problems: Accuracy Aspects
M. Gevers and G. Li

Loop Transfer Recovery: Analysis and Design
A. Saberi, B.M. Chen and P. Sannuti

Markov Chains and Stochastic Stability
S.P. Meyn and R.L. Tweedie

Robust Control: Systems with Uncertain Physical Parameters
J. Ackermann in co-operation with A. Bartlett, D. Kaesbauer, W. Sienel and R. Steinhauser

Optimal Sampled-Data Control Systems
Tongwen Chen and Bruce Francis

Nonlinear Control Systems (3rd edition)
Alberto Isidori

Theory of Robot Control
C. Canudas de Wit, B. Siciliano and G. Bastin (Eds)

Fundamental Limitations in Filtering and Control
María M. Seron, Julio Braslavsky and Graham C. Goodwin

Constructive Nonlinear Control
R. Sepulchre, M. Janković and P.V. Kokotović

A Theory of Learning and Generalization
M. Vidyasagar

Adaptive Control
I.D. Landau, R. Lozano and M.M'Saad

Stabilization of Nonlinear Uncertain Systems
Miroslav Krstić and Hua Deng

Passivity-based Control of Euler-Lagrange Systems
Romeo Ortega, Antonio Loría, Per Johan Nicklasson and Hebertt Sira-Ramírez

Stability and Stabilization of Infinite Dimensional Systems with Applications
Zheng-Hua Luo, Bao-Zhu Guo and Omer Morgul

Nonsmooth Mechanics (2nd edition)
Bernard Brogliato

Nonlinear Control Systems II
Alberto Isidori

L_2-Gain and Passivity Techniques in Nonlinear Control
Arjan van der Schaft

Ali Saberi, Anton A. Stoorvogel and
Peddapullaiah Sannuti

Control of Linear Systems with Regulation and Input Constraints

With 28 Figures

Springer

A. Saberi, Professor
School of Electrical Engineering and Computer Science, Washington State University, Pullman, Washington 99164-2752, USA

A. Stoorvogel, Professor
Department of Mathematics and Computing Science, Eindhoven University of Technology, PO Box 513, 5600 MB Eindhoven, The Netherlands

P. Sannuti, Professor
Department of Electrical and Computer Engineering, Rutgers, The State University of New Jersey, 94 Brett Road, Piscataway, New Jersey, 08854-8058, USA

Series Editors

E.D. Sontag • M. Thoma

ISBN-13: 978-1-4471-1189-4 e-ISBN-13: 978-1-4471-0727-9
DOI: 10.1007/978-1-4471-0727-9

British Library Cataloguing in Publication Data
A catalogue record for this book is available from the British Library.

Library of Congress Cataloging-in-Publication Data
Saberi, Ali
Control of linear systems with regulation and input constraints / A. Saberi, A. Stoorvogel, and P. Sannuti.
p. cm. - (Communications and control engineering)
ISBN-13: 978-1-4471-1189-4
1. Linear control systems. I. Title. II. Series III. Stoorvogel, Anton. IV. Sannuti, Peddapullaiah, 1941-

TJ220.S23 1999
629.8'32 21—dc21

99-043051

Softcover reprint of the hardcover 1st edition 2000

Typesetting: Camera ready by authors

69/3830-543210 Printed on acid-free paper SPIN 10709923

This book is dedicated to:

Ann, Ingmar, and Ula	(Ali Saberi)
Suseela, Aruna, Mohan, Arun, and Pavan	(Pedda Sannuti)

Preface

Ever since the beginnings of mankind eons ago, the desire to control, regulate, and track even under persistent disturbances has been a dominating influence in the development of human civilization. It is so also in the development of *Automatic Control Theory and its Applications*. The subject of output regulation occupies a central theme in all endeavors of theoreticians and practitioners alike. Yet there is no book or monograph that brings all essential modern developments on output regulation under a single cover. This book is intended to fill this void.

Main topics that are brought together in this book include among others, classical exact output regulation of linear systems along with its different facets of well-posedness, internal model principle, and structural stability; output regulation of linear systems with input amplitude and rate saturation constraints; output regulation with transient performance specifications; performance issues (such as H_2, H_∞, L_1 and others) with an output regulation constraint; generalized output regulation in which the set of tracking signals as well as the set of disturbances that act on the plant are broadened beyond those that are common in classical output regulation, and thus permitting us to deal with exact as well as almost output regulation under a variety of controllers, etc.

This book is designed to meet the needs of a variety of audiences including practicing control engineers, graduate students, and researchers in control engineering. Also, it is written to be suitable for self reading and appropriate as a textbook or prime reference when teaching a first or second year graduate course in control theory. Recommended background for this book is (a) linear algebra and matrix theory, (b) linear differential equations, (c) a course in linear systems and state-space methods, and (d) a course in linear control theory.

No work of this magnitude can be undertaken without any sacrifices. Our families are the ones who sacrificed the time we could have spent with them, thus allowing us to complete this work. We owe a debt of gratitude to our families, and it is natural that we dedicate this book to them.

Ali Saberi
Washington State University, Pullman, Washington, U.S.A.

Anton A. Stoorvogel
Eindhoven University of Technology, Eindhoven, The Netherlands

Peddapullaiah Sannuti
Rutgers University, Piscataway, New Jersey, U.S.A.

This book was typeset by the authors using LaTeX. The simulations and numerical computations were developed in *MATLAB*™.

Contents

Chapter 1

Introduction

1.1 Introduction

The subject of output regulation occupies a central role in modern as well as classical control theory. The basic problem dealt with in output regulation is to design a feedback controller which internally stabilizes a given linear time-invariant plant such that the output of the resulting closed-loop system converges to, or tracks, a certain reference signal of known frequencies in the presence of external disturbances of known frequencies. The reference signals and external disturbances are modeled by a system known as exosystem (or exogenous system). The subject of output regulation has many facets associated with it. These facets include but are not limited to internal model principle, well posedness and structural stability. All these have been the subject of a number of studies during the late sixties, seventies and thereafter. These studies are scattered throughout the literature. In spite of all such work, and in spite of the significance of output regulation which resides in the core of control system engineering, *there exists so far no textbook that deals with all essential aspects of the subject matter*. As such one of our goals is to fill this void. More importantly, we would like to couple the subject of output regulation with several recent developments in modern control theory, e.g. output regulation in the presence of omnipresent amplitude as well as rate saturation of actuators, output regulation with optimal transient behavior of the error signal, and output regulation in the presence of uncertainties and under demanding performance requirements. In this regard we observe that the performance and robustness issues have been providing the basic thrust behind most of the modern control developments during the last two decades.

Let us expand and enumerate below our specific goals and attained results:

(*i*) Several modern developments on the subject of output regulation are scattered throughout the literature. We would like to bring all the significant developments into focus in a single folder. While doing so, we would like to elevate or enhance the subject by generalizing several aspects of it. For instance, most of the available literature on output regulation deals with the case where the measured output of the plant that is used as the input to the feedback controller is simply the tracking error signal, i.e. the difference between the desired and the actual output of the plant. This restriction is often unnecessary. Thus, our first goal is to reexamine the literature on output regulation while utilizing a measured output for feedback which is not necessarily the same as the error signal. Such a reexamination enhances the entire subject including the topics such as internal model principle, well-posedness, and structural stability.

(*ii*) In any practical application, one of the important limiting factors is the saturation of actuators. That is, in practical actuators, the amplitudes and rates of change of signals are limited to certain maximum levels. In the face of such limitations, we would like to reexamine the *output regulation theory* so as to come up with appropriate design of controllers which take into account right at the onset of design the physical limiting characteristics of actuators. The subject of output regulation in the presence of saturating actuators has been thoroughly studied recently in a series of papers by the authors and their coworkers. These studies successfully came up with a variety of sound analysis as well as design methodologies. Thus, our next goal in this book is to integrate into a single folder all such methods of analysis and design for output regulation in the presence of saturating actuators.

(*iii*) One of the shortcomings of classical output regulation is that it gives importance only to steady state tracking error, namely rendering it exactly to zero. The natural engineering issues regarding the transient behavior of the error signal are not addressed at all. Such issues can include minimizing the over-shoot or under-shoot of the error signal, or more generally appropriate shaping of the error signal. In this regard, for instance, one may like to impose in the statement of output regulation problem certain requirements on the transient performance so that one can shape appropriately the transient behavior of tracking error in addition to the desired steady state requirements. One way to do so is to achieve asymptotic tracking while minimizing a certain weighted energy of the error signal. This leads to optimal and suboptimal output regulation problems with the performance measure equal to the energy of the error signal. Several issues related to optimal and suboptimal output regulation are discussed. These issues include among others the determination of an expression for the infimum of the performance measure, solvability conditions for the posed optimal and suboptimal output regulation problems, construction of optimal and suboptimal output regulators, relationships between the optimal performance and structural properties of the given system, perfect output regulation etc.

(*iv*) In modern control, one seeks a desired performance as measured by certain properties of a transfer function from a certain input signal to a certain output signal. A common formulation in this regard is to seek a controller that results either in minimizing a certain norm (H_2, H_∞, and L_1 norms are popular) of the transfer function under focus or in rendering this specified norm of the transfer function under focus less than a priori given value. This leads to formulation of an H_2 or H_∞ or L_1 optimal control problem.

One can unite the objective of achieving the desired performance with the objective of achieving output regulation. Such a unification results in an optimal control problem with output regulation constraint. However, constrained optimal control problems are difficult to solve directly. Here we provide a vehicle by which such problems can be solved both in terms of analyzing them or synthesizing appropriate controllers to solve them. That is, we transform the constrained optimal control problem with output regulation constraint for the given system to an unconstrained optimal control problem for a certain auxiliary system without any output regulation constraint. This is done by constructing an auxiliary system from the given data in such a way that once we have an internally stabilizing controller for the auxiliary system that achieves a certain closed-loop transfer matrix from a certain specified input signal to a certain specified output signal equal to G, then one can easily formulate a controller for the given system that achieves internal stability, achieves asymptotic tracking, and more importantly also renders the closed-loop transfer matrix from a certain specified input signal to a certain specified output signal equal to G. This implies that one can study any optimization problem based on minimizing a certain norm of the closed-loop transfer matrix along with the output regulation constraint for the given system by equivalently studying an unconstrained optimization problem for the auxiliary system. This allows us to formulate the necessary and sufficient conditions under which output regulation as well as a specified performance measure can be obtained for the given system. Also, it allows us to synthesize appropriate controllers.

One fundamental and significant issue that arises in solving an optimization problem with the output regulation constraint is this; does the added output regulation constraint compromise the achievable performance? In this regard, our results show that when the performance is measured by the H_2 norm of a transfer function matrix, there is no loss at all in the achievable performance. On the other hand, when the performance is measured by the H_∞ norm of a transfer function matrix, there is in general a certain loss or decay in the achievable performance. It turns out that a very explicit expression for this decay in performance can be presented.

(v) In the literature, while developing the theory and methods of constructing appropriate controllers, the exosystem is often coupled to a model of the given plant. As mentioned earlier, the exosystem models and generates reference signals as well as external disturbances that act on the given plant. The exosystems considered in the literature are autonomous linear dynamical systems. As a result, the disturbance signals and reference signals generated by such exosystems obviously contain only the frequency components of the exosystems. Thus, the class of reference signals and disturbance signals considered in the traditional approach is severely limited. To mitigate this drawback, and to bring into focus several issues inherent in the context of output regulation, one can model the reference and disturbance signals by the outputs of an exosystem that, unlike the traditional approach, is not autonomous but is driven. Here the reference or disturbance signal is still highly dominated by the fixed frequencies but is nevertheless belongs to a much more general class of signals. This leads us to formulate and study what can be termed as *generalized output regulation problems.*

The generalized output regulation problem has many features. We mention below a few among them:

- It lets us treat almost any arbitrary reference signal.
- It lets us utilize the derivative or feedforward information of reference signals whenever it is available.
- It opens up new avenues to pursue whenever exact output regulation problem is not solvable; for instance, one can study almost output regulation in which one can cast several performance measures one at a time on the asymptotic tracking error, e.g. the supremum of asymptotic tracking error be less than a specified fraction

of a specified norm of the reference signal.

The architecture of the book follows the same chronologically itemized discussion as given above. After this first introductory chapter, Chapter 2 concerns with the *classical exact output regulation*. This chapter includes a complete coverage of classical exact output regulation in a broad framework. It reviews, reexamines, clarifies and extends several facets of output regulation such as well-posedness of the output regulation problem, internal model principle, and structural stability.

Chapters 3, 4, and 5 give a complete development of classical exact output regulation when the actuators are subject to amplitude and rate saturation. In particular, Chapters 3 and 4, for continuous- and discrete-time systems respectively, consider the case when the actuators are subject to only amplitude saturation, while Chapter 5 deals with the general case of both amplitude and rate saturation of inputs for both continuous- and discrete-time systems.

Chapter 6 studies exact output regulation along with optimal transient performance requirements.

Chapters 7, 8, 9, 10, and 11 are concerned with general performance and robustness issues with an output regulation constraint. We first set the stage by developing a key result in Chapter 7. We are interested in the transfer function of the closed loop system from an exogenous input d to an exogenous output z and achieving asymptotically tracking or disturbance rejection of an exogenous signal w (generated by some exosystem) in the error signal e. Here, we formulate an auxiliary system denoted by $\bar{\Sigma}$. For each controller $\bar{\Sigma}_C$ for the auxiliary system, we formulate a corresponding controller Σ_C for the given system Σ, and in so doing we generate a class of controllers for the given system. We note that a controller for $\bar{\Sigma}$ can be related to a controller for Σ and vice versa. Then, we develop our main result. It states that a certain transfer function (say, from $\bar{d}$ to $\bar{z}$) of the auxiliary system with an internally stabilizing controller $\bar{\Sigma}_C$, is exactly the same as the transfer function from d to z in the given system with a controller Σ_C that corresponds to $\bar{\Sigma}_C$. Also, such a Σ_C internally stabilizes Σ, and is such that the error e tends to zero asymptotically. Thus, we basically transform any problem of performance with output regulation constraint for the given system to a similar one for the auxiliary system however without any output regulation constraint. This transformation is valid whatever may be the chosen performance measure (H_2, H_∞, L_1, or any other) as long as it is based on the transfer function from d to z in the given system.

Equipped with the basic result of Chapter 7, for continuous-time systems, Chapters 8 and 10, study H_2 optimal and H_∞ optimal control problems with the output regulation constraint. Chapters 9 and 11 do the same but for discrete-time systems.

In Chapter 12, we consider the notion of robust output regulation which requires output regulation in the presence of structured model uncertainties.

The subject of Chapter 13 is generalized output regulation. Essentially, it generalizes the topic of exact output regulation as dealt with in Chapter 2 in several ways. By introducing a non-autonomous exosystem, the class of reference signals to be tracked in the presence of external disturbance signals is broadened considerably. In fact, almost any arbitrary reference signal can be treated. Also, the derivative or feedforward information of reference signals whenever it is available can be utilized. Next, it introduces and studies almost output regulation in detail. Moreover, both exact and almost output regulation are studied under a variety of controllers.

Chapters 14, 15, and 16 are counter parts of Chapters 3, 4 and 5, and consider generalized output regulation. That is, in Chapters 14, 15, and 16, we deal with generalized output regulation for linear systems with actuators subject to amplitude and rate saturation.

Chapter 17 considers the issue of what can be done if classical output regulation is not possible.

Finally, the Epilogue on page 451 poses some open problems.

The material of Chapter 2 is put together from several papers scattered throughout the literature although several facets of it have been extended in a number of ways. The material of the rest of chapters follows the recent research work of the authors and their coworkers.

1.2 Notation and terminology

Throughout this book we shall adopt the following conventions and notations:

$$
\begin{aligned}
\mathbb{R} &:= \text{the set of real numbers,} \\
\mathbb{R}^+ &:= \text{the set of positive real numbers,} \\
\mathbb{C} &:= \text{the entire complex plane,} \\
\mathbb{C}^- &:= \text{the open left-half complex plane,} \\
\mathbb{C}^0 &:= \text{the imaginary axis,} \\
\mathbb{C}^+ &:= \text{the open right-half complex plane,} \\
\mathbb{C}^{\bigcirc} &:= \text{the unit circle in complex plane,}
\end{aligned}
$$

$$
\begin{aligned}
\mathbb{C}^{\ominus} &:= \text{ the set of complex numbers inside the unit circle,}\\
\mathbb{C}^{\otimes} &:= \text{ the set of complex numbers inside or on the unit circle,}\\
\mathbb{C}^{\oplus} &:= \text{ the set of complex numbers outside the unit circle,}\\
I &:= \text{ an identity matrix,}\\
I_k &:= \text{ the identity matrix of dimension } k\times k,\\
A^{\mathrm{T}} &:= \text{ the transpose of } A,\\
\lambda(A) &:= \text{ the set of eigenvalues of } A,\\
\sigma_{\max}(A) &:= \text{ the largest singular value of } A,\\
\sigma_{\min}(A) &:= \text{ the smallest singular value of } A,\\
r(A) &:= \max_i |\lambda_i(A)| \quad \text{(the spectral radius of } A),\\
\operatorname{trace} A &:= \text{ the trace of } A,\\
\ker A &:= \text{ the null space of } A,\\
\operatorname{im} A &:= \text{ the range space of } A,\\
\langle A \mid \operatorname{im} B\rangle &:= \sum_{i=0}^{n-1} \operatorname{im} A^i B, \quad \text{(the controllability subspace),}\\
\mathcal{C}^n &:= \text{ the set of } n \text{ times continuously differentiable functions.}
\end{aligned}
$$

For a vector $q = (q_1, q_2, \ldots q_k)^{\mathrm{T}}$, we define

$$
\|q\|_\infty = \max_i \, |q_i|, \text{ and}
$$

$$
\|q\| = \|q\|_2 = \left(\sum_{i=1}^{k} x(i)^2\right)^{1/2},
$$

Moreover, for a vector-valued function $w(\cdot)$ and $T \geq 0$, we define

$$
\|w\|_\infty := \sup_t \|w(t)\|_\infty, \qquad \|w\|_{\infty,\mathrm{T}} := \sup_{t\geq T} \|w(t)\|_\infty,
$$

$$
\|w\|_p = (\int_0^\infty \|w(t)\|^p dt)^{1/p}, \qquad \|w\|_{p,[0,T]} := (\int_0^T \|w(t)\|^p dt)^{1/p},
$$

for $p \in [1, \infty)$. An obvious notation is valid for discrete-time signals. For each $p \in [1, \infty]$, the space L_p (or ℓ_p) consists of all measurable vector valued functions $w(\cdot)$ such that $\|w\|_p \leq \infty$.

As usual, the H_2 norm of a transfer function G of a continuous-time system is defined as,

$$
\|G\|_2 := \left(\frac{1}{2\pi}\int_{-\infty}^{\infty} \operatorname{trace} G^{\mathrm{T}}(-i\omega)G(i\omega)d\omega\right)^{1/2},
$$

and the H_2 norm of a transfer function G of a discrete-time system is defined as,

$$\|G\|_2 := \left(\frac{1}{2\pi}\int_{-\pi}^{\pi} \text{trace}\, G^{\mathrm{T}}(e^{-i\omega})G(e^{i\omega})d\omega\right)^{1/2}.$$

The H_∞ norm of a transfer function G of a continuous-time system is defined as,

$$\|G\|_\infty := \sup_{\omega\in[-\infty,\infty)} \sigma_{\max} G(i\omega),$$

where $\sigma_{\max}$ denotes the largest singular value. Similarly, the H_∞ norm of a transfer function G of a discrete-time system is defined as,

$$\|G\|_\infty := \sup_{\omega\in[-\pi,\pi)} \sigma_{\max} G(e^{i\omega}).$$

For a continuous-time system with input w and output z with transfer matrix G we also sometimes need the L_p-L_q induced operator norm defined by,

$$\|G\|_{L_p,L_q} := \sup_{w\neq 0} \frac{\|z\|_q}{\|w\|_p}.$$

Note that the L_2-L_2 induced operator norm is equal to the H_∞ norm while the L_∞-L_∞ induced operator norm is sometimes referred to as the L_1 norm because it is equal to the L_1 norm of the impulse response of the system. Similarly for discrete-time systems we define,

$$\|G\|_{\ell_p,\ell_q} := \sup_{w\neq 0} \frac{\|z\|_q}{\|w\|_p},$$

and also in this case the ℓ_2-ℓ_2 induced operator norm is equal to the H_∞ norm while the ℓ_∞-ℓ_∞ induced operator norm is equal to the ℓ_1 norm of the impulse response of the system.

The solvability conditions for certain output regulation problems posed in the book are formulated in terms of well known invariant subspaces of the geometric control theory whose definitions are recalled next. In these definitions, $\mathbb{C}^g$ denotes a subset of complex plane $\mathbb{C}$. $\mathbb{C}^g$ is closed under conjugation, and for continuous-time systems is an unbounded subset of $\mathbb{C}$. Often $\mathbb{C}^g$ is taken as a stability set, that is for continuous-time systems $\mathbb{C}^g$ is taken as a subset of the open left-half complex plane $\mathbb{C}^-$, and for discrete-time systems $\mathbb{C}^g$ is taken as a set of complex numbers inside the unit circle $\mathbb{C}^\ominus$.

We now recall the following standard definitions.

Definition 1.2.1 *Consider a linear system Σ characterized by a quadruple (A, B, C, D). Then,*

(i) *The stabilizable weakly unobservable subspace $\mathcal{V}^g(A, B, C, D)$ is defined as the maximal subspace of $\mathbb{R}^n$ which is $(A + BF)$-invariant and contained in* $\ker(C + DF)$ *such that the eigenvalues of $(A + BF)|\mathcal{V}^g$ are contained in $\mathbb{C}^g \subseteq \mathbb{C}$ for some F.*

(ii) *The detectable strongly controllable subspace $\mathcal{S}^g(A, B, C, D)$ is defined as the minimal subspace of $\mathbb{R}^n$ which is $(A + KC)$ invariant and contains* $\operatorname{im}(B + KD)$ *such that the eigenvalues of the map which is induced by $(A+KC)$ on the factor space $\mathbb{R}^n/\mathcal{S}^g$ are contained in $\mathbb{C}^g \subseteq \mathbb{C}$ for some K.*

In the case of continuous-time systems, it is of interest to have $\mathbb{C}^g$ representing different sets in the complex plane, namely the entire complex plane $\mathbb{C}$, the open left-half complex plane $\mathbb{C}^-$, the imaginary axis $\mathbb{C}^0$, the closed left-half plane $\mathbb{C}^- \cup \mathbb{C}^0$, or the open right-half complex plane $\mathbb{C}^+$. Whenever $\mathbb{C}^g$ represents respectively the sets $\mathbb{C}$, $\mathbb{C}^-$, $\mathbb{C}^0$, $\mathbb{C}^- \cup \mathbb{C}^0$, and $\mathbb{C}^+$ the superscript g in $\mathcal{V}^g$ and $\mathcal{S}^g$ is replaced by a superscript $*$, $-$, 0, -0, and $+$.

Similarly, in the case of discrete-time systems, it is of interest to have $\mathbb{C}^g$ representing different sets in the complex plane, namely the entire complex plane $\mathbb{C}$, the unit circle $\mathbb{C}^{\bigcirc}$, the set of complex numbers inside the unit circle $\mathbb{C}^{\ominus}$, the set of complex numbers either inside or on the unit circle $\mathbb{C}^{\otimes}$, or the set of complex numbers outside the unit circle $\mathbb{C}^{\oplus}$. Whenever $\mathbb{C}^g$ represents respectively the sets $\mathbb{C}$, $\mathbb{C}^{\ominus}$, $\mathbb{C}^{\bigcirc}$, $\mathbb{C}^{\otimes}$, and $\mathbb{C}^{\oplus}$ the superscript g in $\mathcal{V}^g$ and $\mathcal{S}^g$ is replaced by a superscript $*$, $\ominus$, $\bigcirc$, $\otimes$, and $\oplus$.

Definition 1.2.2 *Consider a linear system Σ characterized by a quadruple (A, B, C, D) with $A \in \mathbb{R}^{n\times n}$ be given. We define the following increasing sequence of subspaces:*

$$\begin{aligned}\mathcal{S}_0^*(A, B, C, D) &= B \ker D \\ \mathcal{S}_{i+1}^*(A, B, C, D) &= B \ker D + A(\mathcal{S}_i^*(A, B, C, D) \cap \ker C)\end{aligned}$$

Note that after a finite number (at most n) steps we have $\mathcal{S}_{i+1}^*(A, B, C, D) = \mathcal{S}_i^*(A, B, C, D)$. Moreover, in that case $\mathcal{S}^*(A, B, C, D) = \mathcal{S}_i^*(A, B, C, D)$.

Definition 1.2.3 *Consider a linear system Σ characterized by a quadruple (A, B, C, D). Let a $\mathbb{C}^g$ be chosen such that it has no common elements with the set of invariant zeros of Σ. Then the corresponding $\mathcal{V}^g(A, B, C, D)$, which is always independent of the particular choice of $\mathbb{C}^g$, is referred to as the controllable weakly unobservable subspace $\mathcal{R}^*(A, B, C, D)$.*

Also, we will use the notation, $C^{-1}\{\operatorname{im} D\} := \{x \mid Cx \in \operatorname{im} D\}$. This is obviously valid even if C is not invertible.

The (invariant) zeros of a system with a realization (A, B, C, D) are those points $\lambda \in \mathbb{C}$ for which

$$\operatorname{rank}\begin{pmatrix} \lambda I - A & -B \\ C & D \end{pmatrix} < \operatorname{normrank}\begin{pmatrix} sI - A & -B \\ C & D \end{pmatrix},$$

where by "normrank" we mean the rank of a matrix with entries in the field of rational functions. A more formal definition (if one is concerned with multiplicities) is given via the Smith-McMillan form. The Smith-McMillan form also allows us to define the invariant polynomials associated to the zero dynamics. For details we refer to for instance [87] and [61].

A linear system Σ characterized by a quadruple (A, B, C, D) is said to be:

(*i*) *minimum phase* if it has all its invariant zeros in $\mathbb{C}^-$ for a continuous-time system (respectively, in $\mathbb{C}^\ominus$ for a discrete-time system),

(*ii*) *weakly minimum phase* if it has all its invariant zeros in $\mathbb{C}^- \cup \mathbb{C}^0$ with a restriction that any invariant zero on $\mathbb{C}^0$ is simple for a continuous-time system (respectively, in $\mathbb{C}^\ominus \cup \mathbb{C}^\circ$ with a restriction that any invariant zero on $\mathbb{C}^\circ$ is simple for a discrete-time system),

(*iii*) *weakly non-minimum phase* if it has all its invariant zeros in $\mathbb{C}^- \cup \mathbb{C}^0$ with at least one invariant zero on $\mathbb{C}^0$ non-simple for a continuous-time system (respectively, in $\mathbb{C}^\ominus \cup \mathbb{C}^\circ$ with at least one invariant zero on $\mathbb{C}^\circ$ non-simple for a discrete-time system),

(*iv*) *non-minimum phase* if it has at least one invariant zero in $\mathbb{C}^+$ for a continuous-time system (respectively, in $\mathbb{C}^\oplus$ for a discrete-time system).

Note that a zero λ is called simple if its multiplicity is equal to

$$\operatorname{normrank}\begin{pmatrix} sI - A & -B \\ C & D \end{pmatrix} - \operatorname{rank}\begin{pmatrix} \lambda I - A & -B \\ C & D \end{pmatrix},$$

and otherwise is referred to as non-simple.

We say that a matrix M is Hurwitz-stable if it has all its eigenvalues in the open left-half plane, or Schur-stable if it has all its eigenvalues inside the unit circle. Similarly, we say that a matrix M is anti-Hurwitz-stable if it has all its eigenvalues in the closed right-half plane, or anti-Schur-stable if it has all the eigenvalues on or outside the unit circle. Along the same lines, we say that a matrix is weakly Hurwitz-stable if all its eigenvalues are in the closed left half plane while any eigenvalue on the imaginary axis has geometric multiplicity equal to its algebraic multiplicity or equivalently the associated Jordan blocks of it are of size 1×1. Similarly, we say that a matrix is weakly Schur-stable if all its eigenvalues are inside or on the unit circle while any eigenvalue on the unit circle has geometric multiplicity equal to its algebraic multiplicity or equivalently the associated Jordan blocks of it are of size 1×1. Finally, we say a matrix is weakly Hurwitz-unstable or weakly Schur-unstable if all the eigenvalues are in the closed left-half plane and in the closed unit disc respectively.

For simplicity of notation, for a mapping f, sometimes $f(t)$ is denoted by f_t.

Chapter 2

Classical exact output regulation

2.1 Introduction

The precise formulation and study of the classical *exact output regulation problem* is the topic of this chapter. We consider a system with an exogenous input and a control input (both might be vector-valued). The exogenous input is generated by an autonomous system (i.e. a system without inputs) which is called the exosystem. The objective is to find a feedback controller such that an output of the system converges to zero as time tends to infinity. This can be used to model asymptotic tracking as well as asymptotic disturbance rejection.

In Section 2.2, we formulate the problem in precise terms. For output regulation, one can consider a controller with state feedback or the more general case of a controller with measurement feedback.

This leads to the formulation of two types of *exact output regulation problems*, (1) the exact output regulation problem with state feedback, and (2) the exact output regulation problem with measurement feedback. These two problems are the subject of Sections 2.3 and 2.4 respectively. There we derive necessary and sufficient conditions under which the problem is solvable. Also, if the problem is solvable, we will present algorithms to compute appropriate controllers. It turns out that the existence of either a state feedback regulator or a measurement feedback regulator requires the solvability of what is known as the regulator equation. Section 2.5 examines the conditions for the solvability of the regulator equation. In the measurement feedback case it turns out that the controller often needs to incorporate a model of the exosystem. This property which is often referred to as the internal model principle is the topic of Section 2.6. Of course the solvability of the regulator equation depends on the model of the given plant and on the exosystem that models the disturbance and the reference signals. In practice such data is not precisely known. This leads to issues regarding the "well-posedness" of output regulation problems. Section 2.7 is concerned with such a discussion. It turns out, however, that although the "well-posedness" of an output regulation problem is very much desired, it by itself is not very useful from a practical point of view. This is because the property of "well-posedness" guarantees merely the existence of a controller that solves the output regulation problem for any system in a given neighborhood of the given plant and the exosystem. However, the design or construction of a regulator itself may need the actual data. This leads to the study of what can be termed as a structurally stable output regulation problem which seeks to find a controller that achieves output regulation for all plants which are in a small neighborhood of the given nominal plant. This problem is studied in Section 2.8.

It will be evident throughout the chapter that our development applies equally well for both continuous- and discrete-time systems. As such, in this chapter, we consider continuous- and discrete-time systems together. Often, we only give the proofs for continuous-time systems, since the corresponding proofs for discrete-time systems require only very minor modifications.

Classical exact output regulation has a long history. The primary and early workers in the field were Davison, Francis, Wonham, and their coworkers (see e.g. [13, 14, 16, 19]), for a later work among many others see the work of Desoer and Wang [15] and Pearson, Shields, and Staats [51]. More recently, Isidori, Byrnes, and their coworkers extended it to non-linear systems (see e.g. [25]).

2.2 Problem formulation for exact output regulation

Suppose that we are given a linear multivariable plant which is subject to three types of input signals and two output signals as denoted in Figure 2.1 on the facing page. We design a controller with measurement signal y which generates the input signal u. We have two objectives:

(*i*) **Tracking**. One exosystem generates a reference signal w_2 and it is our objective to track this reference signal by the output z_c of the system. Our objective is asymptotic tracking. As such by defining the error

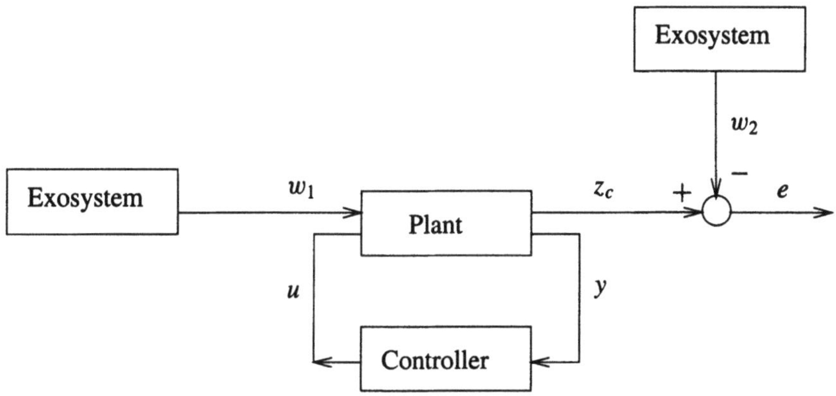

Figure 2.1: Basic setup

signal $e = z_c - w_2$, our objective becomes $\lim_{t\to\infty} e(t) = 0$.

(*ii*) **Disturbance rejection**. The second exosystem generates a disturbance w_1 for the system and our objective is to reduce the effect of the disturbance w_1 on the output z_c. Note that due to linearity of the systems, we can equivalently reduce the effect of w_1 on e. Our objective is again asymptotic, so we have the same objective that $\lim_{t\to\infty} e(t) = 0$.

We will study the above problem and others in detail in this book and all kind of questions such as reducing not only the asymptotic effect but also the transient effect will be discussed in later chapters. Both exosystems are known autonomous systems which generate signals with known frequency but unknown phase and amplitude. Often the exosystem will generate sinusoidal signals as well as step functions.

By combining the two exosystems into one large exosystem we can transform the block diagram in Figure 2.1 into the block diagram depicted in Figure 2.2. This gives us a simpler picture and in many of our derivations it will

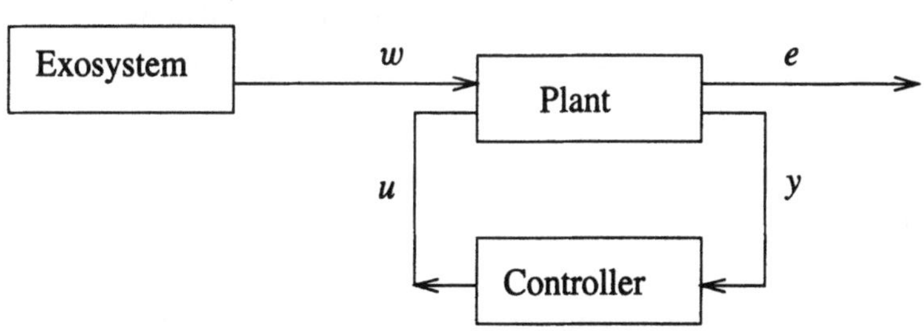

Figure 2.2: Configuration of plant, exosystem, and controller.

simplify our formulae considerably. But in this universal picture we do lose some additional information because tracking and disturbance rejection are not the same objectives. For instance, in tracking, we often know the signal w_1 we want to track while in disturbance rejection it is quite unlikely that we know the disturbance.

The dynamic equations of the plant Σ in Figure 2.2 on the page before are as follows:

$$\Sigma : \begin{cases} \rho x = Ax + Bu + E_w w \\ y = C_y x + D_{yu} u + D_{yw} w \\ e = C_e x + D_{eu} u + D_{ew} w, \end{cases} \tag{2.1}$$

where ρ denotes the time derivative, $\rho x(t) = \frac{dx}{dt}(t)$, for continuous-time systems, and the shift, $(\rho x)(k) = x(k+1)$, for discrete-time systems. The first equation of the above system Σ describes the plant with state $x \in \mathbb{R}^n$ and control input $u \in \mathbb{R}^m$, subject to the effect of an *exogenous disturbance* represented by $E_w w$ where $w \in \mathbb{R}^s$ is the state of an *exosystem* (or exogenous system) Σ_E to be described shortly. The second equation is for the measurement $y \in \mathbb{R}^p$ describing the information available to the controller. The final equation defines the error $e \in \mathbb{R}^q$ between the actual controlled plant output $C_e x$ and the *reference* signal $-D_{ew} w$. If $E_w = 0$ and $D_{yw} = 0$ then we obtain a classical tracking problem where $z_c = C_e x + D_{eu} u$ should track the output of an exogenous system. On the other hand, the case $E_w \neq 0$ or $D_{yw} \neq 0$ is very useful to incorporate disturbance rejection in this setting for those cases where we have specific frequency information about the disturbance.

We note that we can in almost all cases set $D_{yu} = 0$ without loss of generality since it is well known that if we have a controller for the system (2.1) with $D_{yu} = 0$ then it is straightforward to obtain a controller for the original system (2.1) where D_{yu} is not necessarily equal to 0. Therefore in many problems we will set $D_{yu} = 0$ to avoid messy formulae. However, in some cases such as structural stability it will play a more intrinsic role.

The *exosystem* Σ_E is an autonomous system having a state space realization with state $w \in \mathbb{R}^s$,

$$\Sigma_E : \rho w = Sw. \tag{2.2}$$

The above exosystem generates both the exogenous disturbances $E_w w$ and $D_{yw} w$ that acts on the plant as well as the reference signal $-D_{ew} w$ which the plant output is required to track.

We say that we have the state available if both the states x and w are available for feedback, that is

$$y = \begin{pmatrix} x \\ w \end{pmatrix}.$$

When we have the state available for feedback, we consider static feedback controllers of the form,

$$u = Fx + Gw. \tag{2.3}$$

Composing (2.1), (2.2), and (2.3) together yields a closed-loop system

$$\begin{aligned} \rho x &= (A + BF)x + (E_w + BG)w \\ \rho w &= Sw \\ e &= (C_e + D_{eu}F)x + (D_{ew} + D_{eu}G)w. \end{aligned} \tag{2.4}$$

In the general case of any measurement y being available for feedback, we consider dynamic feedback controllers of the form,

$$\Sigma_C : \begin{cases} \rho v = A_c v + B_c y, & v(t) \in \mathbb{R}^{n_c} \\ u = C_c v + D_c y. \end{cases} \tag{2.5}$$

The interconnection of (2.1), (2.2), and (2.5) yields a closed-loop system

$$\begin{aligned} \rho x &= (A + B(I - D_c D_{yu})^{-1} D_c C_y)x + B(I - D_c D_{yu})^{-1} C_c v \\ &\quad + (E_w + B(I - D_c D_{yu})^{-1} D_c D_{yw})w \\ \rho v &= B_c(I - D_{yu} D_c)^{-1} C_y x + (A_c + B_c(I - D_{yu} D_c)^{-1} D_{yu} C_c)v \\ &\quad + B_c(I - D_{yu} D_c)^{-1} D_{yw} w \\ \rho w &= Sw \\ e &= (C_e + D_{eu}(I - D_c D_{yu})^{-1} D_c C_y)x + D_{eu}(I - D_c D_{yu})^{-1} C_c v \\ &\quad + (D_{eu}(I - D_c D_{yu})^{-1} D_c D_{yw} + D_{ew})w. \end{aligned} \tag{2.6}$$

As we discussed earlier, our objective is to find a controller, if possible, that achieves internal stability of the plant as well as output regulation. Internal stability means that, if we disregard the exosystem and set w equal to zero, the closed-loop system comprising of the plant and the controller is asymptotically stable. Output regulation means that for any initial conditions of the closed-loop system, we have $e(t) \to 0$ as $t \to \infty$. Formally, all of this can be summarized in the following two definitions each of which formulates a design problem.

Problem 2.2.1 ***(Exact output regulation problem with state feedback)*** *For a system Σ as given in (2.1), find, if possible, a feedback law of the form (2.3) such that the following conditions hold:*

Internal Stability: *The system $\rho x = (A + BF)x$ is asymptotically stable, i.e. the matrix $A + BF$ is Hurwitz-stable for continuous-time systems and Schur-stable for discrete-time systems.*

Output Regulation: *For all $x(0) \in \mathbb{R}^n$ and $w(0) \in \mathbb{R}^s$, the closed-loop system given in (2.4) satisfies*

$$\lim_{t \to \infty} e(t) = 0.$$

A controller that solves the exact output regulation problem with state feedback is called a state feedback regulator.

Obviously, one needs some natural assumptions on the given system (2.1) in order to solve the above output regulation problem. To start with, the requirement of internal stability of the closed-loop system comprising of the plant and any feedback controller (but excluding the exosystem) requires that the given plant is stabilizable, i.e. the pair (A, B) should be stabilizable. On the other hand, one can assume that the matrix S is anti-Hurwitz-stable for continuous-time systems and anti-Schur-stable for discrete-time systems. This is without loss of generality because, as can be easily seen, the asymptotically stable modes in the exosystem do not affect the regulation of the output as long as the closed-loop system is internally stable (as usual, for internal stability we disregard the exosystem). In fact, if S is Hurwitz-stable for continuous-time systems and Schur-stable for discrete-time systems, the exosystem $\rho w = Sw$ will have a non-trivial stable invariant subspace $\mathcal{S}^-$, and any initial condition $w(0) \in \mathcal{S}^-$ generates an exogenous input exponentially decaying to zero as t tends to infinity. If the closed-loop system is internally stable, as required, the error corresponding to this kind of exogenous input will also exponentially decay to zero as t tends to infinity, for every initial state of the plant and the controller. That is, for any initial condition $w(0) \in \mathcal{S}^-$, the achievement of the property of output regulation is a trivial consequence of the property of the internal stability. This shows that one can assume without any loss of generality that $\mathcal{S}^- = \{0\}$. In Chapter 6 we also look at the transient effect of the exosystem on the error signal e. In that case the stable dynamics do play a role but without loss of generality we can also consider the stable dynamics as a part of the given system.

For ease of referencing, we formulate the above assumptions as follows:

A.1. The pair (A, B) is stabilizable.

A.2. The matrix S is anti-Hurwitz-stable for continuous-time systems and anti-Schur-stable for discrete-time systems.

Next, we define formally the measurement feedback output regulation problem. For clarity of presentation but without loss of generality, we assume that $D_{yu} = 0$.

Problem 2.2.2 ***(Exact output regulation problem with measurement feedback)*** *For a system Σ as given in (2.1), find, if possible, a measurement feedback law of the form (2.5) such that the following conditions hold:*

Internal Stability: *The system*

$$\begin{aligned} \rho x &= (A + BD_c C_y)x + BC_c v \\ \rho v &= B_c C_y x + A_c v \end{aligned}$$

is asymptotically stable, i.e. the matrix

$$\begin{pmatrix} A + BD_c C_y & BC_c \\ B_c C_y & A_c \end{pmatrix}$$

is Hurwitz-stable for continuous-time systems and Schur-stable for discrete-time systems.

Output Regulation: *For all $x(0) \in \mathbb{R}^n$, $v(0) \in \mathbb{R}^{n_c}$, and $w(0) \in \mathbb{R}^s$, the closed-loop system given in (2.6) satisfies*

$$\lim_{t \to \infty} e(t) = 0.$$

A controller that solves the exact output regulation problem with measurement feedback is called a measurement feedback regulator.

Clearly, in order to guarantee the internal stability of the closed-loop system comprising of the plant and any measurement feedback controller, (as usual, for internal stability we disregard the exosystem), we need that (A, B) be stabilizable and (C_y, A) be detectable. To make it easier to refer to, we formulate this assumption and an additional one as follows:

A.3a. The pair (C_y, A) is detectable.

A.3b. For all $\lambda \in \mathbb{C}$ with $\operatorname{Re}\lambda \geqslant 0$ (continuous-time) or $|\lambda| \geqslant 1$ (discrete-time)

$$\ker \begin{pmatrix} \lambda I - A & -E \\ 0 & \lambda I - S \\ C_y & D_{yw} \end{pmatrix} = \ker \begin{pmatrix} \lambda I - A & -E \\ 0 & \lambda I - S \\ C_y & D_{yw} \\ C_e & D_{ew} \end{pmatrix}.$$

The last condition basically states that if an unstable eigenvalue is observable from e then it must also be observable from y. This is obviously necessary for output regulation since otherwise we cannot observe whether output regulation is achieved or not, and then we cannot find a measurement feedback regulator either.

In the next two sections, we discuss in detail the solvability conditions for the output regulation problems formulated here. Also, we construct appropriate controllers that solve the posed output regulation problem whenever it is solvable.

2.3 Exact output regulation with state feedback

Our primary goals in this section are twofold, (1) to establish the necessary and sufficient conditions for the existence of a controller that solves the exact output regulation problem with state feedback, and (2) to construct such a controller whenever it exists. To this end, we have the following theorem.

Theorem 2.3.1 *Consider the exact output regulation problem with state feedback as defined in Problem 2.2.1. Let Assumptions A.1 and A.2 hold. Then, the considered problem is solvable if and only if there exist matrices* Π *and* Γ *which solve the following linear matrix equation (2.7) often called the regulator equation,*

$$\Pi S = A\Pi + B\Gamma + E_w, \tag{2.7a}$$

$$0 = C_e\Pi + D_{eu}\Gamma + D_{ew}. \tag{2.7b}$$

Moreover, a suitable state feedback is then given by

$$u = Fx + (\Gamma - F\Pi)w \tag{2.8}$$

where F *is an arbitrary matrix such that* $A + BF$ *is Hurwitz-stable for continuous-time systems and Schur-stable for discrete-time systems.*

Proof : We only prove this result for continuous-time systems. The proof for discrete-time systems follows along the same lines with minor modifications.

First we prove the property that, under Assumptions A.1 and A.2, if there exists a state feedback law $u = Fx + Gw$ such that $A + BF$ is Hurwitz-stable, then the error $e(t)$ tends to zero as t tends to infinity if and only if there exists a matrix Π which solves the linear equation,

$$\Pi S = (A + BF)\Pi + (E_w + BG), \tag{2.9a}$$

$$0 = C_e\Pi + D_{eu}(F\Pi + G) + D_{ew}. \tag{2.9b}$$

To show this property, we note that equation (2.9a) is a Sylvester equation (see Appendix 2.A). By our assumptions, $\lambda(A + BF) \cap \lambda(S) = \varnothing$ where $\lambda(M)$ denotes the spectrum of matrix M. Therefore, equation (2.9a) has a unique solution Π. Consider now the coordinate transformation,

$$\tilde{x} = x - \Pi w.$$

Then, using (2.4) and in view of Π satisfying (2.9a), we obtain

$$\rho\tilde{x} = (A + BF)\tilde{x}, \tag{2.10}$$

and

$$e = (C_e + D_{eu}F)\tilde{x} + (C_e\Pi + D_{eu}(F\Pi + G) + D_{ew})w.$$

Also, in view of (2.4) and (2.10), we get

$$\tilde{x}(t) = e^{(A+BF)t}\tilde{x}(0), \quad w(t) = e^{St}w(0),$$

and therefore

$$e(t) = (C_e + D_{eu}F)e^{(A+BF)t}\tilde{x}(0) \\ + (C_e\Pi + D_{eu}(F\Pi + G) + D_{ew})e^{St}w(0).$$

Since $A + BF$ is Hurwitz-stable we note that $e(t) \to 0$ as $t \to \infty$, for every $(x(0), w(0))$, if and only if

$$C_e\Pi + D_{eu}(F\Pi + G) + D_{ew} = 0.$$

This completes the proof of the property that the error $e(t)$ tends to zero as t tends to infinity if and only if there exists a matrix Π which solves (2.9).

Now the necessity part of Theorem 2.3.1 is obvious from the above development. Suppose a control law $u = Fx + Gw$ exists solving the exact output regulation problem with state feedback. Then, as discussed above, equations (2.9a) and (2.9b) must hold for some Π. Set

$$\Gamma = F\Pi + G,$$

and note that (2.9a) and (2.9b) reduce to (2.7a) and (2.7b) respectively.

The proof of the sufficiency part of Theorem 2.3.1 is constructive. By Assumption A.1, the pair (A, B) is stabilizable. Therefore, there exists a matrix F such that $A + BF$ is Hurwitz-stable. Suppose Π and Γ are matrices which satisfy (2.7a) and (2.7b), and consider the control law given in (2.8). We claim that this control law solves the problem. Internal stability is guaranteed since $A + BF$ is Hurwitz-stable by the selection of F. Also, we note that the output regulation occurs if F and $G := \Gamma - F\Pi$ are such that (2.9a) and (2.9b) are satisfied for some Π. But the choice of G is such that (2.9a) and (2.9b) are identical to (2.7a) and (2.7b). Therefore we can conclude that the control law given in (2.8) solves the exact output regulation problem with state feedback. ■

The construction described in the necessity part of the proof of Theorem 2.3.1 can be given a simple and expressive geometric interpretation. To this end, we rewrite the dynamic equations of (2.4) in the form

$$\rho x_{cl} = A_{cl} x_{cl}$$

with

$$x_{cl} = \begin{pmatrix} x \\ w \end{pmatrix}, \quad A_{cl} = \begin{pmatrix} A + BF & E_w + BG \\ 0 & S \end{pmatrix}.$$

In the case of continuous-time systems, the $(n+s) \times (n+s)$ matrix A_{cl} has n eigenvalues in the open left half complex plane (the eigenvalues of $A + BF$) and s eigenvalues in the closed right half complex plane (those of S). On the other hand, in the case of discrete-time systems, the matrix A_{cl} has n eigenvalues within the unit circle (the eigenvalues of $A + BF$) and s eigenvalues on or outside the unit circle (those of S). Let $\mathcal{V}^-$ denote the invariant subspace of A_{cl} associated with its eigenvalues in the open left half complex plane for continuous-time systems or its eigenvalues within the unit circle for discrete-time systems. Let $\mathcal{V}^+$ denote the invariant subspace of A_{cl} associated with the rest of its eigenvalues.

It is immediate to realize that $\mathcal{V}^-$ is spanned by the columns of the matrix

$$M^- = \begin{pmatrix} I_n \\ 0 \end{pmatrix}.$$

In fact, the subspace spanned by the columns of M^- is invariant under A_{cl} and the restriction of A_{cl} to this subspace is precisely $A + BF$.

The subspace $\mathcal{V}^+$, being complementary to $\mathcal{V}^-$ in $\mathbb{R}^{n+s}$, will be spanned by the columns of a matrix of the form,

$$M^+ = \begin{pmatrix} X \\ I_s \end{pmatrix}, \tag{2.11}$$

where X is a matrix of appropriate dimensions. It is easy to see that X coincides with the solution Π of (2.9a). In fact, to impose the condition that the subspace spanned by the columns of M^+ is invariant under A_{cl} is equivalent to the requirement that for each $w \in \mathbb{R}^s$ there exists a $\tilde{w} \in \mathbb{R}^s$ satisfying

$$\begin{pmatrix} A + BF & E_w + BG \\ 0 & S \end{pmatrix} \begin{pmatrix} X \\ I_s \end{pmatrix} w = \begin{pmatrix} X \\ I_s \end{pmatrix} \tilde{w}.$$

By expanding the above, one obtains necessarily that

$$(A + BF)X + (E_w + BG) = XS. \tag{2.12}$$

Equation (2.12) for X coincides with (2.9a) for Π. Thus the unique solution Π of (2.9a) is such that the subspace spanned by the columns of (2.11) with $X = \Pi$ is invariant under A_{cl}. Moreover, (2.12) also shows that the restriction of A_{cl} to this invariant subspace is precisely S. Thus the columns of M^+ span the subspace $\mathcal{V}^+$.

In view of the above discussion, for the closed-loop system (2.4), equation (2.9a) expresses the existence of an invariant subspace having the form,

$$\mathcal{V}^+ = \left\{ \begin{pmatrix} x \\ w \end{pmatrix} \in \mathbb{R}^{n+s} \;\middle|\; x = \Pi w \right\}, \tag{2.13}$$

and the restriction of (2.4) reduces to

$$\rho w = Sw.$$

On the other hand, the condition given in (2.9b) expresses the fact that the error map $e = C_e \Pi x + (D_{eu}(F\Pi + G) + D_{ew})w$ is zero at each point of this invariant subspace $\mathcal{V}^+$.

From the above interpretation it is rather easy to deduce why output regulation occurs if and only if the unique solution Π of (2.9a) satisfies (2.9b). The necessity arises from the fact that if the initial condition of (2.4) is in $\mathcal{V}^+$, the corresponding trajectory, which remains in $\mathcal{V}^+$ and is a copy of a trajectory of the exosystem, cannot converge to zero because the matrix S is anti-Hurwitz-stable for continuous-time systems and anti-Schur-stable for discrete-time systems. Thus, the only possibility of output regulation occurring is when $\mathcal{V}^+$ is annihilated by the error map. The sufficiency derives from the fact that all trajectories of (2.4) converge, as t tends to infinity, to $\mathcal{V}^+$ and therefore produce an error which asymptotically decays to zero.

The above discussion leads us to a geometrical interpretation of the equations (2.7a) and (2.7b). We note that the equation (2.7a) expresses the fact that the subspace $\mathcal{V}^+$ given in (2.13) is a *controlled invariant subspace* of the system

$$\begin{pmatrix} \rho x \\ \rho w \end{pmatrix} = \begin{pmatrix} A & E_w \\ 0 & S \end{pmatrix} \begin{pmatrix} x \\ w \end{pmatrix} + \begin{pmatrix} B \\ 0 \end{pmatrix} u,$$

that is, $\mathcal{V}^+$ is rendered invariant by an appropriate choice of a feedback control, in this case, given by

$$u = \Gamma w.$$

Equation (2.7b), as we already observed, is equivalent to the fact that this controlled invariant subspace is annihilated by the error map.

From the above it is also clear that

$$\Gamma w(t) - u(t) \to 0,$$
$$\Pi w(t) - x(t) \to 0$$

as $t \to \infty$. In other words Γ and Π have the interpretation as specifying the asymptotic behavior of the input and the state respectively.

Design of a state feedback regulator:

For clarity, we now give a step by step design of a state feedback regulator.

Step 1 : Find a solution $(\Pi,\ \Gamma)$ of the regulator equation (2.7).

Step 2 : Find a matrix F such that $A + BF$ is Hurwitz-stable for continuous-time systems and Schur-stable for discrete-time systems. Let $G := \Gamma - F\Pi$, and then finally construct the state feedback regulator $u = Fx + (\Gamma - F\Pi)w$ as given in (2.8). □

2.4 Exact output regulation with measurement feedback

In the previous section, we considered the exact output regulation problem for the case of state feedback. In this section, we consider the exact output regulation problem for the case of general measurement feedback. As in the previous case, our primary goals here are two fold, (1) to establish the necessary and sufficient conditions for the existence of a controller that solves the exact output regulation problem with measurement feedback, and (2) to construct such a controller whenever it exists.

We will establish the necessary and sufficient conditions for the existence of a controller that solves the exact output regulation problem with measurement feedback, and then construct such a controller. We first impose an additional assumption:

A.3. The pair $(\tilde{C}, \tilde{A})$ is detectable where

$$\tilde{C} = \begin{pmatrix} C_y & D_{yw} \end{pmatrix}, \text{ and } \tilde{A} = \begin{pmatrix} A & E_w \\ 0 & S \end{pmatrix}. \tag{2.14}$$

Note that this assumption implies that both Assumptions A.3a and A.3b are satisfied. The general case without this additional assumption is discussed after the following theorem. Again, for ease of presentation but without loss of generality, we assume that $D_{yu} = 0$.

Theorem 2.4.1 *Consider the problem of exact output regulation with measurement feedback for the system (2.1) as defined in Problem 2.2.2. Let Assumptions A.1, A.2, and A.3 hold. Then, the considered problem is solvable if and only if there exist matrices Π and Γ which solve the regulator equation (2.7). Moreover, a suitable measurement feedback controller is then given by*

$$\begin{pmatrix} \rho\hat{x} \\ \rho\hat{w} \end{pmatrix} = \begin{pmatrix} A & E_w \\ 0 & S \end{pmatrix} \begin{pmatrix} \hat{x} \\ \hat{w} \end{pmatrix} + \begin{pmatrix} B \\ 0 \end{pmatrix} u + \begin{pmatrix} K_A \\ K_S \end{pmatrix} \left[\begin{pmatrix} C_y & D_{yw} \end{pmatrix} \begin{pmatrix} \hat{x} \\ \hat{w} \end{pmatrix} - y \right] \tag{2.15a}$$

$$u = \begin{pmatrix} F & (\Gamma - F\Pi) \end{pmatrix} \begin{pmatrix} \hat{x} \\ \hat{w} \end{pmatrix}, \tag{2.15b}$$

where F, K_A, and K_S are arbitrary matrices such that the matrices

$$A + BF \quad \textit{and} \quad \begin{pmatrix} A + K_A C_y & E_w + K_A D_{yw} \\ K_S C_y & S + K_S D_{yw} \end{pmatrix}$$

are both Hurwitz-stable for continuous-time systems and Schur-stable for discrete-time systems.

Proof : As in the previous section, the proof is given for continuous-time systems. The proof for discrete-time systems follows along the same lines with minor modifications.

We first prove that a stabilizing measurement feedback law of the form (2.5), i.e. a measurement feedback such that the matrix

$$\begin{pmatrix} A + BD_cC_y & BC_c \\ B_cC_y & A_c \end{pmatrix} \tag{2.16}$$

is Hurwitz-stable, has the property that the error $e(t)$ tends to zero as t tends to infinity if and only if there exist matrices Π and Θ which solve the linear equation,

$$\Pi S = (A + BD_cC_y)\Pi + BC_c\Theta + (E_w + BD_cD_{yw}), \tag{2.17a}$$

$$\Theta S = A_c\Theta + B_c(C_y\Pi + D_{yw}), \tag{2.17b}$$

$$0 = C_e\Pi + D_{ew} + D_{eu}D_c(C_y\Pi + D_{yw}) + D_{eu}C_c\Theta. \tag{2.17c}$$

The two equations (2.17a) and (2.17b) together form the following Sylvester equation,

$$\begin{pmatrix} \Pi \\ \Theta \end{pmatrix} S = \begin{pmatrix} A + BD_cC_y & BC_c \\ B_cC_y & A_c \end{pmatrix} \begin{pmatrix} \Pi \\ \Theta \end{pmatrix} + \begin{pmatrix} E_w + BD_cD_{yw} \\ B_cD_{yw} \end{pmatrix}. \tag{2.18}$$

Assumption A.2 together with the fact that the matrix given in (2.16) is Hurwitz-stable yields that the above Sylvester equation has a unique solution for Π and Θ. We note that this solution is such that the columns of

$$M^+ = \begin{pmatrix} \Pi \\ \Theta \\ I_s \end{pmatrix}$$

span the invariant subspace $\mathcal{V}^+$ of A_{cl} associated with the eigenvalues in the closed right half complex plane $\mathbb{C}^0 \cup \mathbb{C}^+$, where

$$A_{cl} = \begin{pmatrix} A + BD_cC_y & BC_c & E_w + BD_cD_{yw} \\ B_cC_y & A_c & B_cD_{yw} \\ 0 & 0 & S \end{pmatrix}.$$

Consider now the coordinate transformation,

$$\tilde{x} = x - \Pi w, \quad \tilde{v} = v - \Theta w,$$

and note that, in the new coordinates thus defined, the equations which describe the closed-loop system (2.6) assume the form,

$$\begin{aligned}
\rho\tilde{x} &= (A + BD_{\rm c}C_y)\tilde{x} + BC_{\rm c}\tilde{v} \\
\rho\tilde{v} &= B_{\rm c}C_y\tilde{x} + A_{\rm c}\tilde{v} \\
\rho w &= Sw \\
e &= (C_e + D_{eu}D_{\rm c}C_y)\tilde{x} + D_{eu}C_{\rm c}\tilde{v} \\
&\qquad +(C_e\Pi + D_{ew} + D_{eu}D_{\rm c}(C_y\Pi + D_{yw}) + D_{eu}C_{\rm c}\Theta)w.
\end{aligned}$$

Then, using the arguments identical to those used in the first part of the proof of Theorem 2.3.1, it is easy to see that output regulation occurs, i.e. $e(t) \to 0$ as $t \to \infty$, for every $(\tilde{x}(0), \tilde{v}(0), w(0))$, if and only if

$$C_e\Pi + D_{ew} + D_{eu}D_{\rm c}(C_y\Pi + D_{yw}) + D_{eu}C_{\rm c}\Theta = 0.$$

Thus, output regulation occurs if and only if the unique solution Π, Θ of (2.18) satisfies (2.17c).

In view of the above discussion, the necessity part of Theorem 2.4.1 is obvious. Suppose a controller of the form (2.5) exists solving the exact output regulation problem with measurement feedback. Then, as discussed above, equations (2.17a), (2.17b), and (2.17c) must hold for some Π and Θ. Set

$$\Gamma = D_{\rm c}(C_y\Pi + D_{yw}) + C_{\rm c}\Theta,$$

and note that (2.17a) and (2.17c) reduce respectively to (2.7a) and (2.7b).

The proof of the sufficiency part of Theorem 2.4.1 is by the construction of an appropriate controller having an observer based architecture. Assuming the existence of a pair of matrices Π and Γ that satisfy the regulator equation (2.7), we first construct a control law that solves the exact output regulation problem with state feedback. As discussed in Theorem 2.3.1, this can be chosen as

$$u = Fx + (\Gamma - F\Pi)w,$$

where F is an arbitrary matrix such that $A + BF$ is Hurwitz-stable. The existence of such an F is implied by Assumption A.1.

Next, in order to estimate x and w, we construct an observer driven by the measurement y. This observer has the form,

$$\begin{pmatrix} \rho\hat{x} \\ \rho\hat{w} \end{pmatrix} = \begin{pmatrix} A + K_{\rm A}C_y & E_w + K_{\rm A}D_{yw} \\ K_{\rm S}C_y & S + K_{\rm S}D_{yw} \end{pmatrix} \begin{pmatrix} \hat{x} \\ \hat{w} \end{pmatrix} - \begin{pmatrix} K_{\rm A} \\ K_{\rm S} \end{pmatrix} y + \begin{pmatrix} B \\ 0 \end{pmatrix} u,$$

where the pair $(K_{\rm A}, K_{\rm S})$ is such that the matrix

$$\begin{pmatrix} A + K_{\rm A}C_y & E_w + K_{\rm A}D_{yw} \\ K_{\rm S}C_y & S + K_{\rm S}D_{yw} \end{pmatrix}$$

is Hurwitz-stable. The existence of such a pair $(K_{\rm A}, K_{\rm S})$ is implied by Assumption A.3. Finally, we define a measurement feedback controller as in (2.15a) and (2.15b).

Now to prove the sufficiency part of Theorem 2.4.1, we note that the controller given in (2.15) is of the form (2.5) with

$$A_{\rm c} = \begin{pmatrix} A + K_{\rm A}C_y + BF & E_w + K_{\rm A}D_{yw} + B(\Gamma - F\Pi) \\ K_{\rm S}C_y & S + K_{\rm S}D_{yw} \end{pmatrix},$$

$$B_{\rm c} = -\begin{pmatrix} K_{\rm A} \\ K_{\rm S} \end{pmatrix}, \quad C_{\rm c} = \begin{pmatrix} F & \Gamma - F\Pi \end{pmatrix}, \text{ and } D_{\rm c} = 0.$$

Thus the matrix

$$\begin{pmatrix} A + BD_{\rm c}C_y & BC_{\rm c} \\ B_{\rm c}C_y & A_{\rm c} \end{pmatrix}$$

can be rewritten as

$$\left(\begin{array}{c|cc} A & BF & B(\Gamma - F\Pi) \\ \hline -K_{\rm A}C_y & A + K_{\rm A}C_y + BF & E_w + K_{\rm A}D_{yw} + B(\Gamma - F\Pi) \\ -K_{\rm S}C_y & K_{\rm S}C_y & S + K_{\rm S}D_{yw} \end{array}\right).$$

A similarity transformation on the above matrix via

$$T = \begin{pmatrix} I & 0 & 0 \\ -I & I & 0 \\ 0 & 0 & I \end{pmatrix}$$

yields

$$\begin{pmatrix} A + BF & BF & B(\Gamma - F\Pi) \\ 0 & A + K_{\rm A}C_y & E_w + K_{\rm A}D_{yw} \\ 0 & K_{\rm S}C_y & S + K_{\rm S}D_{yw} \end{pmatrix}.$$

This matrix is block diagonal and the eigenvalues of the diagonal blocks are, by construction, in $\mathbb{C}^-$. Thus the closed-loop system comprising of the plant and the controller is internally stable (as usual, for internal stability we disregard the exosystem). Next, in order to prove that output regulation occurs, it suffices to check that the pair Π and Θ with

$$\Theta = \begin{pmatrix} \Pi \\ I \end{pmatrix}$$

is a solution of the equations (2.17a), (2.17b), and (2.17c). Using that Π and Γ satisfy (2.7) this can be checked straightforwardly. ∎

Remark 2.4.1 *It is interesting to observe that under an additional Assumption A.3, the conditions given in Theorem 2.3.1 to solve the exact output regulation problem with state feedback, also guarantee the solution of the exact output regulation problem with measurement feedback.*

Design of an observer based measurement feedback regulator:

For clarity, we now summarize the construction of an observer based measurement feedback regulator for the case when $D_{yu} = 0$.

Step 1 : At first construct a state feedback regulator. That is, find the matrices $(\Pi, \ \Gamma)$ that solve the regulator equation (2.7), find a matrix F such that $A + BF$ is Hurwitz-stable for continuous-time systems and Schur-stable for discrete-time systems, and let $G := \Gamma - F\Pi$.

Step 2 : Design a full order observer so that we can implement the controller with observer based architecture as given in (2.15). That is, find the matrix gains K_A and K_S such that the matrix

$$\begin{pmatrix} A + K_A C_y & E_w + K_A D_{yw} \\ K_S C_y & S + K_S D_{yw} \end{pmatrix}$$

is Hurwitz-stable for continuous-time systems and Schur-stable for discrete-time systems.

Step 3 : Implement the observer based measurement feedback regulator as given in (2.15). □

Remark 2.4.2 *Measurement feedback output regulation as given in (2.15) has a full order observer based architecture. It is easy to see from the proof that one can also use reduced order observer based architecture for the controller. We observe that a full order observer based measurement feedback regulator is strictly proper, where as the reduced order observer based one is proper.*

Theorem 2.4.1 as well as the above development utilize the observer based architecture to construct a measurement feedback regulator. In fact, one does not have to resort to observer based architecture to construct measurement feedback controllers that achieve output regulation. Indeed, under Assumptions A.1, A.2, and A.3, and whenever there exist matrices Π and Γ which solve the regulator equation (2.7), any measurement feedback controller with a desirable architecture (not necessarily observer based one) can be utilized to construct a regulator. We give below a step by step procedure of constructing such a regulator.

Design of a general measurement feedback regulator:

Again, without loss of generality, we assume here that $D_{yu} = 0$.

Step 1 : Consider the auxiliary system $\bar{\Sigma}$ defined by

$$\bar{\Sigma}: \begin{cases} \rho\bar{x} = \bar{A}\bar{x} + \bar{B}\bar{u} \\ \bar{y} = \bar{C}_y\bar{x}, \end{cases} \tag{2.19}$$

where

$$\bar{A} = \begin{pmatrix} A & -B\Gamma \\ 0 & S \end{pmatrix}, \quad \bar{B} = \begin{pmatrix} 0 & B \\ I & 0 \end{pmatrix}, \quad \bar{C}_y = \begin{pmatrix} C_y & D_{yw} + C_y\Pi \end{pmatrix}.$$

We note that the auxiliary system $\bar{\Sigma}$ is constructed from the data of the given system Σ and the exosystem Σ_E respectively as in (2.1) and (2.2), and the matrices Π and Γ that solve the regulator equation (2.7).

Step 2 : Consider for $\bar{\Sigma}$ a general class of measurement feedback controllers with state space representation $(\bar{A}_c, \bar{B}_c, \bar{C}_c, \bar{D}_c)$ and having any desirable architecture. Let $\bar{\Sigma}_C$ be given by

$$\bar{\Sigma}_C: \begin{cases} \rho\bar{v} = \bar{A}_c\bar{v} + \bar{B}_c\bar{y} \\ \bar{u} = \bar{C}_c\bar{v} + \bar{D}_c\bar{y}. \end{cases} \tag{2.20}$$

Design $\bar{\Sigma}_C$ such that the closed-loop system comprising $\bar{\Sigma}$ and $\bar{\Sigma}_C$ is internally stable.

Step 3 : Construct a controller Σ_C having the structure,

$$\Sigma_C: \begin{cases} \rho v_1 = Sv_1 + \bar{C}_{c,1}v_2 + \bar{D}_{c,1}(y + (D_{yw} + C_y\Pi)v_1) \\ \rho v_2 = \bar{A}_c v_2 + \bar{B}_c(y + (D_{yw} + C_y\Pi)v_1) \\ u = -\Gamma v_1 + \bar{C}_{c,2}v_2 + \bar{D}_{c,2}(y + (D_{yw} + C_y\Pi)v_1), \end{cases} \tag{2.21}$$

where $\bar{C}_{c,1}$, $\bar{C}_{c,2}$, $\bar{D}_{c,1}$, and $\bar{D}_{c,2}$ are obtained by partitioning $\bar{C}_c$ and $\bar{D}_c$ in conformity with the partitioning of $\bar{A}$,

$$\bar{C} = \begin{pmatrix} \bar{C}_{c,1} \\ \bar{C}_{c,2} \end{pmatrix} \text{ and } \bar{D} = \begin{pmatrix} \bar{D}_{c,1} \\ \bar{D}_{c,2} \end{pmatrix}.$$

Then, it follows from Theorem 7.3.1 of Chapter 7 that the above controller Σ_C when applied to the given system Σ internally stabilizes it, and moreover achieves output regulation as well. □

As discussed in Section 2.2, in order to solve the measurement feedback output regulation problem, we need the Assumptions A.1, A.2, A.3a and A.3b. However in Theorem 2.4.1, instead of A.3a, we used a stronger Assumption than A.3a and A.3b, namely A.3.

To see that A.3 is stronger than A.3a, one can utilize the so called Hautus' test for detectability. Hautus' test states that a pair (C_y, A) is detectable if and only if the matrix

$$\begin{pmatrix} A - \lambda I \\ C_y \end{pmatrix}$$

has full column-rank for all λ in the closed right half plane for continuous-time systems, or for all λ on or outside the unit circle for discrete-time systems. Now, it is immediate to observe that if the pair (C_y, A) does not pass the Hautus' test for detectability, then necessarily also the pair $(\tilde{C}, \tilde{A})$ does not pass such a test. This is because the matrix

$$\begin{pmatrix} A - \lambda I & E_w \\ 0 & S - \lambda I \\ C_y & D_{yw} \end{pmatrix}$$

will then not have full column-rank for some λ in the closed right half plane for continuous-time systems, or for some λ on or outside the unit circle for discrete-time systems. As a consequence, Assumption A.3 implies A.3a. It is also easy to see that A.3 implies A.3b.

At this time, it is natural to enquire why one needs the stronger Assumption A.3 rather than A.3a and A.3b. After all the latter two assumptions are necessary for output regulation while Assumption A.3 is not necessary. In fact, one does not really need to replace A.3a and A.3b by A.3. However, as it will become evident in a moment, Assumption A.3 is rather convenient, has a natural motivation, and above all, as will be shown later, it does not involve a loss of generality in solving the exact output regulation problem with measurement feedback.

The motivation for using A.3 rather than A.3a and A.3b is simple. We already know, from Section 2.3, how to solve the exact output regulation problem by a control law of the form (2.8) which presupposes the availability of the state x of the plant Σ and the state w of the exosystem Σ_E. If the states x and w are not available for measurement, then in order to implement the control law (2.8), it is reasonable to expect that we need to be able to obtain at least an asymptotic estimate of these states x and w from the available measurement y. The needed assumption, under which one can obtain an asymptotic estimate of the states x and w from y, is indeed A.3. The reason why Assumption A.3 does not involve loss of generality is somewhat technical.

As it turns out, if A.3a and A.3b hold but not A.3, then it is always possible to reduce the state space of the exosystem Σ_E to that part which actually affects the error; and the reduced exosystem thus obtained along with the given plant satisfies the condition A.3. Clearly, without loss of generality, one can delete the states that do not affect the error. To substantiate this discussion, we can use the following proposition (see [16]).

Proposition 2.4.1 *Suppose that Assumptions A.3a and A.3b hold but not Assumption A.3. Consider the augmented system*

$$\begin{aligned} \rho x^a &= \tilde{A}x^a + \tilde{B}u \\ y^a &= \tilde{C}x^a + \tilde{D}u \end{aligned} \tag{2.22}$$

with $\tilde{A}$, $\tilde{B}$, $\tilde{C}$ and $\tilde{D}$ defined by

$$\tilde{A} = \begin{pmatrix} A & E_w \\ 0 & S \end{pmatrix}, \qquad \tilde{B} = \begin{pmatrix} B \\ 0 \end{pmatrix},$$
$$\tilde{C} = \begin{pmatrix} C_y & D_{yw} \\ C_e & D_{ew} \end{pmatrix}, \qquad \tilde{D} = \begin{pmatrix} D_{yu} \\ D_{eu} \end{pmatrix}.$$

Then, there exists a coordinate transformation

$$\tilde{x}^a = T^a x^a$$

such that, in the new coordinates, $\tilde{A}$, $\tilde{B}$, and $\tilde{C}$ assume the form,

$$\tilde{A}^a = T^a \tilde{A}(T^a)^{-1} = \begin{pmatrix} A & \tilde{E}_w \\ 0 & \tilde{S} \end{pmatrix},$$
$$\tilde{B}^a = T^a \tilde{B} = \begin{pmatrix} B \\ 0 \end{pmatrix},$$
$$\tilde{C}^a = \tilde{C}(T^a)^{-1} = \begin{pmatrix} C_y & \tilde{D}_{yw} \\ C_y & \tilde{D}_{ew} \end{pmatrix},$$

with $\tilde{S}$, $\tilde{E}_w$, $\tilde{D}_{yw}$ and $\tilde{D}_{ew}$ having a partitioned structure,

$$\tilde{S} = \begin{pmatrix} S_{11} & 0 \\ S_{21} & S_{22} \end{pmatrix}, \quad \tilde{E}_w = \begin{pmatrix} E_{w,1} & 0 \end{pmatrix},$$
$$\tilde{D}_{yw} = \begin{pmatrix} \tilde{D}_{yw,1} & 0 \end{pmatrix}, \quad \tilde{D}_{ew} = \begin{pmatrix} \tilde{D}_{ew,1} & 0 \end{pmatrix}.$$

Moreover, S_{11}, $E_{w,1}$, $\tilde{D}_{yw,1}$, *and* $\tilde{D}_{ew,1}$ *are such that the pair*

$$\begin{pmatrix} C_y & \tilde{D}_{yw,1} \\ C_e & \tilde{D}_{ew,1} \end{pmatrix}, \begin{pmatrix} A & E_{w,1} \\ 0 & S_{11} \end{pmatrix}$$

is detectable.

Proof : As is well known, if the pair $(\tilde{C}, \tilde{A})$ is not detectable, by appropriately changing the coordinates, one can transform this pair into a pair $(\tilde{C}^a, \tilde{A}^a)$ having the structure,

$$\tilde{C}^a = \begin{pmatrix} \tilde{C}_1 & 0 \end{pmatrix}, \quad \tilde{A}^a = \begin{pmatrix} A_{11} & 0 \\ A_{21} & A_{22} \end{pmatrix},$$

where the pair $(\tilde{C}_1, A_{11})$ is detectable. In the present situation, because of the additional assumption that the pair (C_y, A) is detectable, one can choose the transformation so as to obtain

$$A_{11} = \begin{pmatrix} A & E_{w,1} \\ 0 & S_{11} \end{pmatrix}, \quad A_{21} = \begin{pmatrix} 0 & S_{21} \end{pmatrix},$$

$$A_{22} = S_{22}, \quad \tilde{C}_1 = \begin{pmatrix} C_y & \tilde{D}_{yw,1} \\ C_e & \tilde{D}_{ew,1} \end{pmatrix}.$$

■

We can now use Proposition 2.4.1 to substantiate that the use of Assumption A.3 instead of A.3a and A.3b is indeed without any loss of generality. To do so, we consider a state variable transformation as

$$\begin{pmatrix} \tilde{x} \\ \tilde{w} \end{pmatrix} = T^a \begin{pmatrix} x \\ w \end{pmatrix},$$

where T^a is as defined in Proposition 2.4.1. Also, consistent with the partitioning indicated for $\tilde{S}$, $\tilde{E}_w$, and $\tilde{D}_{yw}$, we partition $\tilde{w}$ into two parts as

$$\tilde{w} = \begin{pmatrix} w_1 \\ w_2 \end{pmatrix}. \tag{2.23}$$

In the new coordinates $\tilde{x}$ and $\tilde{w}$, the original equations which describe the plant are replaced by the equations of the form,

$$\begin{aligned} \rho\tilde{x} &= A\tilde{x} + Bu + E_{w,1}w_1, \\ y &= C_y\tilde{x} + \tilde{D}_{yw,1}w_1, \\ e &= \tilde{C}_e\tilde{x} + \tilde{D}_{ew,1}w_1 + D_{eu}u, \end{aligned} \tag{2.24}$$

for some $\tilde{C}_e$, while those describing the exosystem Σ_E are replaced by the equations of the form,

$$\rho w_1 = S_{11} w_1,$$
$$\rho w_2 = S_{21} w_1 + S_{22} w_2.$$

It is seen that only w_1 affects $\tilde{x}$ and e, i.e. w_2 does not play any role in the plant equations as given in (2.24). Thus, solving the output regulation problem for the original plant Σ and the given exosystem Σ_E is equivalent to solving the output regulation problem for the plant given in (2.24) where the exogenous signal w_1 is generated by the *reduced* exosystem,

$$\rho w_1 = S_{11} w_1. \tag{2.25}$$

We finally note that for the plant (2.24) and exosystem (2.25), Assumption A.3 holds. It is obvious that the system is detectable from y and e. Assumption A.3b then guarantees that the system is also detectable from y only.

Remark 2.4.3 *There may be cases for which, in the notation of (2.23), the component w_1 of the state of the exosystem is vacuous. In these cases, since w has no influence on e, the output regulation problem becomes a trivial problem of finding a controller that internally stabilizes the plant.*

2.5 Solvability conditions for the regulator equation

In the previous sections we saw that when S is anti-Hurwitz-stable for continuous-time systems and anti-Schur-stable for discrete-time systems, and under the stabilizability and detectability Assumptions A.1 and A.3, the necessary and sufficient condition for the existence of a regulator with either a state feedback or a measurement feedback is the solvability of the regulator equation (2.7). In this section we examine the solvability conditions for the regulator equation. This discussion is based on the material in Appendix 2.A which deals with the solvability conditions for general linear matrix equations.

Consider the regulator equation (2.7) which is rewritten here as

$$\Pi S = A\Pi + B\Gamma + E_w, \tag{2.26a}$$
$$0 = C_e\Pi + D_{eu}\Gamma + D_{ew}, \tag{2.26b}$$

where, as seen earlier, $A \in \mathbb{R}^{n\times n}$, $B \in \mathbb{R}^{n\times m}$, $E_w \in \mathbb{R}^{n\times s}$, $C_e \in \mathbb{R}^{q\times n}$, $D_{eu} \in \mathbb{R}^{q\times m}$, $D_{ew} \in \mathbb{R}^{q\times s}$, and $S \in \mathbb{R}^{s\times s}$. The above equation is of the form,

$$A_1 \begin{pmatrix} \Pi \\ \Gamma \end{pmatrix} - A_2 \begin{pmatrix} \Pi \\ \Gamma \end{pmatrix} S = R, \tag{2.27}$$

for

$$A_1 = \begin{pmatrix} A & B \\ C_e & D_{eu} \end{pmatrix}, \quad A_2 = \begin{pmatrix} I & 0 \\ 0 & 0 \end{pmatrix} \quad \text{and} \quad R = -\begin{pmatrix} E_w \\ D_{ew} \end{pmatrix}.$$

Thus, (2.27) is in the form of equation (2.48) of Appendix 2.A with A_1, A_2, and R as given above while $q_1(\lambda) = 1$ and $q_2(\lambda) = -\lambda$.

We have the following corollary of Theorem 2.A.1 of Appendix 2.A regarding the universal solvability of (2.26). We remark that the terms *universal* and *individual* solvability of an equation are defined in Appendix 2.A.

Corollary 2.5.1 *The regulator equation (2.26) is universally solvable for* Π *and* Γ *if and only if the matrix*

$$\begin{pmatrix} A - \lambda I & B \\ C_e & D_{eu} \end{pmatrix}$$

has full row-rank for each λ *which is an eigenvalue of* S.

Proof : Note that in the notation of Theorem 2.A.1, $A(\lambda)$ for the regulator equation (2.26) is given by

$$A(\lambda) = \begin{pmatrix} A & B \\ C_e & D_{eu} \end{pmatrix} - \lambda \begin{pmatrix} I & 0 \\ 0 & 0 \end{pmatrix} = \begin{pmatrix} A - \lambda I & B \\ C_e & D_{eu} \end{pmatrix}.$$

The result is then an immediate consequence of Theorem 2.A.1. ■

Note that Theorem 2.A.1 also implies that if $m = p$ and if the conditions of Corollary 2.5.1 are satisfied then the solution Π and Γ of (2.26) is unique. The work of [22] also quite implicitly gives a general characterization when the solution is unique. The following lemma states this result and gives necessary and sufficient conditions under which the solution of the regulator equation is unique. For the sake of completeness a proof is included.

Lemma 2.5.1 *The regulator equation (2.26) has at most one solution* Π *and* Γ *if and only if the system* (A, B, C_e, D_{eu}) *is left-invertible and its invariant zeros are not eigenvalues of* S.

Proof : Assume the conditions are not satisfied then

$$\begin{pmatrix} A - \lambda I & B \\ C_e & D_{eu} \end{pmatrix} \tag{2.28}$$

does not have full-row rank for at least one eigenvalue λ of S. But then there exists a left eigenvector v, with v a row vector, of S, i.e. we have $vS = \lambda S$ and a vector $(x^{\mathrm{T}}, u^{\mathrm{T}})^{\mathrm{T}}$ in the kernel of the matrix in (2.28). But then it is easy to check that

$$\Pi_0 = xv, \qquad \Gamma_0 = uv$$

satisfies

$$\Pi_0 S = A\Pi_0 + B\Gamma_0 \text{ and } C_e\Pi_0 + D_{eu}\Gamma_0 = 0,$$

but then if Π and Γ is a solution of the regulator equation then also $\Pi + \Pi_0$ and $\Gamma + \Gamma_0$ is a solution of the regulator equation, and hence the solution is not unique.

Conversely, if the system (A, B, C_e, D_{eu}) is square then uniqueness follows immediately from Theorem 2.A.1. If the system is not square, then we have to do some work. Obviously it is sufficient to show that the only solution of the set of equations

$$\Pi S = A\Pi + B\Gamma \text{ and } C_e\Pi + D_{eu}\Gamma = 0 \tag{2.29}$$

is zero. If the system is left-invertible then we can always find a suitable basis for the state space and the output space such that with respect to this basis we have,

$$A + KC = \begin{pmatrix} A_{11} & 0 \\ A_{21} & A_{22} \end{pmatrix}, \quad B + KD = \begin{pmatrix} 0 \\ B_2 \end{pmatrix},$$
$$C = \begin{pmatrix} C_{11} & 0 \\ C_{21} & C_{22} \end{pmatrix}, \quad D = \begin{pmatrix} 0 \\ D_2 \end{pmatrix}$$

with $(A_{22}, B_2, C_{22}, D_2)$ square and invertible. The condition that no invariant zero of (A, B, C_e, D_{eu}) is an eigenvalue of S then implies that

- (C_{11}, A_{11}) has no unobservable eigenvalues which are eigenvalues of S.
- No invariant zero of $(A_{22}, B_2, C_{22}, D_2)$ is an eigenvalue of S.

If we also decompose Π in this new basis,

$$\Pi = \begin{pmatrix} \Pi_1 \\ \Pi_2 \end{pmatrix},$$

then (2.29) expressed in the new basis yields,

$$\Pi_1 S = A\Pi_1, \quad C_{11}\Pi_1 = 0,$$

and due to the fact that (C_{11}, A_{11}) has no unobservable eigenvalues which are eigenvalues of S we find that $\Pi_1 = 0$. Using this result we find the following equations for Π_2 and Γ

$$\Pi_2 S = A_{22}\Pi_2 + B_2\Gamma \text{ and } C_{22}\Pi_2 + D_2\Gamma = 0$$

but since the system $(A_{22}, B_2, C_{22}, D_2)$ is square and its invariant zeros are not eigenvalues of S, we find by applying Theorem 2.A.1 that $\Pi_2 = 0$ and $\Gamma = 0$. ■

Next we would like to examine the individual solvability of the regulator equation (2.26). Before we do this, we first obtain a system Σ_{disc} from the following system,

$$\Sigma : \begin{cases} \rho x = Ax + Bu + E_w w \\ \rho w = Sw \\ e = C_e x + D_{eu} u + D_{ew} w, \end{cases} \tag{2.30}$$

which is the interconnection of the given plant Σ and the exosystem Σ_{E}. The system Σ_{disc} is obtained from Σ by setting $E_w = 0$ and $D_{ew} = 0$, and by disconnecting the interconnections between the plant Σ and the exosystem Σ_{E}. Thus, Σ_{disc} is given by

$$\Sigma_{\text{disc}} : \begin{cases} \rho x = Ax + Bu \\ \rho w = Sw \\ e = C_e x + D_{eu} u. \end{cases} \tag{2.31}$$

Let us also define two polynomial matrices $P_\Sigma(\lambda)$ and $P_{\Sigma,\text{disc}}(\lambda)$,

$$P_\Sigma(\lambda) = \begin{pmatrix} \lambda I - A & -E_w & -B \\ 0 & \lambda I - S & 0 \\ C_e & D_{ew} & -D_{eu} \end{pmatrix},$$

$$P_{\Sigma,\text{disc}}(\lambda) = \begin{pmatrix} \lambda I - A & 0 & -B \\ 0 & \lambda I - S & 0 \\ C_e & 0 & -D_{eu} \end{pmatrix}.$$

Remark 2.5.1 *We note that the matrix $P_\Sigma(\lambda)$ is the Rosenbrock's system matrix [54] for the interconnection of the given plant Σ and the exosystem Σ_{E}. Moreover, we observe that $P_{\Sigma,disc}(\lambda)$ is obtained from $P_\Sigma(\lambda)$ by setting $E_w = 0$ and $D_{ew} = 0$. That is, $P_{\Sigma,disc}(\lambda)$ is the Rosenbrock's system matrix for the system Σ_{disc} given in (2.31).*

We now have the following corollary of Theorem 2.A.2 regarding the individual solvability of (2.26).

Corollary 2.5.2 *The regulator equation (2.26) is individually solvable for* Π *and* Γ *if and only if the polynomial matrices* $P_\Sigma(\lambda)$ *and* $P_{\Sigma,\,disc}(\lambda)$ *are unimodularly equivalent.*

Proof : It follows from Theorem 2.A.2 that the regulator equation (2.26) has a solution Π and Γ for a given D_{eu}, E_w, and D_{ew} if and only if the polynomial matrices $\overline{P}(\lambda)$ and $\overline{P}_{\text{disc}}(\lambda)$ are unimodularly equivalent where

$$\overline{P}(\lambda) = \begin{pmatrix} A-\lambda I & B & -E_w \\ C_e & 0 & -D_{ew} \\ 0 & 0 & \lambda I - S \end{pmatrix},$$

$$\overline{P}_{\text{disc}}(\lambda) = \begin{pmatrix} A-\lambda I & B & 0 \\ C_e & 0 & 0 \\ 0 & 0 & \lambda I - S \end{pmatrix}.$$

By row and column operations, we can now easily reduce $\overline{P}(\lambda)$ and $\overline{P}_{\text{disc}}(\lambda)$ to $P_\Sigma(\lambda)$ and $P_{\Sigma,\,\text{disc}}(\lambda)$. Hence the result. ■

With the help of the above discussion, we can now restate Theorem 2.4.1 as follows.

Theorem 2.5.1 *Consider the problem of exact output regulation with measurement feedback as defined in Problem 2.2.2. Let Assumptions A.1, A.2 and A.3 hold. Then, the considered problem is solvable if and only if the given system* Σ*, given by (2.30), and the disconnected system* Σ_{disc}*, given by (2.31), have the same invariant polynomials. Moreover, a suitable measurement feedback controller is given by (2.15a) and (2.15b) where* F*,* K_A*, and* K_S *are arbitrary matrices such that the matrices*

$$A+BF \quad \textit{and} \quad \begin{pmatrix} A+K_A C_y & E_w + K_A D_{yw} \\ K_S C_y & S + K_S D_{yw} \end{pmatrix}$$

are both Hurwitz-stable for continuous-time systems and Schur-stable for discrete-time systems.

Remark 2.5.2 *Note that it is not sufficient in the above theorem that* Σ *given by (2.30), and the disconnected system* Σ_{disc}*, given by (2.31), have the same invariant zeros. They also need to have the same invariant polynomials.*

Proof : It is known that two polynomial matrices are unimodularly equivalent if and only if they have the same invariant polynomials (see Gantmacher [21, section 6.5]). Then, the result follows from Theorem 2.4.1, Corollary 2.5.2, and Remark 2.5.1. ∎

The result of Theorems 2.4.1 and 2.5.1, under Assumptions A.1, A.2 and A.3, describes a condition which is necessary and sufficient for the existence of a solution to the exact output regulation problem with measurement feedback. From a different point of view, we may regard Assumptions A.1, A.2 and A.3 together with the condition that the regulator equation (2.26) is solvable, as a set of sufficient conditions for the existence of a solution to the exact output regulation problem with measurement feedback. In the following corollary, we provide a slightly different set of sufficient conditions which might be easier to check.

Corollary 2.5.3 *Consider the problem of exact output regulation with measurement feedback as defined in Problem 2.2.2. Let Assumptions A.1, A.2, A.3a and A.3b hold. Then, the considered problem is solvable if the matrix*

$$\begin{pmatrix} A-\lambda I & B \\ C_e & D_{eu} \end{pmatrix}$$

has full row-rank for each λ which is an eigenvalue of S.

Proof : With the help of Proposition 2.4.1, we have seen that if Assumption A.3a and A.3b hold but Assumption A.3 does not hold, without any loss of generality, one can construct a reduced order exosystem such that Assumption A.3 holds when that reduced order exosystem is used. Therefore, in view of Theorems 2.4.1 and 2.5.1, in order to prove the corollary, it suffices to verify that the regulator equation (2.26) is solvable; but this is obvious under the hypotheses of this corollary and in view of Corollary 2.5.1. ∎

2.6 The internal model principle

We would like to examine next the structure of a controller that solves the exact output regulation problem with measurement feedback. It is well known that the controller that solves the output regulation problem copies a part of the dynamics of the exosystem. For instance, in the classical case treated in [16], where $y = e$, it is known that if a controller achieves exact output regulation, then the controller must contain a copy of the exosystem. This is a celebrated result, and is known as the *internal model principle.* Note that this result appeared in [91] where it was connected to structural stability, the concept that requires that output regulation should be preserved under arbitrarily small perturbations of the plant parameters, which we will study in detail in Section 2.8.

Here we will consider whether we need a copy of the exosystem in the controller to achieve output regulation independent of whether or not we have structural stability. This result can be found in [16] but without proof. The need of one or even more copies of the exosystem in the controller to guarantee structural stability will be considered in the next Section 2.8.

We present below an extension of this classical result to the case where y and e are not necessarily equal. A related result for the case that y and e are different but in a frequency domain setting can be found in [50].

Theorem 2.6.1 *Let Assumptions A.1, A.2, and A.3 hold. If the exact output regulation problem via measurement feedback is solved by a certain controller of the form (2.5) with a realization* (A_c, B_c, C_c, D_c)*, then the regulator equation (2.7) has a solution* Π *and* Γ *such that the controller contains a copy of the unobservable dynamics of the pair* $(C_y\Pi + D_{yu}\Gamma + D_{yw}, S)$*, i.e. if there exists* S_o *and an injective matrix* Ψ *such that* $S\Psi = \Psi S_o$ *and* $(C_y\Pi + D_{yu}\Gamma + D_{yw})\Psi = 0$*, then there also exists* $\bar{\Psi}$ *injective such that* $A_c\bar{\Psi} = \bar{\Psi}S_o$.

Proof : For ease of exposition we only prove this result for the case of $D_{yu} = 0$. The general result goes similarly but is much more technical.

Suppose that we have a controller of the form (2.5) with a realization (A_c, B_c, C_c, D_c) that achieves output regulation. Then, it is easy to check that $e(t)$ tends to zero as t tends to infinity for all initial conditions if and only if there exists Π, Θ satisfying the linear equation (2.17).

If we choose $\Gamma = C_c\Theta + D_c(C_y\Pi + D_{yw})$, then it is obvious that we find Π and Γ that satisfy the regulator equation,

$$\begin{aligned} \Pi S &= A\Pi + B\Gamma + E_w, \\ 0 &= C_e\Pi + D_{eu}\Gamma + D_{ew}. \end{aligned}$$

For any eigenvector x of S with associated eigenvalue λ for which $(C_y\Pi + D_{yw})x = 0$, we obtain from (2.17b) that $\lambda\Theta x = \Theta S x = A_c\Theta x$, i.e. Θx is an eigenvector of A_c. This shows that unobservable eigenvalues of $(C_y\Pi + D_{yw}, S)$ are also eigenvalues of A_c as soon as we have shown that $\Theta x \neq 0$. The latter is easily seen by noting that the regulator equation (2.7) implies, after a basis transformation, that the pair

$$\left[\begin{pmatrix} C_y & C_y\Pi + D_{yw} \end{pmatrix}, \begin{pmatrix} A & -B\Gamma \\ 0 & S \end{pmatrix}\right]$$

is detectable. This yields the required result if the matrix S is diagonalizable, i.e. if the state space of the exosystem is spanned by the eigenvectors of S.

In order to prove the general case we define $\mathcal{V}$ to be the subspace spanned by the eigenvectors of S which are unobservable with respect to $C_y\Pi + D_{yw}$. We know that $\Theta|_{\mathcal{V}}$ is injective and that $\Theta\mathcal{V}$ is an invariant subspace of A_c. By choosing a suitable basis for the state space of the exosystem and for the state space of the controller, we can bring the relevant matrices into the following form:

$$\begin{aligned} S &= \begin{pmatrix} S_1 & S_{12} \\ 0 & S_{22} \end{pmatrix}, \quad \mathcal{V} = \operatorname{im}\begin{pmatrix} I \\ 0 \end{pmatrix}, \quad \Theta = \begin{pmatrix} -I & \Theta_{12} \\ 0 & \Theta_{22} \end{pmatrix}, \\ A_c &= \begin{pmatrix} S_1 & A_{c,12} \\ 0 & A_{c,22} \end{pmatrix}, \ B_c = \begin{pmatrix} B_{c,1} \\ B_{c,2} \end{pmatrix}, \quad C_c = \begin{pmatrix} C_{c,1} & C_{c,2} \end{pmatrix}, \\ \Gamma &= \begin{pmatrix} \Gamma_1 & \Gamma_2 \end{pmatrix}, \quad \Pi = \begin{pmatrix} \Pi_1 & \Pi_2 \end{pmatrix}, \quad D_{yw} = \begin{pmatrix} -C_y\Pi_1 & D_{yw2} \end{pmatrix}. \end{aligned}$$

In the basis we have chosen we find that the following controller

$$\begin{aligned} \rho v_2 &= A_{c,22}\, v_2 + B_{c,2}\, y \\ \bar{u} &= \begin{pmatrix} A_{c,12} \\ C_{c,2} \end{pmatrix} v_2 + \begin{pmatrix} B_{c,1} \\ D_c \end{pmatrix} y \end{aligned}$$

solves the output regulation problem for a new system given by

$$\begin{aligned} \rho\bar{x} &= \begin{pmatrix} A & -B\Gamma_1 \\ 0 & S_1 \end{pmatrix}\bar{x} + \begin{pmatrix} 0 & B \\ I & 0 \end{pmatrix}\bar{u} + \begin{pmatrix} -B\Gamma_2 \\ S_{12} \end{pmatrix}\bar{w} \\ \rho\bar{w} &= S_{22}\bar{w} \\ y &= \begin{pmatrix} C_y & 0 \end{pmatrix}\bar{x} + (C_y\Pi_2 + D_{yw2})\bar{w} \\ e &= \begin{pmatrix} C_e & -D_{eu}\Gamma_1 \end{pmatrix}\bar{x} + \begin{pmatrix} 0 & D_{eu} \end{pmatrix}\bar{u} - D_{eu}\Gamma_2\bar{w}. \end{aligned}$$

which can be verified if we decompose the state space of the controller (v_1, v_2) and the exosystem (w_1, w_2) accordingly and choose $\bar{x}_1 = x - \Pi w$, $\bar{x}_2 = v_1 + w_1$ and $\bar{w} = w_2$. Moreover, the above new system satisfies Assumptions A.1, A.2, and A.3. Also, the regulator equation of the type (2.7) is solvable for the new system. Therefore, we can use a similar argument as before.

We define the space spanned by the unobservable eigenvectors of S_{22} and prove that this part of the dynamics of the exosystem must be a part of the controller. After a finitely many steps the exosystem is observable with respect to y (given $\bar{x}$). This then yields that the unobservable part was indeed copied by the controller. ∎

Trivially, if there is only one solution to (2.7) then the controller must contain the unobservable dynamics associated with that unique solution.

Corollary 2.6.1 *Let Assumptions A.1, A.2, and A.3a be satisfied. If there exists a unique solution* (Π, Γ) *of the regulator equation (2.7) and the output regulation problem via measurement feedback is solved by a controller of the form (2.5) with a realization* (A_c, B_c, C_c, D_c)*, then this controller must contain a copy of the unobservable dynamics of the pair* $(C_y\Pi + D_{yu}\Gamma + D_{yw}, S)$*, i.e. if there exists* S_o *and an injective matrix* Ψ *such that* $S\Psi = \Psi S_o$ *and* $(C_y\Pi + D_{yu}\Gamma + D_{yw})\Psi = 0$*, then there also exists* $\bar{\Psi}$ *injective such that* $A_c\bar{\Psi} = \bar{\Psi}S_o$.

Remark 2.6.1 *We note that for an important subset of systems where the subsystem characterized by* (A, B, C_e, D_{eu}) *is left invertible and has no invariant zeros which are eigenvalues of* S*, the solution of the regulator equation (2.7) is unique. Hence in this case the above corollary can be applied.*

In order to state the *internal model principle* for the case of non-unique solutions to the regulator equation, we need to examine the unobservable dynamics of the pair $(C_y\Pi + D_{yu}\Gamma + D_{yw}, S)$ which obviously depends on Π and Γ. In general, the set of all (Π, Γ) satisfying the regulator equation is an affine set and the algebraic multiplicity of each unobservable eigenvalue of the pair $(C_y\Pi + D_{yu}\Gamma + D_{yw}, S)$ is generically independent of the particular solution of the regulator equation, i.e. for an open and dense subset $\mathcal{R}$ of all solutions the multiplicity is independent of the particular solution in $\mathcal{R}$. For non-generic points the algebraic multiplicity can only be higher. We call solutions of the regulator equation in $\mathcal{R}$ *generic solutions of the regulator equation*. We then have the following corollary.

Corollary 2.6.2 *Let Assumptions A.1, A.2, and A.3a be satisfied. Let* (Π, Γ) *be a generic solution of the regulator equation (2.7). If the output regulation problem via measurement feedback is solved by a certain controller of the form (2.5) with a realization* (A_c, B_c, C_c, D_c)*, then this controller has the property that each unobservable eigenvalue of the pair* $(C_y\Pi + D_{yu}\Gamma + D_{yw}, S)$ *is an eigenvalue of* A_c *with at least the same algebraic multiplicity.*

Proof : From Theorem 2.6.1 we know that there exist solutions Π and Γ of the regulator equation such that the controller contains a copy of the unobservable dynamics of the pair $(C_y\Pi + D_{yu}\Gamma + D_{yw}, S)$. The associated algebraic multiplicities are always higher than the algebraic multiplicities of the unobservable dynamics of $(C_y\Pi + D_{yu}\Gamma + D_{yw}, S)$ associated with a generic solution (Π_g, Γ_g) of the regulator equation. ∎

Remark 2.6.2 *To find a* generic *solution, we observe that a random choice among all solutions will be generic with probability* 1. *Therefore, a reasonable method is to randomly pick a number of solutions and look at the smallest unobservable dynamics among the different solutions. The chance that there is one solution for which the unobservable dynamics is too large is already* 0, *so we can be quite certain that we did not pick only exception points.*

Remark 2.6.3 ***(Error feedback and state feedback cases)*** *We would like to observe two special cases of Theorem 2.6.1; one special case corresponds to the error feedback, i.e.* $y = e$*, and the other corresponds to state feedback, i.e. both* x *and* w *are available for feedback. For the first case where* $y = e$*, we have* $C_e = C_y$, $D_{eu} = D_{yu}$*, and* $D_{ew} = D_{yw}$*, and thus* $C_y\Pi + D_{yu}\Gamma + D_{yw} = 0$ *by the definition of* Π*. Hence, in this case the controller must contain a copy of the complete exosystem. For the second case of state feedback, it is easy to check that the pair* $(C_y\Pi + D_{yu}\Gamma + D_{yw}, S)$ *is observable. Hence, the controller does not need to copy any dynamics of the exosystem. In fact, for this special case, as is shown in Theorem 2.3.1, a static state feedback controller can achieve output regulation.*

2.7 Well-posedness of the output regulation problem

In previous sections we considered output regulation problems with full information feedback, and with general measurement feedback. We showed under which conditions such problems are solvable for the given system Σ. Also, whenever such problems are solvable, we constructed appropriate controllers. Basically, the existence of a solution to any output regulation problem is equivalent to the existence of a solution to a special pair of linear matrix equations called together as *regulator equation.* The solvability of the regulator equation depends on the data of the given system as specified by the set of matrices $(A, B, E_w, C_e, D_{eu}, D_{ew}, C_y, D_{yu}, D_{yw})$ and the data of our exosystem characterized by S. Our interest in this section is to study the so called *well-posedness of an output regulation problem* for the given data. We note that a mathematical problem is called *well-posed* if it is solvable and it remains solvable after a small perturbation or variation of the data of the problem. Thus the purpose of this section is to examine how much our ability of solving an output regulation problem is *sensitive to variations of the parameters* which characterize the model of the said controlled system and the associated exosystem. In other words, we would like to examine here to what extent the existence of a solution to the output regulation problem is influenced by the unpredictable and ever present perturbations in the parameters that define such a problem.

Let us next briefly discuss the nature of parameter variations in any given system. As is well documented in the literature, in practical situations, the equations that describe a given plant or system are always approximate for a number of reasons. That is, the parameters of a given system Σ, namely for our discussion the set of matrices (A, B, E_w, C_e, D_{eu}, D_{ew}, C_y, D_{yu}, D_{yw}), are not exactly known. As such one assumes a set of nominal values (A_0, B_0, $E_{w,0}$, $C_{e,0}$, $D_{eu,0}$, $D_{ew,0}$, $C_{y,0}$, $D_{yu,0}$, $D_{yw,0}$) for them. Similarly the parameters of the matrix S are not exactly known and we also assume a nominal value S_0. In order to depict this situation, one can regard (A, B, E_w, C_e, D_{eu}, D_{ew}, C_y, D_{yu}, D_{yw}, S) as an element of a space of parameters

$$\mathcal{P} = \mathbb{R}^{n\times n} \times \mathbb{R}^{n\times m} \times \mathbb{R}^{n\times s} \times \mathbb{R}^{q\times n} \times \mathbb{R}^{q\times m} \\ \times \mathbb{R}^{q\times s} \times \mathbb{R}^{p\times n} \times \mathbb{R}^{p\times m} \times \mathbb{R}^{p\times s} \times \mathbb{R}^{s\times s}.$$

Uncertainty in the values of the parameters, within known intervals around certain nominal values, can be simply expressed by allowing the set (A, B, E_w, C_e, D_{eu}, D_{ew}, C_y, D_{yu}, D_{yw}, S) to vary in some neighborhood $\mathcal{P}_0$ of a nominal set (A_0, B_0, $E_{w,0}$, $C_{e,0}$, $D_{eu,0}$, $D_{ew,0}$, $C_{y,0}$, $D_{yu,0}$, $D_{yw,0}$, S_0) in $\mathcal{P}$. Also, to emphasize that the parameters (A, B, E_w, C_e, D_{eu}, D_{ew}, C_y, D_{yu}, D_{yw}, S) are known only with certain tolerances around the nominal values (A_0, B_0, $E_{w,0}$, $C_{e,0}$, $D_{eu,0}$, $D_{ew,0}$, $C_{y,0}$, $D_{yu,0}$, $D_{yw,0}$, S_0), one can write the equation

$$(A, B, E_w, C_e, D_{eu}, D_{ew}, C_y, D_{yu}, D_{yw}, S) = \\ (A_0 + \delta A, B_0 + \delta B, E_{w,0} + \delta E_w, C_{e,0} + \delta C_e, D_{eu,0} + \delta D_{eu}, \\ D_{ew,0} + \delta D_{ew}, C_{y,0} + \delta C_y, D_{yu} + \delta D_{yu}, D_{yw,0} + \delta D_{yw}, S_0 + \delta S_0),$$

where the set $(\delta A, \delta B, \delta E_w, \delta C_e, \delta D_{eu}, \delta D_{ew}, \delta C_y, \delta D_{yu}, \delta D_{yw}, \delta S)$ represents the variations of (A, B, E_w, C_e, D_{eu}, D_{ew}, C_y, D_{yu}, D_{yw}, S) from the nominal set (A_0, B_0, $E_{w,0}$, $C_{e,0}$, $D_{eu,0}$, $D_{ew,0}$, $C_{y,0}$, $D_{yu,0}$, $D_{yw,0}$, S_0). In this representation, we are assuming that all the parameters of the given plant or system are susceptible to variations. However, in a given situation because of the physics of the plant or system, it may well be that the mathematical set of parameters $(A, B, E_w, C_e, D_{eu}, D_{ew}, C_y, D_{yu}, D_{yw}, S)$ has a special structure which causes these parameters to lie on a given hyper-surface in the parameter space $\mathcal{P}$. In our discussion, we would ignore this kind of special situation, i.e. we always assume that all the parameters of the given system are susceptible to variations. In fact, restricting the indeterminacy to only some, but not all, parameters complicates the analysis, and indeed is not very useful.

In the literature, it is generally assumed that the exosystem characterized by the matrix S is not subject to parameter variations. This is a reasonable assumption when one recognizes that the exosystem is simply an artificial device introduced to model the reference outputs as well as the disturbance inputs affecting the system, and not a real object whose physical parameters cannot exactly be determined. On the other hand, disturbances never behave completely as expected and reference inputs will due to implementation difficulties never be completely equal to the modeled signal. We will show that perturbations of S do not really affect well-posedness. This situation is very different for structural stability as we will see in the next section.

Before we analyze the well-posedness of an output regulation problem, we would like to consider the well-posedness of a set of linear equations. Let $\mathcal{X}$ and $\mathcal{Y}$ be finite dimensional linear spaces and let $A : \mathcal{X} \rightarrow \mathcal{Y}$ be a linear map and $b \in \mathcal{Y}$. Consider the linear equation

$$Ax = b \tag{2.32}$$

in the variable x. We say that equation (2.32) is well-posed at the nominal parameter values (A_0, b_0) if there exists a neighborhood of (A_0, b_0) in the parameter space such that (2.32) is solvable in this neighborhood. The following lemma characterizes the well-posedness of (2.32).

Lemma 2.7.1 *Equation (2.32) in the variable x is well-posed at the nominal parameter values (A_0, b_0) if and only if A_0 is surjective.*

Proof : If A_0 is surjective, it will have a submatrix with non-zero determinant of dimension equal to the number of rows. Since this determinant is a continuous function of the entries of the matrix (in fact, a polynomial) it follows that it will remain non-zero when the entries are perturbed a little bit. Hence A_0 will remain surjective, after a small perturbation. Therefore, equation (2.32) remains solvable for small perturbations of A_0 and b_0. Obviously, b_0 can be perturbed in an arbitrary way and not just locally.

Conversely, if A_0 is not surjective, im A_0 is a proper subspace of $\mathcal{Y}$. Also, equation (2.32) is solvable if and only if $b_0 \in \operatorname{im} A_0$. However, b_0 cannot be an interior point of im A_0, since im A_0 contains no interior points. Consequently, an arbitrary small perturbation of b_0 may take it out of im A_0 and hence destroys the solvability of (2.32). ∎

We now proceed to examine the well-posedness of an output regulation problem the precise meaning of which is given in the following definition.

Definition 2.7.1 *(Well-posedness) For a system Σ as in (2.1), the exact output regulation problem with state feedback or with measurement feedback as defined respectively in Problems 2.2.1 and 2.2.2 is said to be well-posed at (A_0, B_0, $E_{w,0}$, $C_{e,0}$, $D_{eu,0}$, $D_{ew,0}$, $C_{y,0}$, $D_{yu,0}$, $D_{yw,0}$, S_0) if there exists a neighborhood $\mathcal{P}_0$ of (A_0, B_0, $E_{w,0}$, $C_{e,0}$, $D_{eu,0}$, $D_{ew,0}$, $C_{y,0}$, $D_{yu,0}$, $D_{yw,0}$, S_0) in the parameter space $\mathcal{P}$ such that the considered problem is solvable for each element (A, B, E_w, C_e, D_{eu}, D_{ew}, C_y, D_{yu}, D_{yw}, S) of $\mathcal{P}_0$.*

In studying the well-posedness of the exact output regulation problem with either measurement feedback or state feedback, we note that Assumptions A.1, A.3 are invariant under small perturbations. Assumption A.2 is not invariant to small perturbations. However, Assumption A.2 is needed to guarantee that the regulator problem is solvable only if the regulator equations are solvable. If Assumptions A.1 and A.3 (but not necessarily Assumption A.2 are satisfied then solvability of the regulator equation is always a sufficient condition for solvability of the output regulation problem. Therefore, the output regulation problem is well-posed (both in the case of measurement feedback and in the case of state feedback) if and only if the regulator equation (2.7) is solvable not only for the nominal parameters (A_0, B_0, $E_{w,0}$, $C_{e,0}$, $D_{eu,0}$, $D_{ew,0}$, $C_{y,0}$, $D_{yu,0}$, $D_{yw,0}$, S_0) but also for all the parameters (A, B, E_w, C_e, D_{eu}, D_{ew}, C_y, D_{yu}, D_{yw}, S) sufficiently close to (A_0, B_0, $E_{w,0}$, $C_{e,0}$, $D_{eu,0}$, $D_{ew,0}$, $C_{y,0}$, $D_{yu,0}$, $D_{yw,0}$, S_0). In other words, the equation

$$\Pi S - A\Pi - B\Gamma = E_w, \tag{2.33a}$$

$$C_e\Pi + D_{eu}\Gamma = -D_{ew}, \tag{2.33b}$$

with variables Π and Γ should be well-posed at the nominal parameters (A_0, B_0, $E_{w,0}$, $C_{e,0}$, $D_{eu,0}$, $D_{ew,0}$, $C_{y,0}$, $D_{yu,0}$, $D_{yw,0}$, S_0). Now, from Lemma 2.7.1, it follows that this is the case if and only if the linear map

$$\mathcal{L} : (\Pi, \Gamma) \to (\Pi S - A\Pi - B\Gamma,\ C_e\Pi + D_{eu}\Gamma)$$

is surjective at $(A_0, B_0, E_{w,0}, C_{e,0}, D_{eu,0}, D_{ew,0}, C_{y,0}, D_{yu,0}, D_{yw,0})$. In order to check this condition, one can give a matrix representation of this map using tensor products. However, utilizing the material in Appendix 2.A, one can follow a different procedure, based on the general conditions on the solvability of matrix equations. We conclude from this discussion that the well-posedness of (2.33) is equivalent to the requirement that (2.33) be solvable for any E_w and D_{ew} at the nominal values of the parameters (A, B, C_e, D_{eu}, S), i.e. when they take the values $(A_0, B_0, C_{e,0}, D_{eu,0}, S_0)$. In other words, we require the universal solvability of (2.33) at $(A_0, B_0, C_{e,0}, D_{eu,0})$ (see Appendix 2.A for the definition of universal solvability). Now we can utilize the Corollary 2.5.1 to find the conditions for the well-posedness of the output regulation problem. Since state feedback can be considered as a special case of the measurement feedback, the following theorem formalizes the above discussion only for the exact output regulation problem with measurement feedback.

Theorem 2.7.1 *Consider a system Σ as in (2.1) and the exact output regulation problem with measurement feedback as defined in Problem 2.2.2. Let Assumptions A.1, A.2, A.3 hold, for the nominal values*

$$\begin{aligned}(A, B, E_w, C_e, D_{eu}, D_{ew}, C_y, D_{yu}, D_{yw}, S)\\ = (A_0, B_0, E_{w,0}, C_{e,0}, D_{eu,0}, D_{ew,0}, C_{y,0}, D_{yu,0}, D_{yw,0}, S_0).\end{aligned}$$

Then the exact output regulation problem with measurement feedback for Σ is well-posed at $(A_0, B_0, E_{w,0}, C_{e,0}, D_{eu,0}, D_{ew,0}, C_{y,0}, D_{yu,0}, D_{yw,0}, S_0)$ if and only if the matrix

$$\begin{pmatrix} A_0 - \lambda I & B_0 \\ C_{e,0} & D_{eu,0} \end{pmatrix}$$

has full row-rank for each λ which is an eigenvalue of S_0.

Proof : To prove the sufficiency part of the theorem, we note that by continuity there exists a neighborhood $\mathcal{P}_0$ of $(A_0, B_0, E_{w,0}, C_{e,0}, D_{eu,0}, D_{ew,0}, C_{y,0}, D_{yu,0}, D_{yw,0}, S_0)$ such that, for each $(A, B, E_w, C_e, D_{eu}, D_{ew}, C_y, D_{yu}, D_{yw}, S) \in \mathcal{P}_0$, (A, B) is stabilizable, and the matrix

$$\begin{pmatrix} A - \lambda I & B \\ C_e & D_{eu} \end{pmatrix}$$

has full row-rank for each λ that is not an eigenvalue of S. The later property implies by Corollary 2.5.3 that the regulator equation (2.7) is solvable. Therefore, by Theorem 2.4.1, the measurement feedback output regulation problem is solvable for each $(A, B, E_w, C_e, D_{eu}, D_{ew}, C_y, D_{yu}, D_{yw}, S) \in \mathcal{P}_0$.

To prove the necessity, we note that again by Corollary 2.5.3, if the condition fails, the set of all (E_w, D_{ew}) for which the linear equations,

$$\begin{aligned} \Pi S &= A_0\Pi + B_0\Gamma + E_w, \\ 0 &= C_{e,0}\Pi + D_{eu,0}\Gamma + D_{ew}, \end{aligned}$$

are solvable, spans only a proper subspace of $\mathbb{R}^{n\times s} \times \mathbb{R}^{q\times s}$. Thus, the existence condition expressed by Theorem 2.4.1 cannot be satisfied at each $(A, B, E_w, C_e, D_{eu}, D_{ew}, C_y, D_{yu}, D_{yw}, S)$ in a neighborhood $\mathcal{P}_0$ of $(A_0, B_0, E_{w,0}, C_{e,0}, D_{eu,0}, D_{ew,0}, C_{y,0}, D_{yu,0}, D_{yw,0}, S_0)$. ■

2.8 Structural stability of exact output regulation

The previous section considered the issue of well-posedness. Clearly, the property of well-posedness by itself is not very useful from a practical point of view. This is because, as our discussion in the previous section indicates, the property of well-posedness guarantees merely the existence of a controller that solves the exact output regulation problem in a neighborhood of the nominal system. The design or construction of such a controller could very well depend on the actual parameters. This implies that a redesigning of the controller might be needed each time the parameter values change. This is very unrealistic. To add to the difficulty, as discussed earlier, the actual parameter values are seldom known in practice.

An interesting solution to the above predicament would be to seek a *fixed* controller that solves the exact output regulation problem not only for the nominal system but also for systems obtained by small perturbations of the system parameters. Thus, even if the values of the system parameters drift but are confined to the given neighborhood, the same controller always achieves output regulation. In this sense, one fixed controller solves the exact output regulation problem for a *family* of plants, i.e. the family of all plants whose parameters range in a given neighborhood of the nominal point in a parameter space. In the literature many different versions have been studied, see [14, 15, 18, 19, 51]. These papers do not use consistent definitions. They differ in several ways:

- the choice of which system parameters are subject to perturbations,
- whether or not perturbations in the controller parameters are included,

- whether or not the error signal is part of or equal to the complete measurement signal.

A major issue is the exosystem. After almost any perturbations of the matrix S, the closed-loop system will no longer achieve output regulation. This is due to the fact that the system must have a zero from w to e at all frequencies of the exosystem, i.e. for every eigenvalue of S. A system, unless it achieves complete disturbance decoupling, will only have a finite number of zeros. Therefore, any perturbation of S and its eigenvalues will result in a closed-loop system which no longer achieves output regulation. This argument can also be made related to the internal model principle. After all since the controller must generally copy all or part of the exosystem, a perturbation of the exosystem without perturbing the controller will result in a lack of an internal model and hence output regulation is no longer achieved. Therefore in the literature, structural stability is always studied with perturbations of the plant only.

Note that it is easy to see that if we allow for arbitrary perturbations on C_e, D_{eu}, D_{ew} while the other system parameters remain fixed then we can never find one controller that achieves output regulation for the perturbed system. After all if the controller is designed to yield,

$$C_e x + D_{eu} u + D_{ew} w \to 0 \text{ as } t \to \infty,$$

then we find that $C_e x + D_{eu} u + (D_{ew} + \varepsilon I) w \to 0$ as $t \to \infty$ only if $w \to 0$ which is excluded by our basic Assumption A.2 and is any way a trivial case. It will turn out that we need to restrict these changes to preserve a coupling between the perturbations in C_e, D_{eu}, D_{ew} and C_y, D_{yu}, D_{yw} as given later on in (2.36).

The fact that the property of output regulation cannot be preserved under arbitrary perturbations of plant and exosystem parameters is actually not as bad as it may seem. A more natural question would be to ask for continuity with respect to parameter variations:

> Given any nominal values for the system parameters (A_0, B_0, $E_{w,0}$, $C_{e,0}$, $D_{eu,0}$, $D_{ew,0}$, $C_{y,0}$, $D_{yu,0}$, $D_{yw,0}$, S_0) for which the output regulation problem is solvable, find a controller such that for each $\varepsilon > 0$ there exists a neighborhood $\mathcal{P}_0$ of (A_0, B_0, $E_{w,0}$, $C_{e,0}$, $D_{eu,0}$, $D_{ew,0}$, $C_{y,0}$, $D_{yu,0}$, $D_{yw,0}$, S_0) in the parameter space $\mathcal{P}$ such that the interconnection of the given controller with any system with parameters (A, B, E_w, C_e, D_{eu}, D_{ew}, C_y, D_{yu}, D_{yw},

S) in $\mathcal{P}_0$ yields internal stability and

$$\lim_{t\to\infty} \|e(t)\| < \varepsilon.$$

This appears to be a more natural requirement which guarantees reasonable performance in a neighborhood around the nominal plant and this is actually achieved by any controller that achieves output regulation. Obviously additional requirements such as keeping ε small for a large neighborhood of the nominal plant yield interesting problems which at the moment are still unsolved.

We present here the output regulation problem with structural stability for two reasons. One is historical; we want to clarify issues with respect to structural stability which are sometimes quite unclear in the literature. Secondly, it yields a nice sufficient condition; obviously if we can achieve perfect tracking in a neighborhood of the nominal plant with little effort then it is obviously a nice property.

Before defining structural stability, we first state a fundamental structural requirement on plant data which is needed for structural stability. This structural requirement was alluded to before and basically implies that the error signal must be part of the measurements. In [19] the following result has been shown.

Lemma 2.8.1 *Consider a system Σ as in (2.1) with $D_{eu} = 0$ and $D_{yu} = 0$. Let Assumption A.2 hold, i.e. let the matrix S be anti-Hurwitz-stable for continuous-time systems and anti-Schur-stable for discrete-time systems. Assume that there exists a controller of the form (2.5) that achieves internal stability, achieves output regulation and is such that output regulation is preserved for small perturbations of E_w. Then we must have*

$$\ker C_e \supseteq \ker C_y.$$

If output regulation is preserved for small perturbations of E_w and B_c, then we must have

$$\ker \begin{pmatrix} C_e & D_{ew} \end{pmatrix} \supseteq \ker \begin{pmatrix} C_y & D_{yw} \end{pmatrix}. \tag{2.34}$$

In our case, where D_{eu} and D_{yu} are present, the above result cannot be applied but a similar result can be derived.

Lemma 2.8.2 *Consider a system Σ as in (2.1). Let Assumption A.2 hold, i.e. let the matrix S be anti-Hurwitz-stable for continuous-time systems and anti-Schur-stable for discrete-time systems. Assume that there exists a controller of* *the form (2.5) that achieves internal stability, achieves output regulation and is such that output regulation is preserved for small perturbations of E_w, B_c and C_c. Moreover assume that the McMillan degree of the controller is larger than m. Then we must have,*

$$\ker \begin{pmatrix} C_e & D_{eu} & D_{ew} \end{pmatrix} \supseteq \ker \begin{pmatrix} C_y & D_{yu} & D_{yw} \end{pmatrix}. \tag{2.35}$$

Note that (2.35) implies that for a suitable choice of basis we have

$$y = \begin{pmatrix} e \\ y_1 \end{pmatrix}.$$

In other words, essentially the error signal must be part of the measurements. The only discomforting fact about these results is that they are based on perturbations of the controller and system parameters instead of only perturbations of the system parameters. Of course, it is a nice property to be robust against perturbations of controller parameters as well. However, perturbations in controller parameters are strictly limited since the internal model principle tells us that in general the controller must copy a specific part of the exosystem and this is intrinsically not robust against perturbations in the controller. Hence, as is done in [19], we can in a certain case allow for perturbations of the controller input matrix B_c or even C_c as we did in our extension but it is almost never possible to guarantee robustness with respect to the controller state matrix A_c. But, as noted before, some additional structure in the problem is needed since arbitrary perturbations of C_e, D_{eu}, D_{ew}, C_y, D_{yu}, D_{yw} will remove any possibility to achieve output regulation by one controller for all perturbed systems in some neighborhood.

Based on the above discussion, in what follows, we will assume

$$C_y = \begin{pmatrix} C_e \\ C_{y2} \end{pmatrix}, \qquad D_{yu} = \begin{pmatrix} D_{eu} \\ D_{yu2} \end{pmatrix}, \qquad D_{yw} = \begin{pmatrix} D_{ew} \\ D_{yw2} \end{pmatrix}. \tag{2.36}$$

In other words, we assume that the error signal is a part of the measurement signal and by perturbing C_e, D_{eu}, D_{ew}, C_{y2}, D_{yu2}, D_{yw2} this property is preserved. This discussion leads to the following definition of a structurally stable output regulation problem.

Definition 2.8.1 ***(Structurally Stable Output Regulation Problem)*** *Consider a system Σ as in (2.1) with the additional structure given in (2.36). A fixed controller of the form given in (2.5) is said to solve a structurally stable output regulation problem for Σ at (A_0, B_0, $E_{w,0}$, $C_{e,0}$, $D_{eu,0}$, $D_{ew,0}$, $C_{y2,0}$, $D_{yu2,0}$, $D_{yw2,0}$) if it satisfies the following properties:*

(*i*) *The controller solves the exact output regulation problem with measurement feedback as defined in Problem 2.2.2 when the plant in (2.1) is characterized by the nominal set of parameters* (A_0, B_0, $E_{w,0}$, $C_{e,0}$, $D_{eu,0}$, $D_{ew,0}$, $C_{y2,0}$, $D_{yu2,0}$, $D_{yw2,0}$).

(*ii*) *There exist a neighborhood* $\mathcal{P}_0$ *of* (A_0, B_0, $E_{w,0}$, $C_{e,0}$, $D_{eu,0}$, $D_{ew,0}$, $C_{y2,0}$, $D_{yu2,0}$, $D_{yw2,0}$) *such that the controller achieves internal stability and output regulation for each set of perturbed plant parameters* ($A_0 + \delta A$, $B_0 + \delta B$, $E_{w,0} + \delta E_w$, $C_{e,0} + \delta C_e$, $D_{eu,0} + \delta D_{eu}$, $D_{ew,0} + \delta D_{ew}$, $C_{y2,0} + \delta C_{y2}$, $D_{yu2,0} + \delta D_{yu2}$, $D_{yw2,0} + \delta D_{yw2}$) *in* $\mathcal{P}_0$

The above definition obviously implies that, for the existence of a regulator that solves the structurally stable output regulation problem, the exact output regulation problem must necessarily be well-posed (with the obvious modifications implied by (2.36) and the lack of perturbations in S).

As shown below, it turns out that the necessary and sufficient condition given in Theorem 2.7.1 for the well-posedness of the exact output regulation problem with measurement feedback is indeed also the necessary and sufficient condition for the existence of a regulator that solves the structurally stable output regulation problem.

Theorem 2.8.1 *Consider a system* Σ *as in (2.1) with the structural constraint (2.36) and the output regulation problem with measurement feedback as defined in Problem 2.2.2. Let Assumptions A.1, A.2 and A.3 be satisfied.*

Then, there exists a regulator that solves the structurally stable output regulation problem for Σ *at* (A_0, B_0, $E_{w,0}$, $C_{e,0}$, $D_{eu,0}$, $D_{ew,0}$, $C_{y2,0}$, $D_{yu2,0}$, $D_{yw2,0}$) *if and only if the matrix*

$$\begin{pmatrix} A_0 - \lambda I & B_0 \\ C_{e,0} & D_{eu,0} \end{pmatrix} \tag{2.37}$$

has full row-rank for each λ *which is an eigenvalue of* S.

The proof of the necessity part is obvious from Theorem 2.7.1 (although with some care since in well-posedness we perturbed the exosystem and in structural stability we do not). Moreover, the sufficiency part is a direct consequence of the following theorem:

Theorem 2.8.2 *Consider a system* Σ *as in (2.1) with the structural constraint (2.36) and the output regulation problem with measurement feedback as defined in Problem 2.2.2. Let Assumptions A.1, A.2 and A.3 be satisfied.*

Assume the matrix (2.37) has full row-rank for each λ which is an eigenvalue of S. Then, there exists a regulator that solves the output regulation problem for the nominal plant with parameters (A_0, B_0, $E_{w,0}$, $C_{e,0}$, $D_{eu,0}$, $D_{ew,0}$, $C_{y2,0}$, $D_{yu2,0}$, $D_{yw2,0}$). Moreover, this controller achieves output regulation for each set of perturbed plant parameters (A, B, E_w, C_e, D_{eu}, D_{ew}, $C_{y2,0}$, $D_{yu2,0}$, $D_{yw2,0}$) for which the closed-loop system is internally stable.

Remark 2.8.1 *The importance of the above theorem is obviously that, by designing a controller which achieves robust stability for a large uncertainty set and structural stability locally, we are guaranteed that the system will achieve output regulation for any possible system perturbation in this large uncertainty set. So we can not only achieve* local *structural stability but we also obtain a methodology to achieve robust output regulation for systems subject to a large uncertainty set. We will discuss this in more detail in Chapter 12.*

Proof : The proof is by construction of an appropriate controller. A description of the construction is given in this section while a detailed proof is given in Appendix 2.B. ∎

Design of structurally stable measurement feedback regulators:

We will now describe how to find **all** controllers that achieve output regulation with structural stability. Note that this automatically implies that we achieve output regulation for all system parameters which preserve internal stability as considered in Theorem 2.8.2. Obviously we have to assume that the conditions given in Theorem 2.8.1 are satisfied. The detailed construction is divided into the following two parts:

(*i*) Given a plant and an exosystem as in (2.1) and (2.2), we apply a preliminary static output feedback which guarantees that the system we obtain has no poles in common with the exosystem.

(*ii*) We formulate an *auxiliary output regulation problem with measurement feedback* based on the system obtained after the preliminary static output feedback.

It will be shown that there actually exists a one-one correspondence between controllers which achieve output regulation for the auxiliary system and controllers which achieve output regulation with structural stability for the original system. Using this result it will also be shown that we always need q copies of the exosystem in the controller (where q is the dimension of the error signal).

The first step is based on the following lemma which is proven in Appendix 2.B.

Lemma 2.8.3 *Assume*

$$\begin{pmatrix} A_0 - \lambda I & B_0 \\ C_{e,0} & D_{eu,0} \end{pmatrix}$$

has full-row rank for all eigenvalues λ *of* S. *Then there exists a static preliminary feedback* $u = Ne + \tilde{u}$ *such that*

$$\bar{A} = A_0 + B_0 N (I - D_{eu,0} N)^{-1} C_{e,0}$$

has no eigenvalues in common with S.

We apply this preliminary feedback to the system (2.1) where we have the additional structure of (2.36). Note that this preliminary feedback is possible since e is part of the measurements y. Secondly, after we have proved existence of such a static preliminary feedback, it should be noted that any matrix N with proper dimensions generically has the above property, so construction of such an N is much easier than it might seem from the proof of the above lemma. After we apply this preliminary static feedback to the nominal system we obtain the following system

$$\bar{\Sigma} : \begin{cases} \rho x = \bar{A}_0 x + \bar{B}_0 \tilde{u} + \bar{E}_{w,0} w \\ y = \bar{C}_{y,0} x + \bar{D}_{yu,0} \tilde{u} + \bar{D}_{yw,0} w \\ e = \bar{C}_{e,0} x + \bar{D}_{eu,0} \tilde{u} + \bar{D}_{ew,0} w, \end{cases} \tag{2.38}$$

where

$$\begin{aligned}
\bar{A}_0 &= A_0 + B_0 N (I - D_{eu,0} N)^{-1} C_{e,0}, \\
\bar{B}_0 &= B_0 (I - N D_{eu,0})^{-1}, \\
\bar{E}_{w,0} &= E_{w,0} + B_0 N (I - D_{eu,0} N)^{-1} D_{ew,0}, \\
\bar{C}_{e,0} &= (I - D_{eu,0} N)^{-1} C_{e,0}, \\
\bar{D}_{eu,0} &= (I - D_{eu,0} N)^{-1} D_{eu,0}, \\
\bar{D}_{ew,0} &= (I - D_{eu,0} N)^{-1} D_{ew,0}, \\
\bar{C}_{y,0} &= C_{y,0} + D_{yu,0} N (I - D_{eu,0} N)^{-1} C_{e,0}, \\
\bar{D}_{yu,0} &= D_{yu,0} (I - N D_{eu,0})^{-1}, \\
\bar{D}_{yw,0} &= D_{yw,0} + D_{yu,0} N (I - D_{eu,0} N)^{-1} D_{ew,0}.
\end{aligned}$$

Note that perturbations of the above parameters are basically the same as perturbations of the original system (the shape of the neighborhood changes but not the fact that the neighborhood is open). Therefore, achieving structural stability for this new system is intrinsically the same problem as achieving structural stability for the original system. The only thing to remember is that the controllers are slightly different due to the preliminary feedback. Obviously we do have to take care that the perturbations preserve the structure in (2.36). This basically means that for this new system the perturbed parameters must satisfy,

$$\bar{C}_y = \begin{pmatrix} \bar{C}_e \\ \bar{C}_{y2} \end{pmatrix}, \qquad \bar{D}_{yu} = \begin{pmatrix} \bar{D}_{eu} \\ \bar{D}_{yu2} \end{pmatrix}, \qquad \bar{D}_{yw} = \begin{pmatrix} \bar{D}_{ew} \\ \bar{D}_{yw2} \end{pmatrix}. \tag{2.39}$$

The second step involves an extension of the exosystem. Let the plant and the exosystem be given as in (2.1) and (2.2). Without loss of generality, assume that the matrix S which characterizes the exosystem has already been transformed into a block diagonal matrix of the form,

$$S = \begin{pmatrix} S^* & 0 \\ 0 & S_{\min} \end{pmatrix}, \tag{2.40}$$

in which S^* is a certain matrix not of much concern to us, and $S_{\min}$ is a matrix whose characteristic polynomial coincides with the minimal polynomial of S. Moreover, $S_{\min}$ is a cyclic matrix, i.e. its characteristic and minimal polynomials coincide. Note that $S_{\min}$ can be constructed out of S by taking in the Jordan form for each eigenvalue one copy of the largest Jordan block associated with that eigenvalue. The remaining Jordan blocks are dumped into S^*. Let the integer $\tilde{s}$ be such that $S_{\min} \in \mathbb{R}^{\tilde{s}\times\tilde{s}}$.

With $S_{\min}$ as defined above, we define now an *auxiliary* exosystem as

$$\rho\tilde{w} = \tilde{S}_p\tilde{w} \tag{2.41}$$

where $\tilde{w} \in \mathbb{R}^{p\tilde{s}}$ and $\tilde{S}_p$ is a block diagonal matrix given by

$$\tilde{S}_p = \begin{pmatrix} S_{\min} & 0 & \cdots & 0 \\ 0 & S_{\min} & \ddots & \vdots \\ \vdots & \ddots & \ddots & 0 \\ 0 & \cdots & 0 & S_{\min} \end{pmatrix}.$$

We note that the *auxiliary* exosystem given in (2.41) is composed of p identical copies of a constituent exosystem where p is the dimension of the measurement signal y. We decompose $\tilde{S}_p$ as follows:

$$\tilde{S}_p = \begin{pmatrix} \tilde{S}_q & 0 \\ 0 & \tilde{S}_{p-q} \end{pmatrix}$$

where $\tilde{S}_i$ has the obvious interpretation as a block-diagonal matrix composed of i identical copies of $S_{\min}$ and q is the dimension of the error signal.

We define an *auxiliary* system composed of the extended plant (2.1) and the *auxiliary* exosystem (2.41),

$$\tilde{\Sigma}: \begin{cases} \rho\tilde{x} = \bar{A}_0\tilde{x} + \bar{B}_0\tilde{u} \\ \rho\tilde{w} = \tilde{S}_p\tilde{w} \\ y = \bar{C}_{y,0}\tilde{x} + \bar{D}_{yu,0}\tilde{u} + \tilde{D}_{yw}\tilde{w} \\ e = \bar{C}_{e,0}\tilde{x} + \bar{D}_{eu,0}\tilde{u} + \tilde{D}_{ew}\tilde{w} \end{cases} \tag{2.42}$$

and the matrix $\tilde{D}_{ew}$ and $\tilde{D}_{yw}$ are partitioned as

$$\tilde{D}_{yw} = \begin{pmatrix} \tilde{D}_{ew} \\ \tilde{D}_{yw2} \end{pmatrix} = \begin{pmatrix} \tilde{D}_{ew1} & 0 \\ 0 & \tilde{D}_{yw22} \end{pmatrix}$$

where

$$\begin{aligned} \tilde{D}_{ew1} &= \begin{pmatrix} \tilde{D}_{ew1,1} & \tilde{D}_{ew1,2} & \cdots & \tilde{D}_{ew1,q} \end{pmatrix}, \\ \tilde{D}_{yw22} &= \begin{pmatrix} \tilde{D}_{yw22,1} & \tilde{D}_{yw22,2} & \cdots & \tilde{D}_{yw22,p-q} \end{pmatrix}. \end{aligned}$$

Here the matrices $\tilde{D}_{ew1}$ and $\tilde{D}_{yw22}$ are selected so that the pairs of matrices $(\tilde{D}_{ew1}, \tilde{S}_q)$ and $(\tilde{D}_{yw22}, \tilde{S}_{p-q})$ are detectable. Construction of $\tilde{D}_{ew1}$ and $\tilde{D}_{yw22}$ is actually quite easy. Choose any row-vector R such that $(R, S_{\min})$ is detectable which is possible since $S_{\min}$ is cyclic. Then we can choose

$$\tilde{D}_{ew1,i} = e_i R, \qquad \tilde{D}_{yw22,i} = f_i R$$

where e_i $(i = 1, \ldots, q)$ form a basis of $\mathbb{R}^q$ while f_i $(i = 1, \ldots, p-q)$ form a basis of $\mathbb{R}^{p-q}$.

In Appendix 2.B it is shown that the auxiliary system (2.42) satisfies Assumptions A.1, A.2, and A.3 and the regulator equation for this auxiliary system are solvable due to the assumption that (2.37) has full row-rank for each λ which is an eigenvalue of S.

We now proceed with the construction of a regulator that achieves output regulation for the auxiliary system. We follow the design procedure presented on page 29. Let F be an arbitrary matrix such that $A_0 + B_0F$ is Hurwitz-stable for continuous-time systems and Schur-stable for discrete-time systems. Also, let the pair $(\tilde{K}_A, \tilde{K}_S)$ be such that the matrix

$$\begin{pmatrix} \bar{A}_0 + \tilde{K}_A\bar{C}_{y,0} & \tilde{K}_A\tilde{D}_{yw} \\ \tilde{K}_S\bar{C}_{y,0} & \tilde{S}_p + \tilde{K}_S\tilde{D}_{yw} \end{pmatrix}$$

is Hurwitz-stable for continuous-time systems and Schur-stable for discrete-time systems. Then, using the solution Π and Γ of

$$\Pi\tilde{S}_p = \bar{A}_0\Pi + \bar{B}_0\Gamma, \tag{2.43a}$$

$$0 = \bar{C}_{e,0}\Pi + \bar{D}_{eu,0}\Gamma + \tilde{D}_{ew}, \tag{2.43b}$$

we construct the following controller

$$\begin{pmatrix}\rho\hat{x}\\ \rho\hat{w}\end{pmatrix} = \begin{pmatrix}\bar{A}_0 & 0\\ 0 & \tilde{S}_q\end{pmatrix}\begin{pmatrix}\hat{x}\\ \hat{w}\end{pmatrix} + \begin{pmatrix}\bar{B}_0\\ 0\end{pmatrix}$$
$$+\begin{pmatrix}K_A\\ K_S\end{pmatrix}\left[\begin{pmatrix}\bar{C}_{y,0} & \tilde{D}_{yw}\end{pmatrix}\begin{pmatrix}\hat{x}\\ \hat{w}\end{pmatrix} + \bar{D}_{yu,0}u - y\right] \tag{2.44a}$$

$$\tilde{u} = \begin{pmatrix}F & (\Gamma - F\Pi)\end{pmatrix}\begin{pmatrix}\hat{x}\\ \hat{w}\end{pmatrix}. \tag{2.44b}$$

This controller then solves the output regulation problem for the auxiliary system.

It is shown in Appendix 2.B that any controller of the form (2.5) which solves the output regulation problem for the extended plant (2.42) and the exosystem (2.41) solves the structurally stable output regulation problem for the system (2.38) with exosystem (2.2).

Moreover, assume a controller of the form (2.5) is given and is characterized by (A_c, B_c, C_c, D_c) such that it achieves structurally stable output regulation for the system (2.38). Since this system is obtained by a preliminary feedback $u = Ne + \tilde{u}$, we obtain that a controller which achieves structurally stable output regulation for the original system (2.1) is parameterized by

$$(A_c, B_c, C_c, D_c + \begin{pmatrix}N\\ 0\end{pmatrix}).$$

One consequence of this result is given in the following corollary. This is an extension of an earlier result from [14] which considered the case $y = w$.

Corollary 2.8.1 *Any controller which achieves structural stability for the system (2.1) must contain at least q copies of the reduced exosystem*

$$\rho x_e = S_{\min} x_e.$$

Proof : The results in this section show that a controller $\tilde{\Sigma}_c$ with parameters (A_c, B_c, C_c, D_c) achieves structural stability for the auxiliary system (2.42) if and only if the controller Σ_C with parameters

$$(A_c, B_c, C_c, D_c + \begin{pmatrix} N \\ 0 \end{pmatrix}).$$

achieves output regulation for the original system Σ. By the internal model principle the controller $\tilde{\Sigma}_c$ must contain a copy of the unobservable dynamics of $(\bar{C}_{y,0}\Pi + \bar{D}_{yu,0}\Gamma + \tilde{D}_{yw}, \tilde{S}_p)$ where Π and Γ satisfy (2.43). Since we know that $\bar{C}_{e,0}\Pi + \bar{D}_{eu,0}\Gamma + \tilde{D}_{ew} = 0$, we obtain that the controller must contain the unobservable dynamics of $(\bar{C}_{y2,0}\Pi + \bar{D}_{yu,0}\Gamma + \tilde{D}_{yw2}, \tilde{S}_p)$. However, since $\bar{C}_{y2,0}\Pi + \bar{D}_{yu,0}\Gamma + \tilde{D}_{yw2}$ has only $p - q$ rows and $\tilde{S}_p$ contains p identical copies of a cyclic matrix which is anti-Hurwitz-stable for continuous-time systems and anti-Schur-stable for discrete-time systems, we find that at least q copies of $S_{\min}$ must be unobservable. ■

Remark 2.8.2 *Corollary 2.8.1 really points out a bad property of structural stability. In order to achieve structural stability, the existence conditions are not that strong but one really has to blow up the dimension of the controller to actually do it. In the paper [15] it was shown that (for the case $y = e$) even if we only perturb A, B, C_e then we already need to have q copies of the exosystem in the controller. Even in the case of tracking when we have the tracking signal w available for measurement, in order to achieve structural stability we need these multiple copies. This is why alternative setups such as continuity with respect to parameter variations as briefly outlined in the beginning of this section are of interest since these setups do not require these multiple copies of the exosystem in the controller.*

As already indicated, there are many different setups for structural stability which often vary in precisely which parameters are subject to perturbation. A general setup would be as depicted in Figure 2.3 on the facing page. We basically want to preserve output regulation when we vary Δ over a particular set. If this set is very small and can perturb a large subset of the system parameters then we are close in spirit with structural stability. But if the class over which Δ varies is large and only affects a small subset of the parameters then we are closer in spirit with robust control where one asks to guarantee performance for a particular set of model uncertainty. For details we refer to Chapter 12.

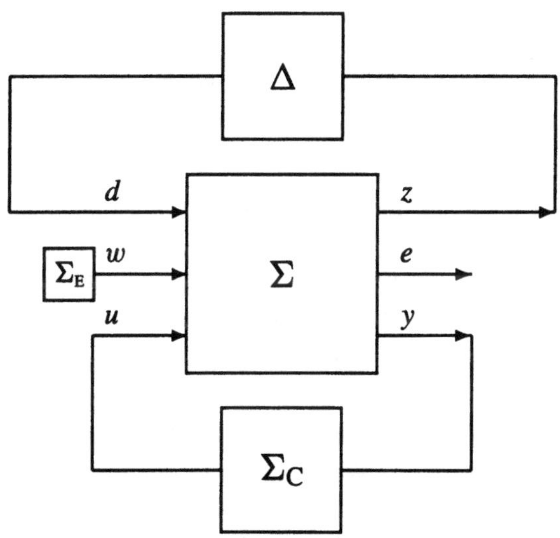

Figure 2.3: Output regulation with model uncertainty

2.A Linear matrix equations

In this appendix, we consider the solvability of certain linear matrix equations encountered often. The equations are of the form,

$$\sum_{i=1}^{k} A_i X S_i = R, \tag{2.45}$$

where A_i, S_i, and R are given matrices and X is an unknown. Hautus [23, 24] presents a detailed discussion on such equations while recalling historical origins of them. Our exposition here is an extract of [23,24]. Equation (2.45) is said to be *universally* solvable if it has a solution for every R; on the other hand, it is said to be *individually* solvable if it has a solution for a particular given R.

A well known example of (2.45) is what is known as *Sylvester Equation*,

$$AX - XS = R \tag{2.46}$$

where A and S are square matrices. It is seen that (2.46) is in the form of (2.45) with $A_1 = A$, $S_1 = I$, $A_2 = I$, and $S_2 = -S$. As proved in 1884 by Sylvester [82], equation (2.46) is universally solvable if and only if the matrices A and S have no eigenvalues in common.

A result for the general equation (2.45), in the same spirit as that of the Sylvester Equation, is not known. Thus, we restrict ourselves here to the case where the matrices S_i are of the form $S_i = q_i(S)$ for certain given polynomials q_i and a fixed square matrix S. We recall the following result.

Theorem 2.A.1 *Let $A_i \in \mathbb{R}^{\bar{n}\times\bar{m}}$, $S \in \mathbb{R}^{\bar{s}\times\bar{s}}$, and $R \in \mathbb{R}^{\bar{n}\times\bar{s}}$. Also, let $q_i(\lambda)$ be polynomials for $i = 1, \cdots, k$. Consider a matrix polynomial in the variable λ,*

$$A(\lambda) := \sum_{i=1}^{k} A_i q_i(\lambda). \tag{2.47}$$

Then the equation

$$\sum_{i=1}^{k} A_i X q_i(S) = R \tag{2.48}$$

is universally solvable if and only if the matrix $A(\lambda)$ has full row-rank for each λ which is an eigenvalue of S. If this is the case and $A(\lambda)$ is square, then the solution X is unique.

Theorem 2.A.2 *Let A_i, S, R, and $q_i(\lambda)$ be as in Theorem 2.A.1. Then the equation (2.48) is individually solvable if and only if the polynomial matrices*

$$\begin{pmatrix} A(\lambda) & R \\ 0 & \lambda I - S \end{pmatrix} \quad \textit{and} \quad \begin{pmatrix} A(\lambda) & 0 \\ 0 & \lambda I - S \end{pmatrix}$$

are unimodularly equivalent[1].

It is easy to see that the Sylvester equation (2.46) is a special case of equation (2.48) for $A_1 = A$, $q_1(\lambda) = 1$, $A_2 = I$, and $q_2(\lambda) = -\lambda$. Thus, $A(\lambda) = A - \lambda I$ for the Sylvester equation, and hence we can conclude from Theorem 2.A.1 that the Sylvester equation has a solution for every R if and only if the matrices A and S have no eigenvalues in common.

2.B The construction of a structurally stable regulator

We first present a proof of Lemma 2.8.3.

Proof of Lemma 2.8.3 : Obviously since $D_{eu,0}$ is a wide matrix there exists a small perturbation of $D_{eu,0}$ say $\tilde{D}_{eu,0}$ such that $\tilde{D}_{eu,0}$ is surjective with a right-inverse $\tilde{D}^{\mathrm{R}}_{eu,0}$ and such that

$$\begin{pmatrix} A_0 - \lambda I & B_0 \\ C_{e,0} & \tilde{D}_{eu,0} \end{pmatrix}$$

has full-row rank for all eigenvalues λ of S. Then it is easy to check that $A_0 - B_0 \tilde{D}^{\mathrm{R}}_{eu,0} C_{e,0}$ has no eigenvalues in common with S. Let $\alpha < 1$ be such that $A_0 - \alpha B_0 \tilde{D}^{\mathrm{R}}_{eu,0} C_{e,0}$ also has no eigenvalues in common with S and such that $I - \alpha D_{eu,0} \tilde{D}^{\mathrm{R}}_{eu,0}$ is invertible. Then

$$N = -\alpha (I - \alpha D_{eu,0} \tilde{D}^{\mathrm{R}}_{eu,0})^{-1} \tilde{D}^{\mathrm{R}}_{eu,0}$$

has the desired property. ∎

Our next observation is that, if all the assumptions of Theorem 2.8.1 hold, the auxiliary system (2.42) satisfies Assumptions A.1, A.2, and A.3. In fact, we obviously have that $\tilde{S}$ is anti-Hurwitz-stable for continuous-time systems

[1] Two polynomial matrices $M(\lambda)$ and $N(\lambda)$ are called unimodularly equivalent if there exist unimodular polynomial matrices $P(\lambda)$ and $Q(\lambda)$ such that $M(\lambda)P(\lambda) = Q(\lambda)N(\lambda)$.

and anti-Schur-stable for discrete-time systems, and that the pair $(\bar{A}_0, \bar{B}_0)$ is stabilizable. Moreover, the pair

$$\left[\begin{pmatrix}\bar{C}_{y,0} & \tilde{D}_{yw}\end{pmatrix}\begin{pmatrix}\bar{A}_0 & 0\\ 0 & \tilde{S}_p\end{pmatrix}\right]$$

is detectable since $(\bar{C}_{y,0}, \bar{A}_0)$ and $(\tilde{D}_{yw}, \tilde{S}_p)$ are detectable and $\bar{A}_0$ and S have no eigenvalues in common. On the other hand, the regulator equation (2.7) for the auxiliary output regulation problem with error feedback assumes the form (2.43). Now, if we partition Π and Γ as

$$\Pi = \begin{pmatrix}\Pi_1 & \Pi_2 & \cdots & \Pi_p\end{pmatrix} \text{ and } \Gamma = \begin{pmatrix}\Gamma_1 & \Gamma_2 & \cdots & \Gamma_p\end{pmatrix} \tag{2.49}$$

in conformity with the partitioning of $\tilde{S}_p$, $\tilde{D}_{ew}$, and $\tilde{D}_{yw}$, we can rewrite (2.43a) and (2.43b) into a set of p equations of the form,

$$\Pi_i S_{\min} = \bar{A}_0 \Pi_i + \bar{B}_0 \Gamma_i, \tag{2.50a}$$

$$0 = \bar{C}_{e,0}\Pi_i + \bar{D}_{eu,0}\Gamma_i + \tilde{D}_{ew1,i} \tag{2.50b}$$

where $\tilde{D}_{ew1,i} = 0$ for $i > q$. The solvability of the above set of equations is guaranteed by the hypothesis that the matrix in (2.37) has full row-rank for each λ which is an eigenvalue of S (thus, in particular, an eigenvalue of $S_{\min}$). It is then obvious that a controller as constructed in Section 2.8 will achieve output regulation for the auxiliary plant.

Our aim now is to show that any controller of the form (2.5) which solves the output regulation problem for the extended plant (2.42) and the exosystem (2.41) solves the structurally stable output regulation problem for the system (2.38) with exosystem (2.2). For simplicity we assume from now on that $\bar{D}_{eu} = 0$ and $\bar{D}_{yu} = 0$. The general result is equally valid but the formulae get much more complex. Let n_c be the order of the controller.

We first check that the controller (2.5) stabilizes the nominal plant. In fact, the interconnection of the nominal plant (2.38) and the controller is equal to the interconnection of the auxiliary plant (2.42) and the controller when w and $\tilde{w}$ are set to zero. Thus, their stability properties are the same. Since the controller stabilizes the auxiliary plant by construction, it does so as well for the original nominal plant.

In order to check the controller (2.5) indeed solves the structurally stable output regulation problem for the original plant, we need to show that the error converges asymptotically to zero no matter how the plant parameters are perturbed as long as the perturbation is such that the corresponding closed-loop system is internally stable (as usual, for internal stability we disregard the exosystem). Note that S is not subject to perturbation and, by assumption, the perturbations preserve (2.36).

In view of the proof of Theorem 2.4.1, we can conclude that the proposed controller indeed solves the structurally stable output regulation problem for the extended plant (2.38) if the equation,

$$\Pi S = (\bar{A} + \bar{B} D_c \bar{C}_{y,0})\Pi + \bar{B} C_c \Theta + \bar{E}_w + \bar{B} D_c \bar{D}_{yw}, \tag{2.51a}$$

$$\Theta S = A_c \Theta + B_c(\bar{C}_y \Pi + \bar{D}_{yw}), \tag{2.51b}$$

$$0 = \bar{C}_e \Pi + \bar{D}_{ew}, \tag{2.51c}$$

has a solution for each set of perturbed plant parameters ($\bar{A}$, $\bar{B}$, $\bar{E}_w$, $\bar{C}_e$, $\bar{D}_{ew}$, $\bar{C}_{y2}$, $\bar{D}_{yw2}$), with the known structure (2.39), for which the closed-loop system is internally stable, i.e. when the matrix

$$\begin{pmatrix} \bar{A} + \bar{B} D_c \bar{C}_y & \tilde{B}\tilde{C}_c \\ \tilde{B}_c \tilde{C}_e & \tilde{A}_c \end{pmatrix} \tag{2.52}$$

is Hurwitz-stable for continuous-time systems and Schur-stable for discrete-time systems. We note further that since the proposed output regulator stabilizes the nominal plant, as discussed earlier, the set of perturbed plant parameters ($\bar{A}$, $\bar{B}$, $\bar{E}_w$, $\bar{C}_e$, $\bar{D}_{ew}$, $\bar{C}_{y2}$, $\bar{D}_{yw2}$) for which the closed-loop system is internally stable (as usual, for internal stability we disregard the exosystem) is a subset of $\mathscr{P}$ having ($\bar{A}_0$, $\bar{B}_0$, $\bar{E}_0$, $\bar{C}_{e,0}$, $\bar{D}_{ew,0}$, $\bar{C}_{y2,0}$, $\bar{D}_{yw2,0}$) as an interior point.

Our goal is now to show that the set of equations given by (2.51) has a solution. To do so, we recall that by hypothesis the matrix S which characterizes the exosystem has all its eigenvalues in the closed-right half plane for continuous-time systems, or on or outside the unit circle for discrete-time systems. Therefore, for each set of perturbed plant parameters for which the closed-loop system is internally stable (as usual, for internal stability we disregard the exosystem), the Sylvester equation,

$$\begin{pmatrix} \bar{A} + \bar{B} D_c \bar{C}_y & \bar{B}\bar{C}_c \\ B_c \bar{C}_y & A_c \end{pmatrix} \begin{pmatrix} \Pi \\ \Theta \end{pmatrix} - \begin{pmatrix} \Pi \\ \Theta \end{pmatrix} S = \begin{pmatrix} X \\ Y \end{pmatrix}, \tag{2.53}$$

is solvable for Π and Θ whatever may be the values of X and Y are on the right hand side. Choosing

$$\begin{pmatrix} X \\ Y \end{pmatrix} = \begin{pmatrix} -\bar{E}_w - \bar{B} D_c \bar{D}_{yw} \\ -B_c \bar{D}_{ew} \end{pmatrix},$$

we obtain the equations,

$$(\bar{A} + \bar{B}D_c\bar{C}_y)\Pi + \bar{B}C_c\Theta + \bar{E}_w + \bar{B}D_c\bar{D}_{yw} = \Pi S, \tag{2.54}$$

and

$$A_c\Theta - \Theta S = -B_c(\bar{C}_y\Pi + \bar{D}_{yw}). \tag{2.55}$$

Equation (2.54) is exactly the same as (2.51a). If we prove that this Π, Θ satisfy (2.51c), the proof that the proposed regulator indeed solves the structurally stable output regulation problem follows obviously. So, our effort now is to prove this implication. With this goal in mind, we define the following two linear mappings:

$$\begin{aligned} \mathcal{A} &: \mathbb{R}^{(n_c+p-q)\times s} \to \mathbb{R}^{n_c\times s} \\ &: \Theta \mapsto \mathcal{A}(\Theta, V) = A_c\Theta - \Theta S + B_{c,2}V\,, \\ \mathcal{B} &: \mathbb{R}^{q\times s} \to \mathbb{R}^{n_c\times s} \\ &: Z \mapsto \mathcal{B}(Z) = B_{c,1}Z \end{aligned}$$

where n_c is the order of the controller and B_c is partitioned conformably with the partition of y into e and y_1,

$$B_c y = B_c \begin{pmatrix} e \\ y_1 \end{pmatrix} = B_{c,1}e + B_{c,2}y_1.$$

With this notation, equation (2.55) can be rewritten as

$$\mathcal{A}(\Theta, C_{y2}\Pi + D_{yw2}) = -\mathcal{B}(C_e\Pi + D_{ew}). \tag{2.56}$$

Now, suppose we can prove that the images of $\mathcal{A}$ and $\mathcal{B}$ intersect at $\{0\}$. Then, (2.56) would imply that

$$\mathcal{A}(\Theta, C_{y2}\Pi + D_{yw2}) = 0, \tag{2.57}$$

$$\mathcal{B}(\tilde{C}_e\Pi + D_{ew}) = 0. \tag{2.58}$$

If we can also show that $\ker \mathcal{B} = \{0\}$, we can deduce the remaining equation (2.51c).

In summary, the above discussion shows that the proposed regulator indeed solves the structurally stable output regulation problem if we can prove that the mappings $\mathcal{A}$ and $\mathcal{B}$ satisfy the properties,

$$\operatorname{im} \mathcal{A} \cap \operatorname{im} \mathcal{B} = \{0\}, \tag{2.59}$$

$$\ker \mathcal{B} = \{0\}. \tag{2.60}$$

In order to prove the above properties we first need a preliminary lemma.

Lemma 2.B.1 *Let S and $S_{\min}$ be as given in (2.40). Then there exist s independent solutions of the linear matrix equation,*

$$S_{\min} X = XS. \tag{2.61}$$

Proof : If S has a basis of eigenvectors then this property is straightforward. Let $\{v_1, \ldots, v_s\}$ be independent left-eigenvectors with corresponding eigenvalues $\{\lambda_1, \lambda_2, \ldots, \lambda_s\}$. Then because of the fact that $S_{\min}$ is cyclic and the fact the minimal polynomials of $S_{\min}$ and S are equal, we find that there exist unique right eigenvectors $\{u_1, \ldots, u_s\}$ with eigenvalues $\{\lambda_1, \lambda_2, \ldots, \lambda_s\}$. Note that the set $\{u_1, \ldots, u_s\}$ is in general not independent. Then the matrices $X_i = u_i v_i$ satisfy (2.61). Moreover, since the eigenvectors of S are independent, the matrices X_i are independent and we have found precisely s independent solutions.

Assume S does not have a full basis of eigenvectors then it does have a basis of generalized eigenvectors. Assume $v_{i_1}, \ldots, v_{i_j}$ are j generalized left-eigenvectors of S belonging to the same Jordan block, or in other words

$$v_{i_k} S = \lambda v_{i_k} + v_{i_{k-1}}$$

for $k > 1$ and

$$v_{i_1} S = \lambda v_{i_k}.$$

Due to construction of $S_{\min}$ and S, there exist a unique set of generalized eigenvectors $u_{i_1}, \ldots, u_{i_j}$ of $S_{\min}$ such that

$$S_{\min} u_{i_k} = \lambda u_{i_k} + u_{i_{k-1}}$$

for $k > 1$ and

$$S_{\min} v_{i_1} = \lambda v_{i_k}.$$

But then $X_k = v_{i_k} u_{i_1} + v_{i_1} u_{i_k}$ satisfies (2.61) for $k = 1, \ldots, j$ and are obviously independent. Since we can repeat this argument for each Jordan block we obtain s independent solutions. ∎

In order to prove the properties (2.59) and (2.60) the following result plays a central role.

Lemma 2.B.2 *There exist at least ps independent solutions Θ of the equation*

$$\mathcal{A}(\Theta, V) = 0. \tag{2.62}$$

Proof : The fact that a controller achieves output regulation for the auxiliary system (2.42) implies that there exist solutions $\tilde{\Pi}$ and $\tilde{\Theta}$ of the set of equations,

$$\tilde{\Pi}\tilde{S}_p = \bar{A}_0\tilde{\Pi} + \bar{B}_0 C_c\tilde{\Theta} + \bar{B}_0 D_c(\bar{C}_{y,0}\tilde{\Pi} + \tilde{D}_{yw}), \tag{2.63a}$$

$$\tilde{\Theta}\tilde{S}_p = A_c\tilde{\Theta} + B_{c,2}(\bar{C}_{y,0}\tilde{\Pi} + \tilde{D}_{yw}), \tag{2.63b}$$

$$0 = \bar{C}_{e,0}\tilde{\Pi} + \tilde{D}_{ew}. \tag{2.63c}$$

By Lemma 2.B.1 there exist s independent solutions L_i ($i = 1, \ldots, s$) such that

$$S_{\min} L_i = L_i S.$$

Let M_j ($j = 1, \ldots, p$) be the projection on the j'th copy of $S_{\min}$ in $\tilde{S}_p$, i.e.,

$$M_j = \begin{pmatrix} 0_{\tilde{s}} \\ \vdots \\ 0_{\tilde{s}} \\ I_{\tilde{s}} \\ 0_{\tilde{s}} \\ \vdots \\ 0_{\tilde{s}} \end{pmatrix}$$

where $I_{\tilde{s}}$ and $0_{\tilde{s}}$ are the $\tilde{s} \times \tilde{s}$ identity and zero matrix respectively and the identity matrix is at the j'th spot. Note that $\tilde{S}_p M_j L_i = M_j L_i S$. Then it is easy to check that we have

$$\mathcal{A}\left(\tilde{\Theta}M_j L_i, (\bar{C}_{y2,0}\tilde{\Pi} + D_{yw2,0})M_j L_i\right) = 0.$$

In other words we have found ps elements in the kernel of $\mathcal{A}$. What remains to be shown is that these elements are independent. Assume

$$\sum_{i=1}^{r}\sum_{j=1}^{s}\alpha_{ij}\tilde{\Theta}M_j L_i = 0,$$

$$\sum_{i=1}^{r}\sum_{j=1}^{s}\alpha_{ij}(\bar{C}_{y2,0}\tilde{\Pi} + D_{yw2,0})M_j L_i = 0. \tag{2.64}$$

Then from equations (2.63a) and (2.63c) and the fact that $\bar{A}_0$ and S have no eigenvalues in common, we get that

$$\sum_{i=1}^{r}\sum_{j=1}^{s}\alpha_{ij}\tilde{\Pi}M_j L_i = 0,$$

but then we find from (2.63c) and (2.64) that

$$V = \sum_{i=1}^{r} \sum_{j=1}^{\tilde{s}} \alpha_{ij} M_j L_i = 0$$

satisfies $\tilde{D}_{yw} V = 0$ and $\tilde{S}_p V = VS$ which yields a contradiction with the detectability of $(\tilde{D}_{yw}, \tilde{S}_q)$ unless $V = 0$. The latter implies $\alpha_{ij} = 0$ for all i and j which yields the required independence. ■

With the aid of Lemma 2.B.2, we proceed now to prove (2.59) and (2.60). We observe that the existence of ps independent solutions of (2.62) shows that the dimension of ker $\mathcal{A}$ is at least ps. Since $\mathcal{A}$ maps $\mathbb{R}^{(n_c+p-q)\times s}$ into $\mathbb{R}^{n_c \times s}$, we deduce that

$$\dim \operatorname{im} \mathcal{A} \leqslant (n_c - q)s. \tag{2.65}$$

where dim indicates the dimension. Moreover,

$$\dim \operatorname{im} \mathcal{B} \leqslant qs, \tag{2.66}$$

because the dimension of the image of a linear mapping cannot exceed that of its domain (in this case qs).

Returning to (2.53), we use any arbitrary X and Y, denote the corresponding solution by $\overline{\Pi}$ and $\overline{\Theta}$, and note that by construction

$$B_c \bar{C}_y \overline{\Pi} + A_c \overline{\Theta} - \overline{\Theta} S = Y.$$

That is,

$$\mathcal{B}(\bar{C}_e \overline{\Pi}) + \mathcal{A}(\overline{\Theta}, \bar{C}_{y2} \Pi) = Y.$$

Because of the arbitrariness of Y, this relation shows that

$$\operatorname{im} \mathcal{A} + \operatorname{im} \mathcal{B} = \mathbb{R}^{n_c \times s}.$$

This, together with (2.65) and (2.66), yields

$$\dim \operatorname{im} \mathcal{A} = (n_c - q)s,$$
$$\dim \operatorname{im} \mathcal{B} = qs.$$

These relations prove that (2.59) and (2.60) are true and therefore the controller with parameters (A_c, B_c, C_c, D_c) achieves structurally stable output regulation for the system (2.38).

We still would like to show that a controller which achieves structurally stable output regulation for system (2.38) achieves output regulation for the system (2.42). Because then we know a controller achieves structurally stable output regulation if and only if it achieves output regulation for the auxiliary system (2.42).

In order to see this property we will only perturb the matrices $\bar{E}_{w,0}$, $\bar{D}_{ew,0}$ and $\bar{D}_{yw2,0}$. Note that structurally stability requires that we achieve output regulation for $\bar{E}_w = \bar{E}_{w,0}$, $\bar{D}_{ew} = \bar{D}_{ew,0}$ and $\bar{D}_{yw2,0} = \bar{D}_{yw2,0}$ and also for $\bar{E}_w = \bar{E}_{w,0}$, $\bar{D}_{ew} = \bar{D}_{ew,0} + \varepsilon U$ and $\bar{D}_{yw2,0} = \bar{D}_{yw2,0} + \varepsilon V$. Linearity of the system dynamics then guarantees that we also achieve output regulation for $\bar{E}_w = 0$, $\bar{D}_{ew} = U$ and $\bar{D}_{yw2,0} = V$. In other words we achieve output regulation for arbitrary matrices $\bar{D}_{ew}$ and $\bar{D}_{yw2,0}$. Next note that if we achieve output regulation for $\bar{E}_w = 0$, $\bar{D}_{ew} = U_1$ and $\bar{D}_{yw2,0} = V_1$ and also for $\bar{E}_w = 0$, $\bar{D}_{ew} = U_2$ and $\bar{D}_{yw2,0} = V_2$ then we must also achieve output regulation for

$$\bar{E}_w = 0,\ \bar{D}_{ew} = \begin{pmatrix} U_1 & U_2 \end{pmatrix} \text{ and } \bar{D}_{yw2,0} = \begin{pmatrix} V_1 & V_2 \end{pmatrix}$$

with a new exosystem

$$\rho\tilde{w} = \begin{pmatrix} S & 0 \\ 0 & S \end{pmatrix} \tilde{w}.$$

Using these arguments it is easy to see that a controller that achieves structurally stable output regulation for the original system must achieve output regulation for the extended system (2.42).

Chapter 3

Classical output regulation with actuators subject to amplitude saturation – continuous-time systems

3.1 Introduction

Chapter 2 considers exact output regulation for linear systems. As defined there, solving an output regulation problem involves constructing an appropriate state or measurement feedback controller that renders the closed-loop system internally stable (as usual, for internal stability we disregard the exosystem) while the tracking error tends to zero as the time t tends to infinity. In this chapter, we revisit the output regulation problem however for linear continuous-time systems with actuators subject to amplitude saturation.

Since 1990 there has been a great deal of renewed interest in the study of linear systems subject to input saturation, probably due to a wide recognition of the inherent constraints on actuators. Since output regulation inherently requires internal stabilization, let us first briefly review internal stabilization of such systems. A crucial result on the subject of internal stabilization of linear continuous-time systems subject to input amplitude saturation appeared in [68] where it was established that such systems can be globally asymptotically stabilized if and only if the system in the absence of saturation is asymptotically null controllable with bounded controls[1]. Another crucial result related to the stabilization problem is that, in general, *linear* feedback control laws cannot be used for the purpose of global asymptotic stabilization of linear systems subject to input saturation. This "negative result" was first pointed out in [20] and elaborated on in [81]. A particular non-linear feedback law using multiple saturation functions for the global asymptotic stabilization of such systems was initiated in [84] and completed in [80]. In response to the above "negative result", the semi-global view-point for stabilization of asymptotically null controllable (with bounded controls) systems subject to input amplitude saturation was proposed in [35–38]. In [36], it was shown that one can semi-globally exponentially stabilize a linear system subject to input amplitude saturation using *linear* feedback laws if and only if the system is asymptotically null controllable with bounded controls. In other words, the basin of attraction of a linear system subject to input amplitude saturation can be made arbitrarily large using appropriately tuned *linear* feedback laws if the system is asymptotically null controllable with bounded controls.

In this chapter, for continuous-time systems, our focus will be on the semi-global output regulation problem for linear asymptotically null controllable (with bounded controls) systems subject to input amplitude saturation. We will consider both state feedback and measurement feedback controllers. The rationale behind the adoption of a semi-global framework for output regulation problem is two-fold. Firstly, the semi-global framework allows us to use linear feedback laws, which is obviously very appealing; and secondly, the semi-global framework seems to be a natural choice when the global output regulation problem, in general, does not have a solution. Hence, we extend here the output regulation theory for linear systems without input amplitude saturation developed in Chapter 2 to the class of linear systems subject to input amplitude saturation which are asymptotically null controllable with bounded controls. More specifically, we first provide a set of solvability conditions for semi-global output regulation of such systems, and then we show that our solvability conditions are also necessary for a fairly general class of systems. We also show that, under certain weak assumptions, we cannot weaken these solvability conditions by using non-linear feedback controllers. However, when these assumptions are not satisfied, an example shows that a non-linear feedback controller can achieve output regulation when no linear feedback controller can do so. This chapter is based on the research of authors and their coworkers, in particular [33] and [42].

A point that should be emphasized is this. As will be shown, under certain solvability conditions, *linear* feedback controllers can be developed to solve

[1] A linear continuous-time system is asymptotically null controllable with bounded controls if and only if it is stabilizable and all the poles of the open-loop system are in the closed left-half plane.

the posed semi-global regulation problems by using what is called a low-gain design technique. Since low-gain controllers under utilize the available control capacity, often one finds that the convergence of the error signal to zero as time progresses to infinity is rather slow. In these cases, under the same solvability conditions, one could develop an improved technique called low-high-gain design method. Such a design method utilizes the available control capacity in a better way, and thus results in a better performance. It is worth noting that, unlike for discrete-time systems that are to be discussed in the next chapter, the feedback controllers that are developed by the improved design technique remain *linear*.

3.2 Classical global output regulation for linear systems subject to input amplitude saturation

As we said above, our primary focus here is the output regulation of linear systems subject to amplitude saturation. Thus we start by redefining the classical output regulation problem. To do so, we consider a time-invariant multivariable continuous-time system with inputs that are subject to amplitude saturation together with a time-invariant exosystem that generates disturbance and reference signals. That is, consider the system,

$$\begin{aligned} \dot{x} &= Ax + B\sigma(u) + E_w w \\ \dot{w} &= Sw \\ y &= C_y x + D_{yw} w \\ e &= C_e x + D_{eu}\sigma(u) + D_{ew} w, \end{aligned} \tag{3.1}$$

where, as usual, $x \in \mathbb{R}^n$, $w \in \mathbb{R}^s$, $u \in \mathbb{R}^m$, $y \in \mathbb{R}^p$, $e \in \mathbb{R}^q$, and σ is a vector-valued saturation function defined as

$$\sigma(s) = [\bar{\sigma}(s_1), \bar{\sigma}(s_2), \cdots, \bar{\sigma}(s_m)]^{\mathrm{T}} \tag{3.2}$$

with

$$\bar{\sigma}(s) = \begin{cases} s & \text{if } |s| \leq 1 \\ -1 & \text{if } s < -1 \\ 1 & \text{if } s > 1. \end{cases} \tag{3.3}$$

Because of the presence of the saturation function σ, the system (3.1) is nonlinear. Note that we can also treat different saturation levels, even differences between channels, by simple scaling. Compared to the system (2.1) as used in Chapter 2, there is, beside the saturation, one more difference. The matrix D_{yu}

equals 0 (compared with (2.1)). For control purposes this matrix can be easily handled and it does not affect any of our solvability conditions. However, it makes the formulae much more complex and therefore we opted to go for the simple case with $D_{yu} = 0$.

Before we proceed further, we recall the following assumptions which have already been defined earlier on pages 19 and 25.

A.1. The pair (A, B) is stabilizable.

A.2. The matrix S is anti-Hurwitz-stable, i.e. all the eigenvalues of S have non-negative real parts.

A.3. The pair $\left(\begin{pmatrix} C_y & D_{yw} \end{pmatrix}, \begin{pmatrix} A & E_w \\ 0 & S \end{pmatrix}\right)$ is detectable.

The global output regulation problem for a non-linear system of the type (3.1) is similar to the one in Chapter 2. However, one must specify the type of internal stability requirement in the problem formulation. We first choose a natural choice of global internal stability requirement which requires the domain of attraction of the closed loop system while disconnected from the exosystem to be the whole space.

Problem 3.2.1 ***(Global state feedback output regulation problem for linear continuous-time systems subject to input amplitude saturation)*** *For a system Σ as given in (3.1), find, if possible, a feedback law $u = \alpha(x, w)$ such that the following conditions hold:*

(i) ***(Internal Stability)*** *The equilibrium point $x = 0$ of*

$$\dot{x} = Ax + B\sigma(\alpha(x, 0))$$

is globally asymptotically stable and locally exponentially stable.

(ii) ***(Output Regulation)*** *For all $x(0) \in \mathbb{R}^n$ and $w(0) \in \mathbb{R}^s$, the solution of the closed-loop system satisfies*

$$\lim_{t \to \infty} e(t) = 0.$$

Problem 3.2.2 ***(Global measurement feedback output regulation problem for linear continuous-time systems subject to input amplitude saturation)*** *For a system Σ as given in (3.1), find, if possible, a dynamic feedback law $u = \theta(z)$, $\dot{z} = \eta(z, y)$ where $v \in \mathbb{R}^{n_c}$ such that the following conditions hold:*

(*i*) ***(Internal Stability)*** *The equilibrium point* $(x, z) = (0, 0)$ *of*

$$\begin{aligned}\dot{x} &= Ax + B\sigma(\theta(z))\\ \dot{z} &= \eta(z, C_y x)\end{aligned}$$

is globally asymptotically stable and locally exponentially stable.

(*ii*) ***(Output Regulation)*** *For all* $x(0) \in \mathbb{R}^n$, $z(0) \in \mathbb{R}^{n_c}$, *and* $w(0) \in \mathbb{R}^s$, *the solution of the closed-loop system satisfies*

$$\lim_{t\to\infty} e(t) = 0.$$

Remark 3.2.1 *A set of sufficient conditions for the above problem, obviously, include the necessary and sufficient conditions for global stabilization of the plant in the presence of input amplitude saturation as established by [68], i.e.,* (A, B) *is stabilizable and all the eigenvalues of* A *are in the closed left half plane.*

Global output regulation, as defined above, was first formulated by Teel in [83] and is clearly a very desirable property. Unfortunately it turns out that only in very special circumstances one can achieve global output regulation. This can be seen by the following lemma.

Lemma 3.2.1 *Consider a system* Σ *as given in (3.1). Let Assumptions A.1, A.2 and A.3 hold. Also, assume that the eigenvalues of* A *are in the closed left half plane. Moreover, assume that at least one unstable pole of the exosystem is observable from the error signal. Then there exist initial conditions* w_0 *for* w *such that there exists no input* u *or initial condition* $x(0)$ *for which the system* Σ *satisfies* $\lim_{t\to\infty} e(t) = 0$.

Proof : We first study the system (3.1) when rewritten in the form,

$$\begin{aligned}\dot{x} &= Ax + Bv + E_w w\\ \dot{w} &= Sw\\ y &= C_y x + D_{yw} w\\ e &= C_e x + D_{eu} v + D_{ew} w,\end{aligned} \tag{3.4}$$

where v denotes an input signal considered bounded.

Let w_0 with $\|w_0\| = 1$ be an eigenvector of S belonging to an eigenvalue λ of S which is detectable from e. Suppose λ is an eigenvalue with strictly positive real part. Then we can decompose e into three components; one due to a possibly non-zero initial condition x_0, another due to the bounded input v, and the third one due to the initial condition w_0. The first two can only grow polynomially in time since all the eigenvalues of A are in the closed left half plane and since the input signal v is considered to be bounded. On the other hand the effect of w_0 will, owing to the detectability assumption, grow exponentially in time. Therefore e will also grow exponentially in time and hence we cannot achieve output regulation without imposing additional conditions.

If an eigenvalue λ of S does not have a strictly positive real part then, by Assumption A.2, it must lie on the imaginary axis. In that case, w will be periodic and bounded since $w(t) = e^{\lambda t} w_0$. To analyze this situation further, we next consider the minimal amplitude of an input signal which achieves tracking, and then minimize it over all possible initial conditions of the plant. That is, we consider

$$\mathcal{J}(w_0) := \inf_{v, x_0} \{ \, \|v\|_\infty \mid v \text{ is such that } \lim_{t\to\infty} e(t) = 0 \\ \text{where } x(0) = x_0 \text{ and } w(0) = w_0 \, \}.$$

Suppose $\mathcal{J}(w_0) = 0$. We take a minimizing sequence $\{v_i, x_{0,i}\}$ for the above optimization problem. For each v_i there exists a T_i such that $\|e(T_i+t)\| < 1/i$ for all $t > 0$ and $w(T_i) = w_0$ (note that w is periodic). Define

$$\bar{v}_i(t) := v_i(T_i + t).$$

Then $\|\bar{v}_i\|_\infty \leqslant \|v_i\|_\infty \to 0$ as $i \to \infty$. The output $\bar{e}_i$ resulting from input $\bar{v}_i$ and initial conditions

$$\bar{x}(0) = \bar{x}_{0,i} := x(T_i),$$

and $\bar{w}(0) = w_0$ satisfies $\|\bar{e}_i\|_\infty < 1/i$. The latter is straightforward since $\bar{e}_i(t) = e(T_i + t)$. We then pick any $T > 0$. On $[0, T]$ the input $\bar{v}_i$ converges in L_∞ norm to 0. Similarly $\bar{e}_i$ converges to 0 uniformly on $[0, T]$.

Define $f : \mathbb{R}^n \to L_\infty[0, T]$ by $[f(z)](t) = C_e e^{At} z$. We can check that $g \in L_\infty[0, t]$ is in the closure of the image of f, where

$$g(t) := \int_0^t C_e e^{A(t-\tau)} E_w w(\tau) d\tau + D_{ew} w(t).$$

Since f is a finite rank operator we know that the image is closed and hence g is in the image of f, i.e. there exists an $\tilde{x}_0$ such that $f(\tilde{x}_0) = -g$. We find that for $t \in [0, T]$

$$\begin{pmatrix} C_e & D_{ew} \end{pmatrix} \exp\left[\begin{pmatrix} A & E_w \\ 0 & S \end{pmatrix} t \right] \begin{pmatrix} \tilde{x}_0 \\ w_0 \end{pmatrix} = 0. \tag{3.5}$$

This immediately implies that (3.5) holds for all t. However, this contradicts the fact that the eigenvalue λ was observable from e. Therefore we have $\mathcal{J}(w_0) > 0$. Hence we find that for $w(0) = 2w_0/\mathcal{J}(w_0)$ any input v which achieves output regulation satisfies $\|v\|_\infty \geqslant 2$. Therefore $v = \sigma(u)$ will never be able to achieve output regulation, i.e. no input u to (3.1) exists for which $e(t) \to 0$ as $t \to \infty$.

The above involves complex inputs. By working with either the real or complex part of the signals at a time we can avoid this technical problem. ∎

Remark 3.2.2 *As argued before, the Assumptions A.1, A.2, A.3, and the assumption that the eigenvalues of A are in the closed left half plane are basically necessary. Therefore the above lemma tells us that we can only achieve global output regulation if the open-loop system already achieves output regulation and the controller only needs to achieve stability without losing the property of output regulation. This is a very exceptional case and therefore, for all practical purposes, global output regulation is not possible when we have input saturation.*

Note that the classical case $y = e$ implies that Assumptions A.2 and A.3 guarantee that all poles of the exosystem should be observable from e and therefore global output regulation is not possible.

3.3 Classical semi-global output regulation for linear systems subject to input amplitude saturation

Since in general global output regulation problems for a system of the type (3.1) are not solvable, one basic question that arises is what type of initial conditions of exosystem and plant should be considered realistically for output regulation when the input is subject to amplitude saturation. Regarding the initial conditions of the exosystem, the discussion at the end of the last section is clearly in favor of the argument that we should restrict our attention only to initial conditions $w(0)$ lying inside a given compact set. Moreover, regarding the initial conditions of the plant, in the theory of stabilization of linear systems subject to amplitude saturation (e.g. see [35] and [36]), the step from global initial conditions to initial conditions inside an arbitrarily given compact set has already been made. This has been named semi-global stabilization. Thus we need to direct our attention here only to a semi-global setting. This, as we shall see, also yields the well-known advantage that the output regulation can be achieved using only linear feedback controllers. Motivated by this we devote ourselves in this section to *semi-global output regulation.*

This section is split into two parts. In the first part we study the so called classical semi-global linear state feedback output regulation problem where we assume that the states x and w are available for feedback, and hence it suffices to look at only static feedback controllers. In the second part we study the so called classical semi-global measurement feedback output regulation problem where only certain measurement signal is available for feedback, and hence we have to resort to dynamic feedback. In this case, we design observer based controllers.

3.3.1 Linear state feedback output regulation problem

The problem considered in this subsection is formulated as follows.

Problem 3.3.1 ***(Classical semi-global linear state feedback regulator problem for linear systems subject to input amplitude saturation)*** *Consider the system (3.1) and a compact set $\mathcal{W}_0 \subset \mathbb{R}^s$. For any a priori given (arbitrarily large) bounded set $\mathcal{X}_0 \subset \mathbb{R}^n$, find, if possible, a linear static feedback law $u = Fx + Gw$ such that the following conditions hold:*

(i) ***(Internal Stability)*** *The equilibrium point $x = 0$ of*

$$\dot{x} = Ax + B\sigma(Fx) \tag{3.6}$$

is asymptotically stable with $\mathcal{X}_0$ contained in its basin of attraction.

(ii) ***(Output Regulation)*** *For all $x(0) \in \mathcal{X}_0$ and $w(0) \in \mathcal{W}_0$, the solution of the closed-loop system satisfies*

$$\lim_{t\to\infty} e(t) = 0. \tag{3.7}$$

Remark 3.3.1 *We would like to emphasize that our definition of classical semi-global linear state feedback output regulation problem does not view the set of initial conditions of the plant as given data. The set of given data consists of the models of the plant and the exosystem and the set of initial conditions for the exosystem. Therefore, any solvability conditions we obtain* must *be independent of the set of initial conditions of the plant, $\mathcal{X}_0$.*

Before discussing the solvability conditions for the output regulation problem, we would like to recall what is known as the low-gain [36] design method introduced earlier by Lin, Saberi and their coworkers. Such a low-gain design method has been used successfully in connection with linear systems with amplitude saturating actuators, not only for internal stabilization [36], but also for various other problems.

There are two methods for low gain design. One is Riccati based which is discussed next and the other a direct method of design which is considered in Appendix 3.B.

Review of Riccati-based low-gain design for linear systems:

We now recall a low-gain state feedback design algorithm from [36]. The objective is to show that we can find stabilizing feedback control inputs with arbitrarily small magnitude which stabilize a given linear system with all its poles in the closed left-half plane. Such a design algorithm yields a family of state feedback gains, parameterized in ε, which is instrumental in proving our results on semi-global output regulation. There exist in the literature two low-gain design algorithms; one is based on the solution of a continuous-time algebraic Riccati equation, parameterized in ε, and the other is a *direct* construction method based on an eigenvalue assignment method. The Riccati-based method is conceptually appealing although solving the parameterized Riccati equation might be numerically stiff. On the other hand, the alternative direct method of explicit construction is numerically efficient but is somewhat involved in details. For conceptual clarity, we present here the Riccati-based method, and the alternative *direct* method is discussed at the end of this chapter in Appendix 3.B.

Consider the linear system

$$\dot{x} = Ax + Bu \tag{3.8}$$

where the state $x \in \mathbb{R}^n$ and the input $u \in \mathbb{R}^m$. Assume that (A, B) is stabilizable and all the eigenvalues of A are located in the closed left-half plane. Consider the Riccati equation defined as

$$P_\varepsilon A + A^{\mathrm{T}} P_\varepsilon - P_\varepsilon B B^{\mathrm{T}} P_\varepsilon + Q_\varepsilon = 0 \tag{3.9}$$

where $Q : (0, 1] \to \mathbb{R}^{n \times n}$ is a continuously differentiable matrix-valued function such that $Q_\varepsilon > 0$, $\frac{dQ_\varepsilon}{d\varepsilon} > 0$ for any $\varepsilon \in (0, 1]$, and $\lim_{\varepsilon \to 0} Q_\varepsilon = 0$. In what follows, we often take $Q_\varepsilon = \varepsilon I$. We next form a family of low-gain state feedback gain matrices F_ε as

$$F_\varepsilon = -B^{\mathrm{T}} P_\varepsilon. \tag{3.10}$$

The following lemma recalls certain important properties of the Riccati equation (3.9).

Lemma 3.3.1 *Consider the Riccati equation (3.9). Assume that* (A, B) *is stabilizable and all the eigenvalues of* A *are located in the closed left-half plane. Also, let* Q_ε *be a continuously differentiable matrix-valued function such that* $Q_\varepsilon > 0$, $\frac{dQ_\varepsilon}{d\varepsilon} > 0$ *for any* $\varepsilon \in (0, 1]$, *and* $\lim_{\varepsilon \to 0} Q_\varepsilon = 0$. *Then the Riccati equation (3.9) has a unique positive definite solution* P_ε *for any* $\varepsilon \in (0, 1]$. *Moreover, this positive definite solution* P_ε *has the following properties:*

(i) *For any* $\varepsilon \in (0, 1]$, *the unique solution* $P_\varepsilon > 0$ *is such that* $A + BF_\varepsilon$ *is Hurwitz-stable where* $F_\varepsilon = -B^T P_\varepsilon$.

(ii) $\lim_{\varepsilon \to 0} P_\varepsilon = 0$.

(iii) P_ε *is continuously differentiable with respect to* ε *and*

$$\frac{dP_\varepsilon}{d\varepsilon} > 0, \quad \textit{for any} \ \ \varepsilon \in (0, 1].$$

Proof : The existence and uniqueness of a positive semi-definite solution P_ε follows from [61]. It also follows from [61] that P_ε is the unique solution for which $A - BB^{\mathrm{T}} P_\varepsilon$ has all its eigenvalues in the closed left-half plane. For $\varepsilon = 0$, it is trivial to see that the Riccati equation (3.9) has a solution $P(0) = 0$ since by assumption, $A - BB^{\mathrm{T}} P(0) = A$ has all its eigenvalues inside or on the unit circle. The fact that $\lim_{\varepsilon \to 0} P_\varepsilon = 0$ follows from standard continuity arguments. Note that for $\varepsilon > 0$ the solution is actually positive definite and is such that $A - BB^{\mathrm{T}} P_\varepsilon$ has all its eigenvalues in the open left-half plane.

Thus, we need to prove here only part (iii). To do so, we observe that the continuous differentiability of P_ε follows from the fact that the Hamiltonian matrix associated with the Riccati equation (3.9) is a continuously differentiable function of ε (see [28]). In order to show that $\frac{dP_\varepsilon}{d\varepsilon} > 0$, we differentiate the Riccati equation (3.9) to obtain the Lyapunov equation,

$$\frac{dP_\varepsilon}{d\varepsilon}(A - BB^{\mathrm{T}} P_\varepsilon) + (A - BB^{\mathrm{T}} P_\varepsilon)^{\mathrm{T}} \frac{dP_\varepsilon}{d\varepsilon} = -\frac{dQ_\varepsilon}{d\varepsilon}. \tag{3.11}$$

Now, it follows from the above equation that $\frac{dP_\varepsilon}{d\varepsilon} > 0$ since $A - BB^{\mathrm{T}} P_\varepsilon$ is asymptotically stable and $\frac{dQ_\varepsilon}{d\varepsilon} > 0$ for all $\varepsilon > 0$. ■

The family of state feedback gains (3.10) parameterized by ε, has the following property.

Theorem 3.3.1 *Consider the system (3.8). Assume that* (A, B) *is stabilizable and* A *has all its eigenvalues in the closed left-half plane. Then, if we apply the state feedback given by (3.10) to the system (3.8), the resulting closed-loop system*

$$\dot{x} = (A + BF_\varepsilon)x \tag{3.12}$$

is asymptotically stable for all $\varepsilon > 0$. *Moreover, there exist* $\kappa_\varepsilon > 0$, $\zeta_\varepsilon > 0$ *and* $\eta_\varepsilon > 0$ *with* $\kappa_\varepsilon, \zeta_\varepsilon \to 0$ *as* $\varepsilon \to 0$ *such that for all* $\varepsilon \in (0, 1]$,

$$\|F_\varepsilon\| \leq \kappa_\varepsilon, \tag{3.13}$$

$$\|F_\varepsilon e^{(A+BF_\varepsilon)t}\| \leqslant \zeta_\varepsilon e^{-\eta_\varepsilon t}, \tag{3.14}$$

Proof : The internal stability of the closed-loop system (3.12), and the inequality (3.13) follow trivially from Lemma 3.3.1. Next, we need to show (3.14). Using the Riccati equation (3.9) we find that, for $\varepsilon \in (0, 1]$,

$$\begin{aligned}\frac{d}{dt}x^{\mathrm{T}}(t)P_\varepsilon x(t) &= -\|B^{\mathrm{T}}P_\varepsilon x(t)\|^2 - x(t)Q_\varepsilon x(t)\\ &\leqslant -2\eta_\varepsilon\|P_\varepsilon^{1/2}x(t)\|^2\end{aligned}$$

where $\eta_\varepsilon = \|Q_\varepsilon\|\|P(1)\|^{-1}$. Hence

$$\|P_\varepsilon^{1/2}x(t)\| \leqslant e^{-\varepsilon\eta t}\|P_\varepsilon^{1/2}x(0)\|. \tag{3.15}$$

Finally,

$$\begin{aligned}\|F_\varepsilon e^{(A+BF_\varepsilon)t}x(0)\| &= \|B^{\mathrm{T}}P_\varepsilon x(t)\|\\ &\leqslant \|B\|\|P_\varepsilon^{1/2}\|e^{-\varepsilon\eta t}\|P_\varepsilon^{1/2}x(0)\|.\end{aligned} \tag{3.16}$$

Since (3.16) is true for all $x(0) \in \mathbb{R}^n$, it follows trivially that

$$\|F_\varepsilon e^{(A+BF_\varepsilon)t}\| \leq \|B\|\|P_\varepsilon^{1/2}\|^2 e^{-\varepsilon\eta t} = \|B\|\|P_\varepsilon\|e^{-\varepsilon\eta t}. \tag{3.17}$$

The proof is then completed by taking $\zeta_\varepsilon = \|B\|\|P_\varepsilon\|$. ■

We next move on to obtain the solvability conditions for the classical semi-global linear state feedback output regulation problem for linear systems subject to input amplitude saturation. The following theorem presents such conditions.

Theorem 3.3.2 *Consider the system (3.1) and the given compact set $\mathcal{W}_0 \subset \mathbb{R}^s$. The classical semi-global linear state feedback output regulation problem is solvable if the following conditions hold:*

(i) (A, B) is stabilizable and A has all its eigenvalues in the closed left half plane.

(ii) There exist matrices Π and Γ such that

(a) they solve the regulator equation (2.7), i.e.,

$$\Pi S = A\Pi + B\Gamma + E_w, \tag{3.18a}$$

$$0 = C_e\Pi + D_{eu}\Gamma + D_{ew}, \tag{3.18b}$$

and

(b) there exist a $\delta > 0$ and a $T \geq 0$ such that $\|\Gamma w\|_{\infty,T} \leqslant 1 - \delta$ for all w with $w(0) \in \mathcal{W}_0$.

Proof : We prove this theorem by first explicitly constructing a family of linear static state feedback laws, parameterized in ε, and then showing that for each given set $\mathcal{X}_0$, there exists an $\varepsilon^* > 0$ such that for all $\varepsilon \in (0, \varepsilon^*]$, both items (i) and (ii) of Problem 3.3.1 hold. The family of linear static state feedback laws we construct takes the form,

$$u = F_\varepsilon x + (\Gamma - F_\varepsilon \Pi)w, \tag{3.19}$$

where F_ε is a state feedback gain matrix, parameterized in ε and chosen in such a way that $A + BF_\varepsilon$ is Hurwitz-stable for all $\varepsilon > 0$ and

$$\|F_\varepsilon e^{(A+BF_\varepsilon)t}\|_\infty \leq \upsilon_\varepsilon, \tag{3.20}$$

where υ is a positive-valued function satisfying $\lim_{\varepsilon\to 0} \upsilon_\varepsilon = 0$. One way of selecting such a gain matrix F_ε is as in (3.10). With this selection of F_ε, the system (3.6) can be written as

$$\dot{x} = Ax + B\sigma(F_\varepsilon x). \tag{3.21}$$

To show that item (i) of Problem 3.3.1 holds, let us next consider the system (3.21) in the absence of the saturation elements. The system then takes the form

$$\dot{x} = (A + BF_\varepsilon)x. \tag{3.22}$$

It follows then from (3.20) that there exists an $\varepsilon_1^* > 0$ such that for all $\varepsilon \in (0, \varepsilon_1^*]$ we have

$$\|F_\varepsilon x\|_\infty \leq 1, \quad \text{for all } x(0) \in \mathcal{X}_0.$$

This shows that for all $\varepsilon \in (0, \varepsilon_1^*]$ and for all $x(0) \in \mathcal{X}_0$, the system (3.21) operates in the linear regions of the saturation elements and hence we can conclude that the equilibrium point $x = 0$ of the system (3.21) is asymptotically stable with $\mathcal{X}_0$ contained in its basin of attraction.

To show that item (ii) of Problem 3.3.1 holds, let us introduce an invertible coordinate change,

$$\xi = x - \Pi w. \tag{3.23}$$

Then, using (3.18), we have

$$\begin{aligned} \dot{\xi} &= \dot{x} - \Pi \dot{w} \\ &= Ax + B\sigma(u) + E_w w - \Pi S w \\ &= A\xi + B(\sigma(u) - \Gamma w). \end{aligned} \tag{3.24}$$

With the considered family of state feedback laws, the closed-loop system can then be rewritten as

$$\dot{\xi} = A\xi + B\left[\sigma(F_\varepsilon \xi + \Gamma w) - \Gamma w\right]. \tag{3.25}$$

Now by Condition (ii)b) of Theorem 3.3.2, $\|\Gamma w\|_{\infty,\mathrm{T}} < 1 - \delta$. Moreover, $\xi(T)$ belongs to a bounded set independent of ε since $\xi(0)$ is bounded and $\xi(T)$ is determined by a linear differential equation with bounded inputs $\sigma(u)$ and Γw. If we consider the system (3.25), from time T onwards, without saturation element, we obtain

$$\dot{\xi} = (A + BF_\varepsilon)\xi. \tag{3.26}$$

Since $\xi(T)$ is bounded, (3.20) and (3.26) imply that there exists an $\varepsilon_2^* > 0$ such that, for all $\varepsilon \in (0, \varepsilon_2^*]$,

$$\|F_\varepsilon \xi\|_{\infty,\mathrm{T}} \leq \delta.$$

We can conclude then that the system will operate within the linear region of the saturation elements for all $t \geq T$ if $\varepsilon \in (0, \varepsilon_2^*]$. Also, in view of (3.19) and (3.23), we find for $t \geq T$ that

$$e(t) = (C_e + D_{eu} F_\varepsilon)\xi(t) + (C_e \Pi + D_{eu}\Gamma + D_{ew})w(t).$$

However, in view of (3.18b), e is given by $e(t) = (C_e + D_{eu}F_\varepsilon)\xi(t)$, and thus, owing to the stability of $A + BF_\varepsilon$, we find that $e(t) \to 0$ as $t \to \infty$.

Finally, taking $\varepsilon^* = \min\{\varepsilon_1^*, \varepsilon_2^*\}$, we complete our proof. ■

We note that Condition (i) of Theorem 3.3.2 is necessary for the existence of a stabilizing feedback. On the other hand from Chapter 2, we know that under Assumptions A.1 and A.2, Condition (ii)a) of Theorem 3.3.2 is necessary to guarantee the solvability of the output regulation problem for the system in the absence of the saturation element. The crucial condition for the solvability of the classical semi-global linear state feedback output regulation problem with amplitude saturating actuators is Condition (ii)b), which is a sufficient condition. Combining the above theorem with the results from Section 3.4 we obtain the following corollary.

Corollary 3.3.1 *Consider the system (3.1). Let the system characterized by the quadruple (A, B, C_e, D_{eu}) be left-invertible without having invariant zeros on the imaginary axis. Moreover, assume that S is weakly Hurwitz-stable. Then, there exists a set of initial conditions $\mathcal{W}_0$ containing 0 in its interior for which the classical semi-global linear state feedback output regulation problem is solvable if and only if the following conditions hold:*

(i) (A, B) is stabilizable and A has all its eigenvalues in the closed left half plane.

(ii) There exist matrices Π and Γ that satisfy the linear matrix equations (3.18).

Moreover, Γ is uniquely determined by (3.18), and if we define the set

$$\bar{\mathcal{W}} := \left\{ w_0 \in \mathbb{R}^{n_c} \mid \limsup_{t\to\infty} \|\Gamma w(t)\|_\infty \leqslant 1 \right\},$$

then the following properties hold:

(a) If $\mathcal{W}_0$ is contained in the interior of $\bar{\mathcal{W}}$, then the classical semi-global linear state feedback output regulation problem is solvable.

(b) If $\mathcal{W}_0$ is not contained in $\bar{\mathcal{W}}$, then the classical semi-global state feedback output regulation problem is not solvable by linear or non-linear state feedback.

Proof : The first part of this corollary is a direct consequence of the (linear) classical output regulation theorem of Chapter 2. The fact that Γ is uniquely defined is a consequence of Lemma 2.5.1. Finally, property (a) is an immediate consequence of Theorem 3.3.2 and property (b) is a direct consequence of Theorem 3.4.1. ■

Design of a low-gain state feedback regulator:

For clarity, we now give a step by step design of a low-gain state feedback regulator.

Step 1 : Find a solution $(\Pi, \ \Gamma)$ of the regulator equation (3.18).

Step 2 : Find a state feedback gain matrix F_ε parameterized in ε in such a way that $A + BF_\varepsilon$ is Hurwitz-stable for all $\varepsilon > 0$ and such that equation (3.20) holds for a positive-valued function υ satisfying $\lim_{\varepsilon \to 0} \upsilon_\varepsilon = 0$. One way of selecting such a gain matrix F_ε is as in (3.10). Another way of doing so is via an alternative *direct* method discussed at the end of this chapter in Section 3.B of the appendix.

Step 3 : Given the sets $\mathcal{X}_0 \subset \mathbb{R}^n$ and $\mathcal{W}_0 \subset \mathbb{R}^s$, find an ε^* by the procedure given in the proof of Theorem 3.3.2.

Then the state feedback controller given in (3.19) for some $\varepsilon \in (0, \ \varepsilon^*]$ solves the classical semi-global state feedback output regulation problem. □

The following example illustrates the design procedure.

Example 3.3.1 *(State feedback case) We consider the system (3.1) with*

$$A = \begin{pmatrix} 1 & 2 & 1 & 0 \\ 1 & 1 & 1 & 0 \\ -1 & -2 & -1.5 & -0.5 \\ 5 & 6 & 4.5 & -0.5 \end{pmatrix}, \ B = \begin{pmatrix} 1 & 0 \\ 0 & 1 \\ 0 & 0 \\ 0 & 0 \end{pmatrix}, \ E_w = \begin{pmatrix} -1 & 0 \\ 0 & 1 \\ 0 & 0 \\ 0 & 0 \end{pmatrix}$$

$$S = \begin{pmatrix} 0 & 1 \\ -1 & 0 \end{pmatrix}, \ C_e = \begin{pmatrix} 3 & 2 & 1.5 & -0.5 \\ 2 & 3 & 1.5 & -0.5 \end{pmatrix},$$

$$D_{ew} = \begin{pmatrix} -0.5 & 0 \\ 0 & -1 \end{pmatrix}, \ D_{eu} = 0,$$

and $\mathcal{W}_0 = \{w \in \mathbb{R}^2 : \|w\| \leq \sqrt{2}/4\}$. *With this data, it is straightforward to show that the solvability conditions for the classical semi-global linear state feedback output regulation problem are satisfied. More specifically, the matrices,*

$$\Pi = \begin{pmatrix} 1 & -1 \\ 0.5 & 0 \\ -1.25 & 1.75 \\ 3.25 & -0.75 \end{pmatrix}, \quad \textit{and } \Gamma = \begin{pmatrix} 1.25 & 0.25 \\ -0.25 & -1.25 \end{pmatrix}, \tag{3.27}$$

solve the regulator equation (3.18). Also, $T = 0$ *and* $\delta = 0.4697$*, since* $\|\Gamma w\|_\infty \leq 0.5303$ *for all* $w(0) \in \mathcal{W}_0$*. Following the design procedure, and taking the gain matrix* F_ε *as in (3.10), we formulate the state feedback control law as*

$$u = -B^T P_\varepsilon x + [B^T P_\varepsilon \Pi + \Gamma] w. \tag{3.28}$$

With the above choice of control law, for the set given by $\mathcal{X}_0 = \{x \in \mathbb{R}^4 : \|x\| \leq 4\}$*, a choice of* ε^* *is* 4×10^{-7}*. For* $\varepsilon = \varepsilon^*$*, the feedback law (3.19) is given by*

$$u = \begin{pmatrix} -0.01894 & -0.03244 & -0.01033 & 0.00807 \\ -0.03244 & -0.05779 & -0.01711 & 0.01438 \end{pmatrix} x + \begin{pmatrix} 1.24602 & 0.25519 \\ -0.25677 & -1.24171 \end{pmatrix} w.$$

Some simulation results are shown in Figure 3.1 on the next page. Plots (a) and (b) show the error signals on the interval $[0, 1000]$ *which converge to* 0*. Plots (c) and (d) give the corresponding input signals on the interval* $[0, 300]$ *(the inputs clearly settle to sinusoidal signals).*

Design of a low-and-high-gain state feedback regulator:

The simulations in Figure 3.1 point out that the convergence of $e(t)$ to zero is very slow. This is a characteristic of the feedback law (3.28) which utilizes a low-gain for F_ε as given by (3.10). That is, low-gain based designs under utilize the available control capacity. Our next goal is to recall a new design methodology which yields significant improvement to the low-gain design method, and leads to a better utilization of the available control capacity and hence better closed-loop performance. The improved design utilizes a low-and-high gain feedback [33]. Such a control law is given by

$$u = -(\mu + 1) B^T P_\varepsilon x + [(\mu + 1) B^T P_\varepsilon \Pi + \Gamma] w, \quad \mu \geq 0. \tag{3.29}$$

We note that when $\mu = 0$ the low-and-high gain control law as given above reduces to the low-gain based control law as given in (3.28). The control

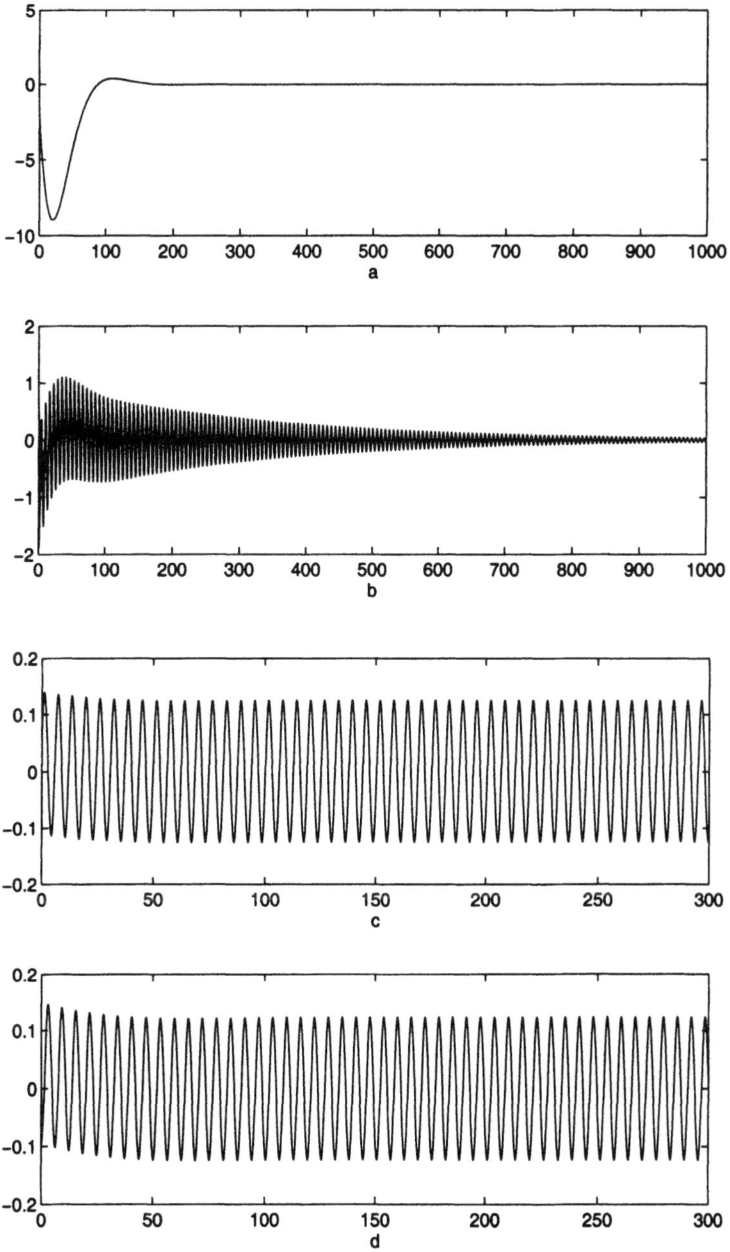

Figure 3.1: a) e_1; b) e_2; c) $\bar{\sigma}(u_1)$; d) $\bar{\sigma}(u_2)$.

law (3.29) is parameterized in two variables, one is ε which is referred to as a low-gain parameter, and the other is μ which is referred to as a high-gain parameter. As will be shown next, for any value of μ, the family of feedback control laws given by (3.29) also solves the classical semi-global linear state feedback output regulation problem under the same solvability conditions as in Theorem 3.3.2.

Theorem 3.3.3 *Consider the system (3.1) and the given compact set $\mathcal{W}_0 \subset \mathbb{R}^s$. Under the same solvability conditions as in Theorem 3.3.2, there exists a controller, among the family of feedback control laws given by (3.29), that solves the classical semi-global linear state feedback output regulation problem. More specifically, for any a priori given (arbitrarily large) bounded set $\mathcal{X}_0 \subset \mathbb{R}^n$, there exists an $\varepsilon^* > 0$ such that for each $\varepsilon \in (0, \varepsilon^*]$ and for each $\mu \geq 0$, the controller in the family (3.29) has the following properties:*

(i) The equilibrium point $x = 0$ of

$$\dot{x} = Ax + B\sigma(-(1+\mu)B^T P_\varepsilon x) \tag{3.30}$$

is asymptotically stable with $\mathcal{X}_0$ contained in its basin of attraction.

(ii) For any $x(0) \in \mathcal{X}_0$ and $w(0) \in \mathcal{W}_0$, the solution of the closed-loop system satisfies

$$\lim_{t\to\infty} e(t) = 0. \tag{3.31}$$

Proof : To show part (i), we rewrite (3.30) as

$$\dot{x} = \tilde{A}x + B\left[\sigma(-(1+\mu)B^{\mathrm{T}} P_\varepsilon x) + B^{\mathrm{T}} P_\varepsilon x\right] \tag{3.32}$$

where

$$\tilde{A} = A - BB^{\mathrm{T}} P_\varepsilon.$$

We can now select the Lyapunov function as

$$V = x^{\mathrm{T}} P_\varepsilon x$$

for the closed-loop system (3.32). From Lemma 3.3.1, it follows that there exists an $\varepsilon^* > 0$ and a constant $c > 0$ such that for all $\varepsilon \in (0, \varepsilon^*]$, we have $\mathcal{X}_0 \subseteq L_V(c)$, and moreover $x \in L_V(c)$ implies that $\|B^{\mathrm{T}} P_\varepsilon x\| \leq 1$, where the level set $L_V(c)$ is defined as

$$L_V(c) = \{x \in \mathbb{R}^n \mid V(x) \leq c\}.$$

Now, the derivative of V along the trajectories of the closed-loop system (3.32) shows that for all $\tilde{x} \in L_V(c)$,

$$\begin{aligned}\dot{V} &= -x^{\mathrm{T}}\left(Q_\varepsilon + P_\varepsilon B B^{\mathrm{T}} P_\varepsilon\right)x \\ &\qquad + 2x P_\varepsilon B\left[\sigma(-(1+\mu)B^{\mathrm{T}}P_\varepsilon x) + B^{\mathrm{T}}P_\varepsilon x\right] \\ &\leqslant -x^{\mathrm{T}}Q_\varepsilon x - 2\sum_{i=1}^{m} v_i\left[\sigma_i((1+\mu)v_i) - \sigma_i(v_i)\right] \\ &\leq -x^{\mathrm{T}}Q_\varepsilon x\end{aligned}$$

where we have defined $v := -B^{\mathrm{T}}P_\varepsilon x$ while v_i is the i-th component of v.

The above development shows that for all $\varepsilon \in (0, \varepsilon^*]$ and $\mu > 0$, the equilibrium point $\tilde{x} = 0$ of the closed-loop system (3.32) is internally stable with $\tilde{\mathcal{X}}_0(\varepsilon) \subseteq L_V(c)$ contained in its domain of attraction. This concludes our proof of part (i).

We next show that there exists an $\varepsilon_2^* \in (0, 1]$ such that for each $\varepsilon \in (0, \varepsilon_2^*]$ and for each $\mu \geq 0$, item (ii) of the theorem holds. To this end, let us introduce an invertible coordinate change $\xi = x - \Pi w$. Using condition (ii)a) of Theorem 3.3.2, we have

$$\dot{\xi} = A\xi + B[\sigma(u) - \Gamma w]. \tag{3.33}$$

With the family of state feedback laws given by (3.29), the closed-loop system can be written as

$$\dot{\xi} = A\xi + B[\sigma(\Gamma w - (\mu+1)B^{\mathrm{T}}P_\varepsilon\xi) - \Gamma w]. \tag{3.34}$$

By condition (ii)a) of Theorem 3.3.2, $\|\Gamma w\|_{\infty,\mathrm{T}} < 1 - \delta$. Moreover, for any $x(0) \in \mathcal{X}_0$ and any $w(0) \in \mathcal{W}_0$, $\xi(T)$ belongs to a bounded set, say $\mathcal{U}_T$, independent of ε since $\mathcal{X}_0$ and $\mathcal{W}_0$ are both bounded and $\xi(T)$ is determined by a linear differential equation with bounded inputs $\sigma(\cdot)$ and Γw.

We next pick a Lyapunov function

$$V(\xi) = \xi^{\mathrm{T}}P_\varepsilon\xi \tag{3.35}$$

and let $c > 0$ be such that

$$c \geq \sup_{\xi\in\mathcal{U}_T,\varepsilon\in(0,1]} \xi^{\mathrm{T}}P_\varepsilon\xi. \tag{3.36}$$

Such a c exists since $\lim_{\varepsilon\to 0} P_\varepsilon = 0$ by Lemma 3.3.1 and $\mathcal{U}_T$ is bounded. Let $\varepsilon_2^* \in (0, 1]$ be such that $\xi \in L_V(c)$ implies that $\|B^{\mathrm{T}}P_\varepsilon\xi\|_\infty \leq \delta$. The existence of such an ε_2^* is again due to the fact that $\lim_{\varepsilon\to 0} P_\varepsilon = 0$. The evaluation of $\dot{V}$ for $t \geq T$, inside the set $L_V(c)$, using (3.9), now shows that

$$\begin{aligned}\dot{V} &= -\xi^{\mathrm{T}}\left(Q_\varepsilon + P_\varepsilon B B^{\mathrm{T}} P_\varepsilon\right)\xi \\ &\qquad + 2\xi^{\mathrm{T}} P_\varepsilon B[\sigma(\Gamma w - (\mu+1)B^{\mathrm{T}} P_\varepsilon \xi) - \Gamma w + B^{\mathrm{T}} P_\varepsilon \xi] \\ &\leq -\xi^{\mathrm{T}} Q_\varepsilon \xi - 2v^{\mathrm{T}}[\sigma((\mu+1)v + \theta) - v - \theta] \\ &= -\xi^{\mathrm{T}} Q_\varepsilon \xi - 2\sum_{i=1}^{m} v_i[\bar{\sigma}((\mu+1)v_i + \theta_i) - v_i - \theta_i]\end{aligned}$$

where we have denoted $v := -B^{\mathrm{T}} P_\varepsilon \xi$ and $\theta := \Gamma w$ with their ith components denoted by v_i and θ_i respectively.

Noting that $\mathrm{sgn}(\bar{\sigma}(s)) = \mathrm{sgn}(s)$ and for $\varepsilon \in (0, \varepsilon_1^*]$, $|v_i + \theta_i| \leq 1$, we observe that,

$$\begin{aligned}&|(\mu+1)v_i + \theta_i| \leq 1 \\ &\qquad \Longrightarrow v_i[\bar{\sigma}((\mu+1)v_i + \theta_i) - v_i - \theta_i] = \mu v_i^2 \geq 0, \\ &(\mu+1)v_i + \theta_i > 1 \Longrightarrow v_i > 0 \\ &\qquad \Longrightarrow v_i[\bar{\sigma}((\mu+1)v_i + \theta_i) - v_i - \theta_i] \geq 0, \\ &(\mu+1)v_i + \theta_i < -1 \Longrightarrow v_i < 0 \\ &\qquad \Longrightarrow v_i[\bar{\sigma}((\mu+1)v_i + \theta_i) - v_i - \theta_i] \geq 0.\end{aligned}$$

Hence, we conclude that

$$\dot{V} \leq -\xi^{\mathrm{T}} Q_\varepsilon \xi. \tag{3.37}$$

This shows that any trajectory of (3.34) starting at $t = 0$ from $\{\xi = x - \Pi w : x \in \mathcal{X}_0, w \in \mathcal{W}_0\}$ remains inside the set $L_V(c)$ and approaches the equilibrium point $\xi = 0$ as $t \to \infty$. It is then easy to see that

$$e = C_e(\xi + \Pi w) + D_{eu}\sigma\left(-(\mu+1)B^{\mathrm{T}} P_\varepsilon)\xi + \Gamma w\right) + D_{ew} w$$

and hence $e(t) \to 0$ as $t \to \infty$ since $\xi(t) \to 0$ and using conditions (3.18b) and (ii)b) of Theorem 3.3.2.

Finally, setting $\varepsilon^* = \min\{\varepsilon_1^*, \varepsilon_2^*\}$, we conclude our proof of Theorem 3.3.3. ■

We now demonstrate the improvement on the closed-loop performance (as the high gain parameter μ increases) by an example.

Example 3.3.2 *Consider the system (3.1) with the matrices A, B, E_w, S, C_e, D_{eu}, and D_{ew} as given in Example 3.3.1. Also, as in Example 3.3.1, let $w(0) \in \mathcal{W}_0$ where $\mathcal{W}_0 = \left\{w : \|w\| < \sqrt{2}/4\right\}$. Then, the solvability conditions for the semi-global linear state feedback output regulation problem are satisfied. Also, Π and Γ are as given in (3.27). Moreover, $T = 0$ and $\delta = 0.4697$, since $\|\Gamma w\|_\infty \leq 0.5303$ for all $w(0) \in \mathcal{W}_0$. Let the set $\mathcal{X}_0$ be given by $\mathcal{X}_0 = \{x : \|x\| \leq 14, x \in \mathbb{R}^4\}$. We choose*

$$Q_\varepsilon = 10^{-4} \times \begin{pmatrix} \varepsilon^2 & 0 & 0 & 0 \\ 0 & \varepsilon^2 & 0 & 0 \\ 0 & 0 & \varepsilon & 0 \\ 0 & 0 & 0 & \varepsilon \end{pmatrix}. \tag{3.38}$$

Then, following the proof of Theorem 3.3.3, a choice of ε^* is 1.5×10^{-4}. For $\varepsilon = \varepsilon^*$, the feedback law (3.29) for $\mu \geq 0$ is given by

$$\begin{aligned} u = &-(\mu+1)10^{-3} \begin{pmatrix} 6.4203 & 11.918 & 3.3628 & -2.9767 \\ 11.918 & 22.623 & 6.113 & -5.6505 \end{pmatrix} x \\ &+ \left((\mu+1)10^{-3} \begin{pmatrix} -2.9799 & 3.3755 \\ -4.2588 & 4.6980 \end{pmatrix} + \begin{pmatrix} 1.25 & 0.25 \\ -0.25 & -1.25 \end{pmatrix}\right) w. \end{aligned}$$

For the initial conditions $x_0 = (7, 7, 7, 7)$, $w_0 = (0.1, 0.1)$, Figures 3.2 on the following page and 3.3 on page 91 show the control action and the closed-loop performance for low-gain feedback ($\mu = 0$) and low-and-high gain feedback ($\mu = 1000$) respectively. These simulation results illustrate that the low-and-high gain feedback regulator (3.29) significantly out-performs the low-gain feedback regulator as given in (3.28).

3.3.2 Dynamic measurement feedback controller

In this section, we consider the classical semi-global linear observer based measurement feedback output regulation problem which can be formulated as follows.

Problem 3.3.2 ***(Classical semi-global linear full-order observer based measurement feedback output regulation problem)*** *Consider the system (3.1) and a compact set $\mathcal{W}_0 \subset \mathbb{R}^s$. For any a priori given (arbitrarily large) bounded sets $\mathcal{X}_0 \subset \mathbb{R}^n$ and $\mathcal{Z}_0 \subset \mathbb{R}^{n+s}$, find, if possible, a measurement feedback law*

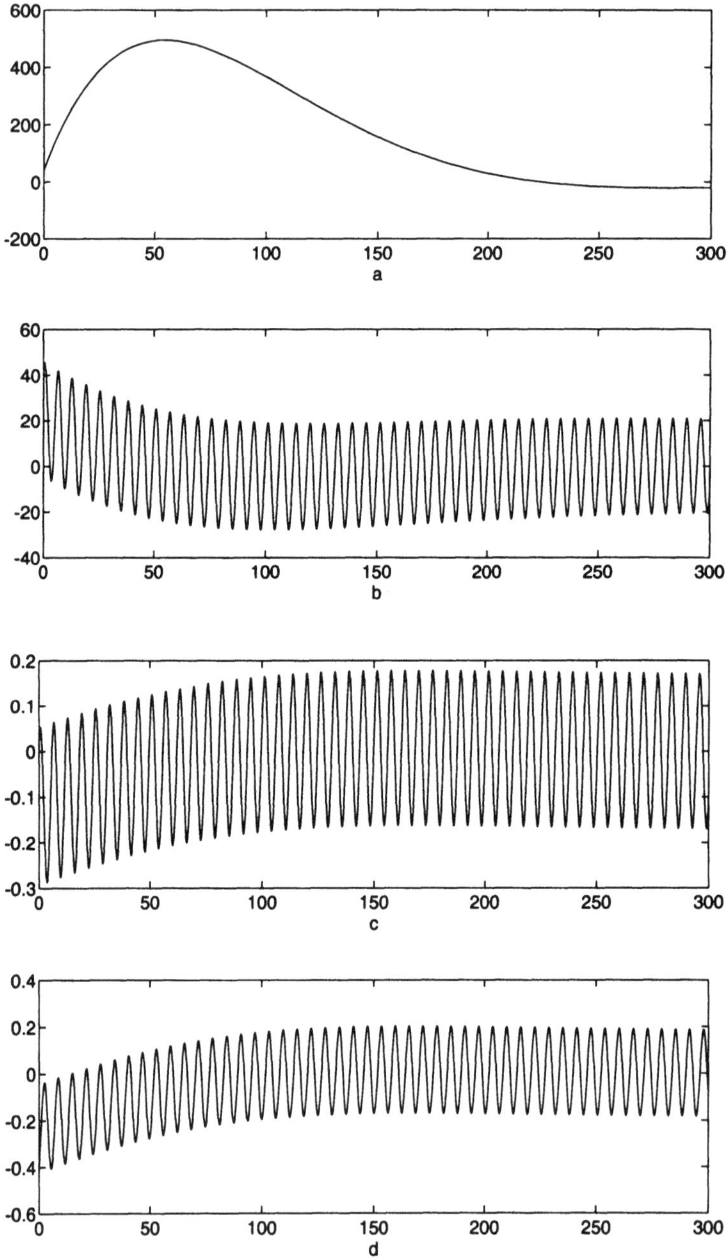

Figure 3.2: $\varepsilon = 1.5 \times 10^{-4}$, $\mu = 0$. a) e_1; b) e_2; c) $\bar{\sigma}(u_1)$; d) $\bar{\sigma}(u_2)$.

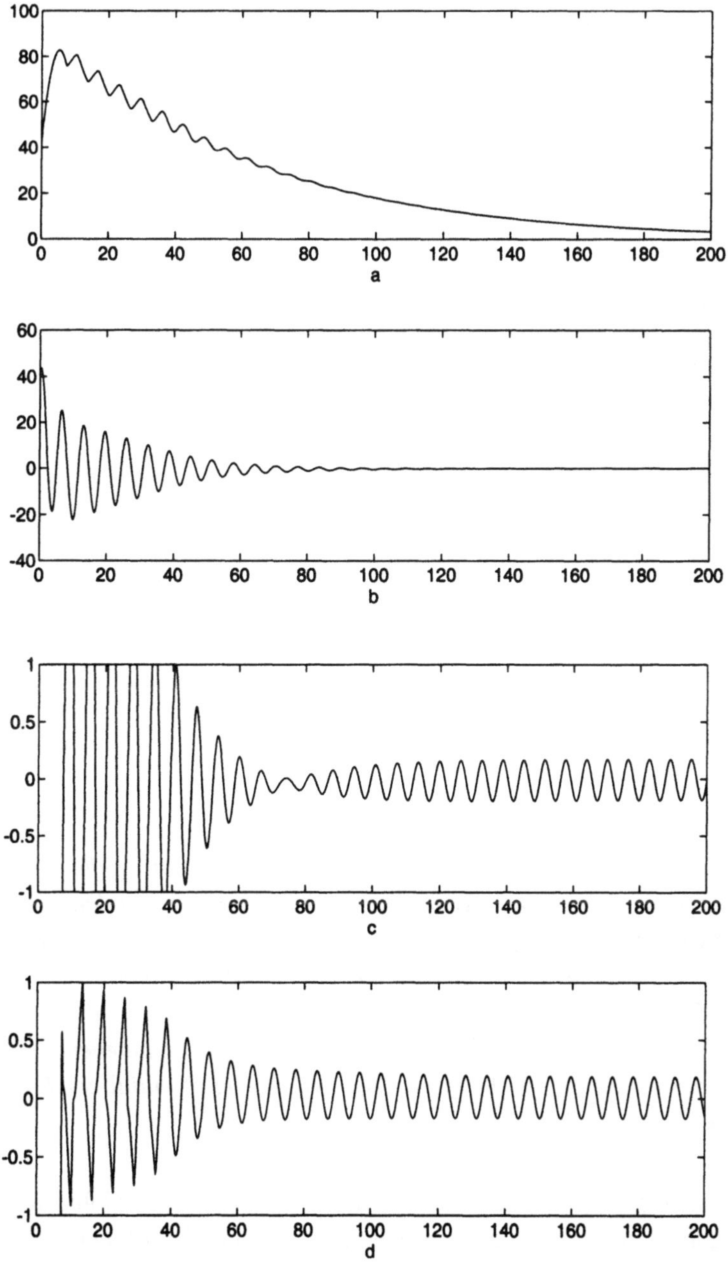

Figure 3.3: $\varepsilon = 1.5 \times 10^{-4}$, $\mu = 1000$. a) e_1; b) e_2; c) $\bar{\sigma}(u_1)$; d) $\bar{\sigma}(u_2)$.

of the form

$$
\begin{aligned}
\begin{pmatrix} \dot{\hat{x}} \\ \dot{\hat{w}} \end{pmatrix} &= \begin{pmatrix} A & E_w \\ 0 & S \end{pmatrix} \begin{pmatrix} \hat{x} \\ \hat{w} \end{pmatrix} + \begin{pmatrix} B \\ 0 \end{pmatrix} \sigma(u) \\
&\quad + \begin{pmatrix} K_A \\ K_S \end{pmatrix} \left(\begin{pmatrix} C_y & D_{yw} \end{pmatrix} \begin{pmatrix} \hat{x} \\ \hat{w} \end{pmatrix} - y \right) \\
u &= F\hat{x} + G\hat{w}
\end{aligned} \tag{3.39}
$$

such that the following conditions hold:

(*i*) ***(Internal Stability)*** *The equilibrium point* $(x, \hat{x}, \hat{w}) = (0, 0, 0)$ *of*

$$
\begin{aligned}
\dot{x} &= Ax + B\sigma(F\hat{x} + G\hat{w}) \\
\begin{pmatrix} \dot{\hat{x}} \\ \dot{\hat{w}} \end{pmatrix} &= \begin{pmatrix} A & E_w \\ 0 & S \end{pmatrix} \begin{pmatrix} \hat{x} \\ \hat{w} \end{pmatrix} + \begin{pmatrix} B \\ 0 \end{pmatrix} \sigma(F\hat{x} + G\hat{w}) \\
&\quad + \begin{pmatrix} K_A \\ K_S \end{pmatrix} \left(\begin{pmatrix} C_y & D_{yw} \end{pmatrix} \begin{pmatrix} \hat{x} - x \\ \hat{w} \end{pmatrix} \right)
\end{aligned} \tag{3.40}
$$

is asymptotically stable with $\mathcal{X}_0 \times \mathcal{Z}_0$ *contained in its basin of attraction.*

(*ii*) ***(Output Regulation)*** *For all* $(x(0), \hat{x}(0)) \in \mathcal{X}_0$, $\hat{w}(0) \in \mathcal{Z}_0$ *and* $w(0) \in \mathcal{W}_0$, *the solution of the closed-loop system satisfies*

$$
\lim_{t \to \infty} e(t) = 0. \tag{3.41}
$$

Remark 3.3.2 *We would like to emphasize again that our definition of the above classical semi-global measurement feedback output regulation problem does not view the set of initial conditions of the plant and the initial conditions of the controller dynamics as given data. The set of given data consists of the models of the plant and the exosystem and the set of initial conditions for the exosystem. Therefore, the solvability conditions* must *be independent of the set of initial conditions of the plant,* $\mathcal{X}_0$, *and the set of initial conditions for the controller dynamics,* $\mathcal{Z}_0$.

The solvability conditions for the above classical semi-global measurement feedback output regulation problem are given in the following theorem.

Theorem 3.3.4 *Consider the system (3.1) and the given compact set* $\mathcal{W}_0 \subset \mathbb{R}^s$. *The classical semi-global measurement feedback output regulation problem is solvable if the following conditions hold:*

(*i*) (A, B) *is stabilizable and* A *has all its eigenvalues in the closed left half plane. Moreover, the pair*

$$\left(\begin{pmatrix} C_y & D_{yw} \end{pmatrix}, \begin{pmatrix} A & E_w \\ 0 & S \end{pmatrix}\right)$$

is detectable.

(*ii*) *There exist matrices* Π *and* Γ *such that*

(*a*) *they solve the regulator equation (3.18), and*

(*b*) *there exists a* $\delta > 0$ *and a* $T \geq 0$ *such that* $\|\Gamma w\|_{\infty,T} \leqslant 1 - \delta$ *for all* w *with* $w(0) \in \mathcal{W}_0$.

Proof : We prove this theorem by first explicitly constructing a family of linear observer based measurement feedback laws of the form (3.39), parameterized in ε, and then showing that for each pair of sets $\mathcal{X}_0 \subset \mathbb{R}^n$ and $\mathcal{Z}_0 \subset \mathbb{R}^{n+s}$, there exists an $\varepsilon^* > 0$ such that for all $\varepsilon \in (0, \varepsilon^*]$, both items (i) and (ii) in Problem 3.3.2 are indeed satisfied. The family of linear observer based measurement feedback laws we construct take the form,

$$\begin{aligned} \dot{\hat{x}} &= A\hat{x} + B\sigma(u) + E_w\hat{w} + K_A C_y(\hat{x} - x) + K_A D_{yw}(\hat{w} - w) \\ \dot{\hat{w}} &= S\hat{w} + K_S C_y(\hat{x} - x) + K_S D_{yw}(\hat{w} - w) \\ u &= F_\varepsilon \hat{x} + (\Gamma - F_\varepsilon \Pi)\hat{w}, \end{aligned} \tag{3.42}$$

where F_ε is a state feedback gain matrix, parameterized in ε, chosen in such a way that $A + BF_\varepsilon$ is Hurwitz-stable and

$$\|F_\varepsilon\| \leq \kappa_\varepsilon, \tag{3.43}$$

$$\|F_\varepsilon e^{(A+BF_\varepsilon)t}\|_\infty \leq \upsilon_\varepsilon e^{-\zeta_\varepsilon t}, \tag{3.44}$$

where $\kappa_\varepsilon, \upsilon_\varepsilon, \zeta_\varepsilon \to 0$ as $\varepsilon \to 0$ and $\zeta_\varepsilon > 0$. The existence and the explicit construction of such an F_ε is established in Theorem 3.3.1. The matrices K_A and K_S are chosen such that the matrix,

$$\bar{A} := \begin{pmatrix} A + K_A C_y & E_w + K_A D_{yw} \\ K_S C_y & S + K_S D_{yw} \end{pmatrix}, \tag{3.45}$$

is Hurwitz-stable. The existence of such K_A and K_S is guaranteed by condition (i) of the theorem.

With this family of feedback laws, the closed-loop system consisting of the system (3.1) and the dynamic measurement feedback laws (3.42) can be written as,

$$\begin{aligned}\dot{x} &= Ax + B\sigma(F_\varepsilon \hat{x} + (\Gamma - F_\varepsilon \Pi)\hat{w}) + E_w w \\ \dot{\hat{x}} &= A\hat{x} + B\sigma(F_\varepsilon \hat{x} + (\Gamma - F_\varepsilon \Pi)\hat{w}) + E_w \hat{w} \\ &\qquad + K_{\mathrm{A}} C_y(\hat{x} - x) + K_{\mathrm{A}} D_{yw}(\hat{w} - w) \\ \dot{\hat{w}} &= S\hat{w} + K_{\mathrm{S}} C_y(\hat{x} - x) + K_{\mathrm{S}} D_{yw}(\hat{w} - w).\end{aligned} \tag{3.46}$$

We then adopt the invertible change of state variables,

$$\begin{aligned}\xi &= x - \Pi w \\ \tilde{x} &= x - \hat{x} \\ \tilde{w} &= w - \hat{w},\end{aligned} \tag{3.47}$$

and then rewrite the closed-loop system (3.46) as

$$\begin{aligned}\dot{\xi} &= A\xi + B\sigma(F_\varepsilon \xi + \Gamma w - \Gamma \tilde{w} - F_\varepsilon \tilde{x} \\ &\qquad + F_\varepsilon \Pi \tilde{w}) + (A\Pi - \Pi S + E_w)w \\ \dot{\tilde{x}} &= (A + K_{\mathrm{A}} C_y)\tilde{x} + (E_w + K_{\mathrm{A}} D_{yw})\tilde{w} \\ \dot{\tilde{w}} &= K_{\mathrm{S}} C_y \tilde{x} + (S + K_{\mathrm{S}} D_{yw})\tilde{w}.\end{aligned} \tag{3.48}$$

To show that item (i) of Problem 3.3.2 holds, we note that (3.39) is equal to (3.46) for $w = 0$. Thus, for $w = 0$, (3.48) reduces to

$$\begin{aligned}\dot{\xi} &= A\xi + B\sigma(F_\varepsilon \xi - \Gamma \tilde{w} - F_\varepsilon \tilde{x} + F_\varepsilon \Pi \tilde{w}) \\ \dot{\tilde{x}} &= (A + K_{\mathrm{A}} C_y)\tilde{x} + (E_w + K_{\mathrm{A}} D_{yw})\tilde{w} \\ \dot{\tilde{w}} &= K_{\mathrm{S}} C_y \tilde{x} + (S + K_{\mathrm{S}} D_{yw})\tilde{w}.\end{aligned} \tag{3.49}$$

Recalling that the matrix $\bar{A}$, defined in (3.45), is Hurwitz-stable, and using (3.43), it readily follows from the last two equations of (3.49) that there exists a $T_1 \geq 0$ such that, for all possible initial conditions $\tilde{x}(0)$, $\tilde{w}(0)$ and for all $\varepsilon \in (0, 1]$, we have

$$\|\Gamma \tilde{w}\|_{\infty,T_1} \leq \tfrac{1}{4}, \quad \|F_\varepsilon \tilde{x}\|_{\infty,T_1} \leq \tfrac{1}{4}, \quad \|F_\varepsilon \Pi \tilde{w}\|_{\infty,T_1} \leq \tfrac{1}{4}. \tag{3.50}$$

We next consider the first equation of (3.49). $\xi(T_1)$ belongs to a bounded set independent of ε since $\xi(0)$ is bounded and since ξ is determined via a linear differential equation with bounded input $\sigma(u)$. Hence there exists an M_1 such that for all possible initial conditions,

$$\|\xi(T_1)\| \leq M_1, \quad \text{for all } \varepsilon \in (0, 1]. \tag{3.51}$$

Let us now assume that, from time T_1 onwards, the saturation elements are non-existent. In this case, the first equation of (3.49) can be written as

$$\dot{\xi} = (A + BF_\varepsilon)\xi - BF_\varepsilon\tilde{x} - B\Gamma\tilde{w} + BF_\varepsilon\Pi\tilde{w}. \tag{3.52}$$

Since $\tilde{x} \to 0$ and $\tilde{w} \to 0$ exponentially with a decay rate independent of ε as $t \to \infty$, it follows trivially from (3.43) and (3.44) that there exist an $\varepsilon_1^* > 0$ and an $M_2 > 0$ such that, for all possible initial conditions and all $\varepsilon \in (0, \varepsilon_1^*]$,

$$\int_{T_1}^{\infty} \|e^{\zeta_\varepsilon \tau} B[F_\varepsilon\tilde{x}(\tau) + \Gamma\tilde{w}(\tau) - F_\varepsilon\Pi\tilde{w}(\tau)]\| d\tau \leq M_2. \tag{3.53}$$

This in turn shows that, for $t \geq T_1$,

$$\begin{aligned}
\|F_\varepsilon\xi(t)\| &= \left\| F_\varepsilon e^{(A+BF_\varepsilon)t}\xi(T_1) - \int_{T_1}^{t} F_\varepsilon e^{(A+BF_\varepsilon)(t-\tau)} Bs(\tau)d\tau \right\| \\
&\leqslant \upsilon_\varepsilon M_1 + \upsilon_\varepsilon \int_{T_1}^{\infty} \|e^{\zeta_\varepsilon \tau} Bs(\tau)\| d\tau \\
&\leqslant \upsilon_\varepsilon (M_1 + M_2).
\end{aligned}$$

where

$$s(\tau) = F_\varepsilon\tilde{x}(\tau) + \Gamma\tilde{w}(\tau) - F_\varepsilon\Pi\tilde{w}(\tau).$$

Choose $\varepsilon_2^* \in (0, \varepsilon_1^*]$ such that, for all $\varepsilon \in (0, \varepsilon_2^*]$,

$$\|F_\varepsilon\xi\|_{\infty,T_1} \leq \frac{1}{4}. \tag{3.54}$$

This, together with (3.50), shows that the system (3.49) will operate linearly after time T_1 and local exponential stability of this linear system follows from the separation principle.

In summary, we have shown that there exists an $\varepsilon_2^* > 0$ such that, for all $\varepsilon \in (0, \varepsilon_2^*]$, the equilibrium point $(0, 0, 0)$ of the system (3.49) is asymptotically stable, with $(\mathcal{X}_0, \mathcal{Z}_0)$ contained in its basin of attraction.

We now proceed to show that item 2 of Problem 3.3.2 also holds. To this end, we consider the closed-loop system (3.48). Recalling that the matrix $\bar{A}$ is Hurwitz-stable, and using (3.43), it readily follows from the last two equations of (3.48) that there exists a $T_2 \geq T$ such that, for all possible initial conditions $(\tilde{x}(0), \tilde{w}(0))$ and for all $\varepsilon \in (0, 1]$

$$\|\Gamma\tilde{w}\|_{\infty,T_2} \leq \tfrac{\delta}{4}, \ \|F_\varepsilon\tilde{x}\|_{\infty,T_2} \leq \tfrac{\delta}{4}, \ \|F_\varepsilon\Pi\tilde{w}\|_{\infty,T_2} \leq \tfrac{\delta}{4}. \tag{3.55}$$

We next consider the first equation of (3.48). $\xi(T_2)$ belongs to a bounded set independent of ε since $\xi(0)$ is bounded and since ξ is determined via a linear differential equation with bounded inputs $\sigma(u)$ and w. Hence there exists an M_3 such that for all possible initial conditions,

$$\|\xi(T_2)\| \leq M_3, \quad \text{for all } \varepsilon \in (0,1]. \tag{3.56}$$

Let us now assume that, from time T_2 onwards, the equation (3.48) operates without the saturation elements. In view of Condition (ii)a), the first equation of (3.48) in the absence of the saturation elements is equal to the first equation of (3.49), and hence also reduces to (3.52) after time T_2. Hence, using a similar argument as above, we can show that there exists an $\varepsilon_3^* > 0$ such that, for all $\varepsilon \in (0, \varepsilon_3^*]$,

$$\|F_\varepsilon \xi\|_{\infty,T_2} < \frac{\delta}{4}. \tag{3.57}$$

This, together with (3.55) and Condition (ii)b), shows that the system (3.48) will operate linearly after time T_2, and thus the exponential stability of this linear system follows from the separation principle.

Next, in view of (3.18b), it is easy to evaluate $e(t)$ for $t \geq T_2$ as

$$e(t) = (C_e + D_{eu}F_\varepsilon)\xi(t) - D_{eu}F_\varepsilon\tilde{x} - D_{eu}(\Gamma - F_\varepsilon\Pi)\tilde{w}.$$

This implies that $e(t) \to 0$ as $t \to \infty$.

Finally, taking $\varepsilon^* = \min\{\varepsilon_2^*, \varepsilon_3^*\}$, we complete our proof. ■

As in the state feedback case, Condition (i) of Theorem 3.3.4 is necessary to guarantee solvability of the measurement feedback output regulation problem for the system in the absence of the saturation element. Condition (ii)a) is also necessary for the existence of a linear feedback controller which semi-globally stabilizes the system (3.1) which is subject to amplitude saturation. Clearly this time we also needed a detectability assumption. Finally, the crucial condition for the solvability of classical semi-global linear observer based measurement feedback output regulation problem is Condition (ii)a). This is a sufficient condition.

Note the surprising fact that, except for the detectability assumption, the solvability conditions for output regulation are the same for both the cases of state feedback and dynamic measurement feedback.

Design of a low-gain measurement feedback regulator:

For clarity, we now summarize the construction of an observer based measurement feedback regulator.

Step 1 : At first construct a state feedback regulator. That is, find the matrices $(\Pi,\ \Gamma)$ that solve the regulator equation (3.18). Also, find a state feedback gain matrix F_ε parameterized in ε in such a way that $A + BF_\varepsilon$ is Hurwitz-stable for all $\varepsilon > 0$ and such that equation (3.20) holds for a positive-valued function υ satisfying $\lim_{\varepsilon\to 0} \upsilon_\varepsilon = 0$.

Step 2 : Design a full order observer so that we can implement the controller with observer based architecture as given in (3.42). That is, find the matrix gains K_A and K_S in such a way that the matrix $\bar{A}$ given in (3.45) is Hurwitz-stable.

Step 3 : Implement the observer based measurement feedback regulator as given in (3.42).

Step 4 : Given the sets $\mathcal{X}_0 \subset \mathbb{R}^n$, $\mathcal{W}_0 \subset \mathbb{R}^s$, and $\mathcal{Z}_0 \subset \mathbb{R}^n$, find an ε^* by the procedure given in the proof of Theorem 3.3.4.

Then for some $\varepsilon \in (0,\ \varepsilon^*]$ the observer based measurement feedback regulator as given in (3.42) solves the classical semi-global measurement feedback output regulation problem. □

The following example illustrates the design procedure.

Example 3.3.3 *(Measurement Feedback Case) Consider the same plant as in Example 3.3.1. However, assume that $y = e$ is available for feedback, instead of the states x and w. Recall that $\mathcal{W}_0 = \{w \in \mathbb{R}^2 : \|w\| \le \sqrt{2}/4\}$, $T = 0$ and $\delta = 0.4697$. Following the design procedure given above, choose K_A and K_S as*

$$K_A = \begin{pmatrix} 1.6021 \times 10^6 & 1.0758 \times 10^5 \\ 8.0176 \times 10^5 & 1.2257 \times 10^6 \\ -5.6686 \times 10^6 & 1.7878 \times 10^6 \\ -4.9298 \times 10^6 & 1.1093 \times 10^7 \end{pmatrix},$$

$$K_S = \begin{pmatrix} 7.4414 \times 10^5 & -1.8083 \times 10^5 \\ -4.2860 \times 10^5 & 1.0281 \times 10^6 \end{pmatrix}.$$

These gains place the eigenvalues of $\bar{A}$ at

$$\{-100, -110, -120, -130, -140, -150\}.$$

Using a low-gain F_ε as in (3.10), for $\mathcal{X}_0 = \{x \in \mathbb{R}^4 : \|x\| \le 1\}$ and for $\mathcal{Z}_0 = \{z \in \mathbb{R}^6 : \|z\| \le 1\}$, a choice of ε^* is $2 \times e^{-8}$. For $\varepsilon = \varepsilon^*$, the linear *observer based feedback law (3.42) is given by*

$$\begin{aligned}
\dot{\hat{x}} &= A\hat{x} + B\sigma(u) + E_w\hat{w} + K_{\mathrm{A}}C_e(\hat{x} - x) + K_{\mathrm{A}}D_{ew}(\hat{w} - w)\\
\dot{\hat{w}} &= S\hat{w} + K_{\mathrm{S}}C_e(\hat{x} - x) + K_{\mathrm{S}}D_{ew}(\hat{w} - w)\\
u &= -10^{-3}\begin{pmatrix} -7.7148 & -14.235 & -4.0450 & 3.5547\\ -14.235 & -26.870 & -7.3084 & 6.7100 \end{pmatrix}\hat{x}\\
&\qquad +10^{-1}\begin{pmatrix} 12.482 & 2.5203\\ -2.5327 & -12.464 \end{pmatrix}\hat{w}.
\end{aligned}$$

Some simulation results are shown in Figure 3.4 on the facing page. Plots a and b show the error signals on the interval $[0, 1000]$ *which converge to* 0. *Plots c and d give the corresponding two input signals on two time intervals* $[0, 0.4]$ *and* $[0.4, 300]$. *The plots of the inputs on the interval* $[0, 0.4]$ *show that the inputs initially saturate. The plots on the interval* $[0.4, 300]$ *show the steady-state behavior of the inputs.*

Design of a low-and-high gain measurement feedback regulator:

The simulations in Figure 3.4 on the next page point out that the convergence of $e(t)$ to zero is very slow. Again, as in the case of state feedback, this is a characteristic of the feedback law (3.42) which utilizes a low-gain for F_ε as given by (3.10). Motivated by this, we recall next from [33] an improved design which utilizes a low-and-high gain feedback.

As before, we first construct here a family of low-and-high gain measurement feedback laws. Such a family of laws is parameterized in a low-gain parameter ε, a high-gain parameter μ, and an observer parameter ℓ. We then show that such a family of low-and-high gain measurement feedback laws solves the semi-global output regulation problem. Significant improvement on the closed-loop performance over the earlier low-gain design (3.42) is again shown by an example. The family of low-and-high gain measurement feedback laws we construct here is linear observer based and is composed by implementing the low-and-high gain state feedback laws developed earlier with the state of a fast observer. More specifically, it takes the following form

$$\begin{aligned}
\dot{\hat{x}} &= A\hat{x} + B\sigma(u) + E_w\hat{w} + K_{\mathrm{A},\ell}\left[C_y\hat{x} + D_{yw}\hat{w} - y\right]\\
\dot{\hat{w}} &= S\hat{w} + K_{\mathrm{S},\ell}\left[C_y\hat{x} + D_{yw}\hat{w} - y\right]\\
u &= -(1+\mu)B^{\mathrm{T}}P_\varepsilon\hat{x} + ((\mu+1)B^{\mathrm{T}}P_\varepsilon\Pi + \Gamma)\hat{w}
\end{aligned}\tag{3.58}$$

where $K_{\mathrm{A},\ell}$ and $K_{\mathrm{S},\ell}$ need to be chosen carefully. There are two ways of designing the observers. One is Riccati-based which we cover here and the other is a direct method which is covered in Appendix 3.C

The following lemma presents some of the basic properties of a dual Riccati equation which we will need to derive our results.

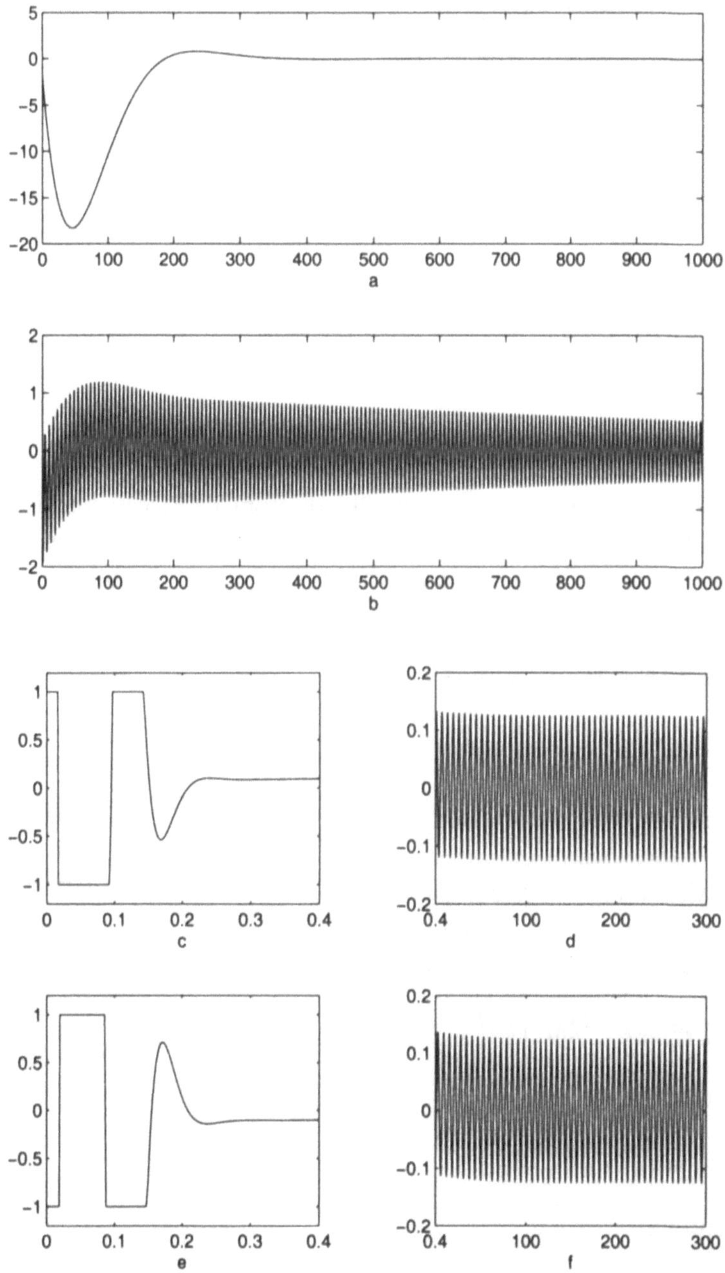

Figure 3.4: a) e_1; b) e_2; c,d) $\bar{\sigma}(u_1)$ (two time-intervals) e,f) $\bar{\sigma}(u_2)$ (two time-intervals)

Lemma 3.3.2 *Let the pair $(H,\ F)$ be observable. Also, let Q_ℓ be the solution of the following dual algebraic Riccati equation,*

$$0 = (F + \ell I)Q + Q(F + \ell I)^T - QHH^TQ + \ell I.$$

Then we have,

$$(F + Q_\ell HH^T)Q_\ell + Q_\ell(F + Q_\ell HH^T)^T \leqslant -\ell I - \ell Q_\ell. \tag{3.59}$$

Moreover for any fixed $t > 0$,

$$\left\| e^{(F+Q_\ell HH^T)t} Q_\ell e^{(F^T+HH^TQ_\ell)t} \right\| \to 0 \tag{3.60}$$

as $\ell \to \infty$.

Proof : The inequality (3.59) follows directly from the Riccati equation. However, to show (3.60) we need to do some work since $Q_\ell \to \infty$ as $\ell \to \infty$. Note that Q_ℓ has an interpretation as the optimal cost of the following optimal control problem:

$$\dot{x} = F^{\mathrm{T}}x + H^{\mathrm{T}}u, \quad x(0) = \xi$$

and

$$\xi^{\mathrm{T}} Q_\ell \xi = \inf_u \int_0^t \ell \|e^{\ell s}x(s)\|^2 + \|e^{\ell s}u(s)\|^2 ds + e^{2\ell t}x(t)^{\mathrm{T}}Q_\ell x(t).$$

There exists a $M > 0$ such that for any ξ there exists an input u such that $x(t/2) = 0$, $\|u\|_2 < M\|\xi\|$ and $\|x\|_2 < M\|\xi\|$. When we choose this suboptimal input in the above optimization problem we find that

$$\xi^{\mathrm{T}} Q_\ell \xi \leqslant e^{\ell t/2} 2M^2 \|\xi\|^2.$$

On the other hand

$$x(t) = e^{(F^{\mathrm{T}}+HH^{\mathrm{T}})t}\xi$$

and therefore

$$\begin{aligned} e^{2\ell t}\xi^{\mathrm{T}} e^{(F+HH^{\mathrm{T}})t} Q_\ell e^{(F^{\mathrm{T}}+HH^{\mathrm{T}})t}\xi &= e^{2\ell t}x(t)^{\mathrm{T}}Q_\ell x(t) \leqslant \xi^{\mathrm{T}}Q_\ell\xi \\ &\leqslant e^{\ell t/2}2M^2\|\xi\|^2 \end{aligned}$$

which implies (3.60). ■

Let R_ℓ be the stabilizing solution of the following dual algebraic Riccati equation:

$$0 = \begin{pmatrix} A+\ell I & E_w \\ 0 & S+\ell I \end{pmatrix} R + R \begin{pmatrix} A+\ell I & E_w \\ 0 & S+\ell I \end{pmatrix}^{\mathrm{T}} - R \begin{pmatrix} C_y^{\mathrm{T}} \\ D_{yw}^{\mathrm{T}} \end{pmatrix} \begin{pmatrix} C_y & D_{yw} \end{pmatrix} R + \ell I. \quad (3.61)$$

On the basics of the above lemma we choose the observer gain as

$$K_\ell = \begin{pmatrix} K_{\mathrm{A},\ell} \\ K_{\mathrm{S},\ell} \end{pmatrix} = -R_\ell \begin{pmatrix} C_y^{\mathrm{T}} \\ D_{yw}^{\mathrm{T}} \end{pmatrix}. \quad (3.62)$$

We then have the following result.

Theorem 3.3.5 *Consider the system (3.1) and the given compact set $\mathcal{W}_0 \subset \mathbb{R}^s$. Let the sufficient conditions given in Theorem 3.3.4 be satisfied. Also, assume that the pair*

$$\left(\begin{pmatrix} C_y & D_{yw} \end{pmatrix}, \begin{pmatrix} A & E_w \\ 0 & S \end{pmatrix} \right)$$

is observable. Then, there exists a controller, among the family of low-and-high gain measurement feedback laws (3.58), that solves the semi-global linear observer based measurement feedback output regulation problem. More specifically, for any a priori given (arbitrarily large) sets $\mathcal{X}_0 \subset \mathbb{R}^n$ and $\mathcal{Z}_0 \subset \mathbb{R}^{n+s}$, there exists an $\varepsilon^* \in (0, 1]$, and for each $\varepsilon \in (0, \varepsilon^*]$ and each $\mu \geq 0$, there exists an $\ell^*(\varepsilon, \mu) > 0$ such that for each $\varepsilon \in (0, \varepsilon^*]$, each $\mu \geq 0$ and each $\ell \geq \ell^*(\varepsilon, \mu)$, the controller in the family (3.58) has the following properties:

(*i*) *The equilibrium point $(x, \hat{x}, \hat{w}) = (0, 0, 0)$ of*

$$\begin{aligned} \dot{x} &= Ax + B\sigma(\Gamma\hat{w} - (1+\mu)B^T P_\varepsilon(\hat{x} - \Pi\hat{w})) \\ \dot{\hat{x}} &= A\hat{x} + B\sigma(\Gamma\hat{w} - (1+\mu)B^T P_\varepsilon(\hat{x} - \Pi\hat{w})) + E_w\hat{w} \\ &\qquad + K_{\mathrm{A},\ell}C_y(\hat{x} - x) + K_{\mathrm{A},\ell}D_{yw}\hat{w} \\ \dot{\hat{w}} &= S\hat{w} + K_{\mathrm{S},\ell}C_y(\hat{x} - x) + K_{\mathrm{S},\ell}D_{yw}\hat{w} \end{aligned} \quad (3.63)$$

is asymptotically stable with $\mathcal{X}_0 \times \mathcal{Z}_0$ contained in its basin of attraction.

(*ii*) *For any $(x(0), \hat{x}(0), \hat{w}(0)) \in \mathcal{X}_0 \times \mathcal{Z}_0$ and $w(0) \in \mathcal{W}_0$, the solution of the closed-loop system satisfies*

$$\lim_{t\to\infty} e(t) = 0. \quad (3.64)$$

Proof : With the family of feedback laws (3.58), the closed-loop system can be written as,

$$\begin{aligned}
\dot{x} &= Ax + B\sigma(\Gamma\hat{w} - (1+\mu)B^{\mathrm{T}}P_\varepsilon(\hat{x} - \Pi\hat{w})) + E_w w \\
\dot{\hat{x}} &= A\hat{x} + B\sigma(\Gamma\hat{w} - (1+\mu)B^{\mathrm{T}}P_\varepsilon(\hat{x} - \Pi\hat{w})) + E_w\hat{w} \\
&\qquad + K_{\mathrm{A},\ell}C_y(\hat{x} - x) + K_{\mathrm{A},\ell}D_{yw}(\hat{w} - w) \\
\dot{\hat{w}} &= S\hat{w} + K_{\mathrm{S},\ell}C_y(\hat{x} - x) + K_{\mathrm{S},\ell}D_{yw}(\hat{w} - w).
\end{aligned} \tag{3.65}$$

We then adopt the invertible change of state variables,

$$\begin{aligned}
\xi &= x - \Pi w, \\
\tilde{x} &= x - \hat{x}, \\
\tilde{w} &= w - \hat{w},
\end{aligned}$$

and then rewrite the closed-loop system (3.65) as

$$\begin{aligned}
\dot{\xi} &= A\xi + B\sigma(-(1+\mu)B^{\mathrm{T}}P_\varepsilon\xi + \Gamma w - \Gamma\tilde{w} + (1+\mu)B^{\mathrm{T}}P_\varepsilon\tilde{x} \\
&\qquad -(1+\mu)B^{\mathrm{T}}P_\varepsilon\Pi\tilde{w}) + (A\Pi - \Pi S + E_w)w \\
\dot{\tilde{x}} &= (A + K_{\mathrm{A},\ell}C_y)\tilde{x} + (E_w + K_{\mathrm{A},\ell}D_{yw})\tilde{w} \\
\dot{\tilde{w}} &= K_{\mathrm{S},\ell}C_y\tilde{x} + (S + K_{\mathrm{S},\ell}D_{yw})\tilde{w}.
\end{aligned} \tag{3.66}$$

We first note that the estimation error

$$\tilde{x}_e(t) = \begin{pmatrix} \tilde{x}(t) \\ \tilde{w}(t) \end{pmatrix}$$

satisfies:

$$\dot{\tilde{x}}_e(t) = \tilde{A}_\ell\tilde{x}_e(t)$$

with

$$\tilde{A}_\ell = \begin{pmatrix} A & E_w \\ 0 & S \end{pmatrix} - R_\ell \begin{pmatrix} C_y^{\mathrm{T}} \\ D_{yw}^{\mathrm{T}} \end{pmatrix} \begin{pmatrix} C_y & D_{yw} \end{pmatrix}$$

and using the properties from Lemma 3.3.2 we find that

$$\frac{d}{dt}\tilde{x}_e^{\mathrm{T}}(t)R_\ell\tilde{x}_e(t) \leqslant -\ell\tilde{x}_e^{\mathrm{T}}(t)R_\ell\tilde{x}_e(t) - \ell\tilde{x}_e^{\mathrm{T}}(t)\tilde{x}_e(t). \tag{3.67}$$

We are now in a position to show the items (i) and (ii) of the theorem. Let T be such that $\|\Gamma w(t)\| \leqslant 1 - \delta$ for all $t > T$.

To show that item (i) of the theorem holds, we note that (3.63) is equal to (3.65) for $w = 0$. Hence (3.65) reduces to

$$\begin{aligned} \dot{\xi} &= A\xi + B\sigma(-(1+\mu)B^{\mathrm{T}}P_{\varepsilon}\xi + (\mu+1)M\tilde{x}_e) \\ \dot{\tilde{x}}_e &= \tilde{A}_{\ell}\tilde{x}_e, \end{aligned} \tag{3.68}$$

where

$$M = \begin{pmatrix} B^{\mathrm{T}}P_{\varepsilon} & 0 \\ 0 & \frac{\Gamma}{\mu+1} - B^{\mathrm{T}}P_{\varepsilon}\Pi. \end{pmatrix},$$

For any $x(0) \in \mathcal{X}_0$, $\xi(T+1)$ belongs to a bounded set, say $\mathcal{U}_T$, independent of ε since $\mathcal{X}_0$ is bounded and $\xi(T+1)$ is determined by a linear differential equation with bounded input $\sigma(\cdot)$. Secondly, using Lemma 3.3.2, we find that there exists an ℓ_1^* such that for all $\ell > \ell_1^*$ we will have $\tilde{x}_e(T+1) \in \mathcal{X}_T$ where

$$\mathcal{X}_T = \left\{ \tilde{z} \in \mathbb{R}^{n+s} \mid \tilde{z}^{\mathrm{T}}R_{\ell}\tilde{z} \leqslant 1 \right\}$$

for any $(x(0), \hat{x}(0), \hat{w}(0)) \in \mathcal{X}_0 \times \mathcal{Z}_0$.

Define a Lyapunov function

$$V(\xi, \tilde{x}_e) = \xi^{\mathrm{T}}P_{\varepsilon}\xi + \tilde{x}_e^{\mathrm{T}}R_{\ell}\tilde{x}_e. \tag{3.69}$$

Let $c_1 > 0$ be such that

$$c_1 \geq \sup_{\xi\in\mathcal{U}_T, \tilde{z}\in\mathcal{X}_T, \varepsilon\in(0,1]} V(\xi, \tilde{z}). \tag{3.70}$$

Such a c_1 exists since $\mathcal{U}_T$ is a bounded set and $\lim_{\varepsilon\to 0} P_{\varepsilon} = 0$ by Lemma 3.3.1. Let $\varepsilon_1^* \in (0,1]$ be such that $\xi \in L_V(c_1)$ implies that $\|B^{\mathrm{T}}P_{\varepsilon}\xi\|_{\infty} \leq \delta$. The existence of such an ε_1^* is again due to the fact that $\lim_{\varepsilon\to 0} P_{\varepsilon} = 0$.

The evaluation of $\dot{V}$, $t \geq T+1$, inside the set $L_V(c_1)$, using (3.9) and (3.67) and following the argument used in proving (3.37), shows that

$$\begin{aligned} \dot{V} &= -\xi^{\mathrm{T}}(Q_{\varepsilon} + P_{\varepsilon}BB^{\mathrm{T}}P_{\varepsilon})\xi - \ell\tilde{x}_e^{\mathrm{T}}R_{\ell}\tilde{x}_e - \ell\tilde{x}_e^{\mathrm{T}}\tilde{x}_e \\ &\quad + 2\xi^{\mathrm{T}}P_{\varepsilon}B[\sigma(-(1+\mu)B^{\mathrm{T}}P_{\varepsilon}\xi + (\mu+1)M\tilde{x}_e) + B^{\mathrm{T}}P_{\varepsilon}\xi] \\ &\leq -\xi^{\mathrm{T}}Q_{\varepsilon}\xi - \ell\tilde{x}_e^{\mathrm{T}}\tilde{x}_e - \xi^{\mathrm{T}}P_{\varepsilon}BB^{\mathrm{T}}P_{\varepsilon}\xi \\ &\quad + 2\xi^{\mathrm{T}}P_{\varepsilon}B[\sigma(-(1+\mu)B^{\mathrm{T}}P_{\varepsilon}\xi \\ &\quad + (\mu+1)M\tilde{x}_e) - \sigma(-(1+\mu)B^{\mathrm{T}}P_{\varepsilon}\xi)] \\ &\leq -\xi^{\mathrm{T}}Q_{\varepsilon}\xi - \ell\|\tilde{x}_e\|^2 + 2\beta(\mu+1)\|M\|\|\xi^{\mathrm{T}}P_{\varepsilon}B\|\|\tilde{x}_e\| \\ &\quad - \|\xi^{\mathrm{T}}P_{\varepsilon}B\|^2 \end{aligned}$$

where β is the Lipschitz constant of the function σ. It is now clear that, for each $\varepsilon \in (0, \varepsilon_1^*]$ and each $\mu \geq 0$, there is an $\ell_2^*(\varepsilon, \mu) > \ell_1^*$ such that, for $\ell \geq \ell_2^*(\varepsilon, \mu)$,

$$(\xi, \tilde{x}_e) \in L_V(c) \Longrightarrow \dot{V} \leq -\xi^{\mathrm{T}} Q_\varepsilon \xi - \frac{1}{2}\ell \tilde{x}_e^{\mathrm{T}} \tilde{x}_e. \tag{3.71}$$

This, in turn, shows that, for any a priori given sets $\mathcal{X}_0$ and $\mathcal{Z}_0$, there exists an $\varepsilon_1^* > 0$ such that for each $\varepsilon \in (0, \varepsilon_1^*]$, $\mu \geq 0$, $\ell \geq \ell_2^*(\varepsilon, \mu)$, the equilibrium point of (3.63) is asymptotically stable with $\mathcal{X}_0 \times \mathcal{Z}_0$ contained in its basin of attraction.

Now, in order to show that item (ii) of the theorem holds, consider (3.65). Again, for any $x(0) \in \mathcal{X}_0$ and any $w(0) \in \mathcal{W}_0$, $\xi(T+1)$ belongs to a bounded set, say $\mathcal{U}_{wT}$, independent of ε since $\mathcal{X}_0$ is bounded and $\xi(T+1)$ is determined by a linear differential equation with bounded inputs $\sigma(\cdot)$ and Γw. Secondly, using Lemma 3.3.2, we find that there exists an $\ell_3^* > 1$ such that for all $\ell > \ell_3^*$ we will have $\tilde{x}_e(T+1) \in \mathcal{X}_T$ for any $(x(0), \hat{x}(0), \hat{w}(0)) \in \mathcal{X}_0 \times \mathcal{Z}_0$ and for any $w(0) \in \mathcal{W}_0$.

For the same Lyapunov function as given by (3.69), we choose $c_2 > 0$ such that

$$c_2 \geq \sup_{\xi \in \mathcal{U}_{wT}, \tilde{z} \in \mathcal{X}_T, \varepsilon \in (0,1]} V(\xi, \tilde{z}). \tag{3.72}$$

Such a c_2 exists since $\mathcal{U}_{wT}$ is a bounded set and $\lim_{\varepsilon \to 0} P_\varepsilon = 0$ by Lemma 3.3.1. Let $\varepsilon_2^* \in (0, 1]$ be such that $\xi \in L_V(c_1)$ implies that $\|B^{\mathrm{T}} P_\varepsilon \xi\|_\infty \leq \delta$. The existence of such an ε_2^* is again due to the fact that $\lim_{\varepsilon \to 0} P_\varepsilon = 0$.

Now, the evaluation of $\dot{V}$, $t \geq T+1$, inside the set $L_V(c_2)$, using the same technique as above, yields

$$\begin{aligned}
\dot{V} &= -\xi^{\mathrm{T}}(Q_\varepsilon + P_\varepsilon B B^{\mathrm{T}} P_\varepsilon)\xi - \ell \tilde{x}_e^{\mathrm{T}} R_\ell \tilde{x}_e - \ell \tilde{x}_e^{\mathrm{T}} \tilde{x}_e \\
&\quad + 2\xi^{\mathrm{T}} P_\varepsilon B[\sigma(-(1+\mu) B^{\mathrm{T}} P_\varepsilon \xi \\
&\quad + (\mu+1) M \tilde{x}_e + \Gamma w) - \Gamma w + B^{\mathrm{T}} P_\varepsilon \xi] \\
&\leq -\xi^{\mathrm{T}} Q_\varepsilon \xi - v^{\mathrm{T}} v - 2v^{\mathrm{T}}[\sigma((\mu+1)v + \theta + \alpha) - v - \theta] - \tilde{x}_e^{\mathrm{T}} \tilde{x}_e \\
&\leq -\xi^{\mathrm{T}} Q_\varepsilon \xi - v^{\mathrm{T}} v - 2v^{\mathrm{T}}[\sigma((\mu+1)v + \theta + \alpha) - v - \theta] - \delta^2 \tilde{x}_e^{\mathrm{T}} \tilde{x}_e \\
&\leq -\xi^{\mathrm{T}} Q_\varepsilon \xi - 2\sum_{i=1}^{m} v_i[\bar{\sigma}((\mu+1)v_i + \theta_i + \alpha_i) - v_i - \theta_i] \\
&\quad - \sum_{i=1}^{m} v_i^2 - \sum_{i=1}^{m} \alpha_i^2,
\end{aligned}$$

where we have denoted $v := -B^{\mathrm{T}}P_\varepsilon\xi$, $\alpha := (\mu+1)M\tilde{x}_e$, and $\theta := \Gamma w$ with their ith components denoted by v_i, α_i and θ_i respectively. We have also chosen $\ell_4^*(\varepsilon,\mu) > \ell_3^*$ such that for each $\varepsilon \in (0,\varepsilon_2^*]$, $\mu \geq 0$, $\ell \geq \ell_4^*(\varepsilon,\mu)$, $\|(\mu+1)M\tilde{x}_e\| < \delta$, which ensures that $|\theta_i + \alpha_i| < 1$.

Noting that $\mathrm{sgn}(\bar{\sigma}(s)) = \mathrm{sgn}(s)$, and for $\varepsilon \in (0,\varepsilon_2^*]$, $\ell \geq \ell_4^*(\varepsilon,\mu)$, $|v_i + \theta_i| \leq 1$ and $|\alpha_i + \theta_i| < 1$, we observe that,

$$
\begin{aligned}
&|(\mu+1)v_i + \theta_i + \alpha_i| \leq 1 \\
&\qquad\Longrightarrow v_i[\bar{\sigma}((\mu+1)v_i + \theta_i + \alpha_i) - v_i - \theta_i] = \mu v_i^2 + v_i\alpha_i, \\
&(\mu+1)v_i + \theta_i + \alpha_i > 1 \Longrightarrow v_i > 0 \\
&\qquad\Longrightarrow v_i[\bar{\sigma}((\mu+1)v_i + \theta_i + \alpha_i) - v_i - \theta_i] \geq 0, \\
&(\mu+1)v_i + \theta_i + \alpha_i < -1 \Longrightarrow v_i < 0 \\
&\qquad\Longrightarrow v_i[\bar{\sigma}((\mu+1)v_i + \theta_i + \alpha_i) - v_i - \theta_i] \geq 0.
\end{aligned}
$$

Now, without loss of generality, let us assume that $|(\mu+1)v_i + \theta_i + \alpha_i| \leq 1, i = 1 \ldots m_\alpha$, for some $0 \leq m_\alpha \leq m$. Hence we conclude that,

$$
\begin{aligned}
\dot{V} &\leq -\xi^{\mathrm{T}}Q_\varepsilon\xi - \sum_{i=1}^{m_\alpha}(v_i + \alpha_i)^2 \\
&\leq -\xi^{\mathrm{T}}Q_\varepsilon\xi.
\end{aligned}
$$

Then, it can be easily seen that, for all $(x(0), \hat{x}(0), \hat{w}(0)) \in \mathcal{X}_0 \times \mathcal{Z}_0$, there exists an $\varepsilon_2^* > 0$, such that for each $\varepsilon \in (0,\varepsilon_2^*]$ and each $\mu \geq 0$, there exists an $\ell_4^*(\varepsilon,\mu)$ such that for each $\varepsilon \in (0,\varepsilon_2^*]$, $\mu \geq 0$, $\ell \geq \ell_4^*(\varepsilon,\mu)$, the solution of the closed-loop system (3.66) satisfies

$$\lim_{t\to\infty} e(t) = 0. \tag{3.73}$$

Finally, taking $\varepsilon^* = \min\{\varepsilon_1^*, \varepsilon_2^*\}$ and $\ell^*(\varepsilon,\mu) = \max\{\ell_2^*(\varepsilon,\mu), \ell_4^*(\varepsilon,\mu)\}$, we complete our proof of Theorem 3.3.5. ■

Following example illustrates our design procedure.

Example 3.3.4 *We consider the same plant and the exosystem as in Example 3.3.1. However, we will utilize $y = e$ for feedback. Let the sets $\mathcal{W}_0$ and $\mathcal{X}_0$ be the same as those in Example 3.3.1. Let the set $\mathcal{Z}_0$, be given by $\mathcal{Z}_0 = \{z : \|z\| \leq 1, z \in \mathbb{R}^6\}$. Following the proof of Theorem 3.3.5, a suitable choice of ε^* is 1.5×10^{-4}. For the observer design we use the direct method as*

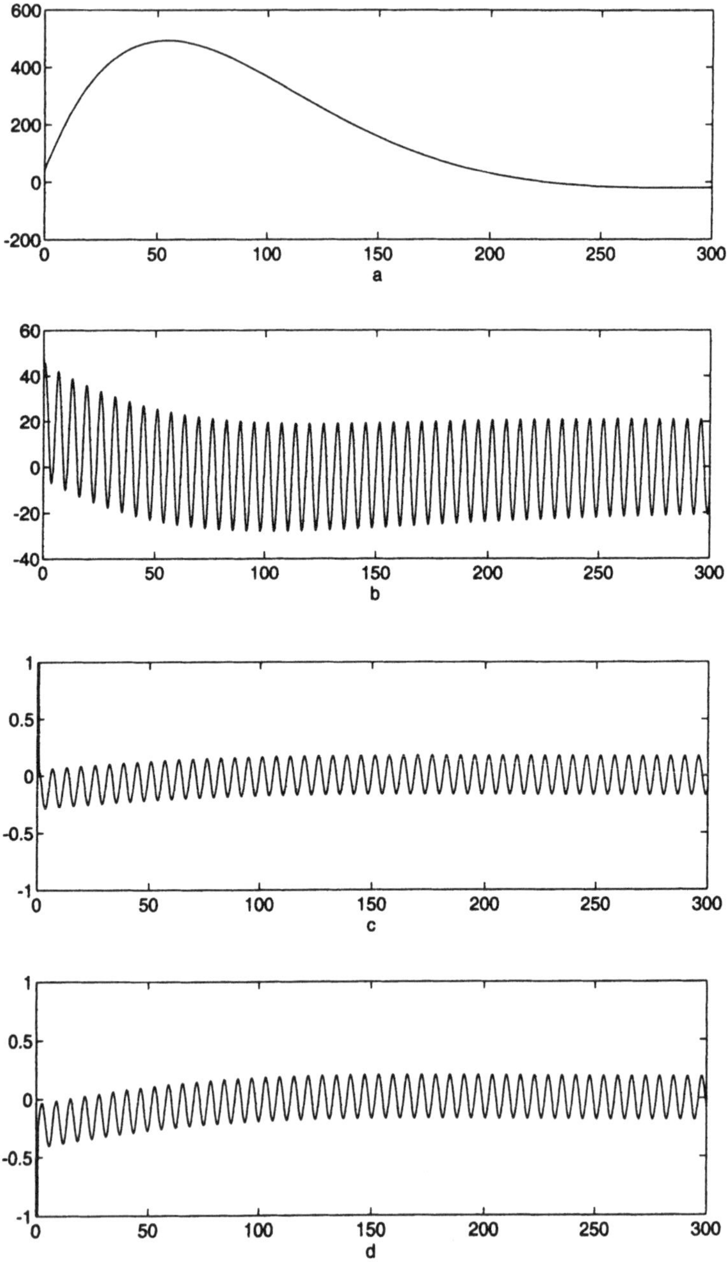

Figure 3.5: $\varepsilon = 1.5 \times 10^{-4}$, $\mu = 0$, $\ell = 5$. a) e_1; b) e_2; c) $\bar{\sigma}(u_1)$; d) $\bar{\sigma}(u_2)$.

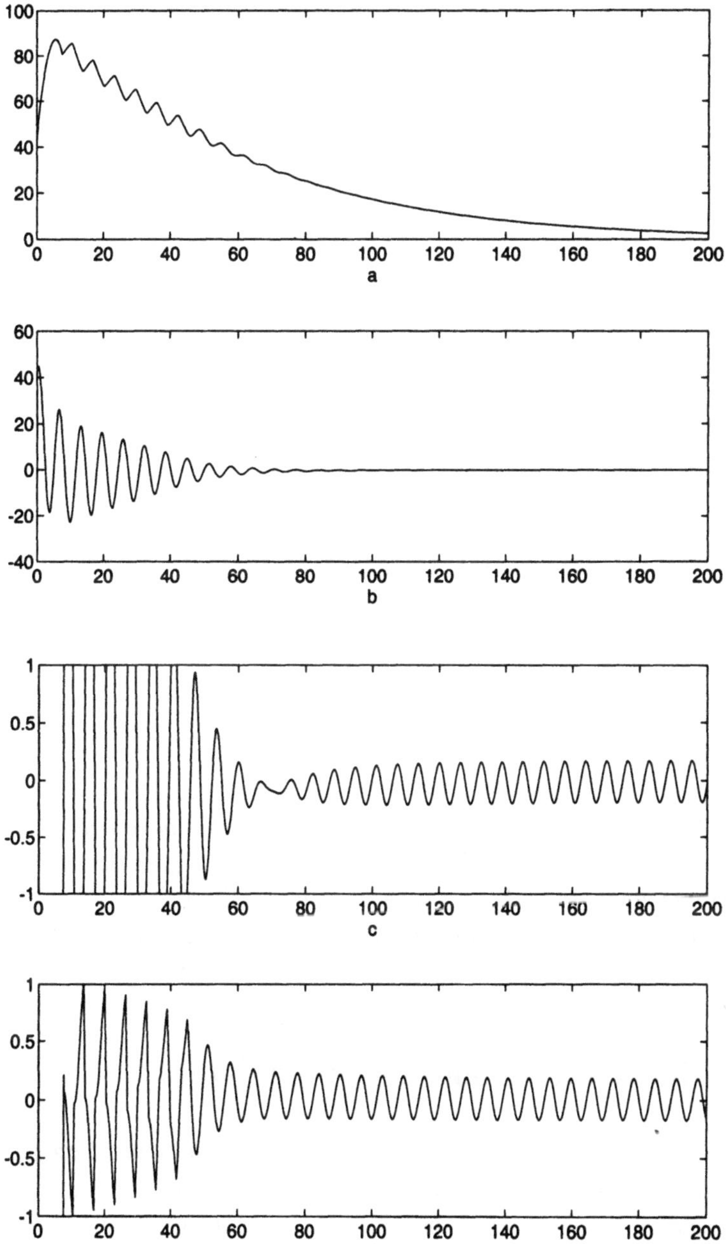

Figure 3.6: $\varepsilon = 1.5 \times 10^{-4}$, $\mu = 1200$, $\ell = 5$. a) e_1; b) e_2; c) $\bar{\sigma}(u_1)$; d) $\bar{\sigma}(u_2)$.

described in Appendix 3.B. Placing all the eigenvalues of $\tilde{A}$ at -4, a suitable choice of ℓ^* for $\mu \leq 1200$ is 5. For $\varepsilon = \varepsilon^*$, $\mu \leq 1200$, and $\ell = \ell^*$, the feedback law (3.58) is given by

$$
\begin{aligned}
\dot{\hat{x}} &= A\hat{x} + B\sigma(u) + E_w\hat{w} + K_{\mathrm{A}}(5)C_e(\hat{x} - x) + K_{\mathrm{A}}(5)D_{ew}(\hat{w} - w) \\
\dot{\hat{w}} &= S\hat{w} + K_{\mathrm{S}}(5)C_e(\hat{x} - x) + K_{\mathrm{S}}(5)D_{ew}(\hat{w} - w) \\
u &= -(\mu + 1)10^{-3}\begin{pmatrix} 6.4203 & 11.918 & 3.3628 & -2.9767 \\ 11.918 & 22/623 & 6.1139 & -5.6505 \end{pmatrix}\hat{x} \\
&\quad +(\mu + 1)10^{-3}\begin{pmatrix} -2.9799 & 3.3755 \\ -4.2588 & 4.6980 \end{pmatrix}\hat{w} \\
&\quad +\begin{pmatrix} 1.25 & 0.25 \\ -0.25 & -1.25 \end{pmatrix}\hat{w}
\end{aligned}
\tag{3.74}
$$

where $\mu \geq 0$ and

$$
K_{\mathrm{A}}(5) = \begin{pmatrix} 6.6148 \times 10^3 & 8.8425 \times 10^2 \\ 2.8146 \times 10^3 & 5.3047 \times 10^3 \\ -2.4153 \times 10^4 & 6.6806 \times 10^3 \\ -2.4947 \times 10^4 & 4.7511 \times 10^4 \end{pmatrix},
$$

$$
K_{\mathrm{S}}(5) = \begin{pmatrix} 3.5526 \times 10^3 & -9.4147 \times 10^2 \\ -2.0833 \times 10^3 & 4.0093 \times 10^3 \end{pmatrix}.
$$

For the initial conditions $x_0 = (7, 7, 7, 7)$, $w_0 = (0.1, 0.1)$, $\hat{x}_0 = (0, 0, 0, 0)$, $\hat{w}_0 = (0, 0)$, Figures 3.5 on page 106 and 3.6 on the page before show the control action and the closed-loop performance for low-gain feedback ($\mu = 0$, $\ell = 5$) and low-and-high gain feedback ($\mu = 1200$, $\ell = 5$) respectively. The simulation results illustrate that the low-and-high gain feedback regulator significantly out-performs the low-gain feedback one given in (3.42).

3.4 Linear versus non-linear regulators

The classical semi-global state feedback output regulation problem as well as the classical semi-global measurement feedback output regulation problem, as defined earlier, require a regulator with a linear structure. That is, linear state feedback laws or the implementation of linear state feedback laws utilizing linear observers are required for output regulation. The sufficient conditions for the existence of such "linear" regulators were given in Section 3.3. We examine in this section the necessity of these conditions. This must be examined in two fronts.

The first is to examine the necessity of the solvability conditions given in Section 3.3 for the existence of "linear" regulators. The second is to examine whether we can weaken the solvability conditions if we allow non-linear regulators in our definitions of the state and the measurement feedback output regulation problems. It turns out that under certain mild conditions our solvability conditions for the existence of "linear" regulators are basically necessary. Moreover, these conditions cannot be weakened by allowing non-linear regulators. We also make an interesting observation that whenever these mild conditions are violated, there *might* be non-linear state feedback controllers that achieve output regulation while no linear state feedback controllers would do so.

The necessary condition for the existence of the classical semi-global state feedback regulator using a general non-linear feedback law is given in the following theorem.

Theorem 3.4.1 *Consider the plant and the exosystem (3.1). Let Assumptions A.1 and A.2 hold. Assume that in the absence of input saturation, the linear state feedback output regulation problem is solvable, i.e., there exist matrices* Π *and* Γ *which solve the regulator equation (3.18). Also assume that the system characterized by the quadruple* (A, B, C_e, D_{eu}) *is left-invertible and has no invariant zeros on the imaginary axis. Then, a necessary condition for the existence of a general, possibly non-linear, state feedback controller that achieves semi-global output regulation for (3.1) is that, for all* $\varepsilon > 0$, *there exists a* $T \geqslant 0$ *such that*

$$\|\Gamma w\|_{\infty,T} \leqslant 1+\varepsilon, \quad \textit{for all } w(0) \in \mathcal{W}_0. \tag{3.75}$$

Proof : The proof of the above theorem depends on certain results described in the Appendix 3.A. We will assume in the proof that the reader is familiar with these results. Consider the system (3.1), however, without the saturation element. That is, let

$$\begin{aligned} \dot{x} &= Ax + Bv + E_w w \\ \dot{w} &= Sw \\ e &= C_e x + D_{eu} v + D_{ew} w. \end{aligned} \tag{3.76}$$

Suppose that we have some arbitrary non-linear feedback $u = \alpha(x, w)$ which achieves output regulation for the system (3.1). Then the feedback $v = \sigma(\alpha(x, w))$ will achieve output regulation for the system (3.76). Note that by our assumption, there exists also a linear feedback,

$$v = Fx + (\Gamma - F\Pi)w,$$

which achieves output regulation for the system (3.76). Moreover,

$$v(t) - \Gamma w(t) \to 0 \text{ as } t \to \infty. \tag{3.77}$$

We have two feedback laws which achieve output regulation for the linear system (3.76). One is non-linear and satisfies a certain amplitude constraint. The other is a linear feedback, of which we have no a priori knowledge regarding its amplitude. Our aim is to show that the linear feedback must necessarily satisfy an amplitude constraint asymptotically as $t \to \infty$. To this end, we define the difference between the two control inputs as

$$s(t) = [Fx + (\Gamma - F\Pi)w](t) - [\sigma(\alpha(x, w))](t).$$

Suppose now that Condition (ii)b) of Theorem 3.3.2 is not true. In that case there exist $\{t_n\}_{n=1}^{\infty}$ and a $\delta > 0$ such that $t_n \to \infty$ as $n \to \infty$ and $\|\Gamma w(t_n)\| > 1 + 2\delta$ for all n. Given the differential equation for w, it is not difficult to see that this implies that there exists an $\varepsilon > 0$ such that, for all n and all $t \in [t_n, t_n + \varepsilon]$, we have $\|\Gamma w(t_n)\| > 1 + \delta$.

If we look at our definition for $s(t)$, we see that the first component asymptotically converges to Γw. The second term is bounded by 1. Combining this with the property for Γw that we just derived, it is easily seen that the vector-valued signal $s(t)$ has a component $s_i(t)$ for which we have, for all n, either

$$s_i(t) > \delta \quad \text{for all } t \in [t_n, t_n + \varepsilon]$$

or

$$s_i(t) < -\delta \quad \text{for all } t \in [t_n, t_n + \varepsilon].$$

In other words $s \in \mathcal{P}$ (as defined in Definition 3.A.1 of the Appendix). Now, if we apply the signal s to the system

$$\begin{aligned} \dot{x} &= Ax + Bs \\ e &= C_e x + D_{eu} s \end{aligned} \tag{3.78}$$

with zero initial conditions, then we have $e(t) \to 0$ (since both the linear feedback and the non-linear saturating feedback achieve output regulation). This is in contradiction with $s \in \mathcal{P}$ according to Theorem 3.A.2. This implies that Condition (ii)b) of Theorem 3.3.2 is satisfied. Thus we find that the existence of a, possibly non-linear, feedback achieving output regulation for the system (3.1) implies that the conditions of Theorem 3.3.2 are satisfied. ∎

Remark 3.4.1 *The necessary conditions given in Theorem 3.4.1 are slightly different from the sufficient conditions given in Theorem 3.3.2. Namely, Condition (3.75) of Theorem 3.4.1 and Condition (ii)b) of Theorem 3.3.2 are not exactly the same, although they are almost equal. Hence, one can conclude that under Assumption A.2 and the assumption that the system characterized by (A, B, C_e, D_{eu}) is left invertible and has no invariant zeros on the imaginary axis, if the classical semi-global state feedback output regulation problem is solvable by a non-linear feedback regulator then it is generally solvable by the linear feedback regulator as well.*

An interesting question that arises now is whether one can weaken the necessary condition given in Theorem 3.4.1 if the system characterized by (A, B, C_e, D_{eu}) is not left invertible and/or has invariant zeros on the $j\omega$ axis. The following example shows that, in fact, this is the case. More significantly, this example shows that if the system characterized by (A, B, C_e, D_{eu}) is not left invertible, non-linear feedback controllers might achieve semi-global output regulation while no linear feedback controller can do so.

Example 3.4.1 *Consider the system*

$$
\begin{aligned}
\dot{x} &= \begin{pmatrix} -1 & 1 & 2 \\ 0 & -2 & 0 \\ 0 & 0 & -1 \end{pmatrix} x + \begin{pmatrix} 1 & 0 & -1 \\ 0 & 1 & 1 \\ -1 & 1 & 0 \end{pmatrix} \sigma(u), \\
\dot{w} &= \begin{pmatrix} 0 & 0 \\ 0 & 0 \end{pmatrix} w, \\
e &= \begin{pmatrix} 0 & 1 & 0 \\ 0 & 0 & 1 \end{pmatrix} x + \begin{pmatrix} -0.5 & 0 \\ 0 & -1 \end{pmatrix} w,
\end{aligned}
\tag{3.79}
$$

with $w(0) \in \mathcal{W}_0$ where

$$
\mathcal{W}_0 = \text{convex hull} \left\{ \begin{pmatrix} 2 \\ 0 \end{pmatrix}, \begin{pmatrix} 0 \\ 2 \end{pmatrix}, \begin{pmatrix} 2 \\ 2 \end{pmatrix} \right\}.
$$

It is straightforward to show that for this example, in the absence of input amplitude saturation, the linear state feedback output regulation problem is solvable. In fact, the matrices Π and Γ that solve the regulator equation (3.18) are given by,

$$
\Pi = \begin{pmatrix} \pi_{11} & \pi_{12} \\ 0.5 & 0 \\ 0 & 1 \end{pmatrix}, \quad \Gamma = \begin{pmatrix} 0.5\pi_{11} + 0.25 & 0.5\pi_{12} - 1.5 \\ 0.5\pi_{11} + 0.25 & 0.5\pi_{12} - 0.5 \\ -0.5\pi_{11} + 0.75 & -0.5\pi_{12} + 0.5 \end{pmatrix},
$$

where π_{11} and π_{12} are any real numbers.

In the presence of input amplitude saturation, however, the sufficient conditions of Theorem 3.3.2 are not satisfied. More specifically, Condition (ii)b) of Theorem 3.3.2 cannot be satisfied. Hence, the design procedure developed in Section 3.3 cannot be applied to this example. It is also evident that the necessary condition (3.75) is not satisfied either. However, since the system characterized by (A, B, C_e, D_{eu}) *in the given plant and the exosystem (3.79) is not left invertible, it does not necessarily imply that there do not exist state feedback laws that achieve semi-global output regulation for (3.79). In fact, in what follows we will establish the following two facts for the plant and the exosystem (3.79):*

(i) *The classical semi-global state feedback output regulation problem is solvable for the system (3.79) even though the condition (3.75) is not satisfied. This implies that the necessary conditions given in Theorem 3.4.1 can be weakened if the system characterized by* (A, B, C_e, D_{eu}) *is not left invertible.*

(ii) *There exist no linear state feedback controllers that can achieve semi-global output regulation for the system (3.79). Yet suitable non-linear controllers do exist. This establishes the important result that if the system characterized by* (A, B, C_e, D_{eu}) *is not left invertible, semi-global output regulation might be achieved via non-linear feedback controllers while no linear feedback controller can do so.*

As the given plant is already asymptotically stable, let us consider a non-linear feedback control law of only the exosystem state; and let it be of the form,

$$u = \begin{pmatrix} 0.5 \\ 0.5 \\ -0.5 \end{pmatrix} f(w) + \begin{pmatrix} 0.5 & -0.5 \\ 0.5 & 0.5 \\ 0.5 & -0.5 \end{pmatrix} w, \tag{3.80}$$

and

$$f(w) = (1 - \alpha - \beta)\begin{pmatrix} -1 & 0 \end{pmatrix} w,$$

where α *and* β *are both* $\geqslant 0$, *and are such that* $\alpha + \beta \leqslant 1$, *and*

$$w(0) = \alpha \begin{pmatrix} 2 \\ 0 \end{pmatrix} + \beta \begin{pmatrix} 0 \\ 2 \end{pmatrix} + (1 - \alpha - \beta) \begin{pmatrix} 2 \\ 2 \end{pmatrix}.$$

Note that for any $w(0) \in \mathcal{W}_0$, $w \equiv w(0)$. *Clearly, since* α *and* β *depend on* w, *the function* f *is non-linear and hence the above controller is non-linear.*

We now study another system,

$$
\begin{aligned}
\dot{x} &= \begin{pmatrix} -1 & 1 & 2 \\ 0 & -2 & 0 \\ 0 & 0 & -1 \end{pmatrix} x + \begin{pmatrix} 1 \\ 0 \\ 0 \end{pmatrix} v_1 + \begin{pmatrix} 0 & 0 \\ 1 & 0 \\ 0 & 1 \end{pmatrix} v_2, \\
\dot{w} &= \begin{pmatrix} 0 & 0 \\ 0 & 0 \end{pmatrix} w, \\
e &= \begin{pmatrix} 0 & 1 & 0 \\ 0 & 0 & 1 \end{pmatrix} x + \begin{pmatrix} -0.5 & 0 \\ 0 & -1 \end{pmatrix} w.
\end{aligned} \tag{3.81}
$$

We point out that the closed-loop system comprising of (3.79) and the control law,

$$
\sigma(u) = \begin{pmatrix} 0.5 \\ 0.5 \\ -0.5 \end{pmatrix} v_1 + \begin{pmatrix} 0.5 & -0.5 \\ 0.5 & 0.5 \\ 0.5 & -0.5 \end{pmatrix} v_2, \tag{3.82}
$$

is indeed given by (3.81). In view of this connection between (3.79) and (3.81), we can try to solve the output regulation problem for the system (3.81) subject to the understanding that v_1 and v_2 thus obtained can be used in (3.82) in order to form a control law for (3.79). A simple look at the output regulation problem for (3.81) indicates that v_1 does not affect e, and hence v_1 can be ignored. Also, we find that the transfer matrix from v_2 to e is left-invertible. Therefore we know that v_2 when achieving output regulation has a unique asymptotic behavior (except for some freedom as described in Theorem 3.A.2, which cannot change the maximal amplitude). We find by solving the regulator equation that v_2 is equal to w at least asymptotically. However for all $w(0) \in \mathcal{W}_0$, equation (3.82) must be solvable. For $w(0) = (2\ \ 0)^T$ and for $w(0) = (0\ \ 2)^T$, we find that v_1 must be 0 when $w = v_2$. If v_1 depends linearly on w, then for $w(0) = (2\ \ 2)^T$, we must also choose $v_1 = 0$ when $w = v_2$. However, in that case (3.82) is not solvable, and hence there exists no linear function from w to v_1 such that output regulation can be achieved for all $w(0) \in \mathcal{W}_0$. On the other hand, setting $v_1 = f(w)$ and $v_2 = w$ results in a solvable equation (3.82) for all possible choices of $w(0) \in \mathcal{W}_0$. Moreover, the controller (3.80) is then the unique solution. Therefore, it is straightforward to check that the non-linear controller (3.80) indeed solves the output regulation problem for (3.79).

We can now move on to consider the classical semi-global measurement feedback output regulation problem, and pose questions similar to those we posed earlier for the semi-global state feedback output regulation problem. In fact, again, there is the question whether the conditions of Theorem 3.3.4 are actually necessary for the solvability of the classical semi-global measurement feedback output regulation problem. But Theorem 3.4.1 basically resolves this question since the conditions which are necessary for the state feedback are clearly also necessary for the case of output feedback on the basis of the error signal only. The only additional assumption we made in Theorem 3.3.4 is the detectability assumption which is clearly necessary for the stabilization of our system.

3.5 Issues of well-posedness and structural stability

We would like to reconsider the problems of well-posedness and structural stability as introduced in the previous chapter but this time for linear systems subject to actuator saturation. First of all note that solvability requires that the eigenvalues of A are in the closed left half plane. This is obviously a property that is not preserved for arbitrary parameter perturbations. But there is another issue to consider. One of our solvability conditions requires

$$\|\Gamma w\|_{\infty,T} < 1 \tag{3.83}$$

for some $T > 0$. If the perturbed exosystem has exponentially unstable modes, then this must occur in the kernel of Γ since otherwise the above condition can never be satisfied. But a small perturbation can obviously change Γ and the unstable dynamics will then become visible in Γw and condition (3.83) will fail for an arbitrarily small perturbation.

Hence in our definition of well-posedness we constrain perturbations of A and S to avoid exponentially unstable eigenvalues. In other words, the perturbed matrices A and S need to be weakly Hurwitz-unstable.

Definition 3.5.1 ***(Well-posedness)*** *For a system Σ as in (3.1), the classical semi-global linear observer based measurement feedback output regulation problem as defined in Problem 3.3.2 is said to be well-posed at (A_0, B_0, $E_{w,0}$, $C_{e,0}$, $D_{eu,0}$, $D_{ew,0}$, $C_{y,0}$, $D_{yu,0}$, $D_{yw,0}$, S_0) if there exists a neighborhood $\mathcal{P}_0$ of (A_0, B_0, $E_{w,0}$, $C_{e,0}$, $D_{eu,0}$, $D_{ew,0}$, $C_{y,0}$, $D_{yu,0}$, $D_{yw,0}$, S_0) in the parameter space $\mathcal{P}$ such that the considered problem is solvable for each element (A, B, E_w, C_e, D_{eu}, D_{ew}, C_y, D_{yu}, D_{yw}, S) of $\mathcal{P}_0$ for which A and S have all its eigenvalues in the closed left-half plane.*

We have the following result.

Theorem 3.5.1 *Consider a system Σ as in (3.1) and the classical semi-global linear observer based measurement feedback output regulation problem as defined in Problem 3.3.2. Let the conditions of Theorem 3.3.4 be satisfied for this system with nominal parameter values,*

$$
\begin{aligned}
&(A, B, E_w, C_e, D_{eu}, D_{ew}, C_y, D_{yu}, D_{yw}, S) \\
&\qquad = (A_0, B_0, E_{w,0}, C_{e,0}, D_{eu,0}, D_{ew,0}, C_{y,0}, D_{yu,0}, D_{yw,0}, S_0).
\end{aligned}
$$

Then the considered problem for Σ is well-posed at $(A_0, B_0, E_{w,0}, C_{e,0}, D_{eu,0}, D_{ew,0}, C_{y,0}, D_{yu,0}, D_{yw,0}, S_0)$ if and only if S is weakly Hurwitz-stable and the matrix

$$
\begin{pmatrix} A_0 - \lambda I & B_0 \\ C_{e,0} & D_{eu,0} \end{pmatrix} \tag{3.84}
$$

has full row-rank for each λ which is an eigenvalue of S_0.

Remark 3.5.1 *In the above theorem, we did not perturb the set of initial conditions for the exosystem $\mathcal{W}_0$ but it is obvious that small perturbations of this set will not affect well-posedness. Note that in the perturbations of A and S we avoid exponentially unstable eigenvalues but it turns out to be necessary that the nominal value for S satisfies the slightly stronger condition of being weakly Hurwitz stable.*

Proof : Let (Π_0, Γ_0) be the solution of the regulator equation for the nominal system,

$$
\begin{aligned}
\Pi_0 S &= A_0 \Pi_0 + B_0 \Gamma_0 + E_{w,0}, \\
0 &= C_{e,0} \Pi_0 + D_{eu,0} \Gamma_0 + D_{ew,0}.
\end{aligned}
$$

such that there exists a $T > 0$ and a $\delta > 0$ such that

$$
\|\Gamma_0 w\|_{\infty,T} < 1 - \delta \tag{3.85}
$$

for all initial conditions $w(0) \in \mathcal{W}_0$. The existence of such a solution of the regulator equation is guaranteed by the fact that the nominal system satisfies the conditions of Theorem 3.3.4.

By the results of Section 2.7 we know that for the perturbed system the regulator equation is solvable. Obviously the solution of the regulator equation need not be unique but is obvious that if the parameter variations are small then there exists a solution (Π, Γ) of the regulator equation for the perturbed system with $\|\Gamma_0 - \Gamma\|$ very small. Since the nominal exosystem is weakly-Hurwitz-stable and the perturbed system has no exponentially unstable modes we know that for small enough perturbations the perturbed exosystem will still be weakly-Hurwitz-stable. But then, from (3.85), it is obvious that for small enough perturbations we will have that $\|\Gamma w\|_{\infty,T} \leqslant 1 - \delta/2$ for all initial conditions $w(0) \in \mathcal{W}_0$. Therefore by Theorem 3.3.4 the classical semi-global measurement feedback output regulation problem is solvable.

From Theorem 2.7.1, we know that (3.84) is necessary to guarantee that for the perturbed system the regulator equation is solvable. Next, assume S is not weakly-Hurwitz-stable. Then there exists arbitrarily small initial conditions for the exosystem such that w is unbounded. But then there exists an arbitrarily small perturbation of Γ, say $\tilde{\Gamma}$, such that $\tilde{\Gamma} w$ is unbounded. Choose $\tilde{E}_w = S\Pi - \Pi A - B\tilde{\Gamma}$ and $\tilde{D}_{ew} = -C_e\Pi - D_{eu}\tilde{\Gamma}$ while the other parameters remain the same. Then for this perturbed system the output regulation problem is not solvable since $\|\tilde{\Gamma} w\|_{\infty,T}$ is equal to infinity and therefore not less than 1 for some T which is required for solvability of the output regulation problem. ■

Next we consider the output regulation problem with structural stability. As already discussed in Section 2.8, we need to restrict our perturbations of the system parameters even more. As already discussed in connection with well-posedness, we need to guarantee that even after perturbation, A still has all its eigenvalues in the closed left-half plane. But based on Section 2.8 we also need to exclude perturbations of the exosystem, i.e. we do not perturb S. Finally we need that the error signal is part of the measurement signal y, i.e. the parameters need to satisfy (2.36).

Definition 3.5.2 ***(Structurally stable output regulation problem)*** *Consider a system Σ as in (3.1) with the additional structure given in (2.36). A fixed controller is said to solve the structurally stable output regulation problem for Σ at $(A_0, B_0, E_{w,0}, C_{e,0}, D_{eu,0}, D_{ew,0}, C_{y2,0}, D_{yu2,0}, D_{yw2,0})$ if it satisfies the following properties:*

(i) The controller solves the classical semi-global linear observer based measurement feedback output regulation problem when the plant in (2.1) is characterized by the nominal set of parameters $(A_0, B_0, E_{w,0}, C_{e,0}, D_{eu,0}, D_{ew,0}, C_{y2,0}, D_{yu2,0}, D_{yw2,0})$.

(ii) There exist a neighborhood $\mathcal{P}_0$ of $(A_0, B_0, E_{w,0}, C_{e,0}, D_{eu,0}, D_{ew,0}, C_{y2,0}, D_{yu2,0}, D_{yw2,0})$ such that the controller achieves internal stability and output regulation for each set of perturbed plant parameters $(A_0 +$

δA, $B_0 + \delta B$, $E_{w,0} + \delta E_w$, $C_{e,0} + \delta C_e$, $D_{eu,0} + \delta D_{eu}$, $D_{ew,0} + \delta D_{ew}$, $C_{y2,0} + \delta C_{y2}$, $D_{yu2,0} + \delta D_{yu2}$, $D_{yw2,0} + \delta D_{yw2}$) in $\mathcal{P}_0$ for which $A_0 + \delta A$ has all its eigenvalues in the closed left-half plane.

In other words, as long as the perturbed parameters remain in $\mathcal{P}_0$, we have $\lim_{t\to\infty} e(t) = 0$ for all $x(0) \in \mathbb{R}^n$, $v(0) \in \mathbb{R}^{n_c}$, and $w(0)) \in \mathbb{R}^s$.

The above definition obviously implies that, for the existence of a regulator that solves the structurally stable output regulation problem, the exact output regulation problem must necessarily be well-posed (with the obvious modification implied by (2.36) and the fact that S is not perturbed).

A main technical complexity is the preliminary static output injection we applied in Section 2.8 to guarantee that A_0 and S have no common eigenvalues. This preliminary feedback is without loss of generality in the case of linear systems but due to the saturation this preliminary output injection changes the structure of the system and makes the analysis complicated. However, first of all in most cases A_0 and S have no eigenvalues in common and then we do not need any preliminary feedback. Secondly, we can make this preliminary feedback arbitrarily small and this allows us to actually resolve all the difficulties associated with this preliminary feedback. Due to its technicality we do not discuss the details.

As shown below, it turns out that the necessary and sufficient condition given in Theorem 3.5.1 for the well-posedness of the exact output regulation problem with measurement feedback is indeed also the necessary and sufficient condition for the existence of a regulator that solves the structurally stable output regulation problem.

Theorem 3.5.2 *Consider a system Σ as in (3.1) with the structural constraint (2.36). Let the conditions of Theorem 3.3.4 be satisfied for this system with nominal parameter values,*

$$(A, B, E_w, C_e, D_{eu}, D_{ew}, C_y, D_{yu}, D_{yw}) \\ = (A_0, B_0, E_{w,0}, C_{e,0}, D_{eu,0}, D_{ew,0}, C_{y,0}, D_{yu,0}, D_{yw,0}).$$

There exists a regulator that solves the structurally stable output regulation problem for Σ at $(A_0, B_0, E_{w,0}, C_{e,0}, D_{eu,0}, D_{ew,0}, C_{y2,0}, D_{yu2,0}, D_{yw2,0})$ if and only if the matrix S is weakly Hurwitz-stable and the matrix

$$\begin{pmatrix} A_0 - \lambda I & B_0 \\ C_{e,0} & D_{eu,0} \end{pmatrix} \tag{3.86}$$

has full row-rank for each λ which is an eigenvalue of S_0.

For linear systems the parameter perturbations could be arbitrarily large as long as stability is preserved. This is not the case here because the perturbations might be such that $\limsup_{t\to\infty} \|\Gamma w(t)\|_\infty > 1$ and then obviously output regulation is no longer possible.

We will only briefly indicate how the above result can be shown and how a suitable controller can be found. For ease of exposition we assume here that A_0 and S have no eigenvalues in common. Define $S_{\min}$, $\tilde{S}_p$, $\tilde{D}_{ew}$ and $\tilde{D}_{yw}$ as in Section 2.8.

We can now define an *auxiliary* system composed of the extended plant (2.1) and the *auxiliary* exosystem (2.41),

$$\tilde{\Sigma}: \begin{cases} \rho\tilde{x} = A_0\tilde{x} + B_0\sigma(\tilde{u}) \\ \rho\tilde{w} = \tilde{S}_p\tilde{w} \\ \tilde{y} = C_{y,0}\tilde{x} + D_{yu,0}\sigma(\tilde{u}) + \tilde{D}_{yw}\tilde{w} \\ \tilde{e} = C_{e,0}\tilde{x} + D_{eu,0}\sigma(\tilde{u}) + \tilde{D}_{ew}\tilde{w}. \end{cases} \tag{3.87}$$

We design a low-gain or a low-high gain measurement feedback controller for this auxiliary system such that the system achieves output regulation for all initial conditions under the condition that $\limsup_{t\to\infty} \|\Gamma\tilde{w}(t)\|_\infty < 1 - \delta/2$ where $\tilde{\Gamma}$ is of course the solution of the regulator equations associated with this system. We can then show that this controller achieves structural stability for the original system if we know that the system $(A_0, B_0, C_{e,0}, D_{eu,0})$ is left-invertible. There are two important issues that need to be clarified at this point:

- A low-gain and a low-high gain measurement feedback controller has the property that when applied to the system (2.42), the controller will asymptotically get out of saturation and therefore we get asymptotically linear behavior where we can use the analysis of Section 2.8.

- We know that the controller achieves output regulation for the perturbed system if we can guarantee that

  $$\limsup_{t\to\infty} \|\Gamma\tilde{w}(t)\|_\infty < 1 - \tfrac{\delta}{2} \tag{3.88}$$

  where Γ satisfies the regulator equation for the perturbed system. Note that if a controller achieves output regulation for a system, then we can associate with that controller a unique solution (Γ, Π) of the regulator equation (because of our left-invertibility assumption). For the nominal system the regulator equation has a solution (Π_0, Γ_0) such that

  $$\|\Gamma_0\tilde{w}\|_{\infty,T} < 1 - \delta.$$

If the perturbations in the system parameters are small enough, we can then guarantee that $\Gamma - \Gamma_0$ is very small and then (3.88) immediately follows.

Next we consider the case where the left-invertibility assumption is not satisfied. We note that when we design our controller for the auxiliary system then this controller is associated with a particular solution $(\tilde{\Pi}, \tilde{\Gamma})$ of the regulator equation for an arbitrary perturbed system. We will then achieve output regulation for this perturbed system for any initial condition which is small enough to guarantee $\|\tilde{\Gamma}\tilde{w}\|_{\infty,T} < 1$. But the question is whether for all initial conditions in $\mathcal{W}_0$ we have $\|\tilde{\Gamma}\tilde{w}\|_{\infty,T} < 1$.

In particular, this controller is then associated with a particular solution $(\tilde{\Pi}_0, \tilde{\Gamma}_0)$ of the regulator equation for the nominal system. But we are not sure whether $\Gamma = \Gamma_0$ and therefore we are not sure that there exists a $T > 0$ such that $\|\tilde{\Gamma}_0\tilde{w}\|_{\infty,T} < 1$ for all $w(0) \in \mathcal{W}_0$. Therefore we are not sure whether we achieve output regulation for all possible initial conditions of the exosystem. This technicality can be removed but due to its complexity we do not discuss the details.

3.A Uniqueness of asymptotic behavior of the input

In this appendix we will prove that under rather weak assumptions the asymptotic behavior of the input is unique given that the output of the system tracks a certain reference signal. This is a result which is used in the proof of Theorem 3.4.1.

The following theorem from [9] is a very powerful tool in the analysis of the asymptotic behavior of the signals.

Theorem 3.A.1 *Let R, S, V and W be matrix valued polynomials. Define k as their maximal degree. The following conditions are then equivalent.*

(*i*) *For any vector-valued function u on $[0, \infty)$ which is k times differentiable the conditions,*

$$r := R\left(\frac{d}{dt}\right)u : \quad r(t) = 0 \qquad \text{for all } t,$$

$$s := S\left(\frac{d}{dt}\right)u : \quad s(t) \to 0 \qquad \text{as } t \to \infty, \text{ and}$$

$$v := V\left(\frac{d}{dt}\right)u : \quad \sup_t v(t) < \infty,$$

imply that

$$w := W\left(\frac{d}{dt}\right)u : \quad w(t) \to 0 \text{ as } t \to \infty.$$

(*ii*) *Defining*

$$M(s)R(s) + N(s)S(s) + L(s)V(s) = W(s), \tag{3.89}$$

we have:

(*a*) *for all $\alpha \in \mathbb{C}^+$ there exist rational matrices M, N and L which do not have poles in α satisfying (3.89);*

(*b*) *for all $\alpha \in \mathbb{C}^0$ there exist rational matrices M, N and L which do not have poles in α satisfying (3.89) and, moreover, $L(\alpha) = 0$.*

(*c*) *there exist rational matrices M, N and L satisfying (3.89) and, moreover, N is a proper and L is a strictly proper rational matrix.*

Proof : This is given in [9]. ■

Remark 3.A.1 *Note that the proof in [9] is set in a distributional setting. Hence the above result still holds true if u is not smooth in which case all the above derivatives are interpreted in a distributional setting. In particular, if $u = \sigma(v)$ where v is smooth and σ is our non-differentiable saturation function, then the above result can still be applied.*

We obtain the following corollary.

Corollary 3.A.1 *Consider the system,*

$$\begin{aligned} \dot{x} &= Ax + Bv \\ y &= Cx + Dv, \end{aligned} \tag{3.90}$$

with $x(0) = 0$. Then, $v = 0$ is the only bounded input for which $y(t) \to 0$ if and only if the system characterized by (A, B, C, D) is left-invertible and has no invariant zeros on the imaginary axis and D is injective. Moreover, $v = 0$ is the only bounded input which has a bounded derivative and for which $y(t) \to 0$ if and only if the system characterized by (A, B, C, D) is left-invertible and has no invariant zeros on the imaginary axis.

Proof : We apply Theorem 3.A.1 with

$$R(s) := [\, sI - A \quad B \,], \tag{3.91}$$

$$S(s) := [\, C \quad D \,], \tag{3.92}$$

$$V(s) := [\, 0 \quad I \,], \text{ and} \tag{3.93}$$

$$W(s) := [\, 0 \quad I \,]. \tag{3.94}$$

■

The extra condition that D must be injective or the derivative of v bounded is really needed. A simple example is $v(t) = \cos(t^2)$ for which the output goes to zero as long as $D = 0$. Basically, v must either go to zero or start oscillating more and more rapidly. To formalize this concept, we define the following class of inputs.

Definition 3.A.1 *We define $\mathcal{P}$ as the set of bounded vector-valued functions v for which there exists a component v_i of v for which there exist $\varepsilon > 0$, $\delta > 0$ and a sequence $\{t_n\}$ ($t_n \to \infty$ as $n \to \infty$) such that for all n, either*

$$v_i(t) > \delta \ \text{ for all } t \in [t_n, t_n + \varepsilon]$$

or

$$v_i(t) < -\delta \ \text{ for all } t \in [t_n, t_n + \varepsilon].$$

Our basic claim is that if $v \in \mathcal{P}$ then it can never result in an output which converges to zero as long as the system is left-invertible and has no invariant zeros on the imaginary axis. This is formalized in the following theorem.

Theorem 3.A.2 *Assume that the system (3.90) is given where it is left-invertible and has no invariant zeros on the imaginary axis. Moreover, assume that v is bounded, and is such that $y(t) \to 0$ as $t \to \infty$. Finally, assume zero initial conditions. In that case $v \notin \mathcal{P}$.*

Proof : We prove this theorem by contradiction. Therefore, assume that ε, δ and $\{t_n\}$ exist satisfying the conditions of the theorem. Let the transfer matrix of (3.90) be given by G. We can factorize $G = \bar{G}H$ where $\bar{G}$ is left-invertible, has no zeros on the imaginary axis, and its direct feedthrough matrix is injective. Moreover, H is square, invertible and has no invariant zeros. The construction of $\bar{G}$ and H can be based on Silverman's algorithm (see [67]). Suppose $w \in \mathbb{R}^m$ is such that Gw is strictly proper. Then, choose k such that $(s+\alpha)^k Gw$ (with $\alpha > 0$) is proper but not strictly proper. Let Π be the projection onto w. Define $G_1 = G(I - \Pi) + G\Pi(s+\alpha)^k$ and $H_1 = (I - \Pi) + (s+\alpha)^{-k}\Pi$ for which we have $G = G_1 H_1$. If the direct feedthrough matrix of G_1 is not yet injective we repeat this procedure on G_1 obtaining G_2 and H_2. We can then repeat the procedure on G_2. After at most m steps (say ℓ steps) this procedure stops and we finally set $\bar{G} = G_\ell$ and $H = H_\ell H_{\ell-1} \cdots H_1$. It is easy to check that these matrices satisfy all our requirements.

Define $s = Hv$. Since H is asymptotically stable and v bounded, we obtain from $s = Hv$ that s is bounded. Moreover, we have $y = \bar{G}s$, and owing to Corollary 3.A.1, we know that $s(t) \to 0$ as $t \to \infty$.

Let the matrices K, L, M, and N characterize H. Due to the way H is constructed it is easy to see that in a suitable basis we have the following additional structure:

$$M = \begin{pmatrix} M_1 \\ 0 \end{pmatrix}, \qquad N = \begin{pmatrix} 0 \\ N_2 \end{pmatrix}.$$

We denote the state of H by p. Define $p_n := p(t_n)$, and define $v_n \in L_2[0,\varepsilon]$ by $v_n(s) := v(t_n + s)$. For all $s \in [0,\varepsilon]$, we have

$$\begin{pmatrix} M_1 \\ 0 \end{pmatrix} \left(e^{Ks} p_n + \int_0^s e^{K(s-\tau)} L v_n(\tau)\, d\tau \right) + \begin{pmatrix} 0 \\ N_2 \end{pmatrix} v_n(s) s(t_n + s) \to 0 \quad \text{as } n \to \infty. \tag{3.95}$$

We know that v_n is bounded, and owing to the stability of H, p_n is also bounded. Hence there exists a subsequence $\{p_{n_r}, v_{n_r}\}$ such that $p_{n_r} \to p_*$ as $r \to \infty$ and $v_{n_r} \overset{w}{\to} v_*$. The latter convergence is a convergence in the weak topology and the existence of the subsequence is based on the fact that the unit ball in $L_2[0, \varepsilon]$ is weakly compact (see e.g. [56]).

Finally, we note that (3.95) then implies that for all $s \in [0, \varepsilon]$, we have

$$Me^{Ks}p_* + \int_0^s Me^{K(s-\tau)}Lv_*(\tau)\,d\tau + Nv_* = 0.$$

Because the system H is invertible and has no zeros, the above implies that $p_* = 0$ and $v_* = 0$.

We complete our proof by showing that $v_* = 0$ contradicts with the requirements on v outlined in the theorem. The conditions in the theorem imply that

$$\left| \int_0^{\varepsilon} v_n(s)\,ds \right| > \varepsilon\delta$$

for all n. But this implies that

$$\left| \int_0^{\varepsilon} v_*(s)\,ds \right| > \varepsilon\delta,$$

which obviously contradicts with $v_* = 0$. Hence, our initial assumption that ε, δ and $\{t_n\}$ exist satisfying the conditions of the theorem was incorrect and our proof is completed. (Note that $v(t) = \cos(t^2)$ would indeed result in v_n which converges to zero in the weak topology.) ■

Remark 3.A.2 *The above theorem basically states that either the amplitude of v gets smaller as $t \to \infty$ or the signal starts oscillating very rapidly as $t \to \infty$. Our previous example $v(t) = \cos(t^2)$ clearly presents the latter behavior. We use this result to show that the difference between two signals achieving output regulation will never be in $\mathcal{P}$. But then we can show that any non-linear feedback can never have asymptotically smaller input than the linear feedback provided both achieve output regulation. Hence we can prove that if we can achieve output regulation via non-linear feedback then we can also achieve output regulation via linear feedback.*

3.B Review of direct low-gain design for linear systems – an explicit construction

We construct here *explicitly* a family of low-gain state feedback gains based on an eigenvalue assignment method.

Consider the linear system

$$\dot{x} = Ax + Bu \tag{3.96}$$

where the state $x \in \mathbb{R}^n$ and the input $u \in \mathbb{R}^m$. Assume that (A, B) is stabilizable and all the eigenvalues of A are located in the closed left-half plane. We have the following direct low-gain state feedback design algorithm.

Step 1 : Find the state transformation T ([10]) such that $(T^{-1}AT,\ T^{-1}B)$ is in the following form

$$T^{-1}AT = \begin{pmatrix} A_1 & A_{12} & \cdots & A_{1p} & A_{1\bar{c}} \\ 0 & A_2 & \cdots & A_{2p} & A_{2\bar{c}} \\ \vdots & \ddots & \ddots & \vdots & \vdots \\ 0 & \cdots & 0 & A_p & A_{p\bar{c}} \\ 0 & \cdots & 0 & 0 & A_{\bar{c}} \end{pmatrix},$$

$$T^{-1}B = \begin{pmatrix} B_1 & 0 & \cdots & 0 & * \\ 0 & B_2 & \ddots & \vdots & \vdots \\ \vdots & \ddots & \ddots & 0 & * \\ 0 & \cdots & 0 & B_p & * \\ 0 & \cdots & 0 & 0 & 0 \end{pmatrix}.$$

Here $*$'s represent submatrices of less interest and for $i = 1, 2, \cdots, p$,

$$A_i = \begin{pmatrix} 0 & 1 & 0 & \cdots & 0 \\ 0 & 0 & 1 & \ddots & \vdots \\ \vdots & \vdots & \ddots & \ddots & 0 \\ 0 & 0 & \cdots & 0 & 1 \\ -a^i_{n_i} & -a^i_{n_i-1} & -a^i_{n_i-2} & \cdots & -a^i_1 \end{pmatrix}, \quad B_i = \begin{pmatrix} 0 \\ \vdots \\ 0 \\ 0 \\ 1 \end{pmatrix}.$$

The transformation T is such that, for all $i = 1, \cdots, p$, (A_i, B_i) is controllable, all the eigenvalues of A_i are in the closed left-half plane, and all the eigenvalues of $A_{\bar{c}}$ are in the open left-half plane.

Step 2 : For each (A_i, B_i), let $F_i(\varepsilon) \in \mathbb{R}^{1\times n_i}$ be the state feedback gain such that

$$\lambda(A_i - B_i F_i(\varepsilon)) = -\varepsilon + \lambda(A_i) \in \mathbb{C}^{-}.$$

Note that $F_i(\varepsilon)$ is unique.

Step 3 : Form a family of low-gain state feedback gain matrices F_ε as

$$F_\varepsilon T = -\begin{pmatrix} F_1(\varepsilon^{\upsilon_2}) & 0 & \cdots & 0 & 0 & 0 \\ 0 & F_2(\varepsilon^{\upsilon_3}) & \ddots & \vdots & \vdots & \vdots \\ \vdots & \ddots & \ddots & 0 & 0 & 0 \\ 0 & \cdots & 0 & F_{p-1}(\varepsilon^{\upsilon_p}) & 0 & 0 \\ 0 & \cdots & 0 & 0 & F_p(\varepsilon) & 0 \\ 0 & \cdots & 0 & 0 & 0 & 0 \end{pmatrix} \tag{3.97}$$

where $\upsilon_i = (r_i + 1)(r_{i+1} + 1)\cdots(r_p + 1)$ and r_i is the largest algebraic multiplicity of the eigenvalues of A_i. □

With the above choice of F_ε, we have the following theorem.

Theorem 3.B.1 *Consider the linear system (3.96). Assume that all the eigenvalues of A are located in the closed left-half plane and that the pair (A, B) is stabilizable. Then, for the state feedback gain given by (3.97), the closed-loop system*

$$\dot{x} = (A + BF_\varepsilon)x \tag{3.98}$$

is asymptotically stable for all $\varepsilon > 0$. Moreover, there exists an $\varepsilon^* > 0$ such that for all $\varepsilon \in (0, \varepsilon^*]$,

$$\|F_\varepsilon\| \le \alpha\varepsilon, \tag{3.99}$$

$$\|F_\varepsilon e^{(A+BF_\varepsilon)t}\| \le \beta\varepsilon e^{-\varepsilon^\gamma t}, \tag{3.100}$$

where α and β are positive constants independent of ε, and γ is a positive integer also independent of ε.

Proof : It is a consequence of [36]. ■

3.C Review of fast observer design – an explicit construction

For the design of a fast observer we have a direct design as an alternative to the Riccati-based approach.

Step 1 : Assuming that the pair

$$\left(\begin{pmatrix} C_y & D_{yw} \end{pmatrix}, \begin{pmatrix} A & E_w \\ 0 & S \end{pmatrix}\right)$$

is observable, we choose a nonsingular state transformation T_S and a nonsingular output transformation T_O such that

$$T_S^{-1}\begin{pmatrix} A & E_w \\ 0 & S \end{pmatrix} T_S = \begin{pmatrix} b_{11} & I_{p_1-1} & b_{21} & 0 & \cdots & b_{k1} & 0 \\ b_{12} & 0 & b_{22} & 0 & \cdots & b_{k2} & 0 \\ b_{13} & 0 & b_{23} & I_{p_2-1} & \cdots & b_{k3} & 0 \\ b_{14} & 0 & b_{24} & 0 & \cdots & b_{k4} & 0 \\ \vdots & \vdots & \vdots & \vdots & \ddots & \vdots & \vdots \\ b_{12k-1} & 0 & b_{22k-1} & 0 & \cdots & b_{k2k-1} & I_{p_k-1} \\ b_{12k} & 0 & b_{22k} & 0 & \cdots & b_{k2k} & 0 \end{pmatrix},$$

and

$$T_O^{-1}\begin{pmatrix} C_y & D_{yw} \end{pmatrix} T_S = \begin{pmatrix} 1 & 0_{1\times p_1-1} & 0 & 0 & \cdots & 0 & 0 \\ 0 & 0 & 1 & 0_{1\times p_2-1} & \cdots & 0 & 0 \\ \vdots & \vdots & \vdots & \vdots & \ddots & \vdots & \vdots \\ 0 & 0 & 0 & 0 & \cdots & 1 & 0_{1\times p_k-1} \\ 0 & 0 & 0 & 0 & \cdots & 0 & 0 \end{pmatrix},$$

where I_{p_i-1} is an identity matrix of dimension $(p_i - 1) \times (p_i - 1)$, and the integers p_i, $i = 1, 2, \cdots, k$, are the so-called observability indices. Such a transformation exists and is usually called the Brunovski transformation.

Step 2 : For each $i = 1$ to k, find an $l_{i1} \in \mathbb{R}^{(p_i-1)\times 1}$ and a scalar l_{i2} such that the following matrix is Hurwitz-stable,

$$\tilde{A}_1 = \begin{pmatrix} -l_{i1} & I_{p_i-1} \\ -l_{i2} & 0 \end{pmatrix}.$$

Step 3 : Compose the observer gain K_ℓ as follows:

$$T_S^{-1} K_\ell T_O =$$
$$\begin{pmatrix} b_{11} + \overline{S}_{p_1-1}(\ell) l_{11} & b_{21} & \cdots & b_{k1} & 0 \\ b_{12} + \ell^{p_1} l_{12} & b_{22} & \cdots & b_{k2} & 0 \\ b_{13} & b_{23} + \overline{S}_{p_2-1}(\ell) l_{21} & \cdots & b_{k3} & 0 \\ b_{14} & b_{23} + \ell^{p_2} l_{22} & \cdots & b_{k4} & 0 \\ \vdots & \vdots & \ddots & \vdots & 0 \\ b_{12k-1} & b_{22k-1} & \cdots & b_{k2k-1} + \overline{S}_{p_k-1}(\ell) l_{k1} & 0 \\ b_{12k} & b_{22k} & \cdots & b_{k2k} + \ell^{p_k} l_{k2} & 0 \end{pmatrix}$$

where for an integer r,

$$\overline{S}_r(\ell) = \begin{pmatrix} \ell & 0 & \cdots & 0 \\ 0 & \ell^2 & \ddots & \vdots \\ \vdots & \ddots & \ddots & 0 \\ 0 & \cdots & 0 & \ell^r \end{pmatrix}.$$

Finally, we partition the matrix K_ℓ and obtain $K_{A,\ell} \in \mathbb{R}^{n\times p}$ and $K_{S,\ell} \in \mathbb{R}^{s\times p}$ as follows

$$K_\ell = \begin{pmatrix} K_{A,\ell} \\ K_{S,\ell} \end{pmatrix}. \qquad \square$$

For this alternative design we find that Theorem 3.3.5 is still valid. For details we refer to [33].

Chapter 4

Classical output regulation with actuators subject to amplitude saturation – discrete-time systems

4.1 Introduction

The previous chapter considers output regulation for continuous-time linear systems with amplitude saturating actuators. This chapter considers the same however for discrete-time linear systems. Although certain technical results differ from similar ones of the previous chapter, there are definite structural similarities between this chapter and the previous one.

Since output regulation inherently requires internal stabilization, let us first briefly review internal stabilization of discrete-time linear systems with amplitude saturating actuators. Analogous to the continuous-time case, Yang [93] established that a linear discrete-time system subject to input amplitude saturation can be globally asymptotically stabilized via feedback if and only if all its poles are located inside or on the unit circle. Also, a non-linear globally stabilizing control law for such a system is explicitly constructed in [93]. Moreover, it is shown in [94] that similar to the continuous-time case, in general, one must resort to *non-linear* control laws for global asymptotic stabilization. On the other hand, in connection with the semi-global framework, it is shown in [37] that one can semi-globally exponentially stabilize a linear system subject to input amplitude saturation using *linear* feedback laws if and only if the given system is stabilizable with bounded controls and has all its poles inside or on the unit circle. In other words, the basin of attraction of a linear system subject to input amplitude saturation can be made arbitrarily large using appropriately tuned *linear* feedback laws if the system is asymptotically null controllable [1] with bounded controls.

Our focus here, as in the case of continuous-time systems, would be on the semi-global output regulation problem for linear asymptotically null controllable (with bounded controls) systems subject to input amplitude saturation. As in the case of continuous-time systems, the rationale behind the adoption of a semi-global framework for output regulation problem is two-fold. Firstly, the semi-global framework allows us to use linear feedback laws, which is obviously very appealing; and secondly, the semi-global framework seems to be a natural choice when we show that the global output regulation problem, in general, does not have a solution. We study here both state feedback and measurement feedback. The dynamic measurement feedback output regulation problem is solved by designing a linear observer based feedback. In this case, although the controller has a linear structure, it has some non-linearity due to input amplitude saturation. We introduce the notion of semi-global output regulation problems for linear systems with amplitude saturating actuators by extending the output regulation theory developed for linear systems in Chapter 2. A set of solvability conditions for these problems is provided and it is shown that for a fairly general class of systems these conditions are necessary. We also show that, under certain weak assumptions, these solvability conditions for semi-global output regulation problems cannot be weakened further even if we resort to non-linear feedback laws. However, when these assumptions are not satisfied, an example shows that a non-linear feedback controller can achieve output regulation when no linear feedback controller can do so. This chapter is based on the research of authors and their coworkers, in particular [41] and [45].

A point that should be emphasized is this. As will be shown, under certain solvability conditions, *linear* feedback controllers can be developed to solve the posed semi-global output regulation problems by using what is familiarly known as a low-gain design technique. Since low-gain controllers under utilize the available control capacity, often one finds that the convergence of the error signal to zero as time progresses to infinity is rather slow. In these cases, under the same solvability conditions, one could develop an improved technique of designing the appropriate feedback controllers. Such a design

[1] A linear discrete-time system is asymptotically null controllable with bounded controls if and only if it is stabilizable and all its poles are inside or on the unit circle.

technique utilizes the available control capacity in a better way, and thus results in a better performance. However, unlike in the continuous-time case, for discrete-time systems, the feedback controllers that are developed by the improved design technique turn out to be *non-linear* when the given system has multiple inputs. For the single input case, we obtain feedback controllers that remain *linear*.

4.2 Classical global output regulation for discrete-time systems subject to input amplitude saturation

As we said above, our primary focus here is output regulation of linear systems subject to amplitude saturation. Thus we start by redefining the classical output regulation problem. To do so, we consider a time-invariant multivariable discrete-time system with inputs that are subject to amplitude saturation together with a time-invariant exosystem that generates disturbance and reference signals. That is, we consider the system,

$$\begin{aligned} x(k+1) &= Ax(k) + B\sigma(u(k)) + E_w w(k) \\ w(k+1) &= Sw(k) \\ y(k) &= C_y x(k) + D_{yw} w(k) \\ e(k) &= C_e x(k) + D_{eu}\sigma(u(k)) + D_{ew} w(k), \end{aligned} \tag{4.1}$$

where as usual $x \in \mathbb{R}^n$, $w \in \mathbb{R}^s$, $u \in \mathbb{R}^m$, $y \in \mathbb{R}^p$, and $e \in \mathbb{R}^q$. Also, as in the continuous-time case, σ is a vector-valued saturation function defined as

$$\sigma(s) = [\bar{\sigma}(s_1), \bar{\sigma}(s_2), \cdots, \bar{\sigma}(s_m)]^{\mathrm{T}} \tag{4.2}$$

with

$$\bar{\sigma}(s) = \begin{cases} s & \text{if } |s| \le 1 \\ -1 & \text{if } s < -1 \\ 1 & \text{if } s > 1. \end{cases} \tag{4.3}$$

Because of the presence of the saturation function σ, the system (4.1) is non-linear. Note that we can also treat different saturation levels, even differences between channels, by simple scaling. The matrix D_{yu} equals 0 (compared with (2.1)). For control purposes this matrix can be easily handled and it does not affect any of our solvability conditions. However, it makes the formulae much more complex and therefore we opted to go for the simple case with $D_{yu} = 0$.

Before we proceed further, for ease of referencing, we formulate the following assumptions:

A.1. The pair (A, B) is stabilizable.

A.2. The matrix S is anti-Schur-stable, i.e. all the eigenvalues of S are on or outside the unit circle.

A.3. The pair $\left(\begin{pmatrix} C_y & D_{yw} \end{pmatrix}, \begin{pmatrix} A & E_w \\ 0 & S \end{pmatrix}\right)$ is detectable.

Note that these assumptions have been defined before on pages 19 and 25.

As discussed in the previous chapter, the global output regulation problem for continuous-time systems was first formulated by Teel in [83]. We now translate Teel's formulation of the global state feedback output regulation problem and the global measurement feedback output regulation problem into the discrete-time language.

Problem 4.2.1 ***(Global state feedback output regulation problem for linear discrete-time systems subject to input amplitude saturation)*** *For a system Σ as given in (4.1), find, if possible, a feedback law $u = \alpha(x, w)$ such that the following conditions hold:*

(*i*) ***(Internal Stability)*** *The equilibrium point $x = 0$ of*

$$x(k+1) = Ax(k) + B\sigma(\alpha(x(k), 0))$$

is globally asymptotically stable.

(*ii*) ***(Output Regulation)*** *For all $x(0) \in \mathbb{R}^n$, and $w(0) \in \mathbb{R}^s$, the solution of the closed-loop system satisfies*

$$\lim_{k \to \infty} e(k) = 0.$$

Problem 4.2.2 ***(Global measurement feedback output regulation problem for linear discrete-time systems subject to input amplitude saturation)*** *For a system Σ as given in (4.1), find, if possible, a dynamic measurement feedback law $u = \theta(z)$, $\rho z = \eta(z, y)$, where $z \in \mathbb{R}^{n_c}$ such that the following conditions hold:*

(*i*) ***(Internal Stability)*** *The equilibrium point $(x, z) = (0, 0)$ of*

$$\begin{aligned} x(k+1) &= Ax(k) + B\sigma(\theta(z(k))) \\ z(k+1) &= \eta(z(k), C_y x(k)) \end{aligned}$$

is globally asymptotically stable.

(*ii*) ***(Output Regulation)*** *For all* $x(0) \in \mathbb{R}^n$, $z(0) \in \mathbb{R}^{n_c}$, *and* $w(0) \in \mathbb{R}^s$, *the solution of the closed-loop system satisfies*

$$\lim_{k\to\infty} e(k) = 0.$$

The global output regulation as defined above is clearly a very desirable property. Unfortunately it turns out that we can achieve global output regulation only under very special circumstances. In fact, the global measurement feedback output regulation problem as formulated in Problem 4.2.2 basically has no solution. This is established in the following lemma.

Lemma 4.2.1 *Consider a system* Σ *as given in (3.1). Let Assumptions A.1, A.2 and A.3 hold. Also, assume that the eigenvalues of A are on or inside the unit circle. Moreover, assume that at least one unstable pole of the exosystem is observable from the error signal. Then there exist initial conditions* w_0 *for* w *such that there exists no input* u *or initial condition* $x(0)$ *for which the system* Σ *satisfies* $\lim_{k\to\infty} e(k) = 0$.

Proof : We study the system (4.1) when it is rewritten as

$$\begin{aligned} x(k+1) &= Ax(k) + Bv(k) + E_w w(k) \\ w(k+1) &= Sw(k) \\ y(k) &= C_y x(k) + D_{yw} w(k) \\ e(k) &= C_e x(k) + D_{eu} v(k) + D_{ew} w(k), \end{aligned} \tag{4.4}$$

where v denotes an input signal considered bounded.

Let w_0 with $\|w_0\| = 1$ be an eigenvector corresponding to an eigenvalue λ of S belonging to an eigenvalue λ of S which is detectable from e. Suppose λ is an eigenvalue with absolute value greater than one. We can decompose e into three components. One due to the possibly non-zero initial condition x_0, one due to the bounded input v, and the other due to the initial condition w_0. The first two can only grow polynomially in time since all the eigenvalues of A are inside or on the unit circle. On the other hand the effect of w_0 will ensure, due to the detectability assumption, that $e(k)$ grows exponentially in time. Therefore we cannot achieve output regulation without imposing additional conditions.

If the absolute value of λ is not greater than one, then by Assumption A.2, it must lie on the unit circle. In that case, w will be bounded since $w(k) = \lambda^k w_0$. Moreover for any natural number K, and $\varepsilon > 0$ there exists a $k > K$ such that $\|w(k) - w(0)\| < \varepsilon$.

To analyze the situation further, we next define the minimal amplitude of an input signal which achieves tracking, and then minimize this over all the possible initial conditions of the plant. That is, we consider

$$\mathcal{J}(w_0) := \inf_{v,x_0}\{\ \|v\|_\infty \mid v \text{ is such that } \lim_{k\to\infty} e(k) = 0 \\ \text{where } x(0) = x_0 \text{ and } w(0) = w_0\ \}.$$

Suppose $\mathcal{J}(w_0) = 0$. We take a minimizing sequence $\{v_i, x_{0,i}\}$ for the above optimization problem. For each v_i there exists a K_i such that $\|e(K_i + k)\| < 1/i$ for all $k > 0$ and $\|w(K_i) - w_0\| < 1/i$. Define

$$\bar{v}_i(k) := v_i(K_i + k).$$

Then $\|\bar{v}_i\|_\infty \leqslant \|v_i\|_\infty \to 0$ as $i \to \infty$. The output $\bar{e}_i$ resulting from input $\bar{v}_i$ and initial conditions

$$\bar{x}(0) = \bar{x}_{0,i} := x(K_i),$$

and $\bar{w}_i(0) = w(K_i)$, satisfies $\|\bar{e}_i\|_\infty < 1/i$. The latter is straightforward since $\bar{e}_i(k) = e(K_i + k)$. We then pick any integer $K > 0$. On $[0, K]$ the input $\bar{v}_i$ converges in L_∞ norm to 0. Similarly $\bar{e}_i$ converges to 0 uniformly on $[0, K]$. Finally $\bar{w}_i(0)$ converges to w_0.

Define $f : \mathbb{R}^n \to L_\infty[0, K]$ by $[f(z)](k) = C_e A^k z$. We can check that $g \in L_\infty[0, K]$ is in the closure of the image of f where

$$g(k) := \sum_{\kappa=0}^{k-1} C_e A^{(k-1-\kappa)} E_w w(\kappa) + D_{ew} w(k).$$

Since f is a finite rank operator we know the image is closed and hence g is in the image of f, i.e. there exists an $\tilde{x}_0$ such that $f(\tilde{x}_0) = -g$. We find that

$$\begin{pmatrix} C_y & D_{yw} \end{pmatrix} \begin{pmatrix} A & E_w \\ 0 & S \end{pmatrix}^k \begin{pmatrix} \tilde{x}_0 \\ w_0 \end{pmatrix} = 0 \tag{4.5}$$

for $k \in [0, K]$. This immediately implies that (4.5) holds for all k. On the other hand $w(k) \not\to 0$ as $k \to \infty$. However this contradicts the fact that the eigenvalue λ was observable from e. Therefore we have $\mathcal{J}(w_0) > 0$.

Hence we find that, for $w(0) = 2w_0/\mathcal{J}(w_0)$, any input v which achieves output regulation satisfies $\|v\|_\infty \geqslant 2$. Therefore $v = \sigma(u)$ will never be able to achieve output regulation, i.e. no input u to (4.1) exists for which $e(k) \to 0$ as $k \to \infty$.

The above involves complex inputs. By working with either the real or complex part of the signals at a time we can avoid this technical problem. ■

Remark 4.2.1 *As argued before, Assumptions A.1, A.2, A.3, and eigenvalues of A being on or inside of unit circle are basically necessary. Therefore the above lemma tells us that we can only achieve global output regulation if the open-loop system already achieves output regulation and the controller only needs to achieve stability without losing the property of output regulation. This is a very exceptional case and therefore, for all practical purposes, global output regulation is not possible when we have input saturation.*

Note that the classical case $y = e$ implies that Assumptions A.2 and A.3 guarantee that all poles of the exosystem should be observable from e and therefore global output regulation is not possible.

4.3 Classical semi-global output regulation for linear systems subject to input amplitude saturation

Since in general global output regulation problems for a system of the type (4.1) are not solvable, one basic question that arises is what type of initial conditions of exosystem and plant should be considered realistic for output regulation when the input is subject to amplitude saturation. Regarding the initial conditions of the exosystem, the discussion at the end of the last section is clearly in favor of the argument that we should restrict our attention only to initial conditions $w(0)$ lying inside a given compact set. Moreover, regarding the initial conditions of the plant, in the theory of stabilization of linear discrete-time systems subject to amplitude saturation, the step from global initial conditions to initial conditions inside a compact set has already been made. This has been named semi-global stabilization. Since, in most cases we have to restrict attention to initial conditions for w inside a compact set, this yields a good motivation anyway to direct our attention to a semi-global setting. Of course this also yields the well-known advantage that we can achieve output regulation with linear feedback controllers. Motivated by this, as in the previous chapter, we devote ourselves here to *semi-global output regulation.*

We split this section into two parts. In the first part we solve the classical semi-global linear state feedback output regulation problem where both the states x and w are available for feedback. In this case, it suffices to look at only static feedback controllers. In the second part we solve the classical semi-global measurement feedback output regulation problem where only certain measurement signal is available for feedback, and hence we need to resort to dynamic feedback. In this case, we design observer based controllers.

4.3.1 State feedback controllers

The problem considered in this subsection is formulated as follows.

Problem 4.3.1 ***(Classical semi-global state feedback output regulation problem for systems subject to input amplitude saturation)*** *Consider the system (4.1) and a compact set $\mathcal{W}_0 \subset \mathbb{R}^s$. For any a priori given (arbitrarily large) bounded set $\mathcal{X}_0 \subset \mathbb{R}^n$, find, if possible, a static feedback law $u = \alpha(x, w)$ such that the following conditions hold:*

(*i*) ***(Internal Stability)*** *The equilibrium point $x = 0$ of*

$$x(k+1) = Ax(k) + B\sigma(\alpha(x(k), 0)) \tag{4.6}$$

is asymptotically stable with $\mathcal{X}_0$ contained in its basin of attraction.

(*ii*) ***(Output Regulation)*** *For all $x(0) \in \mathcal{X}_0$ and $w(0) \in \mathcal{W}_0$, the solution of the closed-loop system satisfies*

$$\lim_{k \to \infty} e(k) = 0. \tag{4.7}$$

Remark 4.3.1 *We would like to emphasize that our definition of classical semi-global state feedback output regulation problem does not view the set of initial conditions of the plant as given data. The set of given data consists of the models of the plant and the exosystem and the set of initial conditions for the exosystem. Therefore, any solvability conditions we obtain* must *be independent of the set of initial conditions of the plant, $\mathcal{X}_0$.*

Before discussing the solvability conditions for the above posed output regulation problem, we would like to recall what is known as a low-gain [37] design method introduced earlier by Lin and Saberi and their coworkers. Such a low-gain design method has successfully been used, in connection with linear systems with amplitude saturating actuators, not only for internal stabilization, but also for various other problems.

Review of Riccati-based low-gain design for linear systems:

We now recall a low-gain state feedback design algorithm from [37]. The objective is to show that we can find stabilizing feedback control inputs with arbitrarily small magnitude which stabilize a given linear system with all its poles located inside or on the unit circle. Such a design algorithm yields a family of state feedback gains, parameterized in ε, which are instrumental in proving our results here on semi-global output regulation. Analogous to the continuous-time case, there exist in the literature two low-gain design algorithms; one is based on the solution of a discrete-time algebraic Riccati equation parameterized in ε, and the other is a *direct* construction method based on an eigenvalue assignment method. The Riccati-based method is conceptually appealing although solving the parameterized Riccati equation might be numerically stiff. On the other hand, the alternative direct method of explicit construction is numerically efficient but is some what involved in details. For conceptual clarity, we present here the Riccati-based method, and the alternative *direct* method is discussed at the end of this chapter in an appendix.

Consider the linear system

$$x(k+1) = Ax(k) + Bu(k), \tag{4.8}$$

where the state $x \in \mathbb{R}^n$ and the input $u \in \mathbb{R}^m$. Assume that (A, B) is stabilizable and all the eigenvalues of A are located inside or on the unit circle. Consider the Riccati equation defined by

$$P_\varepsilon = A^{\mathrm{T}} P_\varepsilon A + Q_\varepsilon - A^{\mathrm{T}} P_\varepsilon B (B^{\mathrm{T}} P_\varepsilon B + I)^{-1} B^{\mathrm{T}} P_\varepsilon A \tag{4.9}$$

where $Q : (0, 1] \to \mathbb{R}^{n \times n}$ is a continuously differentiable matrix-valued function such that $Q_\varepsilon > 0$, $\frac{dQ_\varepsilon}{d\varepsilon} > 0$ for any $\varepsilon \in (0, 1]$, and $\lim_{\varepsilon \to 0} Q_\varepsilon = 0$. Often, Q_ε can be taken as εI, however a judicious choice of Q_ε is some times necessary. We next form a family of low-gain state feedback gain matrices F_ε as

$$F_\varepsilon = -(B^{\mathrm{T}} P_\varepsilon B + I)^{-1} B^{\mathrm{T}} P_\varepsilon A. \tag{4.10}$$

The following lemma recalls from the literature certain important properties of the Riccati equation (4.9).

Lemma 4.3.1 *Consider the Riccati equation given in (4.9). Let (A, B) be stabilizable and all the eigenvalues of A be located inside or on the unit circle. Let Q_ε be a continuously differentiable matrix-valued function such that $Q_\varepsilon > 0$, $\frac{dQ_\varepsilon}{d\varepsilon} > 0$ for any $\varepsilon \in (0, 1]$, and $\lim_{\varepsilon \to 0} Q_\varepsilon = 0$. Then the Riccati equation (4.9) has a unique positive definite solution P_ε for any $\varepsilon \in (0, 1]$. Moreover, this positive definite solution P_ε has the following properties:*

(*i*) *For any $\varepsilon \in (0, 1]$, the unique solution $P_\varepsilon > 0$ is such that $A + BF_\varepsilon$ is Schur-stable where F_ε is as in (4.10).*

(*ii*) $\lim_{\varepsilon \to 0} P_\varepsilon = 0$.

(iii) There exists an $\varepsilon^* > 0$ such that, for $\varepsilon \in (0, \varepsilon^*]$,

$$\|P_\varepsilon^{1/2} A P_\varepsilon^{-1/2}\| \le \sqrt{2}. \tag{4.11}$$

(iv) P_ε is continuously differentiable with respect to ε and

$$\frac{dP_\varepsilon}{d\varepsilon} > 0, \quad \textit{for any } \varepsilon \in (0, 1].$$

Proof : The existence and uniqueness of a positive semi-definite solution P_ε follows from [61]. It also follows from [61] that P_ε is the unique solution for which $A - B(B^{\mathrm{T}}P_\varepsilon B + I)^{-1}B^{\mathrm{T}}P_\varepsilon A$ has all its eigenvalues on or inside the unit circle. For $\varepsilon = 0$, it is trivial to see that the Riccati equation (4.9) has a solution $P(0) = 0$ since by assumption, $A - B(B^{\mathrm{T}}P(0)B + I)^{-1}B^{\mathrm{T}}P(0) = A$ has all its eigenvalues inside or on the unit circle. The fact that $\lim_{\varepsilon\to 0} P_\varepsilon = 0$ follows from standard continuity arguments. Note that for $\varepsilon > 0$ the solution is actually positive definite and is such that $A - B(B^{\mathrm{T}}P_\varepsilon B + I)^{-1}B^{\mathrm{T}}P_\varepsilon A$ has all its eigenvalues inside the unit circle.

To show part (iii), we observe that by pre- and post-multiplying both sides of (4.9) with $P_\varepsilon^{-1/2}$, we obtain

$$V_\varepsilon \left[I - P_\varepsilon^{1/2}B(B^{\mathrm{T}}P_\varepsilon B + I)^{-1}B^{\mathrm{T}}P_\varepsilon^{1/2}\right] V_\varepsilon^{\mathrm{T}} = I - \varepsilon P_\varepsilon^{-1/2}$$

where $V_\varepsilon = P_\varepsilon^{-1/2}A^{\mathrm{T}}P_\varepsilon^{1/2}$. Since $\lim_{\varepsilon\to 0} P_\varepsilon = 0$, it follows from the above equation that there exists an $\varepsilon^* > 0$ such that for all $\varepsilon \in (0, \varepsilon^*]$,

$$P_\varepsilon^{-1/2}A^{\mathrm{T}}P_\varepsilon A P_\varepsilon^{-1/2} \le 2I - 2\varepsilon P_\varepsilon^{-1/2} \le 2I.$$

This implies that

$$\lambda_{\max}\left(P_\varepsilon^{-1/2}A^{\mathrm{T}}P_\varepsilon A P_\varepsilon^{-1/2}\right) \le 2.$$

This completes the proof of part (iii).

To show the last property, we first note that the continuous differentiability of P_ε for $\varepsilon > 0$ follows the fact that the symplectic pencil associated with the Riccati equation (4.9) is a continuously differentiable function of ε (see [28]). Then, in order to show that $\frac{dP_\varepsilon}{d\varepsilon} > 0$, we first observe that

$$\frac{d(B^{\mathrm{T}}P_\varepsilon B + I)^{-1}}{d\varepsilon} = -(B^{\mathrm{T}}P_\varepsilon B + I)^{-1}B^{\mathrm{T}}\frac{dP_\varepsilon}{d\varepsilon}B(B^{\mathrm{T}}P_\varepsilon B + I)^{-1}.$$

Now, differentiating the Riccati equation with respect to ε, and using the above equality, we get

$$\frac{dP_\varepsilon}{d\varepsilon} = (A^{\mathrm{T}} + F_\varepsilon^{\mathrm{T}} B^{\mathrm{T}})\frac{dP_\varepsilon}{d\varepsilon}(A + BF_\varepsilon) + \frac{dQ_\varepsilon}{d\varepsilon}. \tag{4.12}$$

It is now obvious that $\frac{dP_\varepsilon}{d\varepsilon} > 0$. ■

We next move on to obtain the solvability conditions for the classical semi-global linear state feedback output regulation problem for linear systems subject to input amplitude saturation. The following theorem presents such conditions. In fact, the proof of the theorem shows that under the given solvability conditions, a linear low-gain feedback controller can solve the problem.

Theorem 4.3.1 *Consider the system (4.1) and the given compact set $\mathcal{W}_0 \subset \mathbb{R}^s$. The classical semi-global state feedback output regulation problem is solvable if the following conditions hold:*

(i) *(A, B) is stabilizable and A has all its eigenvalues inside or on the unit circle.*

(ii) *There exist matrices Π and Γ such that,*

(a) *they solve the regulator equation (2.7), i.e.,*

$$\Pi S = A\Pi + B\Gamma + E_w, \tag{4.13a}$$
$$0 = C_e\Pi + D_{eu}\Gamma + D_{ew}. \tag{4.13b}$$

(b) *there exist a $\delta > 0$ and a $K \geq 0$ such that $\|\Gamma w\|_{\infty,K} \leqslant 1 - \delta$ for all w with $w(0) \in \mathcal{W}_0$.*

Moreover, a linear state feedback controller of the form, $u = Fx + Gw$, can solve the posed problem.

Proof : We prove this theorem by first explicitly constructing a family of linear static state feedback laws, parameterized in ε, and then showing that for each given set $\mathcal{X}_0$, there exists an $\varepsilon^* > 0$ such that for all $\varepsilon \in (0, \varepsilon^*]$, both items (i) and (ii) of Problem 4.3.1 hold. The family of linear static state feedback laws we construct takes the form,

$$u = F_\varepsilon x + (\Gamma - F_\varepsilon \Pi)w, \tag{4.14}$$

where F_ε is as in (4.10). It then follows from Lemma 4.3.1 that $A + BF_\varepsilon$ is Schur-stable for all $\varepsilon > 0$. With this family of feedback laws, the system (4.6) is written as

$$x(k+1) = Ax(k) + B\sigma(F_\varepsilon x(k)). \tag{4.15}$$

We proceed next to show part (i) of Problem 4.3.1. In view of (4.10), the closed-loop system (4.15) can be rewritten as

$$\begin{aligned} \rho x &= Ax + B\sigma\left[-(B^{\mathrm{T}}P_\varepsilon B + I)^{-1}B^{\mathrm{T}}P_\varepsilon Ax\right] \\ &= \left(A - B(B^{\mathrm{T}}P_\varepsilon B + I)^{-1}B^{\mathrm{T}}P_\varepsilon A\right)x \\ &\qquad\qquad + B[\sigma(u) - u]. \end{aligned} \tag{4.16}$$

Also, it follows from (4.9) that

$$A_{cl}^{\mathrm{T}}P_\varepsilon A_{cl} - P_\varepsilon = -Q_\varepsilon - Q_0 \tag{4.17}$$

where

$$\begin{aligned} A_{cl} &:= A - B(B^{\mathrm{T}}P_\varepsilon B + I)^{-1}B^{\mathrm{T}}P_\varepsilon A \\ Q_0 &:= A^{\mathrm{T}}P_\varepsilon B(B^{\mathrm{T}}P_\varepsilon B + I)^{-2}B^{\mathrm{T}}P_\varepsilon A \geq 0. \end{aligned}$$

We can now select the Lyapunov function as

$$V(x) = x^{\mathrm{T}}P_\varepsilon x,$$

and let c be a strictly positive real number such that

$$c \geq \sup_{x\in\mathcal{X}_0, \varepsilon\in(0,1]} x^{\mathrm{T}}P_\varepsilon x. \tag{4.18}$$

The right hand side is well defined since $\lim_{\varepsilon\to 0} P_\varepsilon = 0$ by Lemma 4.3.1 and $\mathcal{X}_0$ is bounded. Let ε_1^* be such that for all $\varepsilon \in (0, \varepsilon_1^*]$, $x \in L_V(c)$ implies that $\|(B^{\mathrm{T}}P_\varepsilon B + I)^{-1}B^{\mathrm{T}}P_\varepsilon Ax\| \leq 1$, where the level set $L_V(c)$ is defined as

$$L_V(c) = \{x \in \mathbb{R}^n \mid V(x) \leq c\}.$$

Such an ε_1^* exists because of Lemma 4.3.1 and the fact that $\lim_{\varepsilon\to 0} P_\varepsilon = 0$.

The evaluation of the difference of V along the trajectory of the closed-loop system (4.16), using (4.17), shows that for all $x \in L_V(c)$,

$$
\begin{aligned}
V(\rho x) - V(x)& \\
&= -x^{\mathrm{T}}(\varepsilon I + Q_0)x + [\sigma(u) - u]^{\mathrm{T}} B^{\mathrm{T}} P_\varepsilon B[\sigma(u) - u] \\
&\quad + 2x^{\mathrm{T}} A_{cl}^{\mathrm{T}} P_\varepsilon B[\sigma(u) - u] \\
&\le -\varepsilon x^{\mathrm{T}} x - u^{\mathrm{T}} u + [\sigma(u) - u]^{\mathrm{T}} B^{\mathrm{T}} P_\varepsilon B[\sigma(u) - u] \\
&\quad + 2x^{\mathrm{T}} A^{\mathrm{T}} P_\varepsilon B(B^{\mathrm{T}} P_\varepsilon B + I)^{-1}[\sigma(u) - u] \\
&= -\varepsilon x^{\mathrm{T}} x - u^{\mathrm{T}} u + [\sigma(u) - u]^{\mathrm{T}} B^{\mathrm{T}} P_\varepsilon B[\sigma(u) - u] \\
&\quad - 2u^{\mathrm{T}}[\sigma(u) - u] \\
&\le -\varepsilon x^{\mathrm{T}} x - u^{\mathrm{T}} u + (k-1)^2 \lambda_{\max}(B^{\mathrm{T}} P_\varepsilon B) u^{\mathrm{T}} u.
\end{aligned}
$$

Again, recalling that $\lim_{\varepsilon \to 0} P_\varepsilon = 0$, we can easily see that there exists an $\varepsilon^* \in (0, \varepsilon_1^*]$ such that, for all $\varepsilon \in (0, \varepsilon^*]$, we have $(k-1)^2 \lambda_{\max}(B^{\mathrm{T}} P_\varepsilon B) \le 1$. This shows that for any $\varepsilon \in (0, \varepsilon^*]$,

$$x \in L_V(c) \to V(x(k+1)) - V(x(k)) \le -\varepsilon x^{\mathrm{T}}(k)x(k).$$

This in turn shows that, for any $\varepsilon \in (0, \varepsilon^*]$, the equilibrium point $x = 0$ of the closed-loop system is asymptotically stable and its domain of attraction contains the set $L_V(c)$. This completes our proof of part (i) of Problem 4.3.1 since $\mathcal{X}_0 \subseteq L_V(c)$.

Next, we show that there exists an $\varepsilon_2^* \in (0, 1]$ such that for each $\varepsilon \in (0, \varepsilon_2^*]$, item (i) of Problem 4.3.1 holds. To this end, let us introduce an invertible coordinate change,

$$\xi = x - \Pi w. \tag{4.19}$$

Using condition (ii)a), we have

$$
\begin{aligned}
\xi(k+1) &= x(k+1) - \Pi w(k+1) \\
&= Ax(k) + B\sigma(u(k)) + E_w w(k) - \Pi S w(k) \\
&= A\xi(k) + B\left[\sigma(u(k)) - \Gamma w(k)\right].
\end{aligned} \tag{4.20}
$$

With the family of state feedback laws given above, the closed-loop system can be written as

$$\xi(k+1) = A\xi(k) + B\left[\sigma(\Gamma w(k) + F_\varepsilon \xi(k)) - \Gamma w(k)\right]. \tag{4.21}$$

By Condition (ii)b), $\|\Gamma w(k)\|_{\infty,\mathrm{K}} < 1 - \delta$. Moreover, for any $x(0) \in \mathcal{X}_0$ and any $w(0) \in \mathcal{W}_0$, $\xi(K)$ belongs to a bounded set, say $\mathcal{U}_K$, independent of ε since $\mathcal{X}_0$ and $\mathcal{W}_0$ are both bounded and $\xi(K)$ is determined by a linear difference equation with bounded inputs $\sigma(\cdot)$ and Γw.

It follows from (4.9) that

$$(A + BF_\varepsilon)^{\mathrm{T}} P_\varepsilon (A + BF_\varepsilon) - P_\varepsilon = -\varepsilon I - F_\varepsilon^{\mathrm{T}} F_\varepsilon. \tag{4.22}$$

We then pick a Lyapunov function

$$V(\xi) = \xi^{\mathrm{T}} P_\varepsilon \xi, \tag{4.23}$$

and let $c > 0$ be such that

$$c \geq \sup_{\xi \in \mathcal{U}_K, \varepsilon \in (0,1]} \xi^{\mathrm{T}} P_\varepsilon \xi. \tag{4.24}$$

Such a c exists since P_ε and $\mathcal{U}_K$ are bounded. Let $\varepsilon_2^* \in (0, 1]$ be such that $\xi \in L_V(c)$ implies that $\|F_\varepsilon \xi\|_\infty \leq \delta$ where

$$L_V(c) := \left\{\xi \in \mathbb{R}^n \mid V(\xi) < c \right\}.$$

The existence of such an ε_2^* is again due to Lemma 4.3.1. Hence for $k \geq K$, for all $\xi \in L_V(c)$, (4.21) takes the form

$$\xi(k+1) = A\xi(k) + BF_\varepsilon \xi(k). \tag{4.25}$$

The evaluation of the difference of V for $k \geq K$, inside the set $L_V(c)$, using (4.22), shows that for all $\xi \in L_V(c)$,

$$V(\xi(k+1)) - V(\xi(k)) = -\xi(k)^{\mathrm{T}} \left(\varepsilon I + F_\varepsilon^{\mathrm{T}} F_\varepsilon\right) \xi(k). \tag{4.26}$$

This shows that any trajectory of (4.21) starting at $k = 0$ from

$$\{\xi = x - \Pi w : x \in \mathcal{X}_0, w \in \mathcal{W}_0\}$$

remains inside the set $L_V(c)$ and approaches the equilibrium point $\xi = 0$ as $k \to \infty$. Also, in view of (4.14) and (4.19), we find that for $k \geq K$,

$$e(k) = (C_e + D_{eu} F_\varepsilon)\xi(k) + (C_e \Pi + D_{eu}\Gamma + D_{ew})w(k).$$

However, in view of (4.13b), e reduces to $e = (C_e + D_{eu} F_\varepsilon)\xi$, and thus, owing to the stability of $A + BF_\varepsilon$, we find that $e(k) \to 0$ as $k \to \infty$.

Finally, setting $\varepsilon^* = \min\{\varepsilon_1^*, \varepsilon_2^*\}$, we conclude our proof of Theorem 4.3.1. ∎

Remark 4.3.2 *In view of Yang's results ([93]) and the solvability conditions for the state feedback output regulation problem for linear systems in the absence of input saturation as given in Chapter 2, it is obvious to observe that Conditions (i) and (ii)a) of Theorem 4.3.1 are necessary. The crucial condition for the solvability of the classical semi-global linear state feedback output regulation problem with amplitude saturating actuators is Condition (ii)b), which is a sufficient condition. In Section 4.4 we will discuss the necessity of Condition (ii)b).*

Design of a low-gain state feedback regulator:

For clarity, we now give a step by step design of a low-gain state feedback regulator.

Step 1 : Find a solution $(\Pi,\ \Gamma)$ of the regulator equation (4.13).

Step 2 : Find a low-gain state feedback matrix either by the Riccati-based design as in (4.10), or by direct method discussed at the end of this chapter in Appendix 4.B.

Step 3 : Given the sets $\mathcal{X}_0 \subset \mathbb{R}^n$ and $\mathcal{W}_0 \subset \mathbb{R}^s$, find an ε^* by the procedure given in the proof of Theorem 4.3.1.

Then the state feedback controller given in (4.14) for some $\varepsilon \in (0,\ \varepsilon^*]$ solves the classical semi-global state feedback output regulation problem. □

The following example illustrates the design procedure.

Example 4.3.1 *Consider the system,*

$$
\begin{aligned}
x(k+1) &= \begin{pmatrix} 0 & -1 & -1 & 0 \\ 1 & 0 & 0 & 1 \\ 0 & 0 & 0 & -1 \\ 0 & 0 & 1 & 0 \end{pmatrix} x(k) + \begin{pmatrix} 1 & 0 \\ 0 & 1 \\ 1 & 0 \\ 0 & 1 \end{pmatrix} \sigma(u(k)) \\
&\quad + \begin{pmatrix} 0 & 2 \\ -2 & -2 \\ -1 & 2 \\ -2 & -1 \end{pmatrix} w(k) \\
w(k+1) &= \begin{pmatrix} 0 & 1 \\ -1 & 0 \end{pmatrix} w(k) \\
e(k) &= \begin{pmatrix} 1 & 0 & 0 & 0 \\ 0 & 1 & 0 & 0 \end{pmatrix} x(k) + \begin{pmatrix} -1 & 0 \\ 0 & -1 \end{pmatrix} w(k)
\end{aligned}
\tag{4.27}
$$

with $w(0) \in \mathcal{W}_0$, where $\mathcal{W}_0 = \left\{ w \in \mathbb{R}^2 : \|w\| < 0.5 \right\}$. It is straightforward to show that the solvability conditions for the classical semi-global linear *state feedback output regulation problem are satisfied. More specifically, the matrices,*

$$\Pi = \begin{pmatrix} 1 & 0 \\ 0 & 1 \\ 1 & 0 \\ 0 & 1 \end{pmatrix}, \quad \Gamma = \begin{pmatrix} 1 & 0 \\ 0 & 1 \end{pmatrix}, \tag{4.28}$$

solve the regulator equation (4.13). Also, $\delta = 0.5$, *since* $\|\Gamma w\|_\infty \leq 0.5$ *for all* $w(0) \in \mathcal{W}_0$. *Let the set* $\mathcal{X}_0$ *be given by* $\mathcal{X}_0 = \{x \in \mathbb{R}^4 : \|x\| \leq 2\}$.

Then, following the design procedure given above and the proof of Theorem 4.3.1, a choice of ε^* *is* 4.6×10^{-3}. *For* $\varepsilon = \varepsilon^*$, *the feedback law (4.14) is given by*

$$u = -\begin{pmatrix} 0.0529 & -0.0335 & -0.0563 & -0.0340 \\ 0.0335 & -0.0529 & 0.0340 & 0.0563 \end{pmatrix} x + \begin{pmatrix} 0.9966 & -0.0675 \\ 0.0675 & 1.0034 \end{pmatrix} w.$$

For the initial conditions $x(0) = (1, 1, 1, 1)^T$, $w(0) = (0.25, -0.25)^T$, *Figure 4.1 on the facing page shows the control action and the closed-loop performance of the regulator.*

As seen from the above example, and as in the case of continuous-time systems, low-gain based designs under utilize the available control capacity. Our goal next is to recall a new design methodology which incorporates significant improvement to the low-gain design method, and leads to a better utilization of the available control capacity and hence better closed-loop performance.

An improved design of the feedback regulator:

We now construct a family of *non-linear* state feedback laws, parameterized in ε and μ, and then show that such a family of state feedback laws solves the classical semi-global state feedback output regulation problem 4.3.1. This family of non-linear feedback laws reduces to a linear one for the single input case ($m = 1$), that is, the function $\alpha(x, w)$ as in Problem 4.3.1 reduces to the form of $Fx + Gw$.

The new state feedback output regulator design is given as follows: Let P_ε be the solution of the Riccati equation (4.9) with Q_ε satisfying the properties given in Lemma 4.3.1. Assume B injective and consider the control law,

$$u = [F_\varepsilon + \mu\kappa(x, w, \mu)K_\varepsilon](x - \Pi w) + \Gamma w, \quad \mu \in [0, 2], \tag{4.29}$$

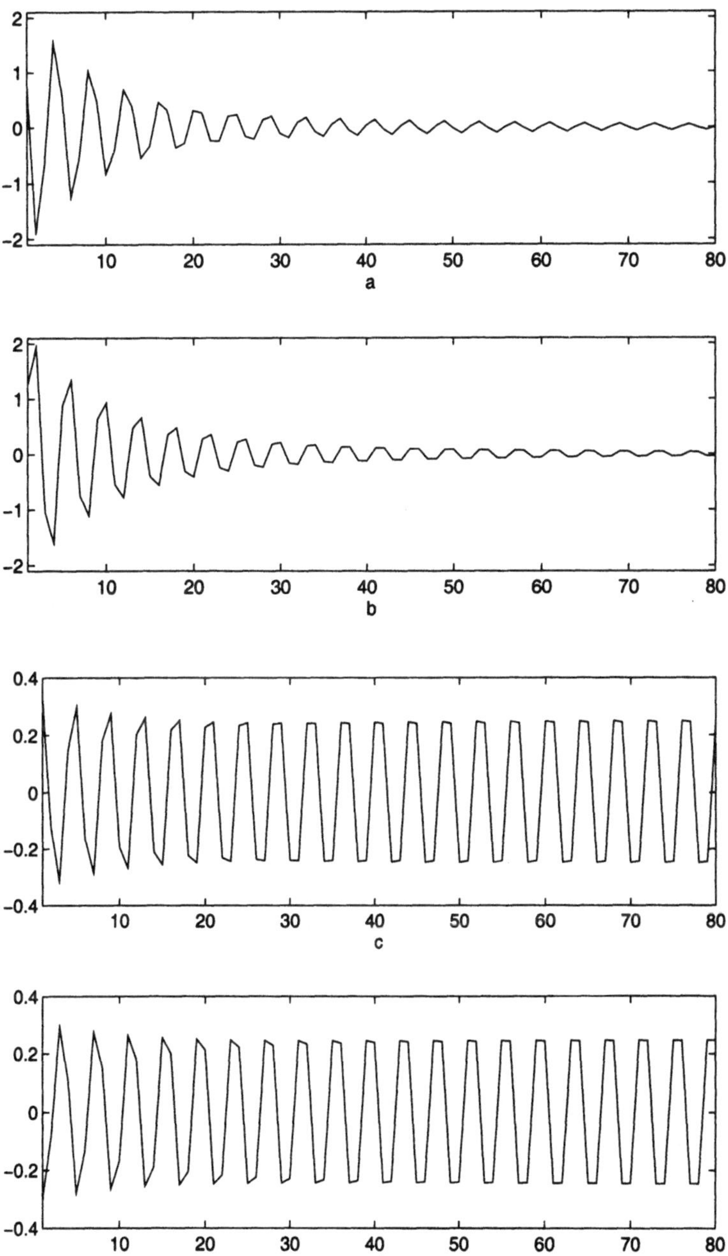

Figure 4.1: $\varepsilon = 4.6 \times 10^{-3}$. a) e_1; b) e_2; c) u_1; d) u_2.

where

$$F_\varepsilon = -(B^{\mathrm{T}} P_\varepsilon B + I)^{-1} B^{\mathrm{T}} P_\varepsilon A,$$
$$K_\varepsilon = -(B^{\mathrm{T}} P_\varepsilon B)^{-1} B^{\mathrm{T}} P_\varepsilon A_c,$$
$$A_c = A + B F_\varepsilon,$$

with $\kappa : \mathbb{R}^n \times \mathbb{R}^s \times [0, 2] \to \mathbb{R}_+$ is defined as,

$$\kappa(x, w, \mu) = \begin{cases} 1 & \text{if } m = 1 \\ z^*(x, w, \mu) & \text{if } m > 1. \end{cases} \tag{4.30}$$

where

$$[z^*(x, w, \mu)](k) = \max_{z \in [0,1]} \{z : \| [F_\varepsilon + \mu z K_\varepsilon](x(k) - \Pi w(k)) + \Gamma w(k) \|_\infty \le 1\}.$$

If, in the above maximization, there exists no z for which the inequality is satisfied then z^* is chosen equal to 0. We note that when $\mu = 0$ the family of state feedback laws as given in (4.29) reduces to the low-gain based family of linear feedback laws as given in (4.14). Thus, as usual, ε is referred to as a low-gain parameter.

As shown next, for any $\mu \in [0, 2]$, the family of state feedback laws (4.29) solves the semi-global state feedback output regulation problem.

Theorem 4.3.2 *Consider the system (4.1) and the given compact set $\mathcal{W}_0 \subset \mathbb{R}^s$. Assume that the sufficient conditions given in Theorem 4.3.1 are satisfied and let B be injective. Then, there exists a controller, among the family of state feedback laws as given in (4.29), that solves the semi-global state feedback output regulation problem. More specifically, for any a priori given (arbitrarily large) bounded set $\mathcal{X}_0 \subset \mathbb{R}^n$, there exists an $\varepsilon^* \in (0, 1]$ such that for each $\varepsilon \in (0, \varepsilon^*]$ and for each $\mu \in [0, 2]$, the controller in the family (4.29) has the following properties:*

(i) The equilibrium point $x = 0$ of

$$\rho x = Ax + B\sigma\left[(F_\varepsilon + \mu\kappa(x, 0, \mu) K_\varepsilon) x\right] \tag{4.31}$$

is asymptotically stable with $\mathcal{X}_0$ contained in its basin of attraction.

(ii) For any $x(0) \in \mathcal{X}_0$ and $w(0) \in \mathcal{W}_0$, the solution of the closed-loop system satisfies

$$\lim_{k \to \infty} e(k) = 0. \tag{4.32}$$

Proof : We prove this theorem by showing that for each given set $\mathcal{X}_0$, there exists an $\varepsilon^* \in (0, 1]$ such that for all $\varepsilon \in (0, \varepsilon^*]$ and all $\mu \in [0, 2]$, both items (i) and (ii) of the theorem hold.

We first show that there exists an $\varepsilon_1^* \in (0, 1]$ such that for each $\varepsilon \in (0, \varepsilon_1^*]$ and for each $\mu \in [0, 2]$, item (i) of the theorem holds. To this end, rewrite (4.31) as

$$\rho x = A_c x + B\left(\sigma\left[(F_\varepsilon + \mu\kappa(x, 0, \mu)K_\varepsilon)x\right] - F_\varepsilon x\right). \tag{4.33}$$

Consider the Lyapunov function

$$V_1(x) = x^{\mathrm{T}} P_\varepsilon x, \tag{4.34}$$

and let $c_1 > 0$ be such that

$$c_1 \geq \sup_{x \in \mathcal{X}_0, \varepsilon \in (0,1]} x^{\mathrm{T}} P_\varepsilon x. \tag{4.35}$$

Such a c_1 exists since $\lim_{\varepsilon \to 0} P_\varepsilon = 0$ by Lemma 4.3.1, and $\mathcal{X}_0$ is bounded. Let ε_1^* be such that for all $\varepsilon \in (0, \varepsilon_1^*]$, $x \in L_{V_1}(c_1)$ implies that $\|F_\varepsilon x\|_\infty \leq 1$, where the level set $L_{V_1}(c_1)$ is defined as $L_{V_1}(c_1) = \{x \in \mathbb{R}^n : V_1(x) \leq c_1\}$. Such an ε_1^* exists because of (4.11) and the fact that $\lim_{\varepsilon \to 0} P_\varepsilon = 0$.

It follows from (4.9) that,

$$A_c^{\mathrm{T}} P_\varepsilon A_c - P_\varepsilon = -\varepsilon I - F_\varepsilon^{\mathrm{T}} F_\varepsilon. \tag{4.36}$$

For the case of multiple input ($m > 1$), by the definition of function κ, (4.30), for all $x \in L_{V_1}(c_1)$, the saturation functions in the closed-loop system (4.33) operate in their respective linear regions, and hence the closed-loop system remains linear. The evaluation of the difference of V_1 along the trajectories of this linear closed-loop system shows that, for $x \in L_{V_1}(c_1)$,

$$\begin{aligned} & V_1(\rho x) - V_1(x) \\ & = -\varepsilon x^{\mathrm{T}} x - x^{\mathrm{T}} F_\varepsilon^{\mathrm{T}} F_\varepsilon x \\ & \quad - \mu\kappa(x, 0, \mu)[2 - \mu\kappa(x, 0, \mu)] x^{\mathrm{T}} A_c^{\mathrm{T}} P_\varepsilon B (B^{\mathrm{T}} P_\varepsilon B)^{-1} B^{\mathrm{T}} P_\varepsilon A_c x \\ & \leq -\varepsilon x^{\mathrm{T}} x. \end{aligned} \tag{4.37}$$

Now for the case of single input ($m = 1$), the evaluation of the difference of V_1 along the trajectories of (4.31) inside the set $L_{V_1}(c_1)$ gives,

$$V_1(\rho x) - V_1(x) = -\varepsilon x^{\mathrm{T}} x - x^{\mathrm{T}} F_\varepsilon^{\mathrm{T}} F_\varepsilon x + \phi_1(\gamma_1), \tag{4.38}$$

where $\phi_1(\gamma_1) = 2x^{\mathrm{T}} A_c^{\mathrm{T}} P_\varepsilon B \gamma_1 + \gamma_1 B^{\mathrm{T}} P_\varepsilon B \gamma_1$ and

$$\gamma_1 = \sigma\left[(F_\varepsilon - \mu(B^{\mathrm{T}} P_\varepsilon B)^{-1} B^{\mathrm{T}} P_\varepsilon A_c)x\right] - F_\varepsilon x.$$

If we denote $\zeta_1 = F_\varepsilon x$ and $\nu_1 = K_\varepsilon$ then $\phi_1(\gamma_1)$ can be written as

$$\begin{aligned}\phi_1(\gamma_1) &= -2\nu_1(B^{\mathrm{T}} P_\varepsilon B)[\sigma(\zeta_1 + \mu\nu_1) - \zeta_1] \\ &\qquad + [\sigma(\zeta_1 + \mu\nu_1) - \zeta_1](B^{\mathrm{T}} P_\varepsilon B)[\sigma(\zeta_1 + \mu\nu_1) - \zeta_1] \\ &= [\sigma(\zeta_1 + \mu\nu_1) - (\zeta_1 + 2\nu_1)](B^{\mathrm{T}} P_\varepsilon B)[\sigma(\zeta_1 + \mu\nu_1) - \zeta_1].\end{aligned}$$

Noting that $\mu \in [0, 2]$, the definition of σ, and $|\zeta_1| \leq 1$ for all $x \in L_{V_1}(c_1)$, we have

$$\begin{aligned}|\zeta_1 + \mu\nu_1| \leq 1 &\Longrightarrow \phi_1(\gamma_1) = -\mu(2-\mu)(B^{\mathrm{T}} P_\varepsilon B)\nu_1^2 \leq 0; \\ \zeta_1 + \mu\nu_1 > 1 &\Longrightarrow \nu_1 > 0,\ \sigma(\zeta_1 + \mu\nu_1) - (\zeta_1 + \mu\nu_1) < 0 \\ &\Longrightarrow \phi_1(\gamma_1) \leq -(2-\mu)(B^{\mathrm{T}} P_\varepsilon B)\nu_1[1 - \zeta_1] \leq 0;\end{aligned}$$

and

$$\begin{aligned}\zeta_1 + \mu\nu_1 < -1 &\Longrightarrow \nu_1 < 0,\ \sigma(\zeta_1 + \mu\nu_1) - (\zeta_1 + \mu\nu_1) > 0 \\ &\Longrightarrow \phi_1(\gamma_1) \leq -(2-\mu)(B^{\mathrm{T}} P_\varepsilon B)\nu_1[-1 - \zeta_1] \leq 0.\end{aligned}$$

We conclude that for all $x \in L_{V_1}(c_1)$, $\phi_1(\gamma_1) \leq 0$ and hence

$$V_1(x(k+1)) - V_1(x(k)) \leq -\varepsilon x(k)^{\mathrm{T}} x(k).$$

At this time, it is appropriate to make a remark. It is clear from the above derivation, the choice of $\kappa(x, w, \mu)$ as in (4.30) prevents the control input from saturating the actuators while increasing the utilization of their capacities. While the avoidance of actuator saturation is essential in establishing (4.37) due to multi-input coupling, the choice of $\kappa(x, w, \mu) \equiv 1$ for single input case allows the control input to saturate the actuators and thus further increases the utilization of their capacities.

So far, we have shown that for both the multiple input and single input cases,

$$V_1(x(k+1)) - V_1(x(k)) \leq -\varepsilon x(k)^{\mathrm{T}} x(k), \quad \forall x(k) \in L_{V_1}(c_1). \tag{4.39}$$

This implies that the closed-loop system (4.31) is locally exponentially stable with $\mathcal{X}_0$ contained in its basin of attraction. We note here that the choice of μ determines the decay rate of $V_1(x(k+1)) - V_1(x(k))$ and hence the freedom in choosing μ can be utilized to ensure fast convergence.

Next, we show that there exists an $\varepsilon_2^* \in (0, 1]$ such that for each $\varepsilon \in (0, \varepsilon_2^*]$, item (ii) of the theorem holds.

To this end, let us introduce an invertible coordinate change $\xi = x - \Pi w$. Using the condition (ii)a) (see Theorem 4.3.1), we have

$$\begin{aligned} \rho\xi &= \rho x - \rho\Pi w \\ &= Ax + B\sigma(u) + E_w w - \Pi S w \\ &= A\xi + B\left[\sigma(u) - \Gamma w\right]. \end{aligned} \tag{4.40}$$

With the family of state feedback laws given by (4.29), the closed-loop system can be written as

$$\begin{aligned} \rho\xi &= A\xi + B\left[\sigma(\Gamma w + (F_\varepsilon + \mu\kappa(\xi + \Pi w, w, \mu)K_\varepsilon)\xi) - \Gamma w\right] \\ &= A_c\xi + B\left[\sigma(\Gamma w + (F_\varepsilon + \mu\kappa(\xi + \Pi w, w, \mu)K_\varepsilon)\xi)\right. \\ &\qquad \left. -\Gamma w - F_\varepsilon\xi\right]. \end{aligned} \tag{4.41}$$

By Condition (ii)b) (see Theorem 4.3.1), $\|\Gamma w\|_{\infty,K} < 1 - \delta$. Moreover, for any $x(0) \in \mathcal{X}_0$ and any $w(0) \in \mathcal{W}_0$, $\xi(K)$ belongs to a bounded set, say $\mathcal{U}_K$, independent of ε since $\mathcal{X}_0$ and $\mathcal{W}_0$ are both bounded and $\xi(K)$ is determined by a linear difference equation with bounded inputs $\sigma(\cdot)$ and Γw.

We then pick a Lyapunov function

$$V_2(\xi) = \xi^{\mathrm{T}} P_\varepsilon \xi, \tag{4.42}$$

and let $c_2 > 0$ be such that

$$c_2 \geq \sup_{\xi \in \mathcal{U}_K, \varepsilon \in (0,1]} \xi^{\mathrm{T}} P_\varepsilon \xi. \tag{4.43}$$

Such a c_2 exists since P_ε and $\mathcal{U}_K$ are bounded. Let $\varepsilon_2^* \in (0, 1]$ be such that $\xi \in L_{V_2}(c_2)$ implies that $\|F_\varepsilon \xi\| \leq \delta$ where

$$L_{V_2}(c_2) = \left\{\xi \in \mathbb{R}^n \mid V_2(\xi) < c_2 \right\}.$$

The existence of such an ε_2^* is again due to (4.11) and the fact that $P_\varepsilon \to 0$ as $\varepsilon \to 0$.

For the case of multiple input ($m > 1$), by the definition of function κ, (4.30), for all $x \in L_{V_1}(c_1)$ and $k \geq K$, the saturation functions in the closed-loop system (4.41) operate in their respective linear regions, and hence the closed-loop system remains linear and hence reduces to

$$\rho\xi = A_c\xi + \mu\kappa(\xi + \Pi w, w, \mu)B(B^{\mathrm{T}}P_\varepsilon B)^{-1}B^{\mathrm{T}}P_\varepsilon A_c\xi.$$

The evaluation of the difference of V_2 along the trajectories of this linear closed-loop system shows that, for $\xi \in L_{V_2}(c_2)$,

$$\begin{aligned}
&V_2(\rho\xi) - V_2(\xi) \\
&= -\varepsilon\xi^{\mathrm{T}}\xi - \xi^{\mathrm{T}}F_\varepsilon^{\mathrm{T}}F_\varepsilon\xi - \mu\kappa(\xi + \Pi w, w, \mu)[2 - \mu\kappa(\xi + \Pi w, w, \mu)]\xi^{\mathrm{T}} \\
&\qquad \times A_c^{\mathrm{T}}P_\varepsilon B(B^{\mathrm{T}}P_\varepsilon B)^{-1}B^{\mathrm{T}}P_\varepsilon A_c\xi \\
&\leq -\varepsilon\xi^{\mathrm{T}}\xi.
\end{aligned}$$

Now for the case of single input ($m = 1$), the evaluation of the difference of V for $k \geq K$, inside the set $L_{V_2}(c_2)$, using (4.36), shows that for all $\xi \in L_{V_2}(c_2)$,

$$V_2(\rho\xi) - V_2(\xi) = -\varepsilon\xi^{\mathrm{T}}\xi - \xi^{\mathrm{T}}F_\varepsilon^{\mathrm{T}}F_\varepsilon\xi + \phi_2(\gamma_2)$$

where $\phi_2(\gamma_2) = 2\xi^{\mathrm{T}}A_c^{\mathrm{T}}P_\varepsilon B\gamma_2 + \gamma_2 B^{\mathrm{T}}P_\varepsilon B\gamma_2$ and

$$\gamma_2 = \sigma(\Gamma w + (F_\varepsilon - \mu(B^{\mathrm{T}}P_\varepsilon B)^{-1}B^{\mathrm{T}}P_\varepsilon A_c)\xi) - \Gamma w - F_\varepsilon\xi.$$

Denoting $\theta_2 = \Gamma w$, $\zeta_2 = F_\varepsilon\xi$ and $\nu_2 = K_\varepsilon\xi$, $\phi_2(\gamma_2)$ can be written as

$$\begin{aligned}
\phi_2(\gamma_2) &= -2\nu_2(B^{\mathrm{T}}P_\varepsilon B)[\sigma(\theta_2 + \zeta_2 + \mu\nu_2) - \theta_2 - \zeta_2] \\
&\qquad + [\sigma(\theta_2 + \zeta_2 + \mu\nu_2) - \theta_2 - \zeta_2](B^{\mathrm{T}}P_\varepsilon B) \\
&\qquad\qquad \times [\sigma(\theta_2 + \zeta_2 + \mu\nu_2) - \theta_2 - \zeta_2] \\
&= [\sigma(\theta_2 + \zeta_2 + \mu\nu_2) - (\theta_2 + \zeta_2 + 2\nu_2)](B^{\mathrm{T}}P_\varepsilon B) \\
&\qquad\qquad \times [\sigma(\theta_2 + \zeta_2 + \mu\nu_2) - \theta_2 - \zeta_2].
\end{aligned}$$

Noting that $\mu \in [0, 2]$, the definition of σ, and $|\theta_2 + \zeta_2| \leq 1$ for all $\xi \in L_{V_2}(c_2)$, we have

$$\begin{aligned}
&|\theta_2 + \zeta_2 + \mu\nu_2| \leq 1 \\
&\qquad \Longrightarrow \phi_2(\gamma_2) = -\mu(2 - \mu)(B^{\mathrm{T}}P_\varepsilon B)\nu_2^2 \leq 0, \\
&\theta_2 + \zeta_2 + \mu\nu_2 > 1 \\
&\qquad \Longrightarrow \nu_2 > 0,\ \sigma(\theta_2 + \zeta_2 + \mu\nu_2) - (\theta_2 + \zeta_2 + \mu\nu_2) < 0 \\
&\qquad \Longrightarrow \phi_2(\gamma_2) \leq -(2 - \mu)(B^{\mathrm{T}}P_\varepsilon B)\nu_2[1 - (\theta_2 + \zeta_2)] \leq 0
\end{aligned}$$

and

$$\begin{aligned}
&\theta_2 + \zeta_2 + \mu\nu_2 < -1 \\
&\qquad \Longrightarrow \nu_2 < 0,\ \sigma(\theta_2 + \zeta_2 + \mu\nu_2) - (\theta_2 + \zeta_2 + \mu\nu_2) > 0 \\
&\qquad \Longrightarrow \phi_2(\gamma_2) \leq -(2 - \mu)(B^{\mathrm{T}}P_\varepsilon B)\nu_2[-1 - (\theta_2 + \zeta_2)] \leq 0.
\end{aligned}$$

We conclude that for all $\xi \in L_{V_2}(c_2)$, $\phi_2(\gamma_2) \le 0$ and hence

$$V_2(\xi(k+1)) - V_2(\xi(k)) \le -\varepsilon\xi(k)^{\mathrm{T}}\xi(k).$$

This shows that any trajectory of the closed-loop system (4.41) starting from $\{\xi = x - \Pi w : x \in \mathcal{X}_0, w \in \mathcal{W}_0\}$ remains inside the set $L_{V_2}(c_2)$ and approaches the equilibrium point $\xi = 0$ as $k \to \infty$. Now, it can easily be seen that $e(k) \to 0$ as $k \to \infty$.

Finally, setting $\varepsilon^* = \min\{\varepsilon_1^*, \varepsilon_2^*\}$, we conclude our proof of Theorem 4.3.1. ∎

We now illustrate the use of the control law (4.29) by an example.

Example 4.3.2 *Consider the system (4.27) of Example 4.3.1 with $w(0) \in \mathcal{W}_0$ where $\mathcal{W}_0 = \{w \in \mathbb{R}^2 : \|w\| < 0.9\}$. It is straightforward to show that the solvability conditions for the semi-global linear state feedback output regulation problem are satisfied. That is, Π and Γ as in (4.28) solve the linear matrix equations (4.13), while $\delta = 0.1$, since $\|\Gamma w\|_\infty \le 0.9$ for all $w(0) \in \mathcal{W}_0$. Let the set $\mathcal{X}_0$ be given by $\mathcal{X}_0 = \{x \in \mathbb{R}^4 : \|x\| \le 10\}$.*

Then, following the proof of Theorem 4.3.2, a suitable choice of ε^* is 6.628×10^{-6}. For $\varepsilon = \varepsilon^*$ and $\mu = 1$, the feedback law (4.29) is given by

$$u(k) = -\begin{pmatrix}-0.3037 & -0.1523 & 0.2403\end{pmatrix} x(k) + \begin{pmatrix}0.9366 & -0.1523\end{pmatrix} w(k).$$

For the initial conditions $x(0) = (6, 6, -2)^T$, $w(0) = (0.45, -0.45)^T$, Figure 4.2 on the next page shows the control action and the closed-loop performance of the regulator for $\mu = 1$. Figure 4.3 on the following page shows the control action and the closed-loop performance of the regulator design in [45], i.e. for $\mu = 0$.

4.3.2 Linear observer based measurement feedback controller

In this section, we consider the classical semi-global linear observer based measurement feedback output regulation problem which can be formulated as follows.

Problem 4.3.2 ***(The classical semi-global linear observer based measurement feedback output regulation problem)*** *Consider the system (4.1) and a*

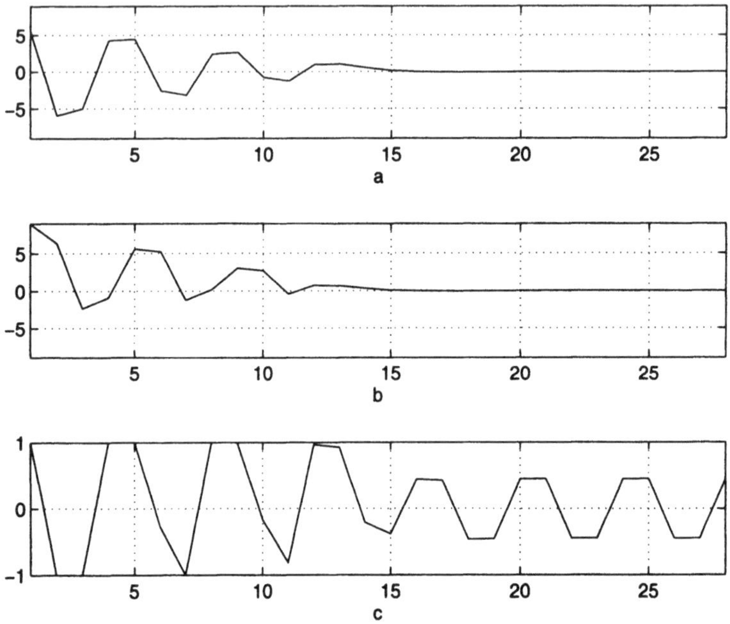

Figure 4.2: $\varepsilon = 6.628 \times 10^{-6}$, $\mu = 1$. a) e_1; b) e_2; c) $\sigma(u)$;

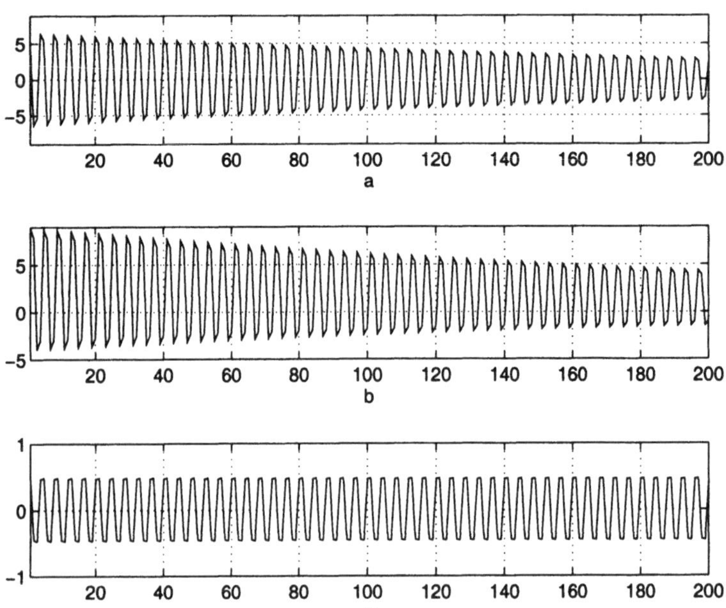

Figure 4.3: $\varepsilon = 6.628 \times 10^{-6}$, $\mu = 0$. a) e_1; b) e_2; c) $\sigma(u)$;

compact set $\mathcal{W}_0 \subset \mathbb{R}^s$. For any a priori given (arbitrarily large) bounded sets $\mathcal{X}_0 \subset \mathbb{R}^n$ and $\mathcal{Z}_0 \subset \mathbb{R}^{n+s}$, find, if possible, a measurement feedback law of the form

$$\begin{aligned} \rho \begin{pmatrix} \hat{x} \\ \hat{w} \end{pmatrix} &= \begin{pmatrix} A & E_w \\ 0 & S \end{pmatrix} \begin{pmatrix} \hat{x} \\ \hat{w} \end{pmatrix} + \begin{pmatrix} B \\ 0 \end{pmatrix} \sigma(u) \\ &\quad + \begin{pmatrix} K_{\mathrm{A}} \\ K_{\mathrm{S}} \end{pmatrix} \left(\begin{pmatrix} C_y & D_{yw} \end{pmatrix} \begin{pmatrix} \hat{x} \\ \hat{w} \end{pmatrix} - y \right) \\ u &= \alpha(\hat{x}, \hat{w}) \end{aligned} \tag{4.44}$$

such that the following conditions hold:

(*i*) ***(Internal Stability)*** *The equilibrium point $(x, \hat{x}, \hat{w}) = (0, 0, 0)$ of*

$$\begin{aligned} \rho x &= Ax + B\sigma(\alpha(\hat{x}, \hat{w})) \\ \rho \begin{pmatrix} \hat{x} \\ \hat{w} \end{pmatrix} &= \begin{pmatrix} A & E_w \\ 0 & S \end{pmatrix} \begin{pmatrix} \hat{x} \\ \hat{w} \end{pmatrix} + \begin{pmatrix} B \\ 0 \end{pmatrix} \sigma(\alpha(\hat{x}, \hat{w})) \\ &\quad + \begin{pmatrix} K_{\mathrm{A}} \\ K_{\mathrm{S}} \end{pmatrix} \begin{pmatrix} C_y & D_{yw} \end{pmatrix} \begin{pmatrix} \hat{x} - x \\ \hat{w} \end{pmatrix} \end{aligned} \tag{4.45}$$

is asymptotically stable with $\mathcal{X}_0 \times \mathcal{Z}_0$ contained in its basin of attraction.

(*ii*) ***(Output Regulation)*** *For all $(x(0), \hat{x}(0), \hat{w}(0)) \in \mathcal{X}_0 \times \mathcal{Z}_0$ and $w(0) \in \mathcal{W}_0$, the solution of the closed-loop system satisfies*

$$\lim_{k \to \infty} e(k) = 0. \tag{4.46}$$

Remark 4.3.3 *We would like to emphasize that our definition of the classical semi-global linear observer based measurement feedback output regulation problem does not view the set of initial conditions of the plant and the initial conditions of the controller dynamics as given data. The set of given data consists of the models of the plant and the exosystem and the set of initial conditions for the exosystem. Therefore, the solvability conditions* must *be independent of the set of initial conditions of the plant, $\mathcal{X}_0$, and the set of initial conditions for the controller dynamics, $\mathcal{Z}_0$.*

The solvability conditions for the classical semi-global linear observer based measurement feedback output regulation problem are given in the following theorem.

Theorem 4.3.3 *Consider the system (4.1) and the given compact set $\mathcal{W}_0 \subset \mathbb{R}^s$. The classical semi-global linear observer based measurement feedback output regulation problem is solvable if the following conditions hold:*

(i) (A, B) *is stabilizable and* A *has all its eigenvalues inside or on the unit circle. Moreover, the pair*

$$\left(\begin{pmatrix} C_y & D_{yw} \end{pmatrix}, \begin{pmatrix} A & E_w \\ 0 & S \end{pmatrix}\right)$$

is detectable.

(ii) There exist matrices Π *and* Γ *such that,*

(a) they solve the regulator equation (4.13),

(b) there exists a $\delta > 0$ *and a* $K \geq 0$ *such that* $\|\Gamma w\|_{\infty,K} \leqslant 1 - \delta$ *for all* w *with* $w(0) \in \mathcal{W}_0$.

Moreover the function $\alpha(\hat{x}(k), \hat{w}(k))$ *in Problem 4.3.2 can be a linear function of* $\hat{x}(k)$ *and* $\hat{w}(k)$.

Proof : We prove this theorem by first explicitly constructing a family of linear observer based measurement feedback laws of the form (4.44), parameterized in ε, and then showing that for each pair of sets $\mathcal{X}_0 \subset \mathbb{R}^n$ and $\mathcal{Z}_0 \subset \mathbb{R}^{n+s}$, there exists an $\varepsilon^* > 0$ such that for all $\varepsilon \in (0, \varepsilon^*]$, both items (i) and (ii) in Problem 4.3.2 are indeed satisfied. The family of linear observer based measurement feedback laws we construct take the form,

$$\begin{aligned} \rho\hat{x} &= A\hat{x} + B\sigma(u) + E_w\hat{w} - K_{\mathrm{A}}y + K_{\mathrm{A}}(C_y\hat{x} + D_{yw}\hat{w}) \\ \rho\hat{w} &= S\hat{w} - K_{\mathrm{S}}y + K_{\mathrm{S}}(C_y\hat{x} + D_{yw}\hat{w}) \\ u &= F_\varepsilon\hat{x} + (\Gamma - F_\varepsilon\Pi)\hat{w}, \end{aligned} \tag{4.47}$$

where $F_\varepsilon := -(B^{\mathrm{T}}P_\varepsilon B + I)^{-1}B^{\mathrm{T}}P_\varepsilon A$ with P_ε being the solution of the Riccati equation (4.9). It follows from Lemma 4.3.1 that $A + BF_\varepsilon$ is Schur-stable for all $\varepsilon > 0$. The matrices K_{A} and K_{S} are chosen such that the following matrix is Schur-stable,

$$\bar{A} := \begin{pmatrix} A + K_{\mathrm{A}}C_y & E_w + K_{\mathrm{A}}D_{yw} \\ K_{\mathrm{S}}C_y & S + K_{\mathrm{S}}D_{yw} \end{pmatrix}. \tag{4.48}$$

With the above family of feedback laws, the closed-loop system consisting of the system (4.1) and the dynamic measurement feedback law (4.47) can be written as (we do not write the equation $w(k+1) = Sw(k)$ explicitly but it is of course always there),

$$\begin{aligned} \rho x &= Ax + B\sigma(\Gamma\hat{w} + F_\varepsilon(\hat{x} - \Pi\hat{w})) + E_w w \\ \rho\hat{x} &= A\hat{x} + B\sigma(\Gamma\hat{w} + F_\varepsilon(\hat{x} - \Pi\hat{w})) + E_w\hat{w} \\ &\qquad + K_{\mathrm{A}}C_y(\hat{x} - x) + K_{\mathrm{A}}D_{yw}(\hat{w} - w) \\ \rho\hat{w} &= S\hat{w} + K_{\mathrm{S}}C_y(\hat{x} - x) + K_{\mathrm{S}}D_{yw}(\hat{w} - w). \end{aligned} \tag{4.49}$$

We then adopt the invertible change of state variables,

$$\xi = x - \Pi w, \qquad \tilde{x} = x - \hat{x}, \qquad \tilde{w} = w - \hat{w}, \tag{4.50}$$

and rewrite the closed-loop system (4.49) as

$$\begin{aligned}
\rho\xi &= A\xi + B\sigma(F_\varepsilon\xi + \Gamma w - \Gamma\tilde{w} - F_\varepsilon\tilde{\xi}) \\
&\qquad +(A\Pi - \Pi S + E_w)w \\
\rho\tilde{x} &= (A + K_{\mathrm{A}}C_y)\tilde{x} + (E_w + K_{\mathrm{A}}D_{yw})\tilde{w} \\
\rho\tilde{w} &= K_{\mathrm{S}}C_y\tilde{x} + (S + K_{\mathrm{S}}D_{yw})\tilde{w}
\end{aligned} \tag{4.51}$$

where we have denoted $\tilde{\xi} = \tilde{x} - \Pi\tilde{w}$.

To show that item (i) of Problem 4.3.2 holds, we note that (4.45) is the same as (4.49) for $w(k) = 0$. We know (4.49) is equivalent to (4.51) which for $w(k) = 0$ reduces to

$$\begin{aligned}
\rho\xi &= A\xi + B\sigma(F_\varepsilon\xi - \Gamma\tilde{w} - F_\varepsilon\tilde{\xi}) \\
\rho\tilde{x} &= (A + K_{\mathrm{A}}C_y)\tilde{x} + (E_w + K_{\mathrm{A}}D_{yw})\tilde{w} \\
\rho\tilde{w} &= K_{\mathrm{S}}C_y\tilde{x} + (S + K_{\mathrm{S}}D_{yw})\tilde{w}.
\end{aligned} \tag{4.52}$$

Denoting $\tilde{m} = [\tilde{x}^{\mathrm{T}}, \tilde{w}^{\mathrm{T}}]^{\mathrm{T}}$, we write (4.52) in the following compact form,

$$\begin{aligned}
\rho\xi &= A\xi + B\left(\sigma\left[F_\varepsilon\xi + M\tilde{m}\right]\right) \\
\rho\tilde{m} &= \bar{A}\tilde{m}.
\end{aligned} \tag{4.53}$$

Also, (4.51) can be written as,

$$\begin{aligned}
\rho\xi &= A\xi + B\left(\sigma\left[F_\varepsilon\xi + M\tilde{m} + \Gamma w\right] - \Gamma w\right) \\
\rho\tilde{m} &= \bar{A}\tilde{m},
\end{aligned} \tag{4.54}$$

where

$$M = \begin{pmatrix} -F_\varepsilon & 0 \\ 0 & (F_\varepsilon\Pi - \Gamma) \end{pmatrix}.$$

Recalling that the matrix $\bar{A}$ defined in (4.48) is Schur-stable, it readily follows from the second equation of (4.53) that there exists a $K_1 \geq 0$ such that, for all possible initial conditions $(\tilde{x}(0), \tilde{w}(0))$,

$$\|M\tilde{m}\|_{\infty,\mathrm{K}_1} \leq \frac{1}{2}, \quad \text{for all } \varepsilon \in (0, 1]. \tag{4.55}$$

For any $x(0) \in \mathcal{X}_0$, $\xi(K_1)$ belongs to a bounded set, say $\mathcal{U}_{K_1}$, independent of ε since $\mathcal{X}_0$ is bounded and $\xi(K_1)$ is determined by a linear difference equation

(4.53) with bounded input $\sigma(\cdot)$. Let $\varepsilon_1^* \in (0, 1]$ be chosen such that for all $\varepsilon \in (0, \varepsilon_1^*]$, $\|F_\varepsilon\| < 1/2$, and $\|F_\varepsilon \Pi\| < 1/2$. This ensures that $\|M\|^2 < (1 + \|\Gamma\|)^2$. Let us define $\beta := (1 + \|\Gamma\|)^2(1 + \|B\|^2)$ for later use. Let $\tilde{P}$ be the unique positive definite solution to the Lyapunov equation

$$\tilde{P} = \bar{A}^{\mathrm{T}} \tilde{P} \bar{A} + I. \tag{4.56}$$

Such a $\tilde{P}$ exists since $\bar{A}$ is Schur-stable.

We next define a Lyapunov function

$$V(\xi, \tilde{m}) = \xi^{\mathrm{T}} P_\varepsilon \xi + (\beta + 1)\tilde{m}^{\mathrm{T}} \tilde{P} \tilde{m}, \tag{4.57}$$

and let $c_1 > 0$ be such that

$$c_1 \geq \sup_{\xi \in \mathcal{U}_{K_1}, \varepsilon \in (0,1]} V(\xi, \tilde{m}). \tag{4.58}$$

Such a c_1 exists since $\mathcal{U}_{K_1}$ is bounded and $\lim_{\varepsilon \to 0} P_\varepsilon = 0$. Let $\varepsilon_2^* \in (0, \varepsilon_1^*]$ be such that $\xi \in L_V(c_1)$ implies that $\|F_\varepsilon \xi\|_\infty \leq \frac{1}{2}$. The existence of such an ε_2^* is due to Lemma 4.3.1. Hence for $k \geq K_1$, (4.53) takes the form

$$\begin{aligned} \xi(k+1) &= (A + BF_\varepsilon)\xi(k) + BM\tilde{m}(k) \\ \tilde{m}(k+1) &= \bar{A}\tilde{m}(k). \end{aligned} \tag{4.59}$$

Now, the evaluation of the difference of V for $k \geq K_1$, inside the set $L_V(c_1)$, using (4.9), shows that for all $\xi \in L_V(c_1)$,

$$\begin{aligned} V(\rho\xi, \rho\tilde{m}) &- V(\xi, \tilde{m}) \\ &= [(A + BF_\varepsilon)\xi + BM\tilde{m}]^{\mathrm{T}} P_\varepsilon [(A + BF_\varepsilon)\xi + BM\tilde{m}] \\ &\qquad - \xi^{\mathrm{T}} P_\varepsilon \xi - (\beta + 1)\tilde{m}^{\mathrm{T}}\tilde{m} \\ &= -\varepsilon\xi^{\mathrm{T}}\xi - \xi^{\mathrm{T}} F_\varepsilon^{\mathrm{T}} F_\varepsilon \xi + 2\tilde{m}^{\mathrm{T}} M^{\mathrm{T}} F_\varepsilon \xi + \tilde{m}^{\mathrm{T}} M^{\mathrm{T}} B^{\mathrm{T}} B M \tilde{m} \\ &\qquad - (\beta + 1)\tilde{m}^{\mathrm{T}}\tilde{m} \\ &\leq -\varepsilon\|\xi\|^2 - \|F_\varepsilon \xi\|^2 + \|F_\varepsilon \xi\|^2 + \|M\|^2\|\tilde{m}\|^2 + \|B\|^2\|M\|^2\|\tilde{m}\|^2 \\ &\qquad - (\beta + 1)\|\tilde{m}\|^2 \\ &\leq -\varepsilon\|\xi\|^2 - \|\tilde{m}\|^2. \end{aligned}$$

Hence we conclude that, for any a priori given sets $\mathcal{X}_0$ and $\mathcal{Z}_0$, there exists an $\varepsilon_2^* \in (0, \varepsilon_1^*]$ such that for each $\varepsilon \in (0, \varepsilon_2^*]$, the equilibrium point (0,0,0) of (4.45) is asymptotically stable with $\mathcal{X}_0 \times \mathcal{Z}_0$ contained in its basin of attraction.

We now proceed to show that item (ii) of Problem 4.3.2 also holds. To this end, we consider the closed-loop system (4.51). Recalling that the matrix $\bar{A}$

is Schur-stable, and using (4.51), which is equivalent to the system (4.54), it readily follows from the last two equations of (4.51) that there exists a $K_2 \geq K$ such that, for all possible initial conditions $(\tilde{x}(0), \tilde{w}(0))$,

$$\|M\tilde{m}\|_{\infty,K_2} \leq \frac{\delta}{2}, \quad \text{for all } \varepsilon \in (0, 1]. \tag{4.60}$$

We then consider the first equation of (4.54). By Condition (ii)b), $\|\Gamma w\|_{\infty,K} < 1 - \delta$. Moreover, for any $x(0) \in \mathcal{X}_0$ and any $w(0) \in \mathcal{W}_0$, $\xi(K_2)$ belongs to a bounded set, say $\mathcal{U}_{K_2}$, independent of ε since $\mathcal{X}_0$ and $\mathcal{W}_0$ are both bounded and $\xi(K_2)$ is determined by a linear difference equation with bounded inputs $\sigma(u)$ and Γw. Then, using the same Lyapunov function as in (4.57), let $c_2 > 0$ be such that

$$c_2 \geq \sup_{\xi \in \mathcal{U}_{K_2}, \varepsilon \in (0,1]} V(\xi, \tilde{m}). \tag{4.61}$$

Such a c_2 exists since $\lim_{\varepsilon \to 0} P_\varepsilon = 0$, and $\mathcal{U}_{K_2}$ is bounded. Let $\varepsilon_3^* \in (0, 1]$ be such that $\xi \in L_V(c_2)$ implies that $\|F_\varepsilon \xi\|_\infty \leq \frac{\delta}{2}$. The existence of such an ε_3^* is again due to Lemma 4.3.1. Hence the system (4.54) operates in the linear region after time K_2. Using the same technique as before, it can then be shown that the system (4.45) is asymptotically stable. Now, it can easily be seen that $e(k) \to 0$ as $k \to \infty$.

Finally, taking $\varepsilon^* = \min\{\varepsilon_2^*, \varepsilon_3^*\}$, we complete our proof. ∎

Remark 4.3.4 *As in the state feedback case, in view of Yang's results ([93]) and the solvability conditions for the measurement feedback output regulation problem for linear systems in the absence of input saturation as given in Chapter 2, it is obvious to observe that conditions (i) and (ii)a) of Theorem 4.3.3 are necessary. The crucial condition for the solvability of the classical semi-global linear observer based measurement feedback output regulation problem with amplitude saturating actuators is Condition (ii)a), which is a sufficient condition. In Section 4.4 we will discuss the necessity of Condition (ii)a).*

Design of a low-gain measurement feedback regulator:

For clarity, we now summarize the construction of an observer based measurement feedback regulator.

Step 1 : At first construct a low-gain state feedback regulator as in (4.10).

Step 2 : Design a full order observer so that we can implement the controller with observer based architecture as given in (4.47). That is, find the matrix gains K_A and K_S in such a way that the matrix $\bar{A}$ given in (4.48) is Schur-stable.

Step 3 : Implement the observer based measurement feedback regulator as given in (4.47).

Step 4 : Given the sets $\mathcal{X}_0 \subset \mathbb{R}^n$, $\mathcal{W}_0 \subset \mathbb{R}^s$, and $\mathcal{Z}_0 \subset \mathbb{R}^n$, find an ε^* by the procedure given in the proof of Theorem 4.3.3.

Then for some $\varepsilon \in (0, \varepsilon^*]$ the observer based measurement feedback regulator as given in (4.47) solves the classical semi-global measurement feedback output regulation problem. □

The following example illustrates the design procedure.

Example 4.3.3 *We consider the same plant and the exosystem as in Example 4.3.1. However, this time, the states x and w are not available for feedback. This forces us to use measurement feedback regulators. Let $y = e$, and also let the sets $\mathcal{W}_0 = \{w \in \mathbb{R}^2 : \|w\| < 0.15\}$ and $\mathcal{X}_0 = \{x \in \mathbb{R}^4 : \|x\| < 0.15\}$. Let the set $\mathcal{Z}_0$, be given by $\mathcal{Z}_0 = \{z \in \mathbb{R}^6 : \|z\| \le 0.2\}$.*

For $\varepsilon = \varepsilon^*$, the feedback laws (4.47) are given by

$$\begin{aligned}
\hat{x}(k+1) &= A\hat{x}(k) + B\sigma(u(k)) + E_w\hat{w}(k) \\
&\quad + K_A C_e(\hat{x}(k) - x(k)) + K_A D_{ew}(\hat{w}(k) - w(k)) \\
\hat{w}(k+1) &= S\hat{w}(k) \\
&\quad + K_S C_e(\hat{x}(k) - x(k)) + K_S D_{ew}(\hat{w}(k) - w(k)) \\
u(k) &= 100\begin{pmatrix} 3.53 & -2.22 & -3.68 & -2.24 \\ 2.22 & -3.53 & 2.24 & 3.68 \end{pmatrix}\hat{x}(k) \\
&\quad + \begin{pmatrix} 0.9986 & -0.0446 \\ 0.0446 & 1.0014 \end{pmatrix}\hat{w}(k).
\end{aligned} \tag{4.62}$$

Following the design procedure given above, and the proof of Theorem 4.3.3, a suitable choice of ε^* is 1.9325×10^{-3}. It can be verified that the matrix $\bar{A}$ as defined in (4.48) is asymptotically stable, if we choose,

$$K_A = \begin{pmatrix} 1.2489 \times 10^{-1} & 1.6643 \times 10^{-2} \\ -9.0244 \times 10^{-1} & -8.7426 \times 10^{-1} \\ -7.5000 \times 10^{-1} & 0.0000 \times 10^{0} \\ -7.7352 \times 10^{-1} & 6.3254 \times 10^{-4} \end{pmatrix} \quad \textit{and}$$

$$K_S = \begin{pmatrix} -3.7468 \times 10^{-1} & -4.9928 \times 10^{-2} \\ -3.8676 \times 10^{-1} & -3.7468 \times 10^{-1} \end{pmatrix}.$$

For the initial conditions $x(0) = (0.075, 0.075, 0.075, 0.075)^T$, $w(0) = (0.1, -0.1)^T$, $\hat{x}(0) = (0, 0, 0, 0)^T$, $\hat{w}(0) = (0, 0)^T$, *Figure 4.4 shows the control action and the closed-loop performance for the dynamic measurement feedback regulator.*

As seen from the above example, and as in the case of state feedback, low-gain based designs under utilize the available control capacity. Our goal next is to recall a new design methodology which incorporates significant improvement to the low-gain design method, and leads to a better utilization of the available control capacity and hence better closed-loop performance.

An improved measurement feedback regulator design:

This section provides an improved design for the measurement feedback regulators. The strategy taken in the new design is to implement the state feedback regulators as constructed in (4.29) with the state of a fast linear observer. The observer is chosen to be of a deadbeat type. Arbitrary fast observers can also be used, however, the use of a deadbeat type observer simplifies our proof drastically since the states of the system will be exactly the same as those of the observer in a finite time. More specifically, the new linear observer based measurement feedback regulator takes the form,

$$\begin{aligned} \rho\hat{x} &= A\hat{x} + B\sigma(u) + E_w\hat{w} - K_A y + K_A(C_y\hat{x} + D_{yw}\hat{w}) \\ \rho\hat{w} &= S\hat{w} - K_S y + K_S(C_y\hat{x} + D_{yw}\hat{w}) \\ u &= (F_\varepsilon + \mu\kappa(\hat{x}, \hat{w}, \mu)K_\varepsilon)(\hat{x} - \Pi\hat{w}) - \Gamma\hat{w} \end{aligned} \tag{4.63}$$

where

$$F_\varepsilon = -(B^T P_\varepsilon B + I)^{-1} B^T P_\varepsilon A,$$
$$K_\varepsilon = -(B^T P_\varepsilon B)^{-1} B^T P_\varepsilon A$$

with P_ε being the solution of the Riccati equation (4.9), and the function κ is as defined by (4.30). The matrices K_A and K_S are chosen such that all eigenvalues of the following matrix

$$\bar{A} = \begin{pmatrix} A + K_A C_y & E_w + K_A D_{yw} \\ K_S C_y & S + K_S D_{yw} \end{pmatrix} \tag{4.64}$$

are at the origin. Here we of course assumed that the pair

$$\left(\begin{pmatrix} C_y & D_{yw} \end{pmatrix}, \begin{pmatrix} A & E_w \\ 0 & S \end{pmatrix} \right) \tag{4.65}$$

is observable.

We have the following results.

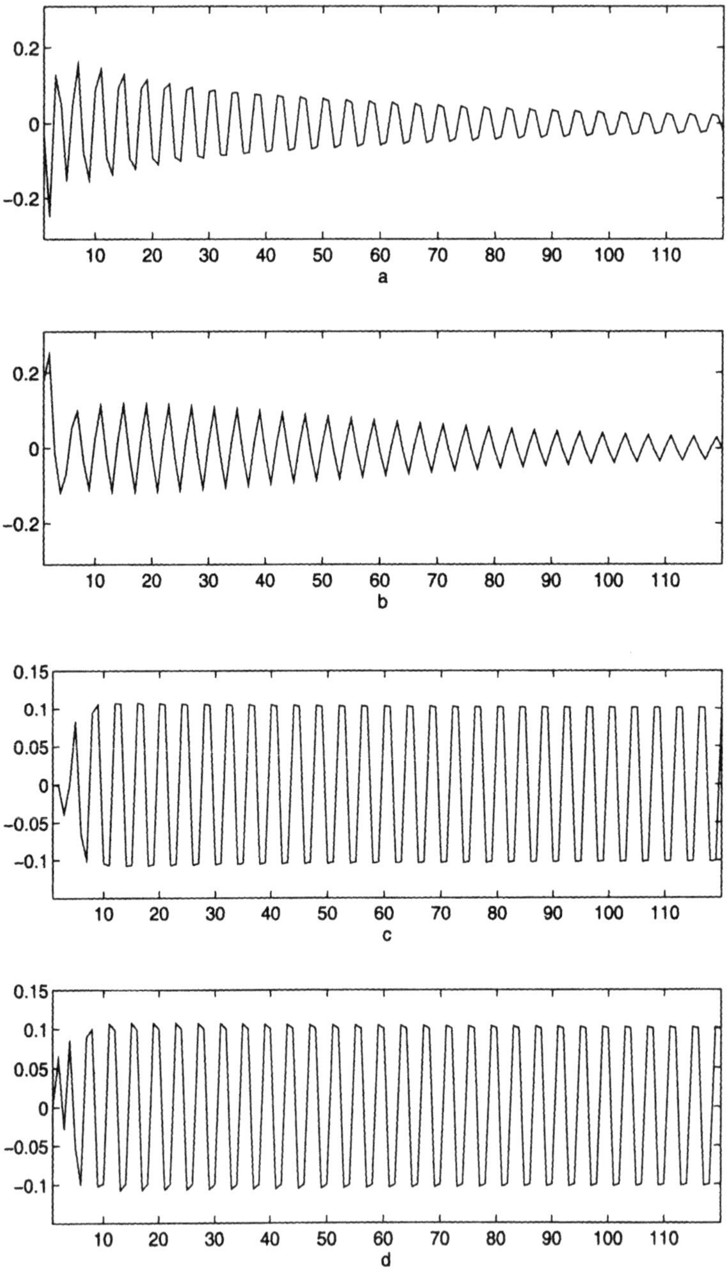

Figure 4.4: $\varepsilon = 1.9325 \times 10^{-3}$. a) e_1; b) e_2; c) u_1; d) u_2.

Theorem 4.3.4 *Consider the system (4.1) and the given compact set $\mathcal{W}_0 \subset \mathbb{R}^s$. Assume the sufficient conditions of Theorem 4.3.3 are satisfied. In addition, assume that the pair (4.65) is observable and that B is injective. Then, there exists a controller, among the family of measurement feedback laws (4.63), that solves the semi-global linear observer based measurement feedback output regulation problem. More specifically, for any a priori given (arbitrarily large) set $\mathcal{X}_0 \subset \mathbb{R}^n$ and $\mathcal{Z}_0 \subset \mathbb{R}^{n+s}$, there exists an $\varepsilon^* \in (0, 1]$, such that for each $\mu \in [0, 2]$, the controller in the family (4.63) has the following properties:*

(*i*) *The equilibrium point $(x, \hat{x}, \hat{w}) = (0, 0, 0)$ of*

$$\begin{aligned} \rho x &= Ax + B\sigma(u) \\ \rho\begin{pmatrix}\hat{x}\\ \hat{w}\end{pmatrix} &= \begin{pmatrix}A & E_w\\ 0 & S\end{pmatrix}\begin{pmatrix}\hat{x}\\ \hat{w}\end{pmatrix} + \begin{pmatrix}B\\ 0\end{pmatrix}\sigma(u) \\ &\quad + \begin{pmatrix}K_{\mathrm{A}}\\ K_{\mathrm{S}}\end{pmatrix}\begin{pmatrix}C_y & D_{yw}\end{pmatrix}\begin{pmatrix}\hat{x}-x\\ \hat{w}\end{pmatrix} \\ u &= F_{lh}\hat{x} - (F_{lh}\Pi - \Gamma)\hat{w} \end{aligned} \tag{4.66}$$

where $F_{lh} = F_\varepsilon + \mu\kappa(\hat{x}, \hat{w}, \mu)K_\varepsilon$, is asymptotically stable with $\mathcal{X}_0 \times \mathcal{Z}_0$ contained in its basin of attraction.

(*ii*) *For any $(x(0), \hat{x}(0), \hat{w}(0)) \in \mathcal{X}_0 \times \mathcal{Z}_0$ and $w(0) \in \mathcal{W}_0$, the solution of the closed-loop system satisfies*

$$\lim_{k\to\infty} e(k) = 0. \tag{4.67}$$

Proof : With the family of feedback laws as given by (4.63), the closed-loop system consisting of the system (4.1) and the dynamic measurement feedback laws (4.63) can be written as (we do not write the equation $w(k+1) = Sw(k)$ explicitly but it is of course always there),

$$\begin{aligned} \rho x &= Ax + B\sigma(u) + E_w w \\ \rho\hat{x} &= A\hat{x} + B\sigma(u) + E_w\hat{w} + K_{\mathrm{A}}C_y(\hat{x} - x) \\ &\qquad + K_{\mathrm{A}}D_{yw}(\hat{w} - w) \\ \rho\hat{w} &= S\hat{w} + K_{\mathrm{S}}C_y(\hat{x} - x) + K_{\mathrm{S}}D_{yw}(\hat{w} - w) \\ u &= (F_\varepsilon + \mu\kappa(\hat{x}, \hat{w}, \mu)K_\varepsilon)\hat{x} \\ &\qquad -((F_\varepsilon + \mu\kappa(\hat{x}, \hat{w}, \mu)K_\varepsilon)\Pi - \Gamma)\hat{w}. \end{aligned} \tag{4.68}$$

We then adopt the invertible change of state variables,

$$\tilde{x} = x - \hat{x}, \qquad \tilde{w} = w - \hat{w}, \tag{4.69}$$

and rewrite the closed-loop system (4.68) as

$$\begin{aligned}
\rho x &= Ax + E_w w + B\sigma\big[\Gamma\hat{w} + (F_\varepsilon \\
&\qquad + \mu\kappa(x - \tilde{x}, w - \tilde{w}, \mu)K_\varepsilon(\hat{x} - \Pi\hat{w})\big] \\
\rho\tilde{x} &= (A + K_{\mathrm{A}}C_y)\tilde{x} + (E_w + K_{\mathrm{A}}D_{yw})\tilde{w} \\
\rho\tilde{w} &= K_{\mathrm{S}}C_y\tilde{x} + (S + K_{\mathrm{S}}D_{yw})\tilde{w}.
\end{aligned} \tag{4.70}$$

Since all eigenvalues of $\bar{A}$ are at the origin, it is easy to verify that for time $k \geq n + s$, $\tilde{x} \equiv 0$ and $\tilde{w} \equiv 0$. As a result, for $k \geq n + s$, $\hat{x} \equiv x$ and $\hat{w} \equiv w$. On the other hand, for any $x(0) \in \mathcal{X}_0$, $(\hat{x}(0), \hat{w}(0)) \in \mathcal{Z}_0$ and $w(0) \in \mathcal{W}_0$, $x(n + s)$ belongs to a bounded set. Hence, the rest of the proof becomes the same as the proof of Theorem 4.3.1. ■

Example 4.3.4 *We consider the same plant and the exosystem as in Example 4.3.2. However, this time, the states x and w are not available for feedback. This forces us to use measurement feedback regulators. Let $y = e$, and also let the sets $\mathcal{W}_0 = \{w \in \mathbb{R}^2 : \|w\| < 0.9\}$ and $\mathcal{X}_0 = \{x \in \mathbb{R}^4 : \|x\| < 10\}$. Let the set $\mathcal{Z}_0$ be given by $\mathcal{Z}_0 = \{z \in \mathbb{R}^6 : \|z\| \leq 2\}$. Following the proof of Theorem 4.3.4, a suitable choice of ε^* is 6.628×10^{-6}. It can be verified that the matrix $\bar{A}$ as defined in (4.64) has all its eigenvalues located at the origin if we choose,*

$$K_{\mathrm{A}} = \begin{pmatrix} -0.3748 & 0.8750 \\ -1.3752 & -0.1250 \\ 0.6252 & 0.8750 \end{pmatrix}, \quad \textit{and} \quad K_{\mathrm{S}} = \begin{pmatrix} -0.25 & 0.25 \\ 0.3752 & 0.125 \end{pmatrix}.$$

For $\varepsilon = \varepsilon^*$ and $\mu = 1$, the feedback laws (4.63) are given by

$$\begin{aligned}
\rho\hat{x} &= A\hat{x} + B\sigma(u) + E_w\hat{w} \\
&\qquad + K_{\mathrm{A}}C_e(\hat{x} - x) + K_{\mathrm{A}}D_{ew}(\hat{w} - w) \\
\rho\hat{w} &= S\hat{w} + K_{\mathrm{S}}C_e(\hat{x} - x) + K_{\mathrm{S}}D_{ew}(\hat{w} - w) \\
u &= -\begin{pmatrix} -0.3037 & -0.1523 & 0.2403 \end{pmatrix}\hat{x} \\
&\qquad + \begin{pmatrix} 0.9366 & -0.1523 \end{pmatrix}\hat{w}.
\end{aligned} \tag{4.71}$$

For the initial conditions $x(0) = (-6, 2, 6)^T$, $w(0) = (0.45, -0.45)^T$, $\hat{x}(0) = (0, 0, 0, 0)^T$, $\hat{w}(0) = (0, 0)^T$, Figure 4.5 on the next page shows the control action and the closed-loop performance for the dynamic measurement feedback regulator for $\mu = 1$. For the purpose of comparison, Figure 4.6 on the facing page shows the same for the case of $\mu = 0$.

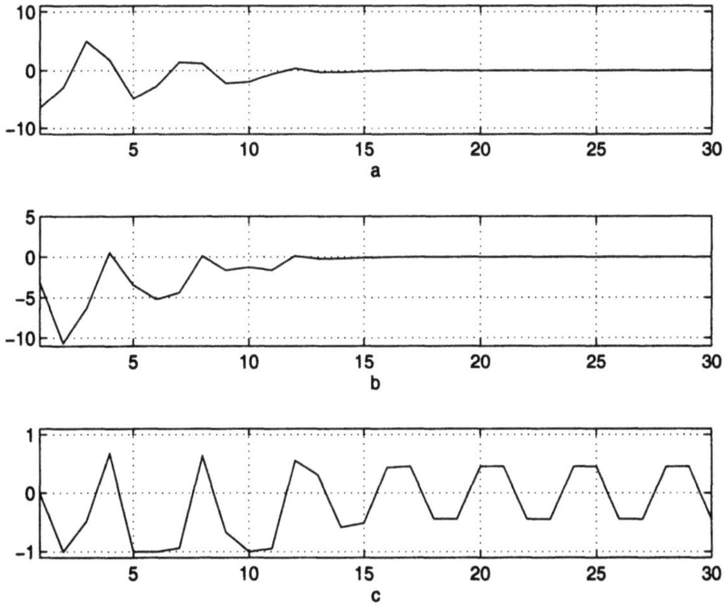

Figure 4.5: $\varepsilon = 6.628 \times 10^{-6}$, $\mu = 1$. a) e_1; b) e_2; c) $\sigma(u)$; .

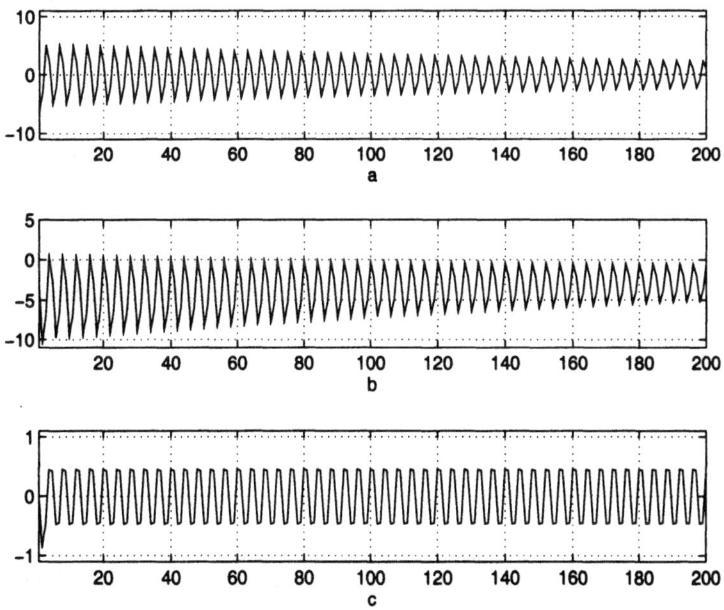

Figure 4.6: $\varepsilon = 6.628 \times 10^{-6}$, $\mu = 0$. a) e_1; b) e_2; c) $\sigma(u)$; .

4.4 Necessary conditions – linear versus non-linear regulator

The classical semi-global state feedback output regulation problem and the classical semi-global measurement feedback output regulation problem as defined in the previous section, can be solved using linear regulators under the sufficient conditions for their existence as given in Theorems 4.3.1 and 4.3.3. In this section we examine the necessity of these conditions. The necessity issue must be examined in two fronts.

The first issue is to examine the necessity of the solvability conditions given in Theorems 4.3.1 and 4.3.3 for the existence of linear regulators. The second issue is to examine whether we can weaken the solvability conditions if we consider non-linear regulators instead of linear regulators. It turns out that under certain mild conditions our solvability conditions for the existence of linear regulators are basically necessary. Moreover, these conditions cannot be weakened by allowing non-linear regulators. We also make an interesting observation that whenever these mild conditions are violated, there *might* be non-linear state feedback controllers that achieve output regulation while no linear state feedback controllers would do so.

The necessary condition for the existence of the classical semi-global state feedback regulator using a general non-linear feedback law is given in the following theorem.

Theorem 4.4.1 *Consider the plant and the exosystem (4.1). Let Assumptions A.1 and A.2 hold. Assume that in the absence of input amplitude saturation, the linear state feedback output regulation problem is solvable, i.e., there exist matrices Π and Γ which solve the regulator equation (4.13). Also assume that (A, B, C_e, D_{eu}) is left-invertible and has no invariant zeros on the unit circle. Then, a necessary condition for the existence of a general, possibly non-linear, state feedback law that achieves semi-global output regulation for (4.1) is that, for all $\varepsilon > 0$, there exists a $K \geqslant 0$ such that*

$$\|\Gamma w\|_{\infty,K} \leqslant 1+\varepsilon, \quad \text{for all } w(0) \in \mathcal{W}_0. \tag{4.72}$$

Proof : The proof of the above theorem depends on a result described in Appendix 3.A. We will assume in the proof that the reader is familiar with these results. Consider the system (4.1) without the saturation element, i.e. let

$$\begin{aligned} x(k+1) &= Ax(k) + Bv(k) + E_w w(k) \\ w(k+1) &= Sw(k) \\ e(k) &= C_e x(k) + D_{eu} v(k) + D_{ew} w(k). \end{aligned} \tag{4.73}$$

Suppose that we have an arbitrary non-linear feedback $u(k) = \alpha(x(k), w(k))$ which achieves output regulation for the system (4.1). Then the feedback $v(k) = \sigma(\alpha(x(k), w(k)))$ will achieve output regulation for the system (4.73).

We note that

$$v = Fx + (\Gamma - F\Pi)w$$

is a linear feedback which achieves output regulation for the system (4.73). Moreover,

$$v(k) - \Gamma w(k) \to 0 \quad \text{as } k \to \infty. \tag{4.74}$$

We have two feedback controllers which achieve output regulation for the linear system (4.73). One is non-linear and satisfies a certain amplitude constraint. The other feedback is a linear feedback of which we have no a priori knowledge regarding its amplitude. Our aim is to show that the linear feedback must necessarily satisfy an amplitude constraint asymptotically as $k \to \infty$. We define the difference between the two control inputs,

$$s = [Fx + (\Gamma - F\Pi)w] - [\sigma(\alpha(x, w))].$$

If we apply s to the system

$$\begin{aligned} x(k+1) &= Ax(k) + Bs(k) \\ e(k) &= C_e x(k) + D_{eu} s(k) \end{aligned} \tag{4.75}$$

with zero initial conditions, then we have $e(k) \to 0$ (since both the linear feedback and the non-linear saturating feedback controllers achieve output regulation). By applying Theorem 4.A.1 we find that $s(k) \to 0$ as $k \to \infty$.

If we look at our definition of $s(k)$, we see that the first component asymptotically converges to Γw. The second term is bounded by 1. Therefore it is easy to see that if $s(k)$ converges to 0 we must necessarily have

$$\limsup_{k\to\infty} \|\Gamma w(k)\|_\infty \leqslant 1.$$

■

Remark 4.4.1 *The necessary conditions given in Theorem 4.4.1 are slightly different from the sufficient solvability conditions given in Theorem 4.3.1. It is necessary that $\|\Gamma w(k)\|_\infty$ is asymptotically less than or equal to* 1 *while in our sufficient conditions we require that $\|\Gamma w(k)\|_\infty$ is asymptotically strictly less than* 1. *Hence, one can conclude that under Assumption A.1 and the condition that (A, B, C_e, D_{eu}) is left invertible and has no invariant zeros on the unit circle, a non-linear feedback regulator cannot do strictly better than a linear feedback regulator.*

An interesting question is whether one can weaken the necessary condition given in Theorem 4.4.1, if (A, B, C_e, D_{eu}) is not left invertible and/or has invariant zeros on the unit circle. The following example shows that, in fact, this is the case. More significantly, this example shows that if (A, B, C_e, D_{eu}) is not left invertible, non-linear feedback controllers might achieve semi-global output regulation while no linear feedback controller can do so.

Example 4.4.1 *Consider the system*

$$\begin{aligned} x(k+1) &= 0.5x(k) + \sigma(u(k)) \\ w(k+1) &= \begin{pmatrix} 0 & 0 & 0 \\ 0 & 0 & 0 \\ 0 & 0 & 0 \end{pmatrix} w(k) \\ e(k) &= \begin{pmatrix} 1 & -1 & 0 & 0 \\ 1 & 0 & -1 & 0 \\ 1 & 0 & 0 & -1 \end{pmatrix} x(k) - \tfrac{2}{7}\begin{pmatrix} 3 & 3 & 6 \\ 6 & 6 & 0 \\ 0 & 6 & 6 \end{pmatrix} w(k) \end{aligned} \tag{4.76}$$

with $w(0) \in \mathcal{W}_0$ where

$$\mathcal{W}_0 = \left\{ w \in \mathbb{R}^3 \mid \|w\|_\infty < 1 \right\}.$$

It is straightforward to show that in the absence of input amplitude saturation the linear state feedback regulator is solvable. In fact, the matrices Π and Γ that solve the linear matrix equation (4.13) are given by,

$$\Gamma = \frac{1}{7}\begin{pmatrix} f_1 & f_2+6 & f_3 \\ f_1-3 & f_2+3 & f_3-6 \\ f_1-6 & f_2 & f_3 \\ f_1 & f_2 & f_3-6 \end{pmatrix}, \quad \text{and} \quad \Pi = 2\Gamma,$$

where f_1, f_2 and f_3 are any real numbers.

In the presence of input amplitude saturation, however, the sufficient conditions of Theorem 4.3.1 are not satisfied. More specifically, Condition (ii)b) of Theorem 4.3.1 cannot be satisfied for any choice of f_1, f_2 and f_3. Hence, the design procedure developed in the previous section cannot be applied to this example. It is also evident that the necessary condition (4.72) is not satisfied either. But, since (A, B, C_e, D_{eu}) in the given plant is not left invertible, Theorem 4.4.1 does not apply anyway. In what follows, we will establish the following two facts for the plant and the exosystem (4.76):

(i) *There exist non-linear feedback controllers that achieve semi-global output regulation. This implies that the necessary condition (4.72) given in Theorem 4.4.1 are not valid if (A, B, C_e, D_{eu}) is not left invertible.*

(ii) *There exist no linear state feedback controllers that can achieve semi-global output regulation. This establishes an important result. That is, if (A, B, C_e, D_{eu}) is not left invertible, the semi-global output regulation might be achieved via non-linear feedback controllers while no linear feedback controllers can do so.*

As the plant is already asymptotically stable, let us consider a non-linear feedback of only the exosystem state of the form,

$$u = \begin{pmatrix} 1 \\ 1 \\ 1 \\ 1 \end{pmatrix} f(w) + \begin{pmatrix} 0 & 3 & 0 \\ -1.5 & 1.5 & -3 \\ -3 & 0 & 0 \\ 0 & 0 & -3 \end{pmatrix} w, \tag{4.77}$$

and

$$f(w) = \frac{6}{7}(1 - \lambda_1 - \lambda_2 - \lambda_3)$$

where $\lambda_1, \lambda_2, \lambda_3$ are such that $|\lambda_1| + |\lambda_2| + |\lambda_3| \leqslant 1$, and

$$w = \lambda_1 \begin{pmatrix} 1 \\ 1 \\ 1 \end{pmatrix} + \lambda_2 \begin{pmatrix} 1 \\ 1 \\ -1 \end{pmatrix} + \lambda_3 \begin{pmatrix} -1 \\ 1 \\ 1 \end{pmatrix} + (1 - \lambda_1 - \lambda_2 - \lambda_3) \begin{pmatrix} 1 \\ -1 \\ 1 \end{pmatrix}. \tag{4.78}$$

In this way the λ_i are not uniquely determined and we make the special choice of choosing those λ_i satisfying (4.78) with minimal λ_1. Note that our feedback is non-linear but continuous. Also, we know that for any $w(0) \in \mathcal{W}_0$, $w \equiv w(0)$. It is then not hard to check that the above non-linear feedback law achieves output regulation.

We still want to show that there does not exist a linear feedback controller that achieves output regulation. Assume that the linear feedback controller $u = Fw$ *achieves output regulation. Define*

$$v = \begin{pmatrix} 1 & -1 & 0 & 0 \\ 1 & 0 & -1 & 0 \\ 1 & 0 & 0 & -1 \end{pmatrix} \sigma(Fw). \tag{4.79}$$

It is easy to see that this feedback controller achieves output regulation for the system

$$\begin{aligned} \rho x &= 0.5x + \begin{pmatrix} 0 & 0 & 0 \\ -1 & 0 & 0 \\ 0 & -1 & 0 \\ 0 & 0 & -1 \end{pmatrix} v \\ \rho w &= \begin{pmatrix} 0 & 0 & 0 \\ 0 & 0 & 0 \\ 0 & 0 & 0 \end{pmatrix} w \\ e &= \begin{pmatrix} 1 & -1 & 0 & 0 \\ 1 & 0 & -1 & 0 \\ 1 & 0 & 0 & -1 \end{pmatrix} x - \tfrac{2}{7} \begin{pmatrix} 3 & 3 & 6 \\ 6 & 6 & 0 \\ 0 & 6 & 6 \end{pmatrix} w \end{aligned} \tag{4.80}$$

with $w(0) \in \mathcal{W}_0$. *However, the following linear feedback controller also achieves output regulation for the system (4.80),*

$$v = \frac{2}{7} \begin{pmatrix} 3 & 3 & 6 \\ 6 & 6 & 0 \\ 0 & 6 & 6 \end{pmatrix} w. \tag{4.81}$$

Since the system from v *to* e *is left-invertible, we know from Theorem 4.A.1 that the asymptotic behavior of signals achieving the output regulation is unique. We have two feedback controllers (4.79) and (4.81) that achieve output regulation. So asymptotically we must have,*

$$\begin{pmatrix} 1 & -1 & 0 & 0 \\ 1 & 0 & -1 & 0 \\ 1 & 0 & 0 & -1 \end{pmatrix} \sigma(Fw) = \frac{2}{7} \begin{pmatrix} 3 & 3 & 6 \\ 6 & 6 & 0 \\ 0 & 6 & 6 \end{pmatrix} w.$$

Using that w *is not dependent on time, and using a simple transformation, we note that the existence of a linear feedback controller achieving regulation requires the existence of a gain* F *satisfying*

$$\left| \begin{pmatrix} 2 \\ 2 \\ 2 \\ 2 \end{pmatrix} Fw + \begin{pmatrix} 0 & 6 & 0 \\ -3 & 3 & -6 \\ -6 & 0 & 0 \\ 0 & 0 & -6 \end{pmatrix} w \right| \leqslant 7.$$

However, it is easily verified (and was explicitly worked out in [72]) that this is not possible for this particular example.

Hence we can note in conclusion that for this example there does exist a suitable non-linear feedback controller but not a suitable linear feedback controller.

Remark 4.4.2 *We can pose similar questions regarding the classical semi-global measurement feedback output regulation problem. Again, there is the question whether the conditions of Theorem 4.3.3 are actually necessary for the solvability of the classical semi-global measurement feedback output regulation problem. But Theorem 4.4.1 basically resolves this question since the conditions which are necessary for the state feedback case are clearly also necessary for the case of measurement feedback on the basis of the error signal only. The only additional assumption we made in Theorem 4.3.3 is the detectability assumption which is clearly necessary for the stabilization of our system.*

4.5 Issues of well-posedness and structural stability

We would like to reconsider the problems of well-posedness and structural stability as introduced in Chapter 2 but this time for discrete-time linear systems subject to actuator saturation. In this case, there are no intrinsic differences between discrete- and continuous-time systems and therefore the results in this section are almost identical with the obvious changes to the discrete-time results of Section 3.5.

Definition 4.5.1 *(Well-posedness) For a system Σ as in (4.1), the classical semi-global linear observer based measurement feedback output regulation problem as defined in Problem 4.3.2 is said to be well-posed at (A_0, B_0, $E_{w,0}$, $C_{e,0}$, $D_{eu,0}$, $D_{ew,0}$, $C_{y,0}$, $D_{yu,0}$, $D_{yw,0}$, S_0) if there exists a neighborhood $\mathcal{P}_0$ of (A_0, B_0, $E_{w,0}$, $C_{e,0}$, $D_{eu,0}$, $D_{ew,0}$, $C_{y,0}$, $D_{yu,0}$, $D_{yw,0}$, S_0) in the parameter space $\mathcal{P}$ such that the considered problem is solvable for each element (A, B, E_w, C_e, D_{eu}, D_{ew}, C_y, D_{yu}, D_{yw}, S) of $\mathcal{P}_0$ for which A and S have all its eigenvalues inside or on the unit circle.*

We have the following result.

Theorem 4.5.1 *Consider a system Σ as in (4.1) and the classical semi-global linear observer based measurement feedback output regulation problem as* *defined in Problem 4.3.2. Let the conditions of Theorem 4.3.3 be satisfied for this system with nominal parameter values,*

$$\begin{aligned}(A, B, E_w, C_e, D_{eu}, D_{ew}, C_y, D_{yu}, D_{yw}, S) \\ = (A_0, B_0, E_{w,0}, C_{e,0}, D_{eu,0}, D_{ew,0}, C_{y,0}, D_{yu,0}, D_{yw,0}, S_0).\end{aligned}$$

Then the considered problem for Σ *is well-posed at* $(A_0, B_0, E_{w,0}, C_{e,0}, D_{eu,0}, D_{ew,0}, C_{y,0}, D_{yu,0}, D_{yw,0}, S_0)$ *if and only if the matrix* S *is weakly Schur-stable and the matrix*

$$\begin{pmatrix} A_0 - \lambda I & B_0 \\ C_{e,0} & D_{eu,0} \end{pmatrix}$$

has full row-rank for each λ *which is an eigenvalue of* S_0.

Remark 4.5.1 *In the above theorem, we did not perturb the set of initial conditions for the exosystem* $\mathcal{W}_0$ *but it is obvious that small perturbations of this set will not affect well-posedness.*

Next we consider the output regulation problem with structural stability. As already discussed in Section 3.5, we need to restrict our perturbations of the system parameters even more. As in the case of well-posedness, we need to guarantee that even after perturbation, A still has all its eigenvalues inside or on the unit circle but we also need to exclude perturbations of the exosystem, i.e. we do not perturb S. Finally we need that the error signal is part of the measurement signal y, i.e. the parameters need to satisfy (2.36).

Definition 4.5.2 ***(Structurally stable output regulation problem)*** *Consider a system* Σ *as in (4.1) with the additional structure given in (2.36). A fixed controller is said to solve a structurally stable output regulation problem for* Σ *at* $(A_0, B_0, E_{w,0}, C_{e,0}, D_{eu,0}, D_{ew,0}, C_{y2,0}, D_{yu2,0}, D_{yw2,0})$ *if it satisfies the following properties:*

(i) The controller solves the classical semi-global linear observer based measurement feedback output regulation problem when the plant in (2.1) is characterized by the nominal set of parameters $(A_0, B_0, E_{w,0}, C_{e,0}, D_{eu,0}, D_{ew,0}, C_{y2,0}, D_{yu2,0}, D_{yw2,0})$.

(ii) There exist a neighborhood $\mathcal{P}_0$ *of* $(A_0, B_0, E_{w,0}, C_{e,0}, D_{eu,0}, D_{ew,0}, C_{y2,0}, D_{yu2,0}, D_{yw2,0})$ *such that the controller achieves internal stability and output regulation for each set of perturbed plant parameters* $(A_0 +$

δA, $B_0 + \delta B$, $E_{w,0} + \delta E_w$, $C_{e,0} + \delta C_e$, $D_{eu,0} + \delta D_{eu}$, $D_{ew,0} + \delta D_{ew}$, $C_{y2,0}+\delta C_{y2}$, $D_{yu2,0}+\delta D_{yu2}$, $D_{yw2,0}+\delta D_{yw2}$) in $\mathcal{P}_0$ for which $A_0+\delta A$ has all its eigenvalues inside or on the unit circle.

In other words, as long as the perturbed parameters remain in $\mathcal{P}_0$, we have $\lim_{t\to\infty} e(t) = 0$ for all $x(0) \in \mathbb{R}^n$, $v(0) \in \mathbb{R}^{n_c}$, and $w(0)) \in \mathbb{R}^s$.

A main technical complexity is the preliminary static output injection we applied in Section 2.8 to guarantee that A_0 and S have no common eigenvalues. These issues can be resolved as already alluded to in Section 3.5. As shown below, it turns out that the necessary and sufficient condition given in Theorem 4.5.1 for the well-posedness of the exact output regulation problem with measurement feedback is indeed also the necessary and sufficient condition for the existence of a regulator that solves the structurally stable output regulation problem.

Theorem 4.5.2 *Consider a system Σ as in (4.1) with the structural constraint (2.36). Let the conditions of Theorem 3.3.4 be satisfied for this system with nominal parameter values,*

$$(A, B, E_w, C_e, D_{eu}, D_{ew}, C_y, D_{yu}, D_{yw}) \\ = (A_0, B_0, E_{w,0}, C_{e,0}, D_{eu,0}, D_{ew,0}, C_{y,0}, D_{yu,0}, D_{yw,0}).$$

Then, there exists a regulator that solves the structurally stable output regulation problem for Σ at $(A_0, B_0, E_{w,0}, C_{e,0}, D_{eu,0}, D_{ew,0}, C_{y2,0}, D_{yu2,0}, D_{yw2,0})$ if and only if the matrix S is weakly Schur-stable and the matrix

$$\begin{pmatrix} A_0 - \lambda I & B_0 \\ C_{e,0} & D_{eu,0} \end{pmatrix} \tag{4.82}$$

has full row-rank for each λ which is an eigenvalue of S_0.

For linear systems the parameter perturbations could be arbitrarily large as long as stability is preserved. This is not the case here because the perturbations might be such that $\limsup_{k\to\infty} \|\Gamma w(k)\|_\infty > 1$ and then obviously output regulation is no longer possible.

We will only briefly indicate how the above result can be shown and how a suitable controller can be found. For ease of exposition we assume here that A_0 and S have no eigenvalues in common. Define $S_{\min}$, $\tilde{S}_p$, $\tilde{D}_{ew}$ and $\tilde{D}_{yw}$ as in Section 2.8.

We can now define an *auxiliary* system composed of the extended plant (2.1) and the *auxiliary* exosystem (2.41),

$$\tilde{\Sigma} : \begin{cases} \rho\tilde{x} = A_0\tilde{x} + B_0\sigma(\tilde{u}) \\ \rho\tilde{w} = \tilde{S}_p\tilde{w} \\ \tilde{y} = C_{y,0}\tilde{x} + D_{yu,0}\sigma(\tilde{u}) + \tilde{D}_{yw}\tilde{w} \\ \tilde{e} = C_{e,0}\tilde{x} + D_{eu,0}\sigma(\tilde{u}) + \tilde{D}_{ew}\tilde{w}. \end{cases} \tag{4.83}$$

We design a low-gain or an improved low-gain measurement feedback controller for this auxiliary system such that the system achieves output regulation for all initial conditions under the condition that $\limsup_{k\to\infty} \|\Gamma\tilde{w}(k)\|_\infty < 1-\delta/2$ where $\tilde{\Gamma}$ is of course the solution of the regulator equations associated with this system. We can then show that this controller achieves structural stability for the original system if we know that the system $(A_0, B_0, C_{e,0}, D_{eu,0})$ is left-invertible.

Extensions to the case where A_0 and S have common eigenvalues and/or where $(A_0, B_0, C_{e,0}, D_{eu,0})$ is not left-invertible are available but due to their technicality have been omitted.

4.A Uniqueness of asymptotic behavior of the input

In this section we will prove that under rather weak assumptions the asymptotic behavior of the input is unique given that the output of the system tracks a certain reference signal. This is a result which is used in the proof of Theorem 4.4.1.

Theorem 4.A.1 *Consider the system,*

$$\begin{aligned} x(k+1) &= Ax(k) + \ \ Bv(k) \\ y(k) \quad &= Cx(k) + Dv(k), \end{aligned} \tag{4.84}$$

with $x(0) = 0$. Assume that (A, B, C, D) is left-invertible, and has no invariant zeros on the unit circle. Moreover, assume that $v(k)$ is bounded, and is such that $y(k) \to 0$ as $k \to \infty$. In that case $v(k) \to 0$ as $k \to \infty$.

Proof : Since we know that the system Σ with the realization (A, B, C, D) is left invertible and has no invariant zeros on the unit circle, we know that there exists a left-inverse Σ_L with input-output operator G_L which has no poles on the unit circle. We split G_L into a stable and an antistable part: $G_L = G_L^+ + G_L{}^-$ where G_L^+ has all poles outside the unit circle and G_L^- has all poles inside the unit circle. We know that $v(k) = Gy(k) = G_L^+ y(k) + G_L^- y(k)$. Clearly since $y(k) \to 0$ as $k \to \infty$, we have that $(G_L^- y)(k) \to 0$ as $k \to \infty$. On the other hand, G_L^+ might not be causal. We write $G_L^+ = G_{L,c}^+ H$ where H consists of only backwards shifts in such a way that $G_{L,c}^+$ has only poles outside the unit circle and is causal. Moreover since H consists of only shifts in time we know that $z(k) = Hy(k)$ satisfies $z(k) \to 0$ as $k \to \infty$. Suppose we have a minimal realization (F, G, H, J) for $G_{L,c}^+$ where F has all its eigenvalues outside the unit circle. Moreover, we know that the output of $G_{L,c}^+$, given the input $z(k)$, is bounded since $v(k)$ is bounded. Since we have a minimal realization this implies that the state $x(k)$ of $G_{L,c}^+$ is bounded. We have,

$$x(k+T) = F^T \left(x(k) + \sum_{i=0}^{T-1} F^{i+1-T} G z(k+i) \right). \tag{4.85}$$

Since F has all its eigenvalues outside the unit circle, we find that

$$\| \sum_{i=0}^{T-1} F^{i+1-T} G z(k+i) \| < M \| z \|_{\infty,k}$$

where M is independent of k and T. Assume that there exists a time k such that

$$\|x(k)\| > M\|z\|_{\infty,k}.$$

Then from (4.85) we find that $x(k)$ is unbounded which yields a contradiction. Therefore,

$$\|x(k)\| < M\|z\|_{\infty,k} \to 0$$

as $k \to \infty$. But since also $z(k) \to 0$ as $k \to \infty$ we find that

$$u(k) = Hx(k) + Jz(k) \to 0$$

as $k \to \infty$. ■

4.B Review of direct low-gain design for linear systems – an explicit construction

We construct here *explicitly* a family of low-gain state feedback gains based on an eigenvalue assignment method.

Consider the linear discrete-time system

$$x(k+1) = Ax(k) + Bu(k) \tag{4.86}$$

where, as usual, the state $x \in \mathbb{R}^n$ and the input $u \in \mathbb{R}^m$. Assume that (A, B) is stabilizable and all the eigenvalues of A are located inside or on the unit circle. We have the following low-gain state feedback design algorithm.

Step 1 : Find the state transformation T ([10]) such that $(T^{-1}AT,\ T^{-1}B)$ is in the following form

$$T^{-1}AT = \begin{pmatrix} A_1 & A_{12} & \cdots & A_{1p} & 0 \\ 0 & A_2 & \cdots & A_{2p} & 0 \\ \vdots & \ddots & \ddots & \vdots & \vdots \\ 0 & \cdots & 0 & A_p & 0 \\ 0 & \cdots & 0 & 0 & A_0 \end{pmatrix},$$

$$T^{-1}B = \begin{pmatrix} B_1 & 0 & \cdots & 0 & * \\ 0 & B_2 & \ddots & \vdots & * \\ \vdots & \ddots & \ddots & 0 & \vdots \\ 0 & \cdots & 0 & B_p & * \\ B_{01} & B_{02} & \cdots & B_{0p} & * \end{pmatrix}.$$

Here $*$'s represent submatrices of less interest, and for $i = 1, 2, \cdots, p$,

$$A_i = \begin{pmatrix} 0 & 1 & 0 & \cdots & 0 \\ 0 & 0 & 1 & \ddots & \vdots \\ \vdots & \vdots & \ddots & \ddots & 0 \\ 0 & 0 & \cdots & 0 & 1 \\ -a^i_{n_i} & -a^i_{n_i-1} & -a^i_{n_i-2} & \cdots & -a^i_1 \end{pmatrix}, \quad B_i = \begin{pmatrix} 0 \\ 0 \\ \vdots \\ 0 \\ 1 \end{pmatrix}.$$

The transformation T is such that, (A_i, B_i) is controllable, and all the eigenvalues of A_i are on the unit circle, while those of A_0 are strictly inside the unit circle.

Step 2 : For each (A_i, B_i), let $F_i(\varepsilon) \in \mathbb{R}^{1 \times n_i}$ be the state feedback gain such that

$$\lambda(A_i - B_i F_i(\varepsilon)) = (1 - \varepsilon)\lambda(A_i), \quad \varepsilon \in (0, 1].$$

Note that $F_i(\varepsilon)$ is unique.

Step 3 : The family of linear low-gain state feedback laws is given as follows,

$$u(k) = F_\varepsilon x(k) \tag{4.87}$$

where the state feedback gain matrix F_ε is given as

$$F_\varepsilon = -\begin{pmatrix} F_1(\varepsilon^{\upsilon_2}) & 0 & \cdots & 0 & 0 & 0 \\ 0 & F_2(\varepsilon^{\upsilon_3}) & \ddots & \vdots & \vdots & \vdots \\ \vdots & \ddots & \ddots & 0 & 0 & 0 \\ 0 & \cdots & 0 & F_{p-1}(\varepsilon^{\upsilon_p}) & 0 & 0 \\ 0 & \cdots & 0 & 0 & F_p(\varepsilon) & 0 \\ 0 & \cdots & 0 & 0 & 0 & 0 \end{pmatrix} T^{-1} \tag{4.88}$$

where $\upsilon_i = (r_i + 1)(r_{i+1} + 1) \cdots (r_p + 1)$ and where r_i is the largest algebraic multiplicity of the eigenvalues of A_i. □

With the above choice of F_ε, we have the following theorem.

Theorem 4.B.1 *Consider the linear system (4.86). Assume that all the eigenvalues of A are located inside or on the unit circle and that the pair (A, B) is stabilizable. Then, for the state feedback gain matrices given by (4.88), the following properties hold:*

(*i*) *For each* $\varepsilon \in (0, 1]$, $A + BF_\varepsilon$ *is Schur-stable.*

(*ii*) *There exists an* $\varepsilon^* \in (0, 1]$ *such that for all* $\varepsilon \in (0, \varepsilon^*]$,

$$\|F_\varepsilon\| \leq \kappa\varepsilon \tag{4.89}$$

$$\|(A + BF_\varepsilon)^k\| \leq \frac{\beta}{\varepsilon^\gamma}(1 - \varepsilon^\gamma)^{k/2} \tag{4.90}$$

$$\|F_\varepsilon(A + BF_\varepsilon)^k\| \leq \alpha\varepsilon(1 - \varepsilon^\gamma)^{k/2} \tag{4.91}$$

where $\gamma \geq 1$ *is an integer dependent only on the system data* (A, B), *and* α, β *and* κ *are some positive constants also independent of* ε.

Proof : It is a consequence of [36]. ■

Chapter 5

Output regulation with actuators subject to amplitude and rate saturation

5.1 Introduction

The previous two chapters considered output regulation of linear systems with actuators subject to amplitude saturation alone. In this chapter, we revisit output regulation of linear systems, however, with actuators subject to both amplitude and rate saturation. Rate saturation refers to the case when actuator outputs cannot change faster than a certain value.

In contrast with the amount of literature that exists when there are only amplitude bounds, not much literature exists when both amplitude as well as rate saturation are present. If we have only rate saturation, we can approach the study of the given system either for stabilization or for output regulation by viewing the derivative of the input signal as a new fictitious input signal which is then bounded only in magnitude; and hence we can readily apply the development given in the previous two chapters. However, if we have simultaneous bounds on the rate as well as on the magnitude, then such a simple approach does not work out. We would like to point out right at the outset that, unlike amplitude saturation which is a static non-linearity, the rate saturation is a dynamic non-linearity.

In modeling amplitude and rate saturation for actuators, one can basically choose one of the following two approaches.

- A natural way is to write down first the linear dynamic equations of an

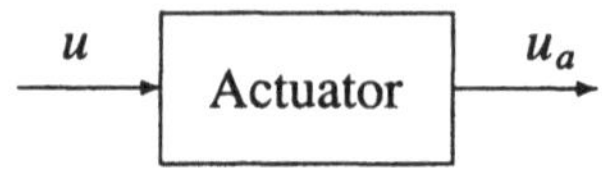

Figure 5.1: Actuator

actuator (see block diagram of Figure 5.1), and then impose on such equations both the amplitude and rate saturation as shown in the following equation,

$$\begin{aligned} \dot{x}_a &= \sigma_\beta(A_a x_a + B_a u) \\ u_a &= \sigma_\alpha(C_a x_a) \end{aligned} \tag{5.1}$$

where $\sigma_\alpha(\cdot)$ and $\sigma_\beta(\cdot)$ are standard saturation functions, say, defined as

$$\sigma_\alpha(s) = \text{sgn}(s)\min\{|s|, \alpha\} \text{ and } \sigma_\beta(s) = \text{sgn}(s)\min\{|s|, \beta\}.$$

Also, $u \in \mathbb{R}^m$ is the input signal and $x_a \in \mathbb{R}^{n_a}$ is the state of the actuator and $u_a \in \mathbb{R}^m$ is the output of the actuator which is applied to the plant.

We would like to point out that it is rather difficult to incorporate the 'nice' external stability behavior of actuators into a state space model characterized by the matrix triple (A_a, B_a, C_a). Anyway, having modeled the actuator as in (5.1), the next obvious step is to augment (5.1) with the plant model in order to obtain the model for both the actuator and the plant. It is well known that it is hard to analyze and design dynamical systems which are modeled as such. However, if one imposes a certain strong mathematical structure on the dynamics of the actuator given in (5.1), one can avoid the complexity in analysis and design. The required mathematical structure is typically to assume that A_a, B_a, and C_a are diagonal matrices. Essentially, such a mathematical structure implies that the actuator dynamics for each component of the input is a first-order or scalar dynamics. This approach is taken in [7, 31, 32, 34]. We note that [7, 34] deal with rate saturation while [31, 32] deal with amplitude as well as rate saturation. It is obvious to see that in [34] with the imposed structure of diagonality of the matrices A_a, B_a, and C_a, the control of linear systems with rate saturation reduces to the control of the augmented plant with only input amplitude saturation. However, we would like to point out an interesting but an undesirable aspect of this approach, namely the necessity of using the state of the actuator for control feedback.

• The second approach is to model the constraints in such a way that they can be incorporated as a part of the controller, and then to design the controller so that its output is always in agreement with the constraints as dictated by the

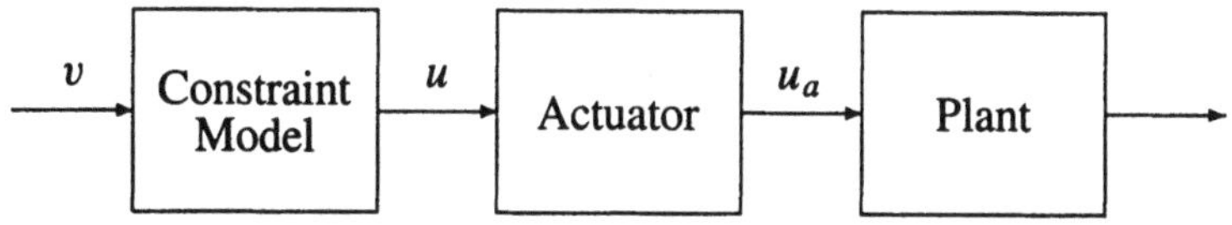

Figure 5.2: Constraint model, actuator, and plant

actuator. Thus, this method avoids overloading of the actuator. By incorporating the actuator constraints in the design of controllers, this method essentially sidesteps the short comings of the first method. The work of [5] takes this approach to model the rate saturation. The block diagram in Figure 5.2 depicts the philosophy of the method.

We take the second approach here, and model the constraints by introducing a non-linear operator that captures both amplitude and rate constraints. We refer to the new non-linear operator as a standard amplitude+rate operator (see Figure 5.3). Such an operator has a 'dead-beat stability' property. With such a property being valid, we study and examine the state space realizations of this operator. It turns out that, although one could obtain a useful state space realization in discrete-time systems, one cannot do so in continuous-time systems. This indicates that a framework that includes functional differential equations for modeling the plant and amplitude as well as rate saturation constraints of the actuator is indeed a natural framework.

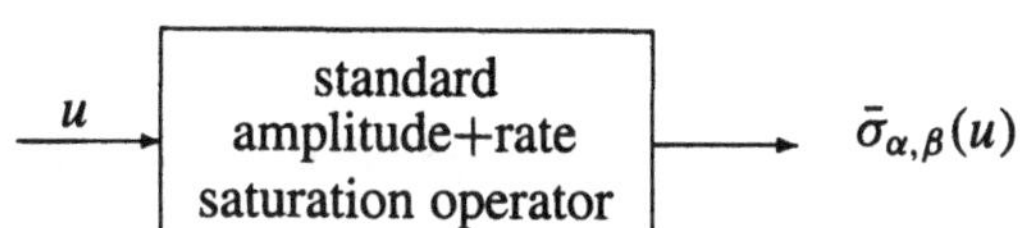

Figure 5.3: Amplitude+rate saturation operator

Utilizing this functional differential framework and knowing that it satisfies a 'dead-beat stability' property, we redefine the notions of semi-global stabilization and output regulation. We then proceed to show that the same low-gain design methodology that was successfully used in the previous two chapters to design controllers for linear systems having only input amplitude saturation can also successfully be used for linear systems with both amplitude and rate saturation on the control input. In fact, we present here explicit controller design methods via low-gain design methodology in order to semi-globally stabilize, and to regulate the output (i.e. to track a desired output and/or to reject exogenous disturbances) of linear systems with both input amplitude as well as rate saturation. As illustrated in the previous two chapters, controllers designed via low-gain design methodology exhibit unacceptably slow transient behavior. For that reason, we also utilize low-and-high gain controllers where the high gain component affects neither the stability and its associated domain of attraction nor the class of exogenous signals for which we can achieve output regulation. On the other hand, the high gain component greatly improves the transient behavior. However, this low-and-high gain controller cannot be obtained in the same way as in the previous two chapters owing to the specific dynamic structure of a rate limiter.

Throughout this chapter, $\sigma_h : \mathbb{R} \to \mathbb{R}$ with $\sigma_h(s) = \operatorname{sgn}(s)\min\{h, |s|\}$ denotes the standard saturation function with amplitude h. We consider both continuous- and discrete-time systems together in this chapter. This chapter is based completely on the recent research work of the authors in [75].

5.2 Modeling issues – standard amplitude + rate saturation operator

In this section, we present modeling aspects of actuators with both amplitude and rate saturation.

5.2.1 Discrete-time

We will first consider a discrete-time system of the form,

$$x(k+1) = Ax(k) + B\sigma_{\alpha,\beta}(u)(k). \tag{5.2}$$

Here $\sigma_{\alpha,\beta}$ is a diagonal operator with identical non-linear elements on the diagonal given by $\bar{\sigma}_{\alpha,\beta}$ which is uniquely defined by the following properties:

- We have

  $$|\bar{\sigma}_{\alpha,\beta}(u)(k)| \leqslant \alpha, \text{ and } |\bar{\sigma}_{\alpha,\beta}(u)(k+1) - \bar{\sigma}_{\alpha,\beta}(u)(k)| \leqslant \beta$$

  for all k.

- If $u(k) > \bar{\sigma}_{\alpha,\beta}(u)(k)$, then either

  ⋆ $u(k) > \alpha$ and $\bar{\sigma}_{\alpha,\beta}(u)(k) = \alpha$, or

- ⋆ $\bar{\sigma}_{\alpha,\beta}(u)(k) = \bar{\sigma}_{\alpha,\beta}(u)(k-1) + \beta$.

- If $u(k) < \bar{\sigma}_{\alpha,\beta}(u)(k)$, then either
  - ⋆ $u(k) < -\alpha$ and $\bar{\sigma}_{\alpha,\beta}(u)(k) = -\alpha$, or
  - ⋆ $\bar{\sigma}_{\alpha,\beta}(u)(k) = \bar{\sigma}_{\alpha,\beta}(u)(k-1) - \beta$.

Obviously $\bar{\sigma}_{\alpha,\beta}$ is a dynamic non-linearity. We can also describe this operator by a state space model,

$$x_\infty(k+1) = \beta\,\mathrm{sgn}(\sigma_\alpha(u(k+1)) - x_\infty(k)), \quad x(0) = \sigma_\alpha(u(0)) \tag{5.3}$$

where $\bar{\sigma}_{\alpha,\beta}(u) = x_\infty$. This is a good model for our amplitude+rate operator which is consistent with our earlier description.

It is easy to see that $\bar{\sigma}_{\alpha,\beta}$ is dynamic and has initial conditions. It is not difficult to see that $\bar{\sigma}_{\alpha,\beta}(u)(k)$ can be viewed as the state of this system at time k. For our purpose, it will in general be sufficient to note that $\bar{\sigma}_{\alpha,\beta}$ is dead-beat. We even have the stronger property that, if u is such that $\|u(k)\|_\infty < \alpha$ and $\|u(k)-u(k-1)\|_\infty < \beta$ for $k > K$, then $\bar{\sigma}_{\alpha,\beta}(u)(k) = u(k)$ for $k > K+\alpha/\beta$. If we refer to arbitrary initial conditions of $\bar{\sigma}_{\alpha,\beta}$ at time 0, then we mean an arbitrary input signal u in the interval $[-\alpha/\beta, 0]$. In our definitions, we will refer to the initial conditions of $\bar{\sigma}_{\alpha,\beta}$ as $\bar{x}_s$, and we will refer to the space of all possible initial conditions for the operator $\bar{\sigma}_{\alpha,\beta}$, namely the space of all signals defined on the interval $[-\alpha/\beta, 0]$, as $\bar{\mathcal{X}}_s$. Similarly, for the operator $\sigma_{\alpha,\beta}$, we denote the initial conditions by x_s, and the space of all initial signals as $\mathcal{X}_s$.

5.2.2 Continuous-time

We consider a continuous-time system of the form,

$$\dot{x}(t) = Ax(t) + B\sigma_{\alpha,\beta}(u)(t). \tag{5.4}$$

Here again $\sigma_{\alpha,\beta}$ is a diagonal operator with identical non-linear elements on the diagonal given by $\bar{\sigma}_{\alpha,\beta}$. We seek an operator $\bar{\sigma}_{\alpha,\beta}$ with the following properties:

- For any continuously differentiable u, $\bar{\sigma}_{\alpha,\beta}(u)$ is differentiable and

  $$|\bar{\sigma}_{\alpha,\beta}(u)(t)| \leqslant \alpha, \text{ and } |\tfrac{d}{dt}\bar{\sigma}_{\alpha,\beta}(u)(t)| \leqslant \beta$$

  for all t.

- If $u(t) > \bar{\sigma}_{\alpha,\beta}(u)(t)$, then either

  - ⋆ $u(t) > \alpha$ and $\bar{\sigma}_{\alpha,\beta}(u)(t) = \alpha$, or
  - ⋆ $\frac{d}{dt}\bar{\sigma}_{\alpha,\beta}(u)(t) = \beta$.

- If $u(t) < \bar{\sigma}_{\alpha,\beta}(u)(t)$, then either
  - ⋆ $u(t) < -\alpha$ and $\bar{\sigma}_{\alpha,\beta}(u)(t) = -\alpha$, or
  - ⋆ $\frac{d}{dt}\bar{\sigma}_{\alpha,\beta}(u)(t) = -\beta$.

However, in the above case of continuous-time systems, it is not clear whether this uniquely determines the operator $\bar{\sigma}_{\alpha,\beta}$. Moreover, for instance due to measurement noise, we might have an input signal which is not smooth but might only be piecewise continuous and for this general class of signals the above definition is clearly not sufficient.

We now consider a different way of looking at $\bar{\sigma}_{\alpha,\beta}$. Consider the class of models,

$$\dot{x}_\lambda = \sigma_\beta(\lambda(\sigma_\alpha(u) - x_\lambda)), \quad x_\lambda(0) = \sigma_\alpha(u(0)). \tag{5.5}$$

It is well known that this differential equation has a unique solution for any measurable input signal u. Next, we define $\bar{\sigma}_{\alpha,\beta}$ by,

$$\bar{\sigma}_{\alpha,\beta}(u) = \lim_{\lambda\to\infty} x_\lambda. \tag{5.6}$$

The following lemma shows that the operator as defined by (5.6) has all the required properties.

Lemma 5.2.1 *For any piecewise continuous function u, the limit in (5.6) exists in L_∞, and the limit $\bar{\sigma}_{\alpha,\beta}(u)$ has an L_∞ norm less than α and is Lipschitz-continuous with Lipschitz constant β.*

Proof : Let $\lambda^* > 0$. For any $\lambda_1, \lambda_2 > \lambda^*$ we have $x_{\lambda_1}(0) - x_{\lambda_2}(0) = 0$. Moreover, if $x_{\lambda_1}(t) - x_{\lambda_2}(t) > 2\beta/\lambda^*$ we have $\dot{x}_{\lambda_1}(t) - \dot{x}_{\lambda_2}(t) \leqslant 0$. After all, we have only two possibilities:

- $x_{\lambda_1}(t) > \sigma_\alpha(u)(t) + \beta/\lambda^*$ in which case $\dot{x}_{\lambda_1}(t) = -\beta$ and $\dot{x}_{\lambda_2}(t) \geqslant -\beta$. Therefore $\dot{x}_{\lambda_1}(t) - \dot{x}_{\lambda_2}(t) \leqslant 0$.

- $x_{\lambda_2}(t) < \sigma_\alpha(u)(t) - \beta/\lambda^*$ in which case $\dot{x}_{\lambda_1}(t) \leqslant \beta$ and $\dot{x}_{\lambda_2}(t) = \beta$. Therefore $\dot{x}_{\lambda_1}(t) - \dot{x}_{\lambda_2}(t) \leqslant 0$.

Similarly, if $x_{\lambda_1}(t) - x_{\lambda_2}(t) < -2\beta/\lambda^*$, we have $\dot{x}_{\lambda_1}(t) - \dot{x}_{\lambda_2}(t) \geqslant 0$. This shows that for all $\lambda_1, \lambda_2 > \lambda^*$ we have

$$\|x_{\lambda_1} - x_{\lambda_2}\|_\infty \leqslant \tfrac{2\beta}{\lambda^*}.$$

Therefore, by definition, $\{x_\lambda\}$ is a Cauchy sequence and it has a limit which we call $x_\infty \in L_\infty$.

We know that $x_\lambda(t) \geqslant \alpha$ implies that $\dot{x}_\lambda(t) \leqslant 0$, and $x_\lambda(t) \leqslant -\alpha$ implies that $\dot{x}_\lambda(t) \geqslant 0$. Combined with $\|x_\lambda(0)\| \leqslant \alpha$, we find then that $\|x_\lambda\|_\infty \leqslant \alpha$. This obviously implies that

$$\|x_\infty\| = \lim_{\lambda\to\infty} \|x_\lambda\| \leqslant \alpha.$$

Finally, we have $\|x_\lambda(t_2) - x_\lambda(t_1)\| \leqslant \beta|t_2 - t_1|$ for any $t_1, t_2 > 0$. By letting $\lambda \to \infty$ we find that

$$\|x_\infty(t_2) - x_\infty(t_1)\| \leqslant \beta|t_2 - t_1| \qquad \text{for all } t_1, t_2 > 0,$$

and hence x_∞ is Lipschitz continuous with Lipschitz constant β. ■

Note that a Lipschitz continuous function is absolutely continuous and hence it is easy to see that $\bar{\sigma}_{\alpha,\beta}(u)$ is differentiable almost everywhere, and there exists an L_∞ function w with L_∞ norm less than β such that

$$\bar{\sigma}_{\alpha,\beta}(u)(t) = \bar{\sigma}_{\alpha,\beta}(u)(0) + \int_0^t w(t)\,dt.$$

However, $\bar{\sigma}_{\alpha,\beta}(u)$ need not be differentiable everywhere. An example is given by the function,

$$u(t) = \begin{cases} 0 & t = 0 \\ t\sin\left(\frac{1}{t}\right) & \text{elsewhere,} \end{cases}$$

for which $\bar{\sigma}_{\alpha,\beta}(u)$ is not differentiable in 0. Obviously, with the more precise definition given in (5.6), $\bar{\sigma}_{\alpha,\beta}$ is uniquely determined. Moreover, as soon as u is sufficiently smooth it is easy to verify that the mathematically precise definition given in (5.6) is consistent with our intuitive definition given initially.

Note that we might define the state model for $\bar{\sigma}_{\alpha,\beta}$ as,

$$\dot{x}_\infty = \beta\,\mathrm{sgn}(\sigma_\alpha(u) - x_\infty), \quad x(0) = \sigma_\alpha(u(0)) \tag{5.7}$$

with $\bar{\sigma}_{\alpha,\beta}(u) = x_\infty$. This is consistent with our intuitive description and it looks like the appropriate model given the state space models for x_λ. However, note that if u satisfies the rate and saturation bounds then we expect that $u = x_\infty$ (this can also be formally shown) but then the above differential equation shows that $\dot{x}_\infty = 0$ which obviously need not be the case. Therefore the model given in (5.7) is **incorrect**.

Like in discrete-time, we see that $\bar{\sigma}_{\alpha,\beta}$ is dynamic and has initial conditions. We note that again $\bar{\sigma}_{\alpha,\beta}(u)(t)$ can be viewed as the state of this system at time t. For our purpose, it will in general be sufficient to note that $\bar{\sigma}_{\alpha,\beta}$ is deadbeat. We again have the stronger property that if u is such that $\|u(t)\|_\infty < \alpha$ and $\|\dot{u}(t)\|_\infty < \beta$ for $t > t_1$, then $\bar{\sigma}_{\alpha,\beta}(u)(t) = u(t)$ for $t > t_1 + \alpha/\beta$. If we refer to arbitrary initial conditions of $\bar{\sigma}_{\alpha,\beta}$ at time 0, then we mean an arbitrary input signal u in the interval $[-\alpha/\beta, 0]$. As in discrete-time systems, we will refer to the initial conditions of $\bar{\sigma}_{\alpha,\beta}$ as $\bar{x}_s$, and we will refer to the space of all possible initial conditions for the operator $\bar{\sigma}_{\alpha,\beta}$, namely the space of all signals defined on the interval $[-\alpha/\beta, 0]$, as $\bar{\mathcal{X}}_s$. Similarly, for the operator $\sigma_{\alpha,\beta}$, we denote the initial conditions by x_s, and the space of all initial signals as $\mathcal{X}_s$.

In the literature, there have been many different models for a rate limit in combination with a saturation. All other models use a form of modeling of the form

$$\dot{x}_r = f(x_r, u), \quad x_r \in \mathbb{R}^k.$$

We do not model the actuator with its rate and amplitude limits. We model these limits and constraints and view them as part of the controller. Namely, we use an operator as part of the controller, which guarantees that the control signal satisfies the bounds in the actuator and avoids overloading the actuator. This is the important difference, but also the fact that our operator only has a state space model in discrete-time and can only be approximated by state space models in continuous-time yields differences in the analysis, while leading to the fact that its state space equals $[-\alpha, \alpha]$. Our approach is really different but, as we will see, very powerful since all results we have obtained in the previous two chapters for saturation can be easily extended to the case with rate limits including the low and high gain design which is difficult to analyze for a state space model with $x_r \in \mathbb{R}^k$.

In summary, in view of the above discussions, the functional differential equations (5.2) and (5.4) along with the definition for the constraint operator $\sigma_{\alpha,\beta}$ given in equations (5.3), (5.5) and (5.6), are the appropriate models for discrete- and continuous-time systems respectively whenever the actuators are constrained by both amplitude and rated saturation.

5.3 Semi-global stabilization

We consider semi-global stabilization of given systems modeled by functional differential equations (5.2) and (5.4) along with the definition for the constraint operator $\sigma_{\alpha,\beta}$ given in equations (5.3), (5.5) and (5.6). We first give a precise definition of the concept of semi-global stabilization for such systems.

Problem 5.3.1 *(Semi-global stabilization problem via linear state feedback laws) Consider a continuous- or discrete-time system of the form (5.4) and (5.2) along with the definition for the constraint operator $\sigma_{\alpha,\beta}$ given in equations (5.3), (5.5) and (5.6). The problem of semi-global stabilization is to find, if possible, a family of feedback gains $\{F_\varepsilon\}_{\varepsilon\in\mathbb{R}^+}$ such that for any a priori given (arbitrarily large) bounded set $\mathcal{X}_0 \subset \mathbb{R}^n$, there exists an ε^* such that for all $\varepsilon < \varepsilon^*$ the linear static feedback law $u = F_\varepsilon x$ is such that the equilibrium $x = 0, x_s = 0$ of the continuous-time system*

$$\dot{x}(t) = Ax(t) + B\sigma_{\alpha,\beta}(Fx)(t) \tag{5.8}$$

or the discrete-time system

$$x(k+1) = Ax(k) + B\sigma_{\alpha,\beta}(Fx)(k) \tag{5.9}$$

is locally exponentially stable with $\mathcal{X}_0 \times \mathcal{X}_s$ contained in its basin of attraction.

We have the following necessary and sufficient conditions for the solvability of the above problem. We first present the continuous-time result.

Theorem 5.3.1 *Consider the continuous-time system (5.4) along with the definition for the constraint operator $\sigma_{\alpha,\beta}$ given in equations (5.5) and (5.6). The semi-global stabilization problem via linear state feedback laws is solvable if and only if (A, B) is stabilizable and the eigenvalues of A are in the closed left half plane.*

Moreover, in that case the semi-global stabilization problem via linear state feedback laws is solved by the family of feedback laws

$$u = F_\varepsilon x = -B^T P_\varepsilon x$$

where

$$0 = A^T P_\varepsilon + P_\varepsilon A - P_\varepsilon B B^T P_\varepsilon + Q_\varepsilon, \tag{5.10}$$

and where Q_ε is a continuously differentiable matrix-valued function such that $Q_\varepsilon > 0$, $\frac{dQ_\varepsilon}{d\varepsilon} > 0$ for any $\varepsilon \in (0, 1]$, and $\lim_{\varepsilon\to 0} Q_\varepsilon = 0$.

Proof : As we have seen in Chapter 3, the conditions that, (A, B) is stabilizable and the eigenvalues of A are in the closed left half plane, are necessary for semi-global stabilization if we only have amplitude saturation and no rate limits. Therefore, obviously, they are still necessary when we have amplitude saturation and rate limits.

To prove that these conditions are also sufficient, it is obviously sufficient to verify that the given family of feedback laws has the desired properties. In other words, while utilizing the given family of feedback laws, we need to show that for each given set $\mathcal{X}_0 \times \mathcal{X}_s$, there exists an $\varepsilon^* > 0$ such that, for all $\varepsilon \in (0, \varepsilon^*]$, we have local asymptotic stability with $\mathcal{X}_0 \times \mathcal{X}_s$ contained in its domain of attraction. In view of the known properties for P_ε (see Lemma 3.3.1), it is easy to verify that for $F_\varepsilon = -B^{\mathrm{T}} P_\varepsilon$ we have

$$\begin{aligned} \|F_\varepsilon e^{(A+BF_\varepsilon)t}\|_\infty &\leqslant \upsilon_\varepsilon e^{-\zeta_\varepsilon t}, \\ \|F_\varepsilon(A+BF_\varepsilon)e^{(A+BF_\varepsilon)t}\|_\infty &\leqslant \upsilon_\varepsilon e^{-\zeta_\varepsilon t}, \end{aligned} \tag{5.11}$$

where υ and ζ depend continuously on ε and are positive for $\varepsilon > 0$. Moreover $\upsilon_0 = \zeta_0 = 0$. Also, we can rewrite (5.8) as

$$\dot{x} = Ax + B\sigma_{\alpha,\beta}(F_\varepsilon x). \tag{5.12}$$

In the absence of saturation elements, the above system takes the form,

$$\dot{x} = (A + BF_\varepsilon)x. \tag{5.13}$$

It then follows from (5.11) that there exists an $\varepsilon_1^* > 0$ such that for all $\varepsilon \in (0, \varepsilon_1^*]$ we have

$$\|F_\varepsilon x\|_\infty \leqslant \alpha, \quad \|F_\varepsilon \dot{x}\|_\infty \leqslant \beta \quad \text{for all } x(0) \in \mathcal{X}_0 \text{ and } x_s \in \mathcal{X}_s.$$

This shows that for all $\varepsilon \in (0, \varepsilon_1^*]$ and for all $x(0) \in \mathcal{X}_0$ and $x_s \in \mathcal{X}_s$, system (5.12) operates in the linear regions of saturation elements, and hence we can conclude that the equilibrium $x = 0$ and $x_s = 0$ of the system (5.12) is asymptotically stable with $\mathcal{X}_0 \times \mathcal{X}_s$ contained in its basin of attraction. ∎

Next, we present the discrete-time version of the above result.

Theorem 5.3.2 *Consider the discrete-time system (5.2) along with the definition for the constraint operator $\sigma_{\alpha,\beta}$ given in equation (5.3). The semi-global stabilization problem via linear state feedback laws is solvable if and only if (A, B) is stabilizable and the eigenvalues of A are in the closed unit disc.*

Moreover, in that case the semi-global stabilization problem via linear state feedback is solved by the family of feedback laws

$$u = F_\varepsilon x = -(B^T P_\varepsilon B + I)^{-1} B^T P_\varepsilon A x$$

where

$$P_\varepsilon = A^T P_\varepsilon A - A^T P_\varepsilon B(B^T P_\varepsilon B + I)^{-1} B^T P_\varepsilon A + Q_\varepsilon, \tag{5.14}$$

and where Q_ε *is a continuously differentiable matrix-valued function such that* $Q_\varepsilon > 0$, $\frac{dQ_\varepsilon}{d\varepsilon} > 0$ *for any* $\varepsilon \in (0, 1]$, *and* $\lim_{\varepsilon \to 0} Q_\varepsilon = 0$.

Proof : It follows directly using the same arguments as in the proof of Theorem 5.3.1 and is based on Lemma 4.3.1. ■

As in the previous two chapters, the low gain techniques presented in this section basically avoid saturation by squeezing the gain. This of course results in a very slow transient response. In the context of output regulation we present an additional high gain that greatly improves the transient response. This is also immediately applicable to the stabilization problem but in order to be concise we do not present the details here. Similarly, the measurement feedback case will only be presented for the problem of stabilization and output regulation but the results presented there can be immediately applied to the stabilization problem.

5.4 Output regulation via state feedback control

In this section, we consider output regulation while using state feedback controllers. Consider a system as given below,

$$\begin{aligned} \rho x &= Ax + B\sigma_{\alpha,\beta}(u) + E_w w \\ \rho w &= S w \\ e &= C_e x + D_{eu}\sigma_{\alpha,\beta}(u) + D_{ew} w, \end{aligned} \tag{5.15}$$

where as usual the first equation of these systems describes a plant, with state $x \in \mathbb{R}^n$, and input $u \in \mathbb{R}^m$, subject to the effect of a *disturbance* represented by $E_w w$. The third equation defines the error $e \in \mathbb{R}^q$ between the actual plant output $C_e x + D_{eu}\sigma_{\alpha,\beta}(u)$ and a *reference* signal $-D_{ew} w$ which the plant output is required to track. The second equation describes the *exosystem*, with state

$w \in \mathbb{R}^s$. As usual, the exosystem models the class of disturbance or reference signals taken into consideration.

As usual, a state feedback controller is of the form,

$$u = Fx + Gw, \tag{5.16}$$

and our objective is to achieve internal stability and output regulation. We formally state the following synthesis problem.

Problem 5.4.1 *Consider a system (5.4) along with the definition for the constraint operator $\sigma_{\alpha,\beta}$ given in equations (5.3), (5.5) and (5.6). Also, consider a compact set $\mathcal{W}_0 \subset \mathbb{R}^s$. The* ***semi-global linear state feedback output regulation problem*** *is defined as follows. For any a priori given (arbitrarily large) bounded set $\mathcal{X}_0 \subset \mathbb{R}^n$, find, if possible, a linear static feedback law $u = Fx + Gw$ such that the following hold:*

(i) *The equilibrium $x = 0, x_s = 0$ of the system*

$$\rho x = Ax + B\sigma_{\alpha,\beta}(Fx) \tag{5.17}$$

is locally exponentially stable with $\mathcal{X}_0 \times \mathcal{X}_s$ contained in its basin of attraction.

(ii) *For all $x(0) \in \mathcal{X}_0$, $x_s \in \mathcal{X}_s$ and $w(0) \in \mathcal{W}_0$, the solution of the closed-loop system satisfies*

$$\lim_{t\to\infty} e(t) = 0. \tag{5.18}$$

First we can ask ourselves whether we can achieve global output regulation where both the sets $\mathcal{W}_0$ and $\mathcal{X}_0$ can be chosen arbitrarily large. Unfortunately, as shown earlier in the previous two chapters, global output regulation is not possible (except in very special circumstances) when amplitude saturation of actuators is present. Obviously, global output regulation is not possible either, in general, when both amplitude and rate saturation of actuators are present. Therefore, as in the previous two chapters where only amplitude saturation of actuators is considered, instead of making $\mathcal{W}_0$ large we view it as a part of the a priori given system data. The output regulation problem does not view the set of initial conditions of the plant as given data. The set of given data consists of the models of the plant and the exosystem and the set of initial conditions for the exosystem. Therefore, any solvability conditions we obtain *must* be independent of the set of initial conditions of the plant, $\mathcal{X}_0$.

As usual, we make the following assumptions which among others enable us to solve Problem 5.4.1.

A.1. The pair (A, B) is stabilizable.

A.2. The matrix S is anti-Hurwitz-stable for continuous-time systems and anti-Schur-stable for discrete-time systems.

As discussed earlier, these assumptions are obviously without loss of generality.

The solvability conditions for the semi-global linear state feedback output regulation problem are given in the following theorem.

Theorem 5.4.1 *Consider the system along with the definition for the constraint operator $\sigma_{\alpha,\beta}$ given in equations (5.3), (5.5) and (5.6). Also, consider the given compact set $\mathcal{W}_0 \subset \mathbb{R}^s$. The semi-global linear state feedback output regulation problem is solvable if the following conditions hold:*

(i) (A, B) is stabilizable, and A has all its eigenvalues in the closed left half plane (continuous-time) or in the closed unit disc (discrete-time).

(ii) There exist matrices Π and Γ such that

(a) they solve the regulator equation,

$$\begin{aligned} \Pi S &= A\Pi + B\Gamma + E_w \\ 0 &= C_e\Pi + D_{eu}\Gamma + D_{ew}\,, \quad \text{and} \end{aligned} \tag{5.19}$$

(b) there exist a $\delta > 0$ and a $T \geqslant 0$ such that

$$\|\Gamma w\|_{\infty,T} \leqslant \alpha - \delta$$

and

$$\begin{aligned} \|\Gamma S w\|_{\infty,T} &\leqslant \beta - \delta \qquad \text{(continuous-time)} \\ \|\Gamma (S - I) w\|_{\infty,T} &\leqslant \beta - \delta \qquad \text{(discrete-time)} \end{aligned}$$

for all w with $w(0) \in \mathcal{W}_0$.

Moreover, if these conditions are satisfied, then in continuous-time a suitable family of linear static state feedback laws is given by

$$u = -B^T P_\varepsilon x + (B^T P_\varepsilon \Pi + \Gamma) w \tag{5.20}$$

where P_ε is defined by (5.10) while, in discrete-time, a suitable family of linear static state feedback laws is given by

$$u = -(B^T P_\varepsilon B + I)^{-1} B^T P_\varepsilon A(x - \Pi w) + \Gamma w \tag{5.21}$$

where P_ε is defined by (5.14).

Proof : We prove this theorem only for continuous-time. The discrete-time version can be derived similarly. Consider the family of feedback laws given in (5.20). We know that (5.11) is satisfied for $F_\varepsilon = -B^{\mathrm{T}}P_\varepsilon$ where υ is a positive-valued function satisfying $\lim_{\varepsilon\to 0}\upsilon_\varepsilon = 0$. For $\varepsilon < \varepsilon_1^*$, the stability of the closed-loop system for $w = 0$, as defined in item (i) of Problem 5.4.1, is then a direct consequence of Theorem 5.3.1 and the corresponding proof.

To show that item (ii) of Problem 5.4.1 holds (i.e. the error e goes to zero asymptotically), let us introduce an invertible coordinate change,

$$\xi = x - \Pi w. \tag{5.22}$$

Then, using (5.19), we have

$$\begin{aligned}\dot{\xi} &= \dot{x} - \Pi\dot{w}\\ &= Ax + B\sigma_{\alpha,\beta}(u) + E_w w - \Pi S w\\ &= A\xi + B(\sigma_{\alpha,\beta}(u) - \Gamma w).\end{aligned} \tag{5.23}$$

With the considered family of state feedback laws, the closed-loop system can then be rewritten as

$$\dot{\xi} = A\xi + B\left[\sigma_{\alpha,\beta}(F_\varepsilon\xi + \Gamma w) - \Gamma w\right]. \tag{5.24}$$

Now by condition (ii)b) of Theorem 5.4.1, $\|\Gamma w\|_{\infty,\mathrm{T}} < \alpha - \delta$ and $\|\Gamma\dot{w}\|_{\infty,\mathrm{T}} < \beta - \delta$. Moreover, $\xi(T)$ belongs to a bounded set independent of ε since $\xi(0)$ is bounded and $\xi(T)$ is determined by a linear differential equation with bounded inputs $\sigma_{\alpha,\beta}(u)$ and Γw. If we consider the system (5.24), from time T onwards, without the saturation element, we obtain

$$\dot{\xi} = (A + BF_\varepsilon)\xi. \tag{5.25}$$

Since $\xi(T)$ is bounded, (5.11) and (5.25) imply that there exists an $\varepsilon_2^* > 0$ such that, for all $\varepsilon \in (0, \varepsilon_2^*]$,

$$\|F_\varepsilon\xi\|_{\infty,\mathrm{T}} \leqslant \delta \text{ and } \|F_\varepsilon\dot{\xi}\|_{\infty,\mathrm{T}} \leqslant \delta.$$

We can conclude then that the system (5.24) will operate within the linear region of the saturation elements for all $t \geqslant T$ if $\varepsilon \in (0, \varepsilon_2^*]$. Also, in view of (5.20) and (5.22), we find that for $t \geqslant T$,

$$e(t) = (C_e + D_{eu}F_\varepsilon)\xi(t) + (C_e\Pi + D_{eu}\Gamma + D_{ew})w(t).$$

However, in view of the second equation of (5.19), $e(t)$ reduces to $e(t) = (C_e + D_{eu}F_\varepsilon)\xi(t)$, and thus, owing to the stability of $A + BF_\varepsilon$, we find that $e(t) \to 0$ as $t \to \infty$.

Finally, by taking $\varepsilon^* = \min\{\varepsilon_1^*, \varepsilon_2^*\}$, we complete our proof. ■

Under certain weak assumptions, the conditions given in Theorem 5.4.1 are very close to being necessary. As seen in the previous chapters, the conditions (i) and condition (ii)a) are necessary for output regulation without any kind of saturation in the actuator. Also, the bound on Γw is very close to being necessary as seen in the previous two chapters when amplitude saturation is present in the actuator. It turns out that the bound on $\Gamma S w$ is also close to being necessary when amplitude as well as rate saturation is present in the actuator. This is all precisely formulated in the following theorem.

Theorem 5.4.2 *Consider the system (5.15) along with the definition for the constraint operator $\sigma_{\alpha,\beta}$ given in equations (5.3), (5.5) and (5.6). Also, consider the given compact set $\mathcal{W}_0 \subset \mathbb{R}^s$. Assume that the system characterized by (A, B, C_e, D_{eu}) is left-invertible and has no zeros on the imaginary axis (continuous-time) or on the unit circle (discrete-time). The semi-global linear state feedback output regulation problem is solvable only if there exist Γ and Π satisfying (5.19) and for each $\varepsilon > 0$ there exists a $T > 0$ such that*

$$\|\Gamma w\|_{\infty,T} \leqslant \alpha + \varepsilon \tag{5.26}$$

and

$$\|\Gamma S w\|_{\infty,T} \leqslant \beta + \varepsilon \tag{5.27}$$

in continuous-time, or

$$\|\Gamma(S - I)w\|_{\infty,T} \leqslant \beta + \varepsilon \tag{5.28}$$

in discrete-time for all w with $w(0) \in \mathcal{W}_0$.

In other words the above theorem shows that under some mild assumptions the conditions of Theorem 5.4.1 are very close to being necessary.

Proof : Our development focuses only on continuous-time systems. Similar development is valid for discrete-time systems as well. The need for the existence of a solution (Π, Γ) of (5.19) is obvious. It is shown in Corollary 3.3.1 that Γ is uniquely defined as a direct consequence of the left-invertibility of the system characterized by (A, B, C_e, D_{eu}). Using this result, the necessity of (5.26) was already established in Theorem 3.4.1. In order to establish the necessity of (5.27) in continuous-time we need to do some additional work.

Consider the system (5.15), however, without the saturation element. That is, let

$$\begin{aligned} \dot{x} &= Ax + Bv + E_w w \\ \dot{w} &= Sw \\ e &= C_e x + D_{eu} v + D_{ew} w. \end{aligned} \tag{5.29}$$

Suppose that we have some arbitrary non-linear feedback law $u = \zeta(x, w)$ which achieves output regulation for the system (5.15). Then the feedback $v = \sigma_{\alpha,\beta}(\zeta(x, w))$ will achieve output regulation for the system (5.29). Note that by our assumption, there exists also a linear feedback,

$$v = Fx + (\Gamma - F\Pi)w,$$

which achieves output regulation for the system (5.29). Moreover,

$$v(t) - \Gamma w(t) \to 0 \text{ as } t \to \infty. \tag{5.30}$$

We have two feedback laws which achieve output regulation for the linear system (5.29). One is non-linear and satisfies certain amplitude and rate constraints. The other is a linear feedback, of which we have no a priori knowledge regarding any constraints. Our aim is to show that the linear feedback must necessarily satisfy an amplitude as well as a rate constraint asymptotically as $t \to \infty$. To this end, we define the difference between the two control inputs as

$$s = [Fx + (\Gamma - F\Pi)w] - [\sigma_{\alpha,\beta}(\zeta(x, w))].$$

Although we proved (5.26) in Chapter 3, let us prove it here as well for the purpose of continuity of our arguments. Suppose now that (5.26) is not true. In that case there exist $\{t_n\}_{n=1}^{\infty}$ and a $\delta > 0$ such that $t_n \to \infty$ as $n \to \infty$ and $\|\Gamma w(t_n)\| > 1 + 2\delta$ for all n. Given the differential equation for w, it is not difficult to see that this implies that there exists an $\varepsilon > 0$ such that, for all n and all $t \in [t_n, t_n + \varepsilon]$, we have $\|\Gamma w(t)\| > 1 + \delta$.

If we look at our definition for s, we see that the first component asymptotically converges to Γw. The second term is bounded by α. Combining this with the property for Γw that we just derived, it is easily seen that the vector-valued signal $s(t)$ has a component $s_i(t)$ for which we have, for all n, either

$$s_i(t) > \delta \quad \text{for all } t \in [t_n, t_n + \varepsilon] \tag{5.31}$$

or

$$s_i(t) < -\delta \quad \text{for all } t \in [t_n, t_n + \varepsilon]. \tag{5.32}$$

Similarly, if (5.27) is not satisfied, there exist $\{t_n\}_{n=1}^{\infty}$ and a $\delta > 0$ such that $t_n \to \infty$ as $n \to \infty$ and $\|\Gamma \dot{w}(t_n)\| > 1 + 2\delta$ for all n. Given the differential equation for w, it is not difficult to see that this implies that there exists an $\varepsilon > 0$ such that, for all n and all $t \in [t_n, t_n + \varepsilon]$, we have $\|\Gamma \dot{w}(t)\| > 1 + \delta$. If we look at our definition for s, we see that the derivative of the first component asymptotically converges to $\Gamma \dot{w}$. The second term has a derivative which is bounded by β. Combining this with the property for $\Gamma \dot{w}$ that we just derived, it is easily seen that the vector-valued signal $s(t)$ has a component $s_i(t)$ for which we have, for all n, that there exist a sequence s_n and ε_2 such that

$$\dot{s}_i(t) > \delta \quad \text{for all } t \in [s_n, s_n + \varepsilon_2] \tag{5.33}$$

or

$$\dot{s}_i(t) < -\delta \quad \text{for all } t \in [s_n, s_n + \varepsilon_2]. \tag{5.34}$$

For any n it is easy to see that, if (5.33) is satisfied, then either

$$|s_i(t)| > \delta \quad \text{for all } t \in [s_n, s_n + \varepsilon_2/4]$$

or

$$|s_i(t)| > \delta \quad \text{for all } t \in [s_n + \varepsilon_2/3, s_n + \varepsilon_2].$$

This implies that either for $t_n = s_n$ or for $t_n = s_n + \varepsilon_2/2$ equation (5.31) or (5.32) is satisfied with $\varepsilon = \varepsilon_2/4$.

So we know that, if either (5.26) or (5.27) is not satisfied, then we can find a sequence $\{t_n\}$ and ε for which either (5.31) or (5.32) is satisfied for each n. Moreover, if we apply $s(t)$ to the system

$$\begin{aligned} \dot{x} &= Ax + Bs \\ e &= C_e x + D_{eu} s \end{aligned} \tag{5.35}$$

with zero initial conditions, then we have $e(t) \to 0$ (since both the linear feedback and the non-linear saturating feedback achieve output regulation). This is in contradiction with the results of Theorem 3.A.2, and therefore we show, by contradiction, that (5.26) and (5.27) must be satisfied. ■

Design of a low-gain state feedback regulator:

For clarity, we now give a step by step design of a low-gain state feedback regulator.

Step 1 : Find a solution (Π, Γ) of the regulator equation (5.19).

Step 2 : For continuous-time systems, find a low-gain state feedback matrix $F_\varepsilon = -B^{\mathrm{T}} P_\varepsilon$ where P_ε is defined by (5.10). Similarly, for discrete-time systems, find a low-gain state feedback matrix $F_\varepsilon = -(B^{\mathrm{T}} P_\varepsilon B + I)^{-1} B^{\mathrm{T}} P_\varepsilon A$ where P_ε is defined by (5.14).

Step 3 : Given the sets $\mathcal{X}_0$, $\mathcal{X}_s$, and $\mathcal{W}_0$, find an ε^* by the procedure given in the proof of Theorem 5.4.1.

Then, for some $\varepsilon \in (0, \varepsilon^*]$, the state feedback controller given in (5.20) for continuous-time systems, or by (5.21) for discrete-time systems solves the semi-global state feedback output regulation problem. □

5.5 Output regulation via dynamic measurement feedback control

In this section, we consider the semi-global linear observer based measurement feedback output regulation problem for the system,

$$\begin{aligned} \rho x &= Ax + B\sigma_{\alpha,\beta}(u) + E_w w \\ \rho w &= Sw \\ y &= C_y x + D_{yw} w \\ e &= C_e x + D_{eu}\sigma_{\alpha,\beta}(u) + D_{ew} w \end{aligned} \tag{5.36}$$

Here, as usual, $y \in \mathbb{R}^p$ is a measured output.

Problem 5.5.1 *Consider the system (5.36) along with the definition for the constraint operator $\sigma_{\alpha,\beta}$ given in equations (5.5) and (5.6). Also, consider a compact set $\mathcal{W}_0 \subset \mathbb{R}^s$. The **semi-global linear observer based measurement feedback output regulation problem** is defined as follows. For any a priori given (arbitrarily large) bounded sets $\mathcal{X}_0 \subset \mathbb{R}^n$ and $\mathcal{Z}_0 \subset \mathbb{R}^{n+s}$, find, if possible, a measurement feedback law of the form,*

$$\begin{aligned} \rho \begin{pmatrix} \hat{x} \\ \hat{w} \end{pmatrix} &= \begin{pmatrix} A & E_w \\ 0 & S \end{pmatrix} \begin{pmatrix} \hat{x} \\ \hat{w} \end{pmatrix} + \begin{pmatrix} B \\ 0 \end{pmatrix} \sigma_{\alpha,\beta}(u) \\ &\quad + \begin{pmatrix} K_A \\ K_S \end{pmatrix} \left(\begin{pmatrix} C_y & D_{yw} \end{pmatrix} \begin{pmatrix} \hat{x} \\ \hat{w} \end{pmatrix} - y \right) \\ u &= F\hat{x} + G\hat{w} \end{aligned} \tag{5.37}$$

such that the following conditions hold:

(*i*) *The equilibrium* $(x, x_s, \hat{x}, \hat{w}) = (0, 0, 0, 0)$ *of*

$$\begin{aligned} \rho x &= Ax + B\sigma_{\alpha,\beta}(F\hat{x} + G\hat{w}) \\ \rho \begin{pmatrix} \hat{x} \\ \hat{w} \end{pmatrix} &= \begin{pmatrix} A & E_w \\ 0 & S \end{pmatrix} \begin{pmatrix} \hat{x} \\ \hat{w} \end{pmatrix} + \begin{pmatrix} B \\ 0 \end{pmatrix} \sigma_{\alpha,\beta}(F\hat{x} + G\hat{w}) \\ &\quad + \begin{pmatrix} K_A \\ K_S \end{pmatrix} \begin{pmatrix} C_y & D_{yw} \end{pmatrix} \begin{pmatrix} \hat{x} - x \\ \hat{w} \end{pmatrix} \end{aligned}$$

is asymptotically stable with $\mathcal{X}_0 \times \mathcal{X}_s \times \mathcal{Z}_0$ *contained in its basin of attraction.*

(*ii*) *For all* $(x(0), x_s, \hat{x}(0), \hat{w}(0)) \in \mathcal{X}_0 \times \mathcal{X}_s \times \mathcal{Z}_0$ *and* $w(0) \in \mathcal{W}_0$, *the solution of the closed-loop system satisfies*

$$\lim_{t\to\infty} e(t) = 0. \tag{5.38}$$

Remark 5.5.1 *We would like to emphasize again that our definition of the above semi-global measurement feedback output regulation problem does not view the set of initial conditions of the plant and the initial conditions of the controller dynamics as given data. The set of given data consists of the models of the plant and the exosystem and the set of initial conditions for the exosystem. Therefore, the solvability conditions* must *be independent of the set of initial conditions of the plant* $\mathcal{X}_0 \times \mathcal{X}_s$, *and the set of initial conditions for the controller dynamics,* $\mathcal{Z}_0$.

The solvability conditions for the above semi-global measurement feedback output regulation problem are given in the following theorem.

Theorem 5.5.1 *Consider the system (5.36) along with the definition for the constraint operator* $\sigma_{\alpha,\beta}$ *given in equations (5.3), (5.5) and (5.6). Also, consider the given compact set* $\mathcal{W}_0 \subset \mathbb{R}^s$. *The semi-global measurement feedback output regulation problem is solvable if the following conditions hold:*

(*i*) (A, B) *is stabilizable and* A *has all its eigenvalues in the closed left half plane (continuous-time) or in the closed unit disc (discrete-time). Moreover, the pair*

$$\left(\begin{pmatrix} C_y & D_{yw} \end{pmatrix}, \begin{pmatrix} A & E_w \\ 0 & S \end{pmatrix} \right) \tag{5.39}$$

is detectable.

(*ii*) *There exist matrices* Π *and* Γ *such that* (*a*) *they solve the regulator equation (5.19), and*

(*b*) *there exist a* $\delta > 0$ *and a* $T \geqslant 0$ *such that*

$$\|\Gamma w\|_{\infty,T} \leqslant \alpha - \delta$$

and

$$\begin{aligned} \|\Gamma S w\|_{\infty,T} &\leqslant \beta - \delta && \textit{(continuous-time)} \\ \|\Gamma (S - I) w\|_{\infty,T} &\leqslant \beta - \delta && \textit{(discrete-time)} \end{aligned}$$

for all w *with* $w(0) \in \mathcal{W}_0$.

Moreover, if these conditions are satisfied, then in continuous-time a suitable family of linear static measurement feedback laws is given by,

$$\begin{aligned} \rho\hat{x} &= A\hat{x} + B\sigma_{\alpha,\beta}(u) + E_w\hat{w} + K_A\left[C_y\hat{x} + D_{yw}\hat{w} - y\right] \\ \rho\hat{w} &= S\hat{w} + K_S\left[C_y\hat{x} + D_{yw}\hat{w} - y\right] \\ u &= F_\varepsilon\hat{x} + (\Gamma - F_\varepsilon\Pi)\hat{w}, \end{aligned} \tag{5.40}$$

with $F_\varepsilon = -B^T P_\varepsilon$ *with* P_ε *defined by (5.10) in continuous-time, while in discrete-time* $F_\varepsilon = -(B^T P_\varepsilon B + I)^{-1} B^T P_\varepsilon A$ *with* P_ε *defined by (5.14).*

The gains K_A *and* K_S *are chosen such that the matrix,*

$$\bar{A} := \begin{pmatrix} A + K_A C_y & E_w + K_A D_{yw} \\ K_S C_y & S + K_S D_{yw} \end{pmatrix}, \tag{5.41}$$

is Hurwitz-stable for continuous-time systems and Schur-stable for discrete-time systems.

Proof : We prove this theorem only for continuous-time. The discrete-time version can be derived similarly. Consider the family of feedback laws given in (5.20). We know that (5.11) is satisfied for $F_\varepsilon = -B^T P_\varepsilon$ where υ is a positive-valued function satisfying $\lim_{\varepsilon\to 0} \upsilon_\varepsilon = 0$.

The closed-loop system consisting of the given system (5.15) and the given family of feedback laws can be written as,

$$\begin{aligned} \dot{x} &= Ax + B\sigma_{\alpha,\beta}(F_\varepsilon\hat{x} + (\Gamma - F_\varepsilon\Pi)\hat{w}) + E_w w \\ \dot{\hat{x}} &= A\hat{x} + B\sigma_{\alpha,\beta}(F_\varepsilon\hat{x} + (\Gamma - F_\varepsilon\Pi)\hat{w}) + E_w\hat{w} \\ &\qquad + K_A C_y(\hat{x} - x) + K_A D_{yw}(\hat{w} - w) \\ \dot{\hat{w}} &= S\hat{w} + K_S C_y(\hat{x} - x) + K_S D_{yw}(\hat{w} - w). \end{aligned} \tag{5.42}$$

We then adopt the invertible change of state variables,

$$\begin{aligned} \xi &= x - \Pi w \\ \tilde{x} &= x - \hat{x} \\ \tilde{w} &= w - \hat{w}, \end{aligned} \tag{5.43}$$

and then rewrite the closed-loop system (5.42) as

$$\begin{aligned} \dot{\xi} &= A\xi + B\sigma_{\alpha,\beta}(F_\varepsilon\xi + \Gamma w - \Gamma\tilde{w} - F_\varepsilon\tilde{x} + F_\varepsilon\Pi\tilde{w}) \\ &\qquad +(A\Pi - \Pi S + E_w)w \\ \dot{\tilde{x}} &= (A + K_\mathrm{A}C_y)\tilde{x} + (E_w + K_\mathrm{A}D_{yw})\tilde{w} \\ \dot{\tilde{w}} &= K_\mathrm{S}C_y\tilde{x} + (S + K_\mathrm{S}D_{yw})\tilde{w}. \end{aligned} \tag{5.44}$$

To show that item (i) of Problem 5.5.1 holds, we note that (5.37) is equal to (5.42) for $w = 0$. Thus, for $w = 0$, (5.44) reduces to

$$\begin{aligned} \dot{\xi} &= A\xi + B\sigma_{\alpha,\beta}(F_\varepsilon\xi - \Gamma\tilde{w} - F_\varepsilon\tilde{x} + F_\varepsilon\Pi\tilde{w}) \\ \dot{\tilde{x}} &= (A + K_\mathrm{A}C_y)\tilde{x} + (E_w + K_\mathrm{A}D_{yw})\tilde{w} \\ \dot{\tilde{w}} &= K_\mathrm{S}C_y\tilde{x} + (S + K_\mathrm{S}D_{yw})\tilde{w}. \end{aligned} \tag{5.45}$$

Recalling that the matrix $\bar{A}$, defined in (5.41), is Hurwitz-stable, it readily follows from the last two equations of (5.45) that there exists a $T_1 \geqslant 0$ such that, for all possible initial conditions $(\tilde{x}(0), x_s, \tilde{w}(0))$,

$$\|\Gamma\tilde{w}\|_{\infty,\mathrm{T}_1} \leqslant \frac{\alpha}{4}, \quad \|F_\varepsilon\tilde{x}\|_{\infty,\mathrm{T}_1} \leqslant \frac{\alpha}{4}, \quad \|F_\varepsilon\Pi\tilde{w}\|_{\infty,\mathrm{T}_1} \leqslant \frac{\alpha}{4}, \tag{5.46}$$

$$\|\Gamma\dot{\tilde{w}}\|_{\infty,\mathrm{T}_1} \leqslant \frac{\beta}{4}, \quad \|F_\varepsilon\dot{\tilde{x}}\|_{\infty,\mathrm{T}_1} \leqslant \frac{\beta}{4}, \quad \|F_\varepsilon\Pi\dot{\tilde{w}}\|_{\infty,\mathrm{T}_1} \leqslant \frac{\beta}{4}, \tag{5.47}$$

for all $\varepsilon \in (0, 1]$. We next consider the first equation of (5.45). Note that $\xi(T_1)$ belongs to a bounded set independent of ε since $\xi(0)$ is bounded and since ξ is determined via a linear differential equation with bounded input $\sigma_{\alpha,\beta}(u)$. Hence there exists an M_1 such that for all possible initial conditions,

$$\|\xi(T_1)\| \leqslant M_1, \quad \text{for all } \varepsilon \in (0, 1]. \tag{5.48}$$

Let us now assume that, from time T_1 onwards, the saturation elements are non-existent. In this case, the first equation of (5.45) can be written as

$$\dot{\xi} = (A + BF_\varepsilon)\xi - BF_\varepsilon\tilde{x} - B\Gamma\tilde{w} + BF_\varepsilon\Pi\tilde{w}. \tag{5.49}$$

Since $\tilde{x} \to 0$ and $\tilde{w} \to 0$ exponentially with a decay rate independent of ε as $t \to \infty$, it follows trivially from (5.11) that there exist an $\varepsilon_1^* > 0$ and an

$M_2 > 0$ such that, for all possible initial conditions $\tilde{x}(0)$ and $\tilde{w}(0)$ and all $\varepsilon \in (0, \varepsilon_1^*]$,

$$\int_{T_1}^{\infty} \|e^{\zeta_\varepsilon \tau} B[F_\varepsilon \tilde{x}(\tau) + \Gamma \tilde{w}(\tau) - F_\varepsilon \Pi \tilde{w}(\tau)]\| d\tau \leqslant M_2. \tag{5.50}$$

This in turn shows that, for $t \geqslant T_1$,

$$\begin{aligned}
\|F_\varepsilon \xi(t)\| &= \Big\| F_\varepsilon e^{(A+BF_\varepsilon)t} \xi(T_1) \\
&\quad - \int_{T_1}^{t} F_\varepsilon e^{(A+BF_\varepsilon)(t-\tau)} B \left[F_\varepsilon \tilde{x}(\tau) + \Gamma \tilde{w}(\tau) - F_\varepsilon \Pi \tilde{w}(\tau)\right] d\tau \Big\| \\
&\leqslant \upsilon_\varepsilon M_1 + \upsilon_\varepsilon \int_{T_1}^{\infty} \|e^{\zeta_\varepsilon \tau} B[F_\varepsilon \tilde{x}(\tau) + \Gamma \tilde{w}(\tau) - F_\varepsilon \Pi \tilde{w}(\tau)]\| d\tau \\
&\leqslant \upsilon_\varepsilon (M_1 + M_2).
\end{aligned}$$

Choose $\varepsilon_2^* \in (0, \varepsilon_1^*]$ such that, for all $\varepsilon \in (0, \varepsilon_2^*]$,

$$\|F_\varepsilon \xi\|_{\infty, T_1} \leqslant \frac{\alpha}{4}. \tag{5.51}$$

Similarly, we can show that there exists an $\varepsilon_3^* \in (0, \varepsilon_2^*]$ such that, for all $\varepsilon \in (0, \varepsilon_3^*]$,

$$\|F_\varepsilon \dot{\xi}\|_{\infty, T_1} \leqslant \frac{\beta}{4}. \tag{5.52}$$

These two bounds, together with (5.46), show that the system (5.45) will operate linearly after time T_1 and local exponential stability of this linear system follows from the separation principle.

In summary, we have shown that there exists an $\varepsilon_3^* > 0$ such that, for all $\varepsilon \in (0, \varepsilon_3^*]$, the equilibrium point $(0, 0, 0, 0)$ of the system (5.45) is asymptotically stable, with $(\mathcal{X}_0, \mathcal{X}_s, \mathcal{Z}_0)$ contained in its basin of attraction.

We now proceed to show that item (ii) of Problem 5.5.1 also holds. To this end, we consider the closed-loop system (5.44). Recalling that the matrix $\bar{A}$ is Hurwitz-stable, it readily follows from the last two equations of (5.44) that there exists an $T_2 \geqslant T$ such that, for all possible initial conditions $(\tilde{x}(0), x_s, \tilde{w}(0))$,

$$\|\Gamma \tilde{w}\|_{\infty, T_2} \leqslant \frac{\delta}{4}, \quad \|F_\varepsilon \tilde{x}\|_{\infty, T_2} \leqslant \frac{\delta}{4}, \quad \|F_\varepsilon \Pi \tilde{w}\|_{\infty, T_2} \leqslant \frac{\delta}{4}, \tag{5.53}$$

$$\|\Gamma \dot{\tilde{w}}\|_{\infty, T_2} \leqslant \frac{\delta}{4}, \quad \|F_\varepsilon \dot{\tilde{x}}\|_{\infty, T_2} \leqslant \frac{\delta}{4}, \quad \|F_\varepsilon \Pi \dot{\tilde{w}}\|_{\infty, T_2} \leqslant \frac{\delta}{4}, \tag{5.54}$$

for all $\varepsilon \in (0, 1]$. We next consider the first equation of (5.44). Note that $\xi(T_2)$ belongs to a bounded set independent of ε since $\xi(0)$ is bounded and since ξ is determined via a linear differential equation with bounded inputs $\sigma_{\alpha,\beta}(u)$ and w. Hence there exists an M_3 such that for all possible initial conditions,

$$\|\xi(T_2)\| \leqslant M_3, \quad \text{for all } \varepsilon \in (0, 1]. \tag{5.55}$$

Let us now assume that, from time T_2 onwards, the equation (5.44) operates without the saturation elements. In view of condition (ii)a) of the theorem, the first equation of (5.44) in the absence of the saturation elements is equal to the first equation of (5.45), and hence also reduces to (5.49) after time T_2. Hence, using a similar argument as above, we can show that there exists an $\varepsilon_3^* > 0$ such that, for all $\varepsilon \in (0, \varepsilon_3^*]$,

$$\|F_\varepsilon \xi\|_{\infty,T_2} < \frac{\delta}{4}, \ \|F_\varepsilon \dot{\xi}\|_{\infty,T_2} < \frac{\delta}{4}. \tag{5.56}$$

This, together with (5.53) and condition (ii)b) of the theorem, shows that the system (5.44) will operate linearly after time T_2, and thus the exponential stability of this linear system follows from the separation principle.

Next, in view of the second equation of (5.19), it is easy to evaluate $e(t)$ for $t \geq T_2$ as

$$e(t) = (C_e + D_{eu}F_\varepsilon)\xi(t) - D_{eu}F_\varepsilon\tilde{x} - D_{eu}(\Gamma - F_\varepsilon\Pi)\tilde{w}.$$

This implies that $e(t) \to 0$ as $t \to \infty$. This completes our proof. ■

Design of a low-gain measurement feedback regulator:

For clarity, we now summarize the construction of an observer based measurement feedback regulator.

Step 1 : At first construct a low-gain state feedback gain matrix F_ε as was done in connection with state feedback.

Step 2 : Design a full order observer so that we can implement the controller with observer based architecture as given in (5.40). That is, find the matrix gains K_A and K_S in such a way that the matrix given in (5.41) is Hurwitz stable for continuous-time systems or Schur-stable for discrete-time systems.

Step 3 : Implement the observer based measurement feedback regulator as given in (5.40).

Step 4 : Given the sets $\mathcal{X}_0$, $\mathcal{X}_s$, $\mathcal{W}_0$, and $\mathcal{Z}_0$, find an ε^* by the procedure given in the proof of Theorem 5.5.1 .

Then for some $\varepsilon \in (0, \varepsilon^*]$ the observer based measurement feedback regulator as given in (5.40) solves the semi-global measurement feedback output regulation problem. □

5.6 Low-and-high-gain feedback regulator design

As already seen in the previous two chapters, low-gain based designs underutilize the available control capacity and the resulting convergence of $e(t)$ to zero is very slow. Clearly, the feedback law (5.20) utilizes a low-gain since P_ε as given by (5.10) converges to zero as ε becomes small. The same holds true for the measurement feedback designs which are based on the same low-gain state feedback. Our next goal is to recall a new design methodology which incorporates a significant improvement to the low-gain design method, and leads to a better utilization of the available control capacity and hence better closed-loop performance.

The improved design utilizes the concepts of low-and-high gain feedback as presented in the previous two chapters. If we just consider stability, then the low gain feedback $u = -B^{\mathrm{T}} P_\varepsilon x$ for the continuous-time system (5.4) achieves stability and the resulting domain of attraction is arbitrarily large for ε small enough. In order to improve the transient performance, the modified feedback $u = -(\mu + 1)B^{\mathrm{T}} P_\varepsilon x$ was shown to achieve stability and the same domain of attraction for any $\mu > 0$ in the case of amplitude saturation. However, this cannot be applied here because it is easy to construct examples to show that, in the case of rate limits, for μ large the domain of attraction can become arbitrarily small. The main problem is the fact that the rate limiter has memory and hence if u was large for some time then it takes a while before the input can become negative again and this delay causes the instability for large μ. Therefore we first present a different low-gain design which is more suited for this low-high gain methodology.

Consider the continuous-time system (5.15). We introduce the following modified system,

$$\begin{aligned} \dot{x}(t) &= Ax(t) + Bu(t) + E_w w(t) \\ \dot{u}(t) &= \sigma_\beta(v(t)) \\ \dot{w}(t) &= Sw(t) \\ e(t) &= C_e x(t) + D_{eu} u(t) + D_{ew} w(t). \end{aligned} \tag{5.57}$$

Then, for this system we derive a low-gain state feedback that solves the output regulation problem. Let $P_{\varepsilon,\zeta}$ be the solution of the following algebraic Riccati equation

$$0 = \begin{pmatrix} A & B \\ 0 & 0 \end{pmatrix}^{\mathrm{T}} P + P \begin{pmatrix} A & B \\ 0 & 0 \end{pmatrix} - P \begin{pmatrix} 0 & 0 \\ 0 & I \end{pmatrix} P + \begin{pmatrix} Q_\varepsilon & 0 \\ 0 & \zeta^2 I \end{pmatrix}. \tag{5.58}$$

Then, the feedback law

$$v = -(\mu + 1)\begin{pmatrix} 0 & I \end{pmatrix} P_{\varepsilon,\zeta} \begin{pmatrix} x \\ u \end{pmatrix} + \left((\mu + 1)\begin{pmatrix} 0 & I \end{pmatrix} P_{\varepsilon,\zeta} \begin{pmatrix} \Pi \\ \Gamma \end{pmatrix} + \Gamma S \right) w \tag{5.59}$$

for any fixed ζ solves the output regulation problem for the system (5.57) in the sense that the closed-loop system is stable while for ε small enough the domain of attraction can be chosen arbitrarily large and we achieve output regulation for any $w(0) \in \mathcal{W}_0$. Moreover, $\mathcal{W}_0$ will remain inside the domain of attraction for any $\mu \geqslant 0$. Hence, by choosing μ large we can improve the transient performance without affecting the domain of attraction. Then, we can apply the following dynamic state feedback to the original system (5.15),

$$\begin{aligned} \dot{u} = \sigma_\beta \Bigg(& -(\mu + 1)\begin{pmatrix} 0 & I \end{pmatrix} P_{\varepsilon,\zeta} \begin{pmatrix} x \\ u \end{pmatrix} \\ & + \left((\mu + 1)\begin{pmatrix} 0 & I \end{pmatrix} P_{\varepsilon,\zeta} \begin{pmatrix} \Pi \\ \Gamma \end{pmatrix} + \Gamma S \right) w \Bigg). \end{aligned} \tag{5.60}$$

Obviously, this interconnection works beautifully if we only have rate constraints and no amplitude constraints because the interconnection of (5.15) and (5.60) for $\alpha = \infty$ is equal to the interconnection of (5.57) and (5.59). In general this is not the case because of the amplitude saturation. However, we will show that for a suitable choice for ε and ζ the above feedback (5.60) has the desired properties when applied to the given system (5.15).

Theorem 5.6.1 *Consider the continuous-time system (5.15) along with the definition for the constraint operator $\sigma_{\alpha,\beta}$ given in equations (5.5) and (5.6). Also, consider the given compact set $\mathcal{W}_0 \subset \mathbb{R}^s$. Under the same solvability conditions as in Theorem 5.4.1, there exists, for any a priori given (arbitrarily large) bounded set $\mathcal{X}_0 \subset \mathbb{R}^n$ and any $\zeta \in (0, 1)$, an $\varepsilon^* > 0$ such that for each $\varepsilon \in (0, \varepsilon^*]$ and for each $\mu \geqslant 0$, the controller in the family (5.60) has the following properties:*

(i) The interconnection of (5.15) and (5.60) under the constraint $w \equiv 0$ is locally exponentially stable with $\mathcal{X}_0 \times \mathcal{X}_s \times [-\alpha, \alpha]^m$ contained in its basin of attraction.

(ii) For any $x(0) \in \mathcal{X}_0$, $x_s \in \mathcal{X}_s$, $u(0) \in [-\alpha, \alpha]^m$, $w(0) \in \mathcal{W}_0$ and arbitrary initial conditions for the rate limiter, the solution of the closed-loop system satisfies

$$\lim_{t \to \infty} e(t) = 0. \tag{5.61}$$

Remark 5.6.1 *Note that for $\mu = 0$, $\zeta \to 0$, and for a fixed ε, the controller converges to the low-gain feedback as presented in Theorem 5.3.1.*

Proof : First we show asymptotic stability and a suitable domain of attraction for the interconnection with $w = 0$. There exists a compact set $\mathcal{X}_1$ such that $x(t) \in \mathcal{X}_1$ for all $t \leqslant \alpha/\beta$, any $x(0) \in \mathcal{X}_0$, any input u and any initial condition for the rate limiter (because the input to the system is bounded).

Choose the Lyapunov function,

$$V_\varepsilon(x,u) = \begin{pmatrix} x^{\mathrm{T}} & u^{\mathrm{T}} \end{pmatrix} P_{\varepsilon,\zeta} \begin{pmatrix} x \\ u \end{pmatrix},$$

and let $c \geqslant 0$ be such that

$$\sup \left\{ V_\varepsilon(x,u) \mid x \in \mathcal{X}_1, u \in [-\alpha,\alpha]^m, \varepsilon \in (0,1] \right\} \leqslant c.$$

Next we note that there exists an $\varepsilon_1^* \in (0,1]$ such that

$$\left\| \begin{pmatrix} 0 & I \end{pmatrix} P_{\varepsilon,\zeta} \begin{pmatrix} x \\ u \end{pmatrix} \right\|_\infty < \alpha$$

for all x, u such that $V(x,u) < c$ and

$$\left[\begin{pmatrix} 0 & I \end{pmatrix} P_{\varepsilon,\zeta} \begin{pmatrix} x \\ u \end{pmatrix} \right]_i > 0$$

for any $i = 1, \ldots, m$ and all x, u such that $V(x,u) < c$ and such that $uq_i \geqslant \alpha$ where $[\cdot]_i$ denotes the i'th element of a vector. Because of symmetry we then also have,

$$\left[\begin{pmatrix} 0 & I \end{pmatrix} P_{\varepsilon,\zeta} \begin{pmatrix} x \\ u \end{pmatrix} \right]_i < 0$$

for any $i = 1, \ldots, m$ and all x, u such that $V(x,u) < c$ and such that $u_i \leqslant -\alpha$.

The existence of ε_1^* is guaranteed by the fact that

$$P_{\varepsilon,\zeta} \to \begin{pmatrix} 0 & 0 \\ 0 & \zeta I \end{pmatrix} \tag{5.62}$$

as $\varepsilon \to 0$.

The above properties imply that $u(0) \in [-\alpha,\alpha]^m$ and $V(x(t),u(t)) < c$ for all $t \in [0,T]$ guarantee that $u(T) \in [-\alpha,\alpha]^m$ for any $T > 0$. After all, if the ith coefficient of u becomes α, then the derivative is negative and if the ith coefficient of u becomes $-\alpha$ then the derivative is positive.

Since $\|u(t)\|_\infty < \alpha$ and $\|\dot{u}(t)\|_\infty < \beta$ for all $t \in [0, \alpha/\beta]$, we find from the characteristics of the rate limiter that $(\sigma_{\alpha,\beta}(u))(\alpha/\beta) = u(\alpha/\beta)$ independent of the initial conditions for u and the rate limiter. We also know that $x(\alpha/\beta) \in \mathcal{X}_1$.

Consider the interconnection of (5.57) and (5.59) with the same initial conditions as the interconnection of (5.15) and (5.60) at time $t = \alpha/\beta$. We can prove easily that the interconnection of (5.57) and (5.59) is stable and that $V(x(t), u(t)) < c$ for all $t \in [\alpha/\beta, \infty)$. Therefore the interconnection of (5.57) and (5.59) is such that $\|u(t)\|_\infty$ is bounded by α for all $t > \alpha/\beta$. The latter then implies that the solution of the interconnection of (5.15) and (5.60) is equal to the solution of the stable interconnection of (5.15) and (5.60) for all $t > \alpha/\beta$. This clearly implies stability and the required domain of attraction.

Next we need to show that we do achieve tracking. There exists a compact set $\mathcal{X}_2$ such that $x(t) \in \mathcal{X}_2$ for all $t \leqslant T + \alpha/\beta$, for any input u and for any initial conditions $w(0) \in \mathcal{W}_0$, $x(0) \in \mathcal{X}_0$, $u(0) \in [-\alpha, \alpha]^m$ and any initial conditions for the rate limiter (because $\sigma_{\alpha,\beta}(u)$ and Γw are both bounded).

Choose the Lyapunov function,

$$V_\varepsilon(x - \Pi w, u - \Gamma w) = \begin{pmatrix} x - \Pi w \\ u - \Gamma w \end{pmatrix} P_{\varepsilon,\zeta} \begin{pmatrix} x - \Pi w \\ u - \Gamma w \end{pmatrix}$$

and let $c \geqslant 0$ be such that

$$\sup\{\, V_\varepsilon(x - \Pi w, u - \Gamma w) \mid x \in \mathcal{X}_2, u \in [-\alpha, \alpha]^m, \varepsilon \in (0, 1]\,\} \leqslant c.$$

Next we note that there exists an $\varepsilon_2^* \in (0, \varepsilon_1^*]$ such that

$$\left\| \begin{pmatrix} 0 & I \end{pmatrix} P_{\varepsilon,\zeta} \begin{pmatrix} x - \Pi w \\ u - \Gamma w \end{pmatrix} \right\|_\infty < \alpha$$

for all x, u, w such that $V(x - \Pi w, u - \Gamma w) < c$ and

$$\left[\begin{pmatrix} 0 & I \end{pmatrix} P_{\varepsilon,\zeta} \begin{pmatrix} x - \Pi w \\ u - \Gamma w \end{pmatrix} \right]_i > 0$$

for any $i = 1, \ldots, m$ and all x, u such that $V(x - \Pi w, u - \Gamma w) < c$ and such that $[u - \Gamma w]_i \geqslant \delta$.

Note that if $[u]_i(t) \geqslant \alpha$ then $[u - \Gamma w]_i > \delta$ and hence $[\dot{u}]_i(t) < 0$ for all $t > T$. Therefore there exists a $T_1 > T$ such that $\|u(t)\|_\infty$ is bounded by α for all $t \in [T_1, T_1 + \alpha/\beta]$. Obviously, $\|\dot{u}(t)\|_\infty$ is bounded by β for all $t \in [T_1, T_1 + \alpha/\beta]$ and hence for $t = T_1 + \alpha/\beta$ we have $(\sigma_{\alpha,\beta}(u))(t) = u(t)$.

Since $\|\Gamma S w\|_{\infty,T_1} < \beta - \delta$ we can see easily that the interconnection of (5.57) and (5.59) achieves asymptotic tracking. We choose the initial conditions at time $T_1 + \alpha/\beta$ for the interconnection of (5.57) and (5.59) equal to the same initial conditions for the interconnection from (5.15) and (5.60). Since the solution of the interconnection of (5.57) and (5.59) is such that $\|u(t)\|_\infty$ is bounded by α, we find that the solution of the interconnection of (5.15) and (5.60) is equal to the solution of (5.57) and (5.59) and therefore also achieves tracking. ■

Next we focus on discrete-time systems. Consider the discrete-time system (5.15) with the additional assumption that $m = 1$ (single input). A more general result can be obtained following a similar development as in Chapter 4 but is very technical. We introduce the following modified system,

$$\begin{aligned} x(k+1) &= Ax(k) + Bu(k) + E_w w(k) \\ u(k+1) &= u(k) + \sigma_\beta(v(k)) \\ w(k+1) &= Sw(k) \\ e(k) &= C_e x(k) + D_{eu} u(k) + D_{ew} w(k). \end{aligned} \tag{5.63}$$

Then for this system we derive a low-gain state feedback law that solves the output regulation problem. Let $P_{\varepsilon,\zeta}$ be the solution of the algebraic Riccati equation,

$$\begin{aligned} P = &\begin{pmatrix} A & B \\ 0 & 1 \end{pmatrix}^{\mathrm{T}} P \begin{pmatrix} A & B \\ 0 & 1 \end{pmatrix} + \begin{pmatrix} Q_\varepsilon & 0 \\ 0 & \zeta^2 \end{pmatrix} \\ &- \begin{pmatrix} A & B \\ 0 & 1 \end{pmatrix}^{\mathrm{T}} P \begin{pmatrix} 0 \\ 1 \end{pmatrix} \left[\begin{pmatrix} 0 & 1 \end{pmatrix} P \begin{pmatrix} 0 \\ 1 \end{pmatrix} + 1 \right]^{-1} \begin{pmatrix} 0 \\ 1 \end{pmatrix}^{\mathrm{T}} P \begin{pmatrix} A & B \\ 0 & 1 \end{pmatrix}. \end{aligned} \tag{5.64}$$

Then the feedback law

$$v = -(F_{\varepsilon,\zeta} + \mu K_{\varepsilon,\zeta}) \left[\begin{pmatrix} x \\ u \end{pmatrix} - \begin{pmatrix} \Pi \\ \Gamma \end{pmatrix} w \right] + \Gamma S w \tag{5.65}$$

for a fixed ζ where

$$\begin{aligned} F_{\varepsilon,\zeta} &= \left[\begin{pmatrix} 0 & 1 \end{pmatrix} P_{\varepsilon,\zeta} \begin{pmatrix} 0 \\ 1 \end{pmatrix} + 1 \right]^{-1} \begin{pmatrix} 0 \\ 1 \end{pmatrix}^{\mathrm{T}} P_{\varepsilon,\zeta} \begin{pmatrix} A & B \\ 0 & 1 \end{pmatrix}, \\ K_{\varepsilon,\zeta} &= \left[\begin{pmatrix} 0 & 1 \end{pmatrix} P_{\varepsilon,\zeta} \begin{pmatrix} 0 \\ 1 \end{pmatrix} \right]^{-1} \begin{pmatrix} 0 \\ 1 \end{pmatrix}^{\mathrm{T}} P_{\varepsilon,\zeta} F_{\varepsilon,\zeta}, \end{aligned}$$

solves the output regulation problem for the system (5.57) in the sense that the closed-loop system is stable while for ε small enough the domain of attraction can be chosen arbitrarily large and we achieve output regulation for any $w(0) \in \mathcal{W}_0$. Moreover, by choosing $\mu \in [0, 2]$ we can improve the transient performance without affecting the domain of attraction. Then, we can apply the following dynamic state feedback to the original system (5.15),

$$\rho u = u + \sigma_\beta\left(-(F_{\varepsilon,\zeta} + \mu K_{\varepsilon,\zeta})\left[\begin{pmatrix} x \\ u \end{pmatrix} - \begin{pmatrix} \Pi \\ \Gamma \end{pmatrix} w\right] + \Gamma S w\right). \tag{5.66}$$

Obviously, this interconnection works beautifully if we only have rate constraints and no amplitude constraints because the interconnection of (5.15) and (5.66) for $\alpha = \infty$ is equal to the interconnection of (5.63) and (5.65). In general this is not the case because of the amplitude saturation. However, we will show that for a suitable choice for ζ the above feedback (5.66) has the desired properties when applied to the system (5.15).

Theorem 5.6.2 *Consider the discrete-time system (5.15) with $m = 1$ (single input) along with the definition for the constraint operator $\sigma_{\alpha,\beta}$ given in equation (5.3). Also, consider the given compact set $\mathcal{W}_0 \subset \mathbb{R}^s$. Assume B is injective and that the solvability conditions of Theorem 5.4.1 are satisfied. In that case, there exists, for any a priori given (arbitrarily large) bounded set $\mathcal{X}_0 \subset \mathbb{R}^n$ and any $\zeta \in (0, 1)$, an $\varepsilon^* > 0$ such that for each $\varepsilon \in (0, \varepsilon^*]$ and for each $\mu \in [0, 2]$, the controller in the family (5.66) has the following properties:*

(i) *The interconnection of (5.15) and (5.66) under the constraint $w = 0$ is asymptotically stable with $\mathcal{X}_0 \times \mathcal{X}_s \times [-\alpha, \alpha]$ contained in its basin of attraction.*

(ii) *For any $x(0) \in \mathcal{X}_0$, $x_s \in \mathcal{X}_s$, $u(0) \in [-\alpha, \alpha]$, $w(0) \in \mathcal{W}_0$, and arbitrary initial conditions for the rate limiter the solution of the closed-loop system satisfies*

$$\lim_{k\to\infty} e(k) = 0. \tag{5.67}$$

Remark 5.6.2 *Note that for $\mu = 0$, $\zeta \to 0$, and for a fixed ε, the controller converges to the low-gain feedback as presented in Theorem 5.3.2.*

Proof : In view of the proofs of Theorems 4.3.2 and 4.3.3, the proof of this theorem can be obtained using the same kind of arguments as in the proof of Theorem 5.6.1. ■

We now proceed to discuss measurement feedback controllers while using low-and-high-gain feedback. In connection with continuous-time systems, it seems obvious that we can apply the same argument as before to improve the performance of measurement feedback controllers by combining the observer used in (5.40) with the low-high gain state feedback controller presented in (5.60).

We define the following family of linear dynamic measurement feedback laws,

$$\begin{aligned}\dot{\hat{x}} &= A\hat{x} + B\sigma_{\alpha,\beta}(u) + E_w\hat{w} + K_{\mathrm{A}}C_y(\hat{x}-x) + K_{\mathrm{A},\ell}D_{yw}(\hat{w}-w)\\ \dot{\hat{w}} &= S\hat{w} + K_{\mathrm{S},\ell}C_y(\hat{x}-x) + K_{\mathrm{S}}D_{yw}(\hat{w}-w)\\ \dot{u} &= \sigma_\beta\Bigg(-(\mu+1)\begin{pmatrix}0 & I\end{pmatrix}P_{\varepsilon,\zeta}\begin{pmatrix}\hat{x}\\ u\end{pmatrix}\\ &\qquad + \left[(\mu+1)\begin{pmatrix}0 & I\end{pmatrix}P_{\varepsilon,\zeta}\begin{pmatrix}\Pi\\ \Gamma\end{pmatrix} + \Gamma S\right]\hat{w}\Bigg).\end{aligned} \tag{5.68}$$

However, we will see that we need stronger conditions on the observer. It is no longer sufficient to choose a fixed observer such that (5.41) is stable. Therefore we choose the observer gain parameterized by ℓ.

Let R_ℓ be the solution of the dual algebraic Riccati equation,

$$\begin{aligned}0 = \begin{pmatrix}A+\ell I & E_w\\ 0 & S+\ell I\end{pmatrix}R_\ell + R_\ell\begin{pmatrix}A+\ell I & E_w\\ 0 & S+\ell I\end{pmatrix}^{\mathrm{T}}\\ - R_\ell\begin{pmatrix}C_y^{\mathrm{T}}\\ D_{yw}^{\mathrm{T}}\end{pmatrix}\begin{pmatrix}C_y & D_{yw}\end{pmatrix}R_\ell + \ell I.\end{aligned} \tag{5.69}$$

We recall that we have certain properties of this Riccati equation as outlined in Lemma 3.3.2. We choose the following observer gain,

$$K_\ell = \begin{pmatrix}K_{\mathrm{A},\ell}\\ K_{\mathrm{S},\ell}\end{pmatrix} = -R_\ell\begin{pmatrix}C_y & D_{yw}\end{pmatrix}^{\mathrm{T}}.$$

We have the following result.

Theorem 5.6.3 *Consider the continuous-time system (5.36) along with the definition for the constraint operator $\sigma_{\alpha,\beta}$ given in equations (5.5) and (5.6). Also, consider the given compact set $\mathcal{W}_0 \subset \mathbb{R}^s$. Under the solvability conditions of Theorem 5.5.1 and the additional condition that the pair (5.39) is observable, there exists, for any a priori given (arbitrarily large) bounded sets $\mathcal{X}_0 \subset \mathbb{R}^n$ and $\mathcal{Z}_0 \subset \mathbb{R}^{n+s}$ and any $\zeta \in (0,1)$, an $\varepsilon^* > 0$ and a $\ell^*(\varepsilon,\mu)$ such that for each $\varepsilon \in (0,\varepsilon^*]$, $\mu \geqslant 0$ and $\ell > \ell^*(\varepsilon,\mu)$ the controller (5.68) has the following properties:*

(*i*) *The interconnection of (5.36) and (5.68) under the constraint* $w \equiv 0$ *is locally exponentially stable with* $\mathcal{X}_0 \times \mathcal{Z}_0 \times \mathcal{X}_s \times [-\alpha, \alpha]^m$ *contained in its basin of attraction.*

(*ii*) *For any* $x(0) \in \mathcal{X}_0$, $x_s \in \mathcal{X}_s$, $(\hat{x}(0), \hat{w}(0)) \in \mathcal{Z}_0$, $u(0) \in [-\alpha, \alpha]^m$, $w(0) \in \mathcal{W}_0$, *and arbitrary initial conditions for the rate limiter, the solution of the closed-loop system satisfies*

$$\lim_{t\to\infty} e(t) = 0. \tag{5.70}$$

Proof : We first note that the estimation error

$$\xi = \begin{pmatrix} \hat{x} - x \\ \hat{w} - w \end{pmatrix}$$

satisfies

$$\dot{\xi}(t) = \left[\begin{pmatrix} A & E_w \\ 0 & S \end{pmatrix} - R_\ell \begin{pmatrix} C_y^{\mathrm{T}} \\ D_{yw}^{\mathrm{T}} \end{pmatrix} \begin{pmatrix} C_y & D_{yw} \end{pmatrix} \right] \xi(t),$$

and, using the properties from Lemma 3.3.2, we find that

$$\frac{d}{dt} \xi^{\mathrm{T}}(t) R_\ell \xi(t) \leqslant -\ell \xi^{\mathrm{T}}(t) R_\ell \xi(t) - \ell \xi^{\mathrm{T}}(t) \xi(t).$$

Moreover, for any given bounded set Ξ and a $T > 0$ there exists an ℓ^* such that for all $\ell > \ell^*$ we will have

$$\xi^{\mathrm{T}}(T) R_\ell \xi(T) \leqslant 1$$

for any $\xi(0) \in \Xi$. The proof basically uses the same arguments as the proof of Theorem 5.6.1 to relate this result to the result for the measurement feedback case of Theorem 3.3.4. ∎

In connection with discrete-time systems, we use the combination of the observer used in (5.40) with the low-high gain state feedback controller presented in (5.66).

We define the following family of linear dynamic measurement feedback laws,

$$\begin{aligned} \rho \hat{x} &= A\hat{x} + B\sigma_{\alpha,\beta}(u) + E_w \hat{w} + K_{\mathrm{A}} \left(C_y \hat{x} + D_{yw} \hat{w} - y \right) \\ \rho \hat{w} &= S\hat{w} + K_{\mathrm{S}} \left(C_y \hat{x} + D_{yw} \hat{w} - y \right) \\ \rho u &= u + \sigma_\beta \left(-(F_{\varepsilon,\zeta} + \mu K_{\varepsilon,\zeta}) \left[\begin{pmatrix} \hat{x} \\ u \end{pmatrix} - \begin{pmatrix} \Pi \\ \Gamma \end{pmatrix} \hat{w} \right] + \Gamma S \hat{w} \right). \end{aligned} \tag{5.71}$$

We choose the observer gain such that (5.41) has all its eigenvalues in the origin. We have the following result.

Theorem 5.6.4 *Consider the discrete-time system (5.36) with* $m = 1$ *(single input) and the given compact set* $\mathcal{W}_0 \subset \mathbb{R}^s$. *Under the solvability conditions of Theorem 5.5.1 and the additional conditions that the pair (5.39) is observable and that B is injective, there exists, for any a priori given (arbitrarily large) bounded sets* $\mathcal{X}_0 \subset \mathbb{R}^n$ *and* $\mathcal{Z}_0 \subset \mathbb{R}^{n+s}$ *and any* $\zeta \in (0, 1)$, *an* $\varepsilon^* > 0$ *such that for each* $\varepsilon \in (0, \varepsilon^*]$, $\mu \in [0, 2]$ *the controller (5.71) has the following properties:*

(i) *The interconnection of (5.36) and (5.71) under the constraint* $w \equiv 0$ *is locally exponentially stable with* $\mathcal{X}_0 \times \mathcal{Z}_0 \times \mathcal{X}_s \times [-\alpha, \alpha]$ *contained in its basin of attraction.*

(ii) *For any* $x(0) \in \mathcal{X}_0$, $x_s \in \mathcal{X}_s$, $(\hat{x}(0), \hat{w}(0)) \in \mathcal{Z}_0$, $u(0) \in [-\alpha, \alpha]$, $w(0) \in \mathcal{W}_0$ *and arbitrary initial conditions for the rate limiter, the solution of the closed-loop system satisfies*

$$\lim_{k \to \infty} e(k) = 0. \tag{5.72}$$

Proof : This proof can be obtained from Theorem 4.3.3 using the same kind of arguments as presented earlier in this chapter. ■

5.7 Issues of well-posedness and structural stability

We would like to reconsider the problems of well-posedness and structural stability as introduced in Chapter 2 but this time for linear systems subject to actuator amplitude and rate saturation. In this case there are no intrinsic differences between systems subject only to amplitude saturation and systems which are subject to both rate and amplitude saturation and therefore the results in this section are almost identical with the obvious changes to the results of Sections 3.5 and 4.5.

Definition 5.7.1 ***(Well-posedness)*** *For a system* Σ *as in (5.36), the classical semi-global linear observer based measurement feedback output regulation problem as defined in Problem 3.3.2 is said to be well-posed at* $(A_0, B_0, E_{w,0}, C_{e,0}, D_{eu,0}, D_{ew,0}, C_{y,0}, D_{yu,0}, D_{yw,0}, S_0)$ *if there exists a neighborhood* $\mathcal{P}_0$ *of* $(A_0, B_0, E_{w,0}, C_{e,0}, D_{eu,0}, D_{ew,0}, C_{y,0}, D_{yu,0}, D_{yw,0}, S_0)$ *in the parameter space* $\mathcal{P}$ *such that the considered problem is solvable for each element* $(A, B, E_w, C_e, D_{eu}, D_{ew}, C_y, D_{yu}, D_{yw}, S)$ *of* $\mathcal{P}_0$ *for which A and S have all its eigenvalues in the closed left-half plane (continuous-time) or in the closed unit disc (discrete-time).*

We have the following result.

Theorem 5.7.1 *Consider a system Σ as in (5.36) and the classical semi-global linear observer based measurement feedback output regulation problem as defined in Problem 5.5.1. Let the conditions of Theorem 5.5.1 be satisfied for this system with nominal parameter values,*

$$\begin{aligned}(A, B, E_w, C_e, D_{eu}, D_{ew}, C_y, D_{yu}, D_{yw}, S)\\ = (A_0, B_0, E_{w,0}, C_{e,0}, D_{eu,0}, D_{ew,0}, C_{y,0}, D_{yu,0}, D_{yw,0}, S_0).\end{aligned}$$

Then the considered problem for Σ is well-posed at $(A_0, B_0, E_{w,0}, C_{e,0}, D_{eu,0}, D_{ew,0}, C_{y,0}, D_{yu,0}, D_{yw,0}, S_0)$ if and only if the matrix S is weakly Hurwitz-stable for continuous-time systems and weakly Schur-stable for discrete-time systems, and the matrix

$$\begin{pmatrix} A_0 - \lambda I & B_0 \\ C_{e,0} & D_{eu,0} \end{pmatrix}$$

has full row-rank for each λ which is an eigenvalue of S_0.

Remark 5.7.1 *In the above theorem, we did not perturb the set of initial conditions for the exosystem $\mathcal{W}_0$ but it is obvious that small perturbations of this set will not affect well-posedness.*

Next we consider the output regulation problem with structural stability. As already discussed in Section 2.8, we need to restrict our perturbations of the system parameters even more. As already discussed with well-posedness, we need to guarantee that even after perturbation, A still has all its eigenvalues in the closed left-half plane for continuous-time systems and inside or on the unit circle for discrete-time systems. But based on Section 2.8 we also need to exclude perturbations of the exosystem, i.e. we do not perturb S. Finally we need that the error signal is part of the measurement signal y, i.e. the parameters need to satisfy (2.36).

Definition 5.7.2 ***(Structurally stable output regulation problem)*** *Consider a system Σ as in (5.36) with the additional structure given in (2.36). A fixed controller is said to solve a structurally stable output regulation problem for Σ at $(A_0, B_0, E_{w,0}, C_{e,0}, D_{eu,0}, D_{ew,0}, C_{y2,0}, D_{yu2,0}, D_{yw2,0})$ if it satisfies the following properties:*

(*i*) *The controller solves the classical semi-global linear observer based measurement feedback output regulation problem when the plant in (2.1) is characterized by the nominal set of parameters* (A_0, B_0, $E_{w,0}$, $C_{e,0}$, $D_{eu,0}$, $D_{ew,0}$, $C_{y2,0}$, $D_{yu2,0}$, $D_{yw2,0}$).

(*ii*) *There exist a neighborhood* $\mathcal{P}_0$ *of* (A_0, B_0, $E_{w,0}$, $C_{e,0}$, $D_{eu,0}$, $D_{ew,0}$, $C_{y2,0}$, $D_{yu2,0}$, $D_{yw2,0}$) *such that the controller achieves internal stability and output regulation for each set of perturbed plant parameters* ($A_0 + \delta A$, $B_0 + \delta B$, $E_{w,0} + \delta E_w$, $C_{e,0} + \delta C_e$, $D_{eu,0} + \delta D_{eu}$, $D_{ew,0} + \delta D_{ew}$, $C_{y2,0}+\delta C_{y2}$, $D_{yu2,0}+\delta D_{yu2}$, $D_{yw2,0}+\delta D_{yw2}$) *in* $\mathcal{P}_0$ *for which* $A_0+\delta A$ *has all its eigenvalues in the closed left-half plane for continuous-time systems and inside or on the unit circle for discrete-time systems.*

In other words, as long as the perturbed parameters remain in $\mathcal{P}_0$, *we have* $\lim_{t\to\infty} e(t) = 0$ *for all* $x(0) \in \mathbb{R}^n$, $v(0) \in \mathbb{R}^{n_c}$, *and* $w(0)) \in \mathbb{R}^s$.

The above definition obviously implies that, for the existence of a regulator that solves the structurally stable output regulation problem, the exact output regulation problem must necessarily be well-posed (with the obvious modification implied by (2.36) and the fact that S is not perturbed).

A main technical complexity is the preliminary static output injection we applied in Section 2.8 to guarantee that A_0 and S have no common eigenvalues. These issues can be resolved as already alluded to in Section 3.5. As shown below, it turns out that the necessary and sufficient condition given in Theorem 4.5.1 for the well-posedness of the exact output regulation problem with measurement feedback is indeed also the necessary and sufficient condition for the existence of a regulator that solves the structurally stable output regulation problem.

Theorem 5.7.2 *Consider a system* Σ *as in (5.36) with the structural constraint (2.36). Let the conditions of Theorem 5.5.1 be satisfied for this system with nominal parameter values,*

$$(A, B, E_w, C_e, D_{eu}, D_{ew}, C_y, D_{yu}, D_{yw}) \\ = (A_0, B_0, E_{w,0}, C_{e,0}, D_{eu,0}, D_{ew,0}, C_{y,0}, D_{yu,0}, D_{yw,0}).$$

Then, there exists a regulator that solves the structurally stable output regulation problem for Σ *at* (A_0, B_0, $E_{w,0}$, $C_{e,0}$, $D_{eu,0}$, $D_{ew,0}$, $C_{y2,0}$, $D_{yu2,0}$, $D_{yw2,0}$) *if and only if the matrix* S *is weakly Hurwitz-stable for continuous-time systems and weakly Schur-stable for discrete-time systems, and the matrix*

$$\begin{pmatrix} A_0 - \lambda I & B_0 \\ C_{e,0} & D_{eu,0} \end{pmatrix} \tag{5.73}$$

has full row-rank for each λ which is an eigenvalue of S_0.

For linear systems the parameter perturbations could be arbitrarily large as long as stability is preserved. This is not the case here because the perturbations might be such that $\limsup_{t\to\infty} \|\Gamma w(t)\|_\infty > \alpha$ and then obviously output regulation is no longer possible. Similarly, if $\limsup_{t\to\infty} \|\Gamma S w(t)\|_\infty > \beta$ (continuous-time) or $\limsup_{t\to\infty} \|\Gamma(I - S)w(t)\|_\infty > \beta$ (discrete-time) for the perturbed system then output regulation is no longer possible.

We will only briefly indicate how the above result can be shown and how a suitable controller can be found. For ease of exposition we assume here that A_0 and S have no eigenvalues in common. Define $S_{\min}$, $\tilde{S}_p$, $\tilde{D}_{ew}$ and $\tilde{D}_{yw}$ as in Section 2.8.

We can now define an *auxiliary* system composed of the extended plant (2.1) and the *auxiliary* exosystem (2.41),

$$\tilde{\Sigma} : \begin{cases} \rho\tilde{x} = A_0\tilde{x} + B_0\sigma_{\alpha,\beta}(\tilde{u}) \\ \rho\tilde{w} = \tilde{S}_p\tilde{w} \\ \tilde{y} = C_{y,0}\tilde{x} + D_{yu,0}\sigma_{\alpha,\beta}(\tilde{u}) + \tilde{D}_{yw}\tilde{w} \\ \tilde{e} = C_{e,0}\tilde{x} + D_{eu,0}\sigma_{\alpha,\beta}(\tilde{u}) + \tilde{D}_{ew}\tilde{w}. \end{cases} \tag{5.74}$$

We design a low-gain or a low-high gain measurement feedback controller for this auxiliary system such that the system achieves output regulation for all initial conditions such that $\|\tilde{\Gamma}\tilde{w}\| < \alpha - \delta/2$ and $\|\tilde{\Gamma}S\tilde{w}\| < \beta - \delta/2$ (continuous-time) or $\|\tilde{\Gamma}(I - S)\tilde{w}\| < \beta - \delta/2$ (discrete-time) where $\tilde{\Gamma}$ is of course the solution of the regulator equation associated with this system. We can then show that this controller achieves structural stability for the original system if we know that the system $(A_0, B_0, C_{e,0}, D_{eu,0})$ is left-invertible.

Extensions to the case where A_0 and S have common eigenvalues and/or where $(A_0, B_0, C_{e,0}, D_{eu,0})$ is not left-invertible are available but due to their technicality have been ommited.

Chapter 6

Transient performance in classical output regulation

6.1 Introduction

Chapter 2 deals with the classical output regulation problem. One of its shortcomings is that it gives importance only to steady state tracking error, namely rendering it exactly equal to zero. The natural engineering issues regarding the transient behavior of the error signal are not addressed at all. Such issues can include minimizing the over-shoot or under-shoot of the error signal, or more generally appropriate shaping of the error signal. In this regard, for instance, one may like to impose in the statement of output regulation problem certain requirements on the transient performance so that one can shape appropriately the transient behavior of tracking error in addition to the desired steady state requirements. This is what we consider in this chapter.

A related work to this chapter is the work of Qiu and Davison [52] who consider servomechanism problems. However, this work has certain limitations. In particular the work of Qiu and Davison [52] assumes that the external disturbance signals and the signals that need to be tracked are either constant or sinusoidal in nature. Moreover, it considers only state feedback controllers. In this chapter, we rectify these short comings. We pose here several optimal and suboptimal output regulation problems. Next, utilizing both state feedback and measurement feedback controllers, a number of issues related to optimal and suboptimal output regulation are discussed. These issues include among others the determination of an expression for the infimum of performance measure, solvability conditions for the posed optimal and suboptimal output regulation problems, construction of optimal and suboptimal output regulators, relationship between the optimal performance and the structural properties of the given system, perfect output regulation etc. This chapter is based on the recent research work of authors [64].

6.2 Problem formulation

We start with a linear system with state space realization,

$$\Sigma : \begin{cases} \rho x = Ax + Bu + E_w w \\ e = C_e x + D_{eu} u + D_{ew} w \\ y = C_y x + D_{yu} u + D_{yw} w, \end{cases} \tag{6.1}$$

where as usual ρ is an operator indicating the time derivative $\frac{d}{dt}$ for continuous-time systems and a forward unit time shift for discrete-time systems. As before, Σ describes the plant with state $x \in \mathbb{R}^n$, control input $u \in \mathbb{R}^m$, measured output $y \in \mathbb{R}^p$, and tracking error $e \in \mathbb{R}^q$. The exogenous disturbance input $w \in \mathbb{R}^s$ is generated by an exosystem Σ_E with state space realization,

$$\Sigma_E : \{ \rho w = Sw. \tag{6.2}$$

Graphically, the given plant and the exosystem are depicted in Figure 6.1.

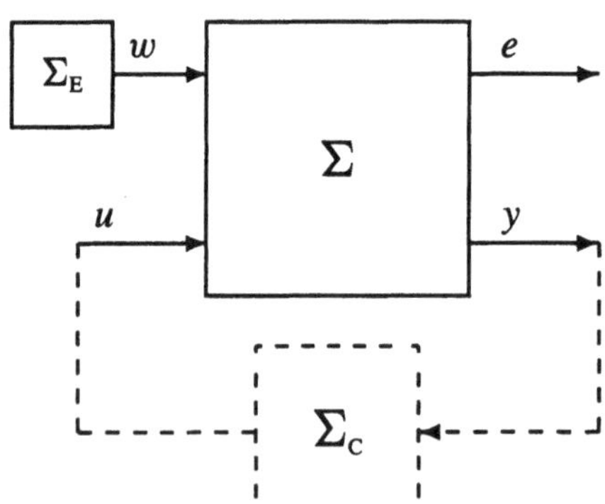

Figure 6.1: Regulation problem

As in the previous chapters, the measurement feedback controllers we seek are of the form,

$$\Sigma_C : \begin{cases} \rho v = A_c v + B_c y, \\ u = C_c v + D_c y. \end{cases} \tag{6.3}$$

As usual, the closed-loop system consisting of the given system Σ and the controller Σ_C is denoted by $\Sigma \times \Sigma_C$.

Although we have not yet formulated the problems we study in this chapter, it is evident from Chapter 2 that in any regulation problem the following assumptions are reasonable and almost necessary.

A.1. The pair (A, B) is stabilizable.

A.2. The matrix S is anti-Hurwitz-stable for continuous-time systems, and it is anti-Schur-stable for discrete-time systems.

A.3. The pair $\left(\begin{pmatrix} C_y & D_{yw} \end{pmatrix}, \begin{pmatrix} A & E_w \\ 0 & S \end{pmatrix} \right)$ is detectable.

A.4. There exists the pair (Π, Γ) that solve the *regulator equation*,

$$\begin{aligned} \Pi S &= A\Pi + B\Gamma + E_w, \\ 0 &= C_e\Pi + D_{eu}\Gamma + D_{ew}. \end{aligned} \tag{6.4}$$

Note that Assumptions A.1, A.2, and A.3 have been defined earlier on pages 19 and 25. Also, as discussed in earlier chapters, the solvability of the regulator equation is necessary to solve any problem with a regulation constraint.

We now recall the following statement of the classical output regulation problem.

Problem 6.2.1 *Consider the given system Σ as in (6.1) and the exosystem Σ_E as in (6.2). Then the* classical output regulation problem *is to find, if possible, a controller Σ_C of the form (6.3) such that the following two properties of internal stability and output regulation as defined in items (i) and (ii) below hold:*

(i) ***(Internal Stability)*** *In the absence of the disturbance w, the closed-loop system $\Sigma \times \Sigma_C$ is internally stable.*

(ii) ***(Output Regulation)*** *For all $x(0) = x_0 \in \mathbb{R}^n$ and $w(0) = w_0 \in \mathbb{R}^s$, the solution of the closed-loop system $\Sigma \times \Sigma_C$ satisfies*

$$\lim_{t \to \infty} e(t) = 0.$$

As usual, any controller Σ_C that solves the above classical output regulation problem is called a regulator.

The conditions under which the above classical output regulation problem can be solved are already known from the previous chapters and will be recalled soon. Clearly, the above problem besides requiring the internal stability of the closed-loop system, requires only that the steady state error e be zero. It does not consider any aspects of the transient behavior of the error signal e. That is, no performance measure is associated with the regulators. To rectify this, we introduce below a performance measure for each regulator.

(Performance Measure) For any given $x(0) = x_0 \in \mathbb{R}^n$ and $w(0) = w_0 \in \mathbb{R}^s$, and for any regulator, let a performance measure be given by

$$J(x_0, w_0, \Sigma_C) = \int_0^\infty e(t)^T Q e(t)\, dt \tag{6.5}$$

for continuous-time systems, and

$$J(x_0, w_0, \Sigma_C) = \sum_{i=0}^{\infty} e(i)^T Q e(i) \tag{6.6}$$

for discrete-time systems, where the matrix Q is positive semi-definite, and as usual ^{T} denotes the matrix transpose.

As we know by now, any regulator we construct is dependent on a solution (Π, Γ) of the regulator equation (6.4). However, in general, (Π, Γ) that solves (6.4) is not unique. In what follows, at first, we consider a class of regulators which are derived based on a given specific solution (Π, Γ) of (6.4). For this given class of regulators and for a given x_0 and w_0, we denote the infimum of the performance measure over all proper (or strictly proper) measurement feedback regulators by J_p^* (and J_{sp}^* respectively). Moreover, for the particular case of static state feedback regulators, we denote the infimum of the performance measure by J_{sf}^*. That is

$$\begin{aligned}
J_p^*(x_0, w_0, \Pi, \Gamma) &= \inf\{J(x_0, w_0, \Sigma_C) \mid \Sigma_C \text{ is a proper regulator} \\
&\qquad\qquad \text{utilizing the given } (\Pi, \Gamma)\}, \\
J_{sp}^*(x_0, w_0, \Pi, \Gamma) &= \inf\{J(x_0, w_0, \Sigma_C) \mid \Sigma_C \text{ is a strictly proper} \\
&\qquad\qquad \text{regulator utilizing the given } (\Pi, \Gamma)\}, \\
J_{sf}^*(x_0, w_0, \Pi, \Gamma) &= \inf\{J(x_0, w_0, \Sigma_C) \mid \Sigma_C \text{ is a state feedback} \\
&\qquad\qquad \text{regulator utilizing the given } (\Pi, \Gamma)\}.
\end{aligned}$$

The above infima of performance measures indeed define the best possible transient performance that could be achieved for a given initial conditions x_0 and w_0 and for a given solution (Π, Γ) of the regulator equation (6.4). In what follows one of the problems we deal with is to find the conditions under which one can obtain the best possible transient performance.

In all the problems we formulate in this chapter we assume that the classical output regulation problem is solvable. Given a solution (Π, Γ) of the regulator equation (6.4), our first problem seeks a proper (or a strictly proper) regulator that attains the optimal performance J_p^* (J_{sp}^* or J_{sf}^*) for a given x_0 and w_0.

Problem 6.2.2 *Consider the given system* Σ *as in (6.1) and the exosystem* Σ_E *as in (6.2). Let* x_0 *and* w_0 *be given. Also, let* (Π, Γ) *be a given solution of the regulator equation (6.4). With this choice of* (Π, Γ) *and the given* x_0 *and* w_0*, consider the corresponding optimal performance* $J_p^*(x_0, w_0, \Pi, \Gamma)$ *(*$J_{sp}^*(x_0, w_0, \Pi, \Gamma)$ *or* $J_{sf}^*(x_0, w_0, \Pi, \Gamma)$*). Then the* optimal output regulation problem for a given x_0 and w_0 *is to find, if possible, a proper (strictly proper or state feedback) regulator that attains the optimal performance* $J_p^*(x_0, w_0, \Pi, \Gamma)$ *(*$J_{sp}^*(x_0, w_0, \Pi, \Gamma)$ *or* $J_{sf}^*(x_0, w_0, \Pi, \Gamma)$*).*

Any regulator Σ_C *that solves the above optimal output regulation problem for a given* x_0 *and* w_0 *is called an optimal regulator for the given* x_0 *and* w_0*.*

It is transparent that the above problem deals with a given x_0 and w_0 at a time. In this regard, we would like to emphasize that the optimal output regulation problem for a given x_0 and w_0 as defined above seeks a controller that achieves output regulation for all initial conditions of the system and exosystem. However, it yields the best transient performance for the given initial conditions x_0 and w_0. A fundamental question that arises immediately is this: does a **fixed** regulator independent of x_0 and w_0 exist solving Problem 6.2.2 for the set of all x_0 and w_0 for which Problem 6.2.2 is solvable? This is formally posed below.

Problem 6.2.3 *Consider the given system* Σ *as in (6.1) and the exosystem* Σ_E *as in (6.2). Let* (Π, Γ) *be a given solution of the regulator equation (6.4). Then the* optimal output regulation problem *is to find, if possible, a* ***fixed*** *proper (strictly proper or state feedback) regulator that attains the optimal performance* $J_p^*(x_0, w_0, \Pi, \Gamma)$ *(*$J_{sp}^*(x_0, w_0, \Pi, \Gamma)$ *or* $J_{sf}^*(x_0, w_0, \Pi, \Gamma)$*) for the set of all* x_0 *and* w_0 *for which Problem 6.2.2 is solvable.*

Any regulator Σ_C *that solves the above optimal output regulation problem is called an optimal regulator.*

We would like to emphasize again that the optimal output regulation problem as defined above seeks a fixed controller that achieves output regulation for all initial condition of the system and exosystem. However, it yields the best transient performance for the set of all x_0 and w_0 for which Problem 6.2.2 is solvable.

As will be discussed soon, the above problems are solvable in general only for a set of initial conditions which satisfy some specific conditions. Such conditions may not always be satisfied by the given system Σ. In an attempt to weaken the conditions and thus broaden the set of initial conditions, one would like to construct a "suboptimal" regulator. In the absence of a formal definition of a suboptimal regulator, any regulator can be construed as a suboptimal regulator. For a given solution (Π, Γ) of the regulator equation (6.4) and for a given x_0 and w_0, a good definition of suboptimality can be given through the notion of ensuring that the performance is arbitrarily close to the infimum J_p^* (J_{sp}^* or J_{sf}^*). In this regard, a sequence or a family of proper (strictly proper or state feedback) regulators can be called suboptimal for the given x_0 and w_0 if one can select a regulator from the family such that the resulting performance is within an arbitrarily given value, say ε, from the infimum J_p^* (J_{sp}^* or J_{sf}^*). A formal definition of a suboptimal output regulation problem for a given x_0 and w_0 is given below.

Problem 6.2.4 *Consider the given system Σ as in (6.1) and the exosystem Σ_E as in (6.2). Let x_0 and w_0 be given. Also, let (Π, Γ) be a given solution of the regulator equation (6.4). With this choice of (Π, Γ) and the given x_0 and w_0, consider the corresponding optimal performance $J_p^*(x_0, w_0, \Pi, \Gamma)$ ($J_{sp}^*(x_0, w_0, \Pi, \Gamma)$ or $J_{sf}^*(x_0, w_0, \Pi, \Gamma)$). Then the* suboptimal output regulation problem for the given x_0 and w_0 *is to find, if possible, a sequence of parameterized proper (strictly proper or state feedback) regulators $\{\, \Sigma_C(\varepsilon) \mid \varepsilon > 0 \,\}$ which satisfy the condition that, as $\varepsilon \to 0$, the attained performance measure $J(x_0, w_0, \Sigma_C(\varepsilon))$ tends to the infimum $J_p^*(x_0, w_0, \Pi, \Gamma)$ ($J_{sp}^*(x_0, w_0, \Pi, \Gamma)$ or $J_{sf}^*(x_0, w_0, \Pi, \Gamma)$).*

The sequence of regulators $\{\, \Sigma_C(\varepsilon) \mid \varepsilon > 0 \,\}$ satisfying the above is referred to as a suboptimal regulator sequence for the given x_0 and w_0.

Obviously, the above problem deals with a given x_0 and w_0 at a time. In this regard, we would like to emphasize again that the suboptimal output regulation problem for a given x_0 and w_0 as defined above seeks a sequence of controllers each member of which achieves output regulation for all initial condition for the system and exosystem. However, by selecting an appropriate regulator from the sequence one can come as close to the best transient performance as desired for the given initial conditions x_0 and w_0. It is interesting to enquire whether there exists a **fixed** sequence of regulators independent of x_0

and w_0 which solves Problem 6.2.4 for all x_0 and w_0. This is formally posed below.

Problem 6.2.5 *Consider the given system Σ as in (6.1) and the exosystem Σ_E as in (6.2). Let (Π, Γ) be a given solution of the regulator equation (6.4). With this choice of (Π, Γ), consider the corresponding optimal performance $J_p^*(x_0, w_0, \Pi, \Gamma)$ ($J_{sp}^*(x_0, w_0, \Pi, \Gamma)$ or $J_{sf}^*(x_0, w_0, \Pi, \Gamma)$). Then the* suboptimal output regulation problem *is to find, if possible, a* ***fixed*** *sequence of parameterized proper (strictly proper or state feedback) regulators $\{\, \Sigma_C(\varepsilon) \mid \varepsilon > 0 \}$ which satisfy the condition that, as $\varepsilon \to 0$, the attained performance $J(x_0, w_0, \Sigma_C(\varepsilon))$ tends to the infimum $J_p^*(x_0, w_0, \Pi, \Gamma)$ ($J_{sp}^*(x_0, w_0, \Pi, \Gamma)$ or $J_{sf}^*(x_0, w_0, \Pi, \Gamma)$) uniformly for all x_0 and w_0.*

The sequence of regulators $\{\, \Sigma_C(\varepsilon) \mid \varepsilon > 0 \}$ satisfying the above is referred to as a suboptimal regulator sequence.

Remark 6.2.1 *The above suboptimal output regulation problem via proper (or strictly proper) regulators for the case where $J_p^*(x_0, w_0, \Pi, \Gamma) = 0$ ($J_{sp}^*(x_0, w_0, \Pi, \Gamma) = 0$ or $J_{sf}^*(x_0, w_0, \Pi, \Gamma) = 0$) for all $x(0) = x_0 \in \mathbb{R}^n$ and $w(0) = w_0 \in \mathbb{R}^s$, is referred to as a perfect output regulation problem [26].*

In this chapter, we first establish that $J_p^*(x_0, w_0, \Pi, \Gamma)$, $J_{sp}^*(x_0, w_0, \Pi, \Gamma)$, and $J_{sf}^*(x_0, w_0, \Pi, \Gamma)$ are finite, and then give explicit expressions for them. Next, we develop the conditions under which each of the above defined optimal and suboptimal output regulation problems is solvable. Whenever such conditions are satisfied, we also develop explicit methods for constructing an optimal regulator or a suboptimal regulator sequence that solves a specified problem. Finally, we examine the relationship between the infima $J_{sf}^*(x_0, w_0, \Pi, \Gamma)$, $J_p^*(x_0, w_0, \Pi, \Gamma)$, or $J_{sp}^*(x_0, w_0, \Pi, \Gamma)$ and the structural properties of the given system Σ. This chapter is organized as follows. Section 6.3 develops certain preliminaries we need, while Sections 6.4 and 6.5 respectively deal with state and measurement feedback controllers.

6.3 Preliminaries

Before we discuss the solution of the problems posed in the previous subsection, we need to develop certain preliminaries. It is well known that an L_2 optimal control problem can be rewritten as an H_2 optimal control problem. Hence we can rewrite the optimal output regulation problem as an H_2 optimal control problem with the output regulation constraint. To do so, let us construct an auxiliary system Σ_a by utilizing the data of the given system Σ and the exosystem Σ_E given by (6.1) and (6.2) respectively along with their initial conditions $x(0) = x_0$ and $w(0) = w_0$,

$$\Sigma_a : \begin{cases} \rho x = & Ax + \quad Bu + \quad E_w w + x_o r \\ \rho w = & Sw + w_o r \\ e \;\; = & C_e x + D_{eu} u + D_{ew} w \\ y \;\; = & C_y x + D_{yu} u + D_{yw} w \\ z \;\; = & Q^{1/2} e = C_z x + D_{zu} u + D_{zw} w \end{cases} \tag{6.7}$$

where z is a controlled output while

$$C_z = Q^{1/2} C_e \ , D_{zu} = Q^{1/2} D_{eu} \ , D_{zw} = Q^{1/2} D_{ew},$$

and r is a fictitious scalar disturbance signal. Then, it can easily be seen that the optimal output regulation problem for the given system Σ is equivalent to an H_2 optimal control problem for Σ_a with an output regulation constraint. That is, the optimal output regulation problem for Σ is equivalent to finding a controller for Σ_a which minimizes the H_2 norm from r to z over all controllers that internally stabilizes Σ and yield $e(t) \to 0$ as $t \to \infty$ for arbitrary initial conditions of the system Σ_a.

6.4 Optimal output regulation problem via state feedback

In this section, we study the optimal and suboptimal output regulation problems posed earlier by utilizing feedback from both the state vectors x and w. In other words, we assume the state is available for feedback. That is, the matrices C_y and D_{yw} take the form

$$C_y = \begin{pmatrix} I \\ 0 \end{pmatrix} \text{ and } D_{yw} = \begin{pmatrix} 0 \\ I \end{pmatrix}. \tag{6.8}$$

The solvability conditions for the posed optimal output regulation problems via state feedback will be expressed shortly in terms of the solution of certain linear matrix inequalities. We recall below the needed linear matrix inequalities. Consider first a continuous-time linear matrix inequality,

$$F(\tilde{P}) := \begin{pmatrix} A^{\mathrm{T}} \tilde{P} + \tilde{P} A + C_z^{\mathrm{T}} C_z & \tilde{P} B + C_z^{\mathrm{T}} D_{zu} \\ B^{\mathrm{T}} \tilde{P} + D_{zu}^{\mathrm{T}} C_z & D_{zu}^{\mathrm{T}} D_{zu} \end{pmatrix} \geqslant 0. \tag{6.9}$$

As shown in [61] and as explained further in Appendix 6.A, whenever the pair (A, B) is stabilizable, there exists a unique symmetric semi-stabilizing solution $\tilde{P}$ of the linear matrix inequality given in (6.9). Moreover, such a solution $\tilde{P}$ is positive semi-definite and is the largest among all symmetric solutions.

Consider next a discrete-time linear matrix inequality,

$$F(\tilde{P}) := \begin{pmatrix} A^{\mathrm{T}}\tilde{P}A - \tilde{P} + C_z^{\mathrm{T}}C_z & A^{\mathrm{T}}\tilde{P}B + C_z^{\mathrm{T}}D_{zu} \\ B^{\mathrm{T}}\tilde{P}A + D_{zu}^{\mathrm{T}}C_z & B^{\mathrm{T}}\tilde{P}B + D_{zu}^{\mathrm{T}}D_{zu} \end{pmatrix} \geqslant 0. \tag{6.10}$$

Again, as shown in [61] and as explained further in Appendix 6.B, whenever the pair (A, B) is stabilizable, there exists a unique symmetric semi-stabilizing and strongly rank minimizing solution $\tilde{P}$ of the linear matrix inequality given in (6.10). Moreover, such a solution $\tilde{P}$ is positive semi-definite, and is the largest among all such strongly rank minimizing symmetric solutions.

Utilizing the unique symmetric semi-stabilizing solution $\tilde{P}$ of (6.9) for continuous-time systems or the unique symmetric semi-stabilizing and strongly rank minimizing solution $\tilde{P}$ of (6.10) for discrete-time systems, we can define a pair of matrices C_{P} and D_{P} as

$$\begin{pmatrix} C_{\mathrm{P}}^{\mathrm{T}} \\ D_{\mathrm{P}}^{\mathrm{T}} \end{pmatrix} \begin{pmatrix} C_{\mathrm{P}} & D_{\mathrm{P}} \end{pmatrix} = F(\tilde{P}). \tag{6.11}$$

6.4.1 Determination of J_{sf}^*

Under the natural Assumptions A.1, A.2, and A.4 of Section 6.2, the following theorem shows that the infimum $J_{sf}^*(x_0, w_0, \Pi, \Gamma)$ is finite and provides an expression for it.

Theorem 6.4.1 *Consider the optimal output regulation problem for a given x_0 and w_0 (i.e., Problem 6.2.2) via state feedback for the system Σ and the exosystem Σ_{E} given by (6.1) and (6.2) respectively. Let Assumptions A.1, A.2, and A.4 be satisfied. Also, let (Π, Γ) be a given solution of the regulator equation (6.4). Moreover, let $\tilde{P}$ be the unique symmetric semi-stabilizing solution of (6.9) for continuous-time systems or the unique semi-stabilizing and strongly rank minimizing solution of (6.10) for discrete-time systems. Then, $J_{sf}^*(x_0, w_0, \Pi, \Gamma)$ is finite and is given by*

$$J_{sf}^*(x_0, w_0, \Pi, \Gamma) = (x_0 - \Pi w_0)^T \tilde{P} (x_0 - \Pi w_0). \tag{6.12}$$

Proof : Consider the given x_0 and w_0, and (Π, Γ) which is a given solution of the regulator equation (6.4). Also, let $\bar{x} = x - \Pi w$ and $\bar{u} = u - \Gamma w$. Then

Σ_a given in (6.7) can be rewritten as

$$\bar{\Sigma} : \begin{cases} \rho\bar{x} = A\bar{x} + B\bar{u} + (x_0 - \Pi w_0)r \\ e = C_e\bar{x} + D_{eu}\bar{u} \\ z = C_z\bar{x} + D_{zu}\bar{u}. \end{cases} \tag{6.13}$$

The dynamics $\rho w = Sw + w_0 r$ are omitted since in this new basis w has no direct effect on e and z.

Now, in view of the results of Theorem 2.3.1, we note that any internally stabilizing state feedback controller $\bar{\Sigma}_{\mathrm{C}}$,

$$\bar{\Sigma}_{\mathrm{C}} : \left\{ \bar{u} = F\bar{x}, \right. \tag{6.14}$$

for $\bar{\Sigma}$ defines a controller

$$u = Fx + (\Gamma - F\Pi)w$$

that achieves exact output regulation for the system Σ given in (6.1). Obviously, $J^*_{sf}(x_0, w_0, \Pi, \Gamma)$ is then the square of the infimum of the H_2 norm of the transfer function from r to z over all possible internally stabilizing state feedback controllers for the system $\bar{\Sigma}$. Then, the expression (6.12) for $J^*_{sf}(x_0, w_0, \Pi, \Gamma)$ immediately follows from [61]. ■

We would like to comment that the expression $J^*_{sf}(x_0, w_0, \Pi, \Gamma)$ given in (6.12) is equal to the minimal possible transient energy over all state feedback controllers that achieve output regulation. As such it points out a fundamental limitation or characteristic of the given system Σ and the exosystem Σ_{E}.

6.4.2 Solvability conditions

In this subsection we develop the solvability conditions for the posed optimal and suboptimal output regulation problems via state feedback.

We have the following result for continuous-time systems.

Theorem 6.4.2 *For continuous-time systems, consider Problems 6.2.2 and 6.2.3 via state feedback for the system Σ and the exosystem Σ_{E} given by (6.1) and (6.2) respectively. Let Assumptions A.1, A.2, and A.4 of Section 6.2 hold. Also, let (Π, Γ) be a given solution of the regulator equation (6.4). Then, the optimal output regulation problem for a given x_0 and w_0 (i.e. Problem 6.2.2) is solvable via state feedback if and only if*

$$\mathrm{im}[x_0 - \Pi w_0] \subseteq \mathcal{V}^-(A, B, C_{\mathrm{P}}, D_{\mathrm{P}}) \tag{6.15}$$

where $\tilde{P}$ is the unique semi-stabilizing solution of the linear matrix inequality (6.9) and the matrices C_P and D_P are as in (6.11).

*Moreover, the optimal output regulation problem (i.e. Problem 6.2.3) is solvable via state feedback. That is, there exist a **fixed** state feedback regulator that attains the optimal performance $J_p^*(x_0, w_0, \Pi, \Gamma)$ (or respectively $J_{sp}^*(x_0, w_0, \Pi, \Gamma)$) for all x_0 and w_0 for which Problem 6.2.2 is solvable.*

Proof : It is obvious from the proof of Theorem 6.4.1 that a controller $u = Fx + Gw$ solves the optimal output regulation problem for a given x_0 and w_0 via state feedback if and only if $G = \Gamma - F\Pi$, and the matrix F is such that the feedback $\bar{u} = F\bar{x}$ solves the H_2 optimal control problem for the system $\bar{\Sigma}$ given in (6.13). Then, the results in H_2 optimal control theory [61, see Theorem 7.2.2] tell us that such a controller $\bar{u} = F\bar{x}$ exists if and only if (6.15) is satisfied. Moreover, the results of Theorem 7.2.2 of [61] guarantee that there exists a fixed state feedback regulator that solves Problem 6.2.3. ■

The following theorem is the analog of Theorem 6.4.2, however it deals with discrete-time systems.

Theorem 6.4.3 *For discrete-time systems, consider Problems 6.2.2 and 6.2.3 via state feedback for Σ and the exosystem Σ_E given by (6.1) and (6.2) respectively. Let Assumptions A.1, A.2, and A.4 of Section 6.2 hold. Also, let (Π, Γ) be a given solution of the regulator equation (6.4). Then, the optimal output regulation problem for a given x_0 and w_0 (i.e. Problem 6.2.2) is solvable via state feedback if and only if*

$$\operatorname{im}[x_0 - \Pi w_0] \subseteq \mathcal{V}^{\ominus}(A, B, C_P, D_P) \tag{6.16}$$

where $\tilde{P}$ is the unique semi-stabilizing and strongly rank minimizing solution of the linear matrix inequality (6.10) and the matrices C_P and D_P are as in (6.11).

*Moreover, the optimal output regulation problem (i.e. Problem 6.2.3) is solvable via state feedback. That is, there exist a **fixed** state feedback regulator that attains the optimal performance measure $J_{sf}^*(x_0, w_0, \Pi, \Gamma)$ for all x_0 and w_0 for which Problem 6.2.2 is solvable.*

Proof : It follows along the same lines as the proof of Theorem 6.4.2. ■

In order to ascertain the solvability of the optimal output regulation problem via state feedback, one needs to ascertain whether there exists a solution (Π, Γ) of the regulator equation (6.4) such that the condition given in Theorem 6.4.2 or in Theorem 6.4.3 holds. Secondly, we have to ask ourselves whether the optimal performance depends on the specific choice for the solutions (Π, Γ) of the regulator equation (6.4).

Regarding the second issue we note that $\tilde{P}$ is independent of Π and Γ. Moreover, the set of solutions (Π, Γ) of the regulator equation (6.4) is an affine set. Therefore minimizing (6.12) over the set of solutions (Π, Γ) of the regulator equation is straightforward. It can actually been shown that there exists *one* solution (Π, Γ) of the regulator equation which minimizes this criterion for all x_0 and w_0.

Whether or not there exists a solution (Π, Γ) of the regulator equation (6.4) such that the condition given in Theorem 6.4.2 or in Theorem 6.4.3 holds, can be checked in a straightforward way. To see this, let H be a matrix such that $\ker H = \mathcal{V}^-(A, B, C_P, D_P)$ for continuous-time systems, or $\ker H = \mathcal{V}^\ominus(A, B, C_P, D_P)$ for discrete-time systems. Then, the two algebraic equations that define the regulator equation (6.4) and the condition of Theorem 6.4.2 or Theorem 6.4.3 can be merged into three algebraic equations consisting of the two equations of the regulator equation (6.4), and the third as

$$H(x_0 - \Pi w_0) = 0. \tag{6.17}$$

Obviously then, the optimal output regulation problem via state feedback is solvable whenever there exists a solution to the above set of three algebraic equations. Note that $\mathcal{V}^-(A, B, C_P, D_P)$ and $\mathcal{V}^\ominus(A, B, C_P, D_P)$, and thus H do not depend on the solution Π and Γ.

Obviously, we can combine the above issues to try to find a solution (Π, Γ) that minimizes the optimal cost and which is such that the optimal cost can be attained.

In view of Theorems 6.4.2 and 6.4.3, it is clear that in general the optimal output regulation via state feedback is possible only for a subset of initial conditions x_0 and w_0 which satisfy the condition (6.15) for continuous-time systems or the condition (6.16) for discrete-time systems. This leads to a fundamental question: when does there exist a fixed regulator which achieves optimal output regulation for all $x_0 \in \mathbb{R}^n$ and $w_0 \in \mathbb{R}^s$. The following theorem deals with this issue.

Theorem 6.4.4 *Consider the given system Σ and the exosystem Σ_E given by (6.1) and (6.2) respectively. Let Assumptions A.1, A.2 , and A.4 of Section*

6.2 hold. Also, let (Π, Γ) *be any solution of the regulator equation (6.4). Then, there exists a fixed state feedback regulator* $u = Fx$ *that achieves the performance measure* $J^*_{sf}(x_0, w_0, \Pi, \Gamma)$ *for all* $x_0 \in \mathbb{R}^n$ *and* $w_0 \in \mathbb{R}^s$ *if and only if the following conditions hold:*

(*i*) *for continuous-time systems, the matrix* D_{zu} *is injective and the system characterized by the quadruple* (A, B, C_z, D_{zu}) *has no invariant zeros on the imaginary axis,*

(*ii*) *for discrete-time systems, the system characterized by the quadruple* (A, B, C_z, D_{zu}) *is left invertible and has no invariant zeros on the unit circle.*

Proof : In view of the proof of Theorem 6.4.1, a simple look at H_2 optimal control theory (e.g. see [61]) shows that the results of this theorem follow if and only if the H_2 optimal control problem via state feedback for the system $\bar{\Sigma}$ given in (6.13) is regular. The conditions for the H_2 optimal control problem for $\bar{\Sigma}$ to be regular are indeed the conditions given in the theorem. ■

Optimal output regulation seeks to achieve the best possible transient performance. However, as seen above, this demands certain conditions to be satisfied by the given system and the given initial conditions. It is natural then to enquire whether a compromise is possible between the required conditions on the given system or initial conditions and the level of transient performance that is sought. This leads to the study of suboptimal output regulation problems. The following theorem shows that, under some natural assumptions, the suboptimal output regulation problem is solvable via state feedback for both continuous- and discrete-time systems.

Theorem 6.4.5 *For both continuous- and discrete-time systems, consider the Problems 6.2.4 and 6.2.5 via state feedback for* Σ *as in (6.1) and the exosystem* Σ_E *as in (6.2). Let Assumptions A.1, A.2, and A.4 of Section 6.2 hold. Also, let* (Π, Γ) *be any solution of the regulator equation (6.4). Moreover, let* $J^*_{sf}(x_0, w_0, \Pi, \Gamma)$ *be the corresponding optimal performance for a given* x_0 *and* w_0*. Then, the suboptimal regulation problem as formulated in Problem 6.2.4 is always solvable via state feedback for any given* x_0 *and* w_0*. Moreover, the suboptimal regulation problem as formulated in Problem 6.2.5 is also always solvable via state feedback. That is, to be precise, there exists a* ***fixed*** *suboptimal state feedback regulator sequence* $\{\Sigma_C(\varepsilon) \mid \varepsilon > 0\}$ *such that the attained performance measure* $J(x_0, w_0, \Sigma_C(\varepsilon))$ *tends to* $J^*_{sf}(x_0, w_0, \Pi, \Gamma)$ *as* ε *tends to zero for all* $x_0 \in \mathbb{R}^n$ *and* $w_0 \in \mathbb{R}^s$*.*

Proof : The H_2 suboptimal control problem via state feedback for the system $\bar{\Sigma}$ given in (6.13) is solvable whenever Assumption A.1 holds [61]. Thus, the proof of this theorem is obvious when we follow the same lines as in the proof of Theorem 6.4.2. ∎

Obviously, Theorem 6.4.5 is very appealing. It shows that under natural assumptions, there exists a fixed sequence of state feedback regulators such that one can choose from it a regulator that achieves a transient performance arbitrarily close to the optimal transient performance. This is in contrast with Theorems 6.4.2 and 6.4.3 which seek to achieve the optimal transient performance. However, we remark that the suboptimal output regulation is obtained at some cost; for instance, in continuous-time systems in order to obtain suboptimal output regulation, one need to employ high-gain feedback controllers whereas optimal output regulation, whenever it is feasible, does not require high-gain feedback controllers.

6.4.3 Perfect output regulation

Earlier in Remark 6.2.1, we alluded to what is referred to as a perfect output regulation problem. We say that perfect output regulation is attainable via state feedback if we can achieve, uniformly for all $x_0 \in \mathbb{R}^n$ and $w_0 \in \mathbb{R}^s$, transient performance arbitrarily close to zero, i.e. if one can solve the suboptimal output regulation problem (Problem 6.2.5) with the additional property that $J_{sf}^*(x_0, w_0, \Pi, \Gamma)$ equals zero for all $x_0 \in \mathbb{R}^n$ and $w_0 \in \mathbb{R}^s$. In other words, the notion of perfect output regulation seeks the satisfaction of two conditions, (1) having $J_{sf}^*(x_0, w_0, \Pi, \Gamma)$ equal to zero for all $x_0 \in \mathbb{R}^n$ and $w_0 \in \mathbb{R}^s$, and (2) the existence of a fixed sequence $\{\Sigma_C(\varepsilon) \mid \varepsilon > 0\}$ of regulators by selecting a member of which one can render the attained performance $J(x_0, w_0, \Sigma_C(\varepsilon))$ arbitrary small uniformly for all $x_0 \in \mathbb{R}^n$ and $w_0 \in \mathbb{R}^s$. The perfect output regulation problem via state feedback has a long history (e.g. see Kwakernaak and Sivan [27], Francis [17], Kimura [26], Lin et al. [39,40]). A more recent but generalized result is given in [61]. The following theorem gives the conditions under which perfect output regulation via state feedback is possible.

Theorem 6.4.6 *Consider the given system Σ as in (6.1) and the exosystem Σ_E as in (6.2). Let Assumptions A.1, A.2, and A.4 of Section 6.2 hold. Then, perfect output regulation can be achieved via state feedback if and only if the following conditions hold:*

(*i*) *For continuous-time systems, the system characterized by the quadruple* (A, B, C_z, D_{zu}) *is right invertible and has all its invariant zeros in the closed left-half plane, i.e. in* $\mathbb{C}^- \cup \mathbb{C}^0$ *(the system it is either minimum phase, weakly minimum phase, or weakly non-minimum phase).*

(*ii*) *For discrete-time systems, the system characterized by the quadruple* (A, B, C_z, D_{zu}) *is right invertible, has all its invariant zeros inside or on the unit circle, i.e. in* $\mathbb{C}^{\otimes}$ *(the system is either minimum phase, weakly minimum phase, or weakly non-minimum phase), and has no infinite zeros of order greater than or equal to one.*

Proof : As discussed in the proof of Theorem 6.4.1, $J^*_{sf}(x_0, w_0, \Pi, \Gamma)$ is the square of the infimum of the H_2 norm of the transfer function from r to z over all possible internally stabilizing state feedback controllers for the system $\bar{\Sigma}$ given in (6.13). Then, the results of this theorem are a consequence of Theorem 5.8.1 (for continuous-time systems) and Theorem 6.8.1 (for discrete-time systems) of [61]. ■

Note that perfect regulation is possible if and only if $\tilde{P}$ in Theorem 6.4.1 is equal to zero. Since $\tilde{P}$ is independent of the specific solution Π, Γ of the regulator equations, we find that perfect regulation is independent of the specific solution of the regulator equations.

6.4.4 Construction of optimal and suboptimal state feedback regulators

In this subsection, we give a procedure of constructing the optimal and suboptimal regulators whenever the solvability conditions developed in the previous subsections are satisfied. In fact, Theorems 6.4.1 and 6.4.2, and their proofs suggest the following step by step procedure of constructing an optimal state feedback regulator.

Construction of an optimal state feedback regulator:

Step 1 : Choose a solution (Π, Γ) of the regulator equation (6.4) such that there exists a matrix H satisfying the algebraic equation (6.17).

Step 2 : Let $\check{E}$ be such that $\operatorname{im} \check{E} = \mathcal{V}^-(A, B, C_P, D_P)$ for continuous-time systems, and $\operatorname{im} \check{E} = \mathcal{V}^{\ominus}(A, B, C_P, D_P)$ for discrete-time systems. Consider the system,

$$\begin{aligned}
\rho\bar{x} &= A\bar{x} + B\bar{u} + \check{E}r \\
e &= C_e\bar{x} + D_{eu}\bar{u} \\
z &= C_z\bar{x} + D_{zu}\bar{u}.
\end{aligned} \tag{6.18}$$

Find an H_2 optimal state feedback controller $\bar{u} = F\bar{x}$ for the above system using the COGFMDZ algorithm for continuous-time systems and DOGFMDZ algorithm for discrete-time systems as given in [61].

Step 3 : Form the state feedback controller,

$$u = Fx + (\Gamma - F\Pi)w.$$

The controller constructed above is an optimal state feedback regulator for the given system Σ. □

Similarly, we can give the following procedure to construct a suboptimal regulator sequence.

Construction of a suboptimal state feedback regulator sequence:

Step 1 : Choose a solution (Π, Γ) of the regulator equation (6.4).

Step 2 : Construct a sequence of state feedback gains $\{ F_\varepsilon \mid \varepsilon > 0 \}$ such that the sequence of state feedback controllers of the type $\bar{u} = F_\varepsilon\bar{x}$ is a sequence of H_2 suboptimal controllers for the system given in (6.18).

Step 3 : Form the sequence of state feedback controllers,

$$u = F_\varepsilon x + (\Gamma - F_\varepsilon\Pi)w.$$

The sequence of controllers constructed above is a suboptimal state feedback regulator sequence for the given system Σ. □

We remark that an algorithm based on a perturbation method is developed in [61] to construct a sequence of state feedback gains $\{ F_\varepsilon \mid \varepsilon > 0 \}$ such that the sequence of state feedback controllers of the type $u = F_\varepsilon x$ is a sequence of H_2 suboptimal state feedback controllers for any given system. However, perturbation methods although conceptually simple are extremely sensitive to numerical errors. An alternative direct method of computing that is insensitive to numerical errors is developed in [39, 40] to construct a similar sequence of state feedback gains $\{ F_\varepsilon \mid \varepsilon > 0 \}$. This alternative method enables us to construct a sequence of H_2 suboptimal state feedback controllers.

We would like to state that constructing a sequence of state feedback regulators that achieve perfect output regulation is exactly the same as constructing a sequence of suboptimal state feedback regulators.

6.4.5 Relationships between J_{sf}^* and the structural properties of Σ

In the previous subsection we derived an expression for the optimal performance $J_{sf}^*(x_0, w_0, \Pi, \Gamma)$ as is given by (6.12). In this subsection, we would like to identify the part of the dynamics of the given system that is responsible for $J_{sf}^*(x_0, w_0, \Pi, \Gamma)$. In view of the proof of Theorem 6.4.1, we remark that the expression for $J_{sf}^*(x_0, w_0, \Pi, \Gamma)$ given in (6.12) is obtained by studying the H_2 optimal state feedback control problem for $\bar{\Sigma}$ given in (6.13) which has been obtained from the system Σ_a of (6.7) by the change of variables, $\bar{x} = x - \Pi w$ and $\bar{u} = u - \Gamma w$. We now rewrite $\bar{\Sigma}$ by setting the fictitious disturbance signal r to zero,

$$\bar{\Sigma} : \begin{cases} \rho\bar{x} = A\bar{x} + B\bar{u} \\ z = C_z\bar{x} + D_{zu}\bar{u}. \end{cases} \tag{6.19}$$

To proceed with our development, we observe that, other than the initial conditions, there are two matrices that contribute to $J_{sf}^*(x_0, w_0, \Pi, \Gamma)$ and these are (1) $\tilde{P}$ the unique symmetric semi-stabilizing solution of (6.9) for continuous-time systems or the unique semi-stabilizing and strongly rank minimizing solution of (6.10) for discrete-time systems, and (2) Π which is a solution of the regulator equation (6.4). Of course, both of these contributing matrices depend on the dynamics of the given system. It turns out that we can separate a part of the dynamics of the given system which determines $\tilde{P}$. It has been shown earlier (e.g. see [59]) that under cheap control certain state and output variables of the given system Σ attain their equilibrium values relatively fast while others do so relatively slowly. That is, under cheap control, a given system Σ exhibits multiple time-scale structure. As discussed below, the dynamics that exhibits the slowest time-scale structure under cheap control (the so called slow dynamics) influences heavily the structure and magnitude of $\tilde{P}$.

We proceed now to extract the subsystem of the given system that exhibits the slowest time-scale structure under cheap control. To do so, we need to rewrite the system given in (6.19) in a special coordinate basis (s.c.b) [60, 66] that can explicitly display the finite and infinite zero structures of any linear system. For this purpose, let us assume without loss of generality that B and C_z are of maximal rank. Also, without loss of generality, assume that

$$D_{zu} = \begin{pmatrix} I_{m_0} & 0 \\ 0 & 0 \end{pmatrix}$$

where I_{m_0} is an identity matrix of order $m_0 \times m_0$. Partition B and C_z in conformity with the partitioning of D_{zu},

$$B = \begin{pmatrix} B_0 & B_1 \end{pmatrix}, \quad \text{and} \quad C_z = \begin{pmatrix} C_0 \\ C_1 \end{pmatrix}.$$

Then it follows from [60,66] that there exist non-singular matrices $\Gamma_s \in \mathbb{R}^{n\times n}$, $\Gamma_o \in \mathbb{R}^{p\times p}$, and $\Gamma_i \in \mathbb{R}^{m\times m}$ such that

$$\Gamma_s^{-1}(A - B_0C_0)\Gamma_s$$
$$= \begin{pmatrix} A_{aa}^- & 0 & 0 & L_{ab}^-C_b & 0 & L_{ad}^-C_d \\ 0 & A_{aa}^0 & 0 & L_{ab}^0C_b & 0 & L_{ad}^0C_d \\ 0 & 0 & A_{aa}^+ & L_{ab}^+C_b & 0 & L_{ad}^+C_d \\ 0 & 0 & 0 & A_{bb} & 0 & L_{bd}C_d \\ B_cE_{ca}^- & B_cE_{ca}^0 & B_cE_{ca}^+ & B_cE_{cb} & A_{cc} & L_{cd}C_d \\ B_dE_{da}^- & B_dE_{da}^0 & B_dE_{da}^+ & B_dE_{db} & B_dE_{dc} & A_d \end{pmatrix}, \tag{6.20}$$

$$\Gamma_s^{-1}\begin{pmatrix} B_0 & B_1 \end{pmatrix}\Gamma_i = \begin{pmatrix} B_{a0}^- & 0 & 0 \\ B_{a0}^0 & 0 & 0 \\ B_{a0}^+ & 0 & 0 \\ B_{b0} & 0 & 0 \\ B_{c0} & 0 & B_c \\ B_{d0} & B_d & 0 \end{pmatrix}, \tag{6.21}$$

and

$$\Gamma_o^{-1}\begin{pmatrix} C_0 \\ C_1 \end{pmatrix}\Gamma_s = \begin{pmatrix} C_{0a}^- & C_{0a}^0 & C_{0a}^+ & C_{0b} & C_{0c} & C_{0d} \\ 0 & 0 & 0 & 0 & 0 & C_d \\ 0 & 0 & 0 & C_b & 0 & 0 \end{pmatrix}, \tag{6.22}$$

where the pair (C_b, A_{bb}) is observable, and the pair (A_{cc}, B_c) is controllable. Also, the eigenvalues of A_{aa}^-, A_{aa}^0, and A_{aa}^+ are the *invariant* zeros of the system given in (6.19), and they are in the open left-half complex plane, on the imaginary axis, and in the open right-half complex plane respectively for continuous-time systems, while they are inside the unit circle, on the unit circle, and outside the unit circle respectively for discrete-time systems. Moreover, the system characterized by the quadruple $(A_d, B_d, C_d, 0)$ is invertible and is free of any invariant zeros. Furthermore, $C_dC_d^{\mathrm{T}} = I_{\bar{p}-m_0}$ with $\bar{p}$ being the normal rank of $C_z(sI - A)^{-1}B + D_{zu}$.

Shortly, we also need the decomposition of the state $\bar{x}$ of the system given in (6.19) in accordance with the special coordinate basis (s.c.b). Let

$$\bar{x} := x - \Pi w := \Gamma_s \left((x_a^-)^{\mathrm{T}} \quad (x_a^0)^{\mathrm{T}} \quad (x_a^+)^{\mathrm{T}} \quad x_b^{\mathrm{T}} \quad x_c^{\mathrm{T}} \quad x_d^{\mathrm{T}} \right)^{\mathrm{T}}. \tag{6.23}$$

Based on the above development, we can now study $\tilde{P}$ which is the unique symmetric semi-stabilizing solution of (6.9) for continuous-time systems or the unique semi-stabilizing and strongly rank minimizing solution of (6.10) for discrete-time systems. There are structurally profound differences between the solution $\tilde{P}$ of (6.9) and that of (6.10). As such, we treat continuous- and discrete-time systems separately.

For continuous-time systems, in view of Property 4.3.2 (page 124) of [61], we see that the unique semi-stabilizing solution $\tilde{P}$ of the linear matrix inequality (6.9) can be written in the form,

$$\tilde{P} = (\Gamma_s^{-1})^{\mathrm{T}} \begin{pmatrix} 0 & 0 & 0 & 0 & 0 \\ 0 & 0 & 0 & 0 & 0 \\ 0 & 0 & P_s & 0 & 0 \\ 0 & 0 & 0 & 0 & 0 \\ 0 & 0 & 0 & 0 & 0 \end{pmatrix} \Gamma_s^{-1} \tag{6.24}$$

where the symmetric positive semi-definite matrix P_s is the solution of the algebraic Riccati equation,

$$\begin{aligned} P_s A_s + A_s^{\mathrm{T}} P_s + C_s^{\mathrm{T}} C_s & \\ - (P_s B_s + C_s^{\mathrm{T}} D_s)(D_s^{\mathrm{T}} D_s)^{-1} &(B_s^{\mathrm{T}} P_s + D_s^{\mathrm{T}} C_s) = 0, \end{aligned} \tag{6.25}$$

and where

$$A_s := \begin{pmatrix} A_{aa}^+ & L_{ab}^+ C_b \\ 0 & A_{bb} \end{pmatrix}, \quad B_s := \begin{pmatrix} B_{a0}^+ & L_{ad}^+ \\ B_{b0} & L_{bd} \end{pmatrix}, \tag{6.26}$$

$$C_s := \Gamma_o \begin{pmatrix} 0 & 0 \\ 0 & 0 \\ 0 & C_b \end{pmatrix}, \qquad D_s := \Gamma_o \begin{pmatrix} I_{m_0} & 0 \\ 0 & I_{\bar{p}-m_0} \\ 0 & 0 \end{pmatrix}. \tag{6.27}$$

Remark 6.4.1 *Note that (6.24) is an explicit way of describing the fact that*

$$\ker P = \mathcal{V}^{-0}(A, B, C_z, D_z) + \mathcal{S}^*(A, B, C_z, D_z)$$

which has a direct connection with the solvability of (almost) disturbance decoupling problems as discussed in Section 13.2.

The matrix P_s of (6.25) can be given a physical meaning. Let

$$x_s = \begin{pmatrix} x_a^+ \\ x_b \end{pmatrix}.$$

Also, consider a linear quadratic control problem for a continuous-time system Σ_{sub},

$$\Sigma_{\text{sub}} : \begin{cases} \rho x_s = A_s x_s + B_s u_s \\ z \quad = C_s x_s + D_s u_s, \end{cases} \tag{6.28}$$

with a performance measure

$$J_{\text{sub}} = \int_0^\infty z^{\mathrm{T}} z \, dt. \tag{6.29}$$

The infimum of performance measure J_{sub} over all possible state feedback controllers is given by

$$J_{\text{sub}}^*(x_s(0)) = x_s(0)^{\mathrm{T}} P_s x_s(0). \tag{6.30}$$

Thus the study of $J_{sf}^*(x_0, w_0, \Pi, \Gamma)$ for continuous-time systems is reduced to the study of $J_{\text{sub}}^*(x_s(0))$ in which P_s is the solution of the algebraic Riccati equation (6.25) whose dimension is in general smaller than that of the linear matrix inequality (6.9). In fact, the above development leads to the following lemma.

Lemma 6.4.1 *For continuous-time systems, consider the optimal output regulation problem for a given x_0 and w_0 via state feedback, i.e. Problem 6.2.2. Let Assumptions A.1, A.2, and A.4 of Section 6.2 hold. Also, assume that the system (6.19) which is characterized by the quadruple (A, B, C_z, D_{zu}) is already in the form of the special coordinate basis (s.c.b). Then, we have*

$$J_{sf}^*(x_0, w_0, \Pi, \Gamma) = J_{\text{sub}}^*(x_s(0)).$$

It is important to recognize that, in view of (6.23), the state x_s is a component of $x - \Pi w$, and hence $x_s(0)$ is a component of $x_0 - \Pi w_0$.

For continuous-time systems, the above lemma clearly points out that the dynamics given by the subsystem Σ_{sub} is indeed the one that is responsible for the non-zero $J_{sf}^*(x_0, w_0, \Pi, \Gamma)$. Subsystem Σ_{sub} has two types of dynamics. The first type of dynamics is represented by the state x_a^+, and it is present only when the system $\bar{\Sigma}$ given in (6.19) has invariant zeros in the open right-half plane $\mathbb{C}^+$. In the literature, the dynamics represented by x_a^+ is often called the *unstable zero dynamics*. The second type of dynamics in Σ_{sub} is represented by the state x_b, and it is present only when the system $\bar{\Sigma}$ is not right invertible.

The above development leads to the following important remarks.

Remark 6.4.2 *We remark that the state x_a^+ is non-existent if the system $\bar{\Sigma}$ given in (6.19) has no invariant zeros in the open right-half plane $\mathbb{C}^+$ (i.e. it is either minimum phase, or weakly minimum phase, or weakly non-minimum phase). Similarly, the state x_b is non-existent if the system $\bar{\Sigma}$ is right invertible. Thus, for continuous-time systems, $P_s = 0$ and hence $\tilde{P} = 0$ whenever the system $\bar{\Sigma}$ is right invertible and is either minimum phase, weakly minimum phase, or weakly non-minimum phase. Under these conditions, $J_{sf}^*(x_0, w_0, \Pi, \Gamma)$ is indeed identically zero as already discussed in Theorem 6.4.6.*

Remark 6.4.3 ***(Energy interpretation)*** *In view of Lemma 6.4.1, whenever the system $\bar{\Sigma}$ given in (6.19) is already in the special coordinate basis, we know that $J_{sf}^*(x_0, w_0, \Pi, \Gamma)$ equals $J_{\text{sub}}^*(x_s(0))$. This gives an interesting energy interpretation. If $\bar{\Sigma}$ is right invertible, one can interpret $J_{sf}^*(x_0, w_0, \Pi, \Gamma)$ as the minimum energy required to stabilize the unstable zero dynamics of $\bar{\Sigma}$.*

One can also write $J_{sf}^*(x_0, w_0, \Pi, \Gamma)$ explicitly in terms of the open right half plane zeros of $\bar{\Sigma}$ which is characterized by the quadruple (A, B, C_z, D_{zu}). Let us illustrate this for the case when the reference/disturbance signal is a step function, i.e., w is a constant. For simplicity, we also assume that $\bar{\Sigma}$ is right invertible and is already in the form of the special coordinate basis. For this case, $A_s = A_{aa}^+$ and is an invertible matrix. The steady state value of x_s, denoted by $\bar{x}_s$, must satisfy

$$A_{aa}^+\bar{x}_s + B_s\bar{u}_s = 0.$$

Using the properties of the special coordinate basis, it is easy to see that

$$\bar{u}_s = \bar{z}$$

where $\bar{z}$ is the steady state value of z. Hence

$$\bar{x}_s = (A_{aa}^+)^{-1}B_s\bar{z}.$$

Let $x(0) = x_0 = 0$. Then, whenever output regulation is attained, one can see that $x_s(0) = \bar{x}_s$. Thus for the particular initial condition $x(0) = 0$, we have

$$\begin{aligned} J_{sf}^*(x_0, w_0, \Pi, \Gamma) &= J_{\text{sub}}^*(\bar{x}_s) = J_{\text{sub}}^*((A_{aa}^+)^{-1}B_s\bar{z}) \\ &= \left[(A_{aa}^+)^{-1}B_s\bar{z}\right]^{\mathrm{T}} P_s \left[(A_{aa}^+)^{-1}B_s\bar{z}\right] = \bar{z}^{\mathrm{T}}M\bar{z} \end{aligned} \tag{6.31}$$

where $M = [(A_{aa}^{+})^{-1}B_s]^{\mathrm{T}}P_s[(A_{aa}^{+})^{-1}B_s]$. But for the case we are considering, Riccati equation (6.25) reduces to

$$P_s A_{aa}^{+} + (A_{aa}^{+})^{\mathrm{T}} P_s - P_s B_s B_s^{\mathrm{T}} P_s = 0. \tag{6.32}$$

By utilizing the properties of the trace operator and (6.32), it can be seen that

$$\operatorname{trace} M = 2\operatorname{trace}(A_{aa}^{+})^{-1} = 2\sum_{i=1}^{\ell}\frac{1}{\zeta_i} \tag{6.33}$$

where ζ_i, $i = 1, 2, \cdots, \ell$ are the eigenvalues of A_{aa}^{+} or equivalently the open right half plane zeros of $\bar{\Sigma}$ which is characterized by the matrix quadruple (A, B, C_z, D_{zu}). This result was obtained earlier by [52].

Remark 6.4.4 *The equations (6.31) and (6.33) lead to an interesting insight. They show that the closer the open right half plane invariant zero is to the imaginary axis the larger is its contribution to the minimum possible transient energy $J_{sf}^{*}(x_0, w_0, \Pi, \Gamma)$. There also exists a curious phenomena which points out a discontinuity. It is clear that the invariant zeros on the imaginary axis do not contribute to $J_{sf}^{*}(x_0, w_0, \Pi, \Gamma)$. However, the open right half plane invariant zeros close to the imaginary axis can lead to arbitrarily large transient error.*

We consider now discrete-time systems. Following Property 4.4.2 (page 140) of [61], we see that the unique semi-stabilizing and strongly rank minimizing solution $\tilde{P}$ of linear matrix inequality (6.10) can be written in the form,

$$\tilde{P} = (\Gamma_s^{-1})^{\mathrm{T}} \begin{pmatrix} 0 & 0 & 0 \\ 0 & 0 & 0 \\ 0 & 0 & \breve{P}_s \end{pmatrix} \Gamma_s^{-1} \tag{6.34}$$

where the symmetric positive semi-definite matrix $\breve{P}_s$ is the solution of the discrete algebraic Riccati equation,

$$\begin{aligned} &\breve{A}_s^{\mathrm{T}} \breve{P}_s \breve{A}_s - \breve{P}_s + \breve{C}_s^{\mathrm{T}} \breve{C}_s \\ &\quad - (\breve{A}_s^{\mathrm{T}} \breve{P}_s \breve{B}_s + \breve{C}_s^{\mathrm{T}} \breve{D}_s)(\breve{B}_s^{\mathrm{T}} \breve{P}_s \breve{B}_s + \breve{D}_s^{\mathrm{T}} \breve{D}_s)^{-1}(\breve{B}_s^{\mathrm{T}} \breve{P}_s \breve{A}_s + \breve{D}_s^{\mathrm{T}} \breve{C}_s) = 0, \end{aligned} \tag{6.35}$$

and where

$$\breve{A}_s = \begin{pmatrix} A_{aa}^+ & L_{ab}^+ C_b & L_{ad}^+ C_d \\ 0 & A_{bb} & L_{bd} C_d \\ B_d E_{da}^+ & B_d E_{db} & A_d \end{pmatrix}, \quad \breve{B}_s = \begin{pmatrix} B_{a0}^+ & 0 \\ B_{b0} & 0 \\ B_{d0} & B_d \end{pmatrix}, \tag{6.36}$$

$$\breve{C}_s = \Gamma_o \begin{pmatrix} 0 & 0 & 0 \\ 0 & 0 & C_d \\ 0 & C_b & 0 \end{pmatrix}, \qquad \breve{D}_s = \Gamma_o \begin{pmatrix} I_{m_0} & 0 \\ 0 & 0 \\ 0 & 0 \end{pmatrix}. \tag{6.37}$$

Remark 6.4.5 *Note that (6.34) is an explicit way of describing the fact that*

$$\ker \tilde{P} = \mathcal{V}^{\otimes}(A, B, C_z, D_{zu}).$$

The above expression has a direct connection with the solvability of (almost) disturbance decoupling problems as discussed in Section 13.2.

Again, the matrix $\breve{P}_s$ of (6.35) can be given a physical meaning. Let $\breve{x}_s = ((x_a^+)^{\mathrm{T}} \;\; x_b^{\mathrm{T}} \;\; x_d^{\mathrm{T}})^{\mathrm{T}}$. Consider a linear quadratic control problem for a discrete-time system $\breve{\Sigma}_{\mathrm{sub}}$,

$$\breve{\Sigma}_{\mathrm{sub}} : \begin{cases} \rho \breve{x}_s = \breve{A}_s \breve{x}_s + \breve{B}_s \breve{u}_s \\ z \;\;\, = \breve{C}_s \breve{x}_s + \breve{D}_s \breve{u}_s, \end{cases} \tag{6.38}$$

with a performance measure

$$\breve{J}_{\mathrm{sub}} = \sum_{i=0}^{\infty} z^{\mathrm{T}}(i) z(i). \tag{6.39}$$

The infimum of possible performance measure $\breve{J}_{\mathrm{sub}}$ over all possible state feedback controllers is then given by

$$\breve{J}_{\mathrm{sub}}^*(\breve{x}_s(0)) = \breve{x}_s(0)^{\mathrm{T}} \breve{P}_s \breve{x}_s(0). \tag{6.40}$$

Thus the study of $J_{sf}^*(x_0, w_0, \Pi, \Gamma)$ for discrete-time systems is reduced to the study of $\breve{J}_{\mathrm{sub}}^*(\breve{x}_s(0))$ in which $\breve{P}_s$ is the solution of the algebraic Riccati equation (6.35) whose dimension is in general smaller than that of the linear matrix inequality (6.10). In fact, as in the case of continuous-time systems, the above development leads to the following lemma.

Lemma 6.4.2 *For discrete-time systems, consider the optimal output regulation problem for a given x_0 and w_0 via state feedback, i.e. Problem 6.2.2. Let*

Assumptions A.1, A.2, and A.4 of Section 6.2 hold. Also, assume that the system $\bar{\Sigma}$ of (6.19) which is characterized by the quadruple (A, B, C_z, D_{zu}) is already in the form of the s.c.b. Then, we have

$$J_{sf}^*(x_0, w_0, \Pi, \Gamma) = \breve{J}_{\text{sub}}^*(\breve{x}_s(0)).$$

Again, it is important to recognize that, in view of (6.23), the state x_s is a component of $x - \Pi w$, and hence $\breve{x}_s(0)$ is a component of $x_0 - \Pi w_0$.

For discrete-time systems, the above lemma clearly points out that the dynamics given by the subsystem $\breve{\Sigma}_{\text{sub}}$ is indeed the one that is responsible for non-zero $J_{sf}^*(x_0, w_0, \Pi, \Gamma)$. Subsystem $\breve{\Sigma}_{\text{sub}}$ has three types of dynamics. The first type of dynamics is represented by the state x_a^+, and it is present only when the system $\bar{\Sigma}$ given in (6.19) has invariant zeros in $\mathbb{C}^{\oplus}$ (outside the unit circle). Again, the dynamics represented by x_a^+ is often called *unstable zero dynamics*. The second type of dynamics in $\breve{\Sigma}_{\text{sub}}$ is represented by the state x_b, and it is present only when $\bar{\Sigma}$ is not right invertible. Finally, the third type of dynamics in $\breve{\Sigma}_{\text{sub}}$ is represented by the state x_d, and it is present only when $\bar{\Sigma}$ has infinite zeros of order greater than or equal to one.

The above development leads to the following important remarks.

Remark 6.4.6 *As we said earlier, the state x_a^+ is non-existent if the system $\bar{\Sigma}$ given in (6.19) has no invariant zeros in $\mathbb{C}^{\oplus}$ (i.e. it is either minimum phase, or weakly minimum phase, or weakly non-minimum phase). Similarly, the state x_b is non-existent if $\bar{\Sigma}$ is right invertible. Also, the state x_d is non-existent if $\bar{\Sigma}$ has no infinite zeros of order greater than or equal to one. Thus, for discrete-time systems, $\breve{P}_s = 0$ and hence $\tilde{P} = 0$ whenever the system $\bar{\Sigma}$ is right invertible, has no infinite zeros of order greater than or equal to one, and is of either minimum phase, or weakly minimum phase, or weakly non-minimum phase. Under these conditions, $J_{sf}^*(x_0, w_0, \Pi, \Gamma)$ is indeed identically zero as already discussed in Theorem 6.4.6.*

Remark 6.4.7 ***(Energy interpretation)*** *In view of Lemma 6.4.2, whenever the system $\bar{\Sigma}$ given in (6.19) is already in the special coordinate basis, we know that $J_{sf}^*(x_0, w_0, \Pi, \Gamma)$ equals $\breve{J}_{\text{sub}}^*(\breve{x}_s(0))$. As before, this gives an interesting energy interpretation for discrete-time systems as well. If $\bar{\Sigma}$ is right invertible and has no infinite zeros of order greater than or equal to one, one can interpret $J_{sf}^*(x_0, w_0, \Pi, \Gamma)$ as the minimum energy required to stabilize the unstable zero dynamics of $\bar{\Sigma}$. We would like to stress a profound difference between continuous- and discrete-time systems. Unlike in the continuous-time case, for discrete-time systems, in order to have the above interpretation, the* *subsystem $\check{\Sigma}$ while being right invertible must also have no infinite zeros of order greater than or equal to one.*

6.5 Optimal output regulation via measurement feedback

In this section, we study again the optimal output regulation problems posed in Section 6.2 however via measurement feedback rather than state feedback. That is, we consider the system Σ_a as given in (6.7) and along with a measurement feedback controller of the form as given in (6.3). Throughout this section, for simplicity, we assume that $D_{yu} = 0$.

In order to study the optimal output regulation problem and similarly constrained problems, our philosophy here (as well as in the upcoming chapters) follows the work of [77]. We first assume that a solution (Π, Γ) of the regulator equation (6.4) exists. Given (Π, Γ), we formulate an auxiliary system denoted by $\bar{\Sigma}$.

There exists a $1 - 1$ relationship between an internally stabilizing controller $\bar{\Sigma}_C$ for the auxiliary system $\bar{\Sigma}$ and an internally stabilizing controller Σ_C for the original system Σ which achieves output regulation and is associated to the particular solution Π and Γ of the regulator equation. In other words, for each internally stabilizing controller $\bar{\Sigma}_C$ for the auxiliary system $\bar{\Sigma}$, we can formulate a corresponding controller Σ_C for the system Σ which achieves internal stability, output regulation, and is associated to Π, Γ. Conversely any controller Σ_C for the system Σ which achieves internal stability, output regulation and is associated to Π, Γ yields an internally stabilizing controller $\bar{\Sigma}_C$ for the auxiliary system $\bar{\Sigma}$.

In fact, we construct $\bar{\Sigma}$ in such a way that a certain transfer function (say, from $\bar{r}$ to $\bar{z}$) of $\bar{\Sigma}$ with an internally stabilizing controller $\bar{\Sigma}_C$, is exactly the same as the transfer function from r to z in Σ_a with a controller Σ_C that corresponds to $\bar{\Sigma}_C$. Recall that the H_2 norm of the closed loop transfer function from r to z in Σ_a with a controller Σ_C is equal to the transient tracking error for given initial conditions x_0 and w_0.

This lets us to transform easily the optimal output regulation problem for Σ via measurement feedback to an H_2 optimal control problem for $\bar{\Sigma}$ via measurement feedback without any regulation constraint. We follow this philosophy in the upcoming chapters as well where different performance measures are used.

Let us consider an auxiliary system defined by

$$\bar{\Sigma}: \begin{cases} \rho\bar{x} = \bar{A}\bar{x} + \bar{B}\bar{u} + \bar{E}\bar{r} \\ \bar{y} = \bar{C}_y\bar{x} \\ \bar{z} = \bar{C}_z\bar{x} + \bar{D}_{zu}\bar{u} \end{cases} \tag{6.41}$$

where

$$\begin{aligned} &\bar{A} = \begin{pmatrix} A & -B\Gamma \\ 0 & S \end{pmatrix}, \quad \bar{B} = \begin{pmatrix} 0 & B \\ I & 0 \end{pmatrix}, \quad \bar{E} = \begin{pmatrix} x_0 - \Pi w_0 \\ w_0 \end{pmatrix} \\ &\bar{C}_z = \begin{pmatrix} C_z & -D_{zu}\Gamma \end{pmatrix}, \quad \bar{D}_{zu} = \begin{pmatrix} 0 & D_{zu} \end{pmatrix}, \\ &\bar{C}_y = \begin{pmatrix} C_y & (D_{yw} + C_y\Pi) \end{pmatrix}. \end{aligned} \tag{6.42}$$

The motivation behind considering the above auxiliary system will be evident in the next chapter. Let us next consider a controller for the above auxiliary system in the form,

$$\bar{\Sigma}_C: \begin{cases} \rho\bar{v} = \bar{A}_c\bar{v} + \bar{B}_c\bar{y} \\ \bar{u} = \bar{C}_c\bar{v} + \bar{D}_c\bar{y}. \end{cases} \tag{6.43}$$

Suppose (Π, Γ) solve the regulator equation (6.4). Then a controller Σ_C for the original system Σ induced by $\bar{\Sigma}_C$ takes the following form,

$$\Sigma_C: \begin{cases} \rho v_1 = Sv_1 + \bar{C}_{c,1}v_2 + \bar{D}_{c,1}(y + (D_{yw} + C_y\Pi)v_1) \\ \rho v_2 = \bar{A}_c v_2 + \bar{B}_c(y + (D_{yw} + C_y\Pi)v_1) \\ u = -\Gamma v_1 + \bar{C}_{c,2}v_2 + \bar{D}_{c,2}(y + (D_{yw} + C_y\Pi)v_1) \end{cases} \tag{6.44}$$

where $\bar{C}_{c,1}$, $\bar{C}_{c,2}$, $\bar{D}_{c,1}$, and $\bar{D}_{c,2}$ are obtained by partitioning $\bar{C}_c$ and $\bar{D}_c$ in conformity with the partitioning of $\bar{A}$,

$$\bar{C}_c = \begin{pmatrix} \bar{C}_{c,1} \\ \bar{C}_{c,2} \end{pmatrix} \text{ and } \bar{D}_c = \begin{pmatrix} \bar{D}_{c,1} \\ \bar{D}_{c,2} \end{pmatrix}.$$

To establish the connection between the interconnection of Σ and Σ_C and the interconnection of $\bar{\Sigma}$ and $\bar{\Sigma}_C$ we used the following basis transformation: $\bar{x}_1 = x - \Pi w$, $\bar{x}_2 = v_1 + w$, $\bar{u}_2 = \Gamma v_1 + u$, and $\bar{v} = v_2$, where $\bar{x}$ and $\bar{u}$ are partitioned in accordance with the partitioning of $\bar{A}$ and $\bar{B}$,

$$\bar{x} = \begin{pmatrix} \bar{x}_1 \\ \bar{x}_2 \end{pmatrix} \text{ and } \bar{u} = \begin{pmatrix} \bar{u}_1 \\ \bar{u}_2 \end{pmatrix}.$$

We now state the following key result.

Lemma 6.5.1 *Consider the given system Σ as in (6.1) and the exosystem Σ_{E} as in (6.2) or equivalently consider Σ_a as given in (6.7). Let Assumptions A.1, A.2, A.3, and A.4 be satisfied. Also, let a solution (Π, Γ) of the regulator equation (6.4) be given. Then, there exists a controller of the form Σ_{C} which, when applied to Σ and Σ_{E}, solves the classical output regulation problem, and which achieves a closed-loop transfer matrix G from r to z when applied to Σ_a if and only if the following condition holds:*

> *There exist a controller $\bar{\Sigma}_{\mathrm{C}}$ of the form (6.43) which internally stabilizes the auxiliary system $\bar{\Sigma}$ given in (6.41). Moreover, when such an internally stabilizing controller $\bar{\Sigma}_{\mathrm{C}}$ is applied to $\bar{\Sigma}$, the achieved closed-loop transfer matrix from $\bar{r}$ to $\bar{z}$ equals G.*

Furthermore, given any such controller $\bar{\Sigma}_C$ for the auxiliary system $\bar{\Sigma}$, the correspondingly constructed controller Σ_{C} given in (6.44) internally stabilizes the given system Σ, achieves output regulation when applied to (6.1) and (6.2), and also yields a closed-loop transfer matrix from r to z equal to G.

Proof : The proof of this lemma follows more or less that of Theorem 7.3.1 given in the next chapter. ∎

Remark 6.5.1 *It is important to note that in the connection that we obtain between Σ_{C} and $\bar{\Sigma}_{\mathrm{C}}$ that the controller Σ_{C} is strictly proper if and only if the controller $\bar{\Sigma}_{\mathrm{C}}$ of the form (6.43) has the property that*

$$\bar{D}_{\mathrm{c}} = \begin{pmatrix} \bar{D}_{\mathrm{c},1} \\ 0 \end{pmatrix}$$

or, in other words, $\bar{D}_{\mathrm{c},2} = 0$. Therefore the problem of minimizing the transient error over the class of strictly proper controllers can also be solved using the above lemma but for the auxiliary system we have a slightly nonstandard H_2 control problem because of the constraint on its direct feedthrough matrix.

Obviously, for our present purpose, the above lemma transforms the optimal output regulation problem for Σ_a (which is a constrained optimization problem) to an unconstrained optimization problem, namely an H_2 optimal control problem for $\bar{\Sigma}$ where in one seeks a controller to minimize the H_2 norm of the closed-loop transfer matrix from r to $\bar{z}$. In other words, the optimal output regulation problem for Σ_a can be examined in detail through a study of the H_2 optimal control problem for $\bar{\Sigma}$.

Analogous to the case of state feedback, the solvability conditions for the optimal output regulation problem via measurement feedback will be expressed shortly in terms of the solutions of certain linear matrix inequalities. We recall below for both continuous- and discrete-time systems the needed linear matrix inequalities.

In connection with continuous-time systems, we first introduce two continuous-time linear matrix inequalities,

$$\bar{F}(\bar{P}) := \begin{pmatrix} \bar{A}^{\mathrm{T}}\bar{P} + \bar{P}\bar{A} + \bar{C}_z^{\mathrm{T}}\bar{C}_z & \bar{P}\bar{B} + \bar{C}_z^{\mathrm{T}}\bar{D}_{zu} \\ \bar{B}^{\mathrm{T}}\bar{P} + \bar{D}_{zu}^{\mathrm{T}}\bar{C}_z & \bar{D}_{zu}^{\mathrm{T}}\bar{D}_{zu} \end{pmatrix} \geqslant 0, \tag{6.45}$$

and

$$\bar{G}(\bar{Q}) := \begin{pmatrix} \bar{A}\bar{Q} + \bar{Q}\bar{A}^{\mathrm{T}} + \bar{E}\bar{E}^{\mathrm{T}} & \bar{Q}\bar{C}_y^{\mathrm{T}} \\ \bar{C}_y\bar{Q} & 0 \end{pmatrix} \geqslant 0. \tag{6.46}$$

Again, as shown in [61] and as explained further in Appendix 6.A, whenever the pair $(\bar{A}, \bar{B})$ is stabilizable (or equivalently (A, B) is stabilizable), there exists a unique semi-stabilizing solution $\bar{P}$ of the linear matrix inequality given in (6.45). Similarly, whenever the pair $(\bar{C}_y, \bar{A})$ is detectable (or equivalently Assumption A.3 is satisfied), there exists a unique semi-stabilizing solution $\bar{Q}$ of the linear matrix inequality given in (6.46). Moreover, such solutions are positive semi-definite and are the largest among all symmetric solutions.

In connection with discrete-time systems, we introduce two discrete-time linear matrix inequalities,

$$\bar{F}(\bar{P}) := \begin{pmatrix} \bar{A}^{\mathrm{T}}\bar{P}\bar{A} - \bar{P} + \bar{C}_z^{\mathrm{T}}\bar{C}_z & \bar{A}^{\mathrm{T}}\bar{P}\bar{B} + \bar{C}_z^{\mathrm{T}}\bar{D}_{zu} \\ \bar{B}^{\mathrm{T}}\bar{P}\bar{A} + \bar{D}_{zu}^{\mathrm{T}}\bar{C}_z & \bar{B}^{\mathrm{T}}\bar{P}\bar{B} + \bar{D}_{zu}^{\mathrm{T}}\bar{D}_{zu} \end{pmatrix} \geqslant 0, \tag{6.47}$$

and

$$\bar{G}(\bar{Q}) := \begin{pmatrix} \bar{A}\bar{Q}\bar{A}^{\mathrm{T}} - \bar{Q} + \bar{E}\bar{E}^{\mathrm{T}} & \bar{A}\bar{Q}\bar{C}_y^{\mathrm{T}} \\ \bar{C}_y\bar{Q}\bar{A}^{\mathrm{T}} & \bar{C}_y\bar{Q}\bar{C}_y^{\mathrm{T}} \end{pmatrix} \geqslant 0. \tag{6.48}$$

Again, as shown in [61] and as explained further in Appendix 6.B, whenever the pair $(\bar{A}, \bar{B})$ is stabilizable (or equivalently (A, B) is stabilizable), there exists a unique symmetric semi-stabilizing and strongly rank minimizing solution $\bar{P}$ of the linear matrix inequality given in (6.47). Similarly, whenever the pair $(\bar{C}_y, \bar{A})$ is detectable (or equivalently Assumption A.3 is satisfied), there exists a unique symmetric semi-stabilizing and strongly rank minimizing solution $\bar{Q}$ of the linear matrix inequality given in (6.48). Moreover, such solutions $\bar{P}$ and $\bar{Q}$ are positive semi-definite, and are the largest among all such strongly rank minimizing solutions.

Utilizing the unique symmetric semi-stabilizing solutions $\bar{P}$ and $\bar{Q}$ of (6.45) and (6.46) for continuous-time systems or the unique semi-stabilizing and strongly rank minimizing solutions $\bar{P}$ and $\bar{Q}$ of (6.47) and (6.48) for discrete-time systems, we can define a pair of matrices $\bar{C}_{\mathrm{P}}$ and $\bar{D}_{\mathrm{P}}$ and another pair of matrices $\bar{E}_{\mathrm{Q}}$ and $\bar{D}_{\mathrm{Q}}$ as

$$\begin{pmatrix} \bar{C}_{\mathrm{P}}^{\mathrm{T}} \\ \bar{D}_{\mathrm{P}}^{\mathrm{T}} \end{pmatrix} \begin{pmatrix} \bar{C}_{\mathrm{P}} & \bar{D}_{\mathrm{P}} \end{pmatrix} = \bar{F}(\bar{P}) \text{ and } \begin{pmatrix} \bar{E}_{\mathrm{Q}} \\ \bar{D}_{\mathrm{Q}} \end{pmatrix} \begin{pmatrix} \bar{E}_{\mathrm{Q}}^{\mathrm{T}} & \bar{D}_{\mathrm{Q}}^{\mathrm{T}} \end{pmatrix} = \bar{G}(\bar{Q}). \tag{6.49}$$

It is worth noting that the optimal and suboptimal regulation problems via proper or strictly proper measurement feedback as formulated in Problems 6.2.3 and 6.2.5 are in general not solvable. In other words, **in general there does not exist a fixed optimal regulator** solving the Problem 6.2.3. Also, in general there does not exist a sequence of fixed proper (strictly proper) measurement feedback regulators by selecting a member of which one can attain a transient performance measure as close as desired to the infimum $J_p^*(x_0, w_0, \Pi, \Gamma)$ ($J_{sp}^*(x_0, w_0, \Pi, \Gamma)$) uniformly for all $x_0 \in \mathbb{R}^n$ and $w_0 \in \mathbb{R}^s$.

Obviously in many cases we do not know the initial conditions and this is actually a prime reason for using measurement feedback regulators which in most cases contain an observer to estimate the state. The question is then which controller should be used since there does not exist a measurement feedback regulator which is uniformly optimal or close to optimal for all initial conditions. It seems then reasonable to minimize an average cost. Let e_i ($i = 1, \ldots, n$) and f_i ($i = 1, \ldots, s$) be bases of the state spaces of the system and the exosystem respectively. We can then use the cost function,

$$\tilde{J}(\Pi, \Gamma) = \sum_{i=1}^{n} J(e_i, 0, \Sigma_{\mathrm{C}}) + \sum_{j=1}^{s} J(0, f_i, \Sigma_{\mathrm{C}}),$$

where Σ_{C} is a regulator associated with the solution (Π, Γ) of the regulator equations. Having defined $\tilde{J}(\Pi, \Gamma)$, we can define optimal and suboptimal regulation problems associated with this cost similar to the ones formulated in Problems 6.2.3 and 6.2.5. We will use $\tilde{J}_p^*(\Pi, \Gamma)$ and $\tilde{J}_{sp}^*(\Pi, \Gamma)$ for the infimum of $\tilde{J}(\Pi, \Gamma)$ over all proper and strictly proper controllers respectively which achieve output regulation and are associated with the solutions (Π, Γ) of the regulator equations.

A first question is whether this cost function $\tilde{J}(\Pi, \Gamma)$ depends on the specific bases we choose for the state spaces of the system and the exosystem. It turns out that as long as we take orthonormal bases, this cost is independent of the choice of bases. Actually, if we choose as a basis $e_i = Q_1 \tilde{e}_i$ ($i = 1, \ldots, n$) and $f_i = Q_2 \tilde{f}_i$ ($i = 1, \ldots, s$) with $\tilde{e}_i$ ($i = 1, \ldots, n$) and $\tilde{f}_i$ ($i = 1, \ldots, s$)

as orthonormal bases, then the matrices Q_1 and Q_2 play the role of weighting functions which can be used to incorporate a priori information about the initial conditions. By making Q_1 large in a certain direction, we actually weigh that direction more heavily in the cost function and this can be used to reflect that, on the basis of a priori information, this direction is considered more likely.

Finally note that the classical H_2 norm which is normally interpreted as the trace of the covariance of the output given white noise inputs can also be interpreted in terms of this average cost. Let G be the transfer function with realization $(A, B, C, 0)$. Let $e_1, \ldots, e_m$ $(i = 1, \ldots, m)$ be an orthonormal bases of the input space. Then we have,

$$\|G\|_2^2 = \sum_{i=1}^{m} \|z_i\|_2^2,$$

where z_i is the output of the system with initial condition Be_i and zero input.

6.5.1 Determination of J_p^*, J_{sp}^*, $\tilde{J}_p^*$, $\tilde{J}_{sp}^*$

We are now ready to determine the optimal performance for the given initial conditions $J_p^*(x_0, w_0, \Pi, \Gamma)$ and $J_{sp}^*(x_0, w_0, \Pi, \Gamma)$ as well as the optimal average performance $\tilde{J}_p^*(\Pi, \Gamma)$ and $\tilde{J}_{sp}^*(\Pi, \Gamma)$.

For continuous-time systems, the following theorem presents the conditions under which the infimum $J_p^*(x_0, w_0, \Pi, \Gamma)$ (or $J_{sp}^*(x_0, w_0, \Pi, \Gamma)$) exists and gives an expression for it.

Theorem 6.5.1 *For continuous-time systems, consider the system Σ as in (6.1) and the exosystem Σ_E as in (6.2). Let Assumptions A.1, A.2, A.3, and A.4 of Section 6.2 hold. Also, let a solution (Π, Γ) of the regulator equation (6.4) be given. Moreover, let $\bar{A}$, $\bar{B}$, $\bar{E}$, $\bar{C}_z$, $\bar{D}_{zu}$, and $\bar{C}_y$ be given by (6.42).*

Let $\bar{P}$ and $\bar{Q}$ be the unique symmetric semi-stabilizing solutions of (6.45) and (6.46) respectively. Moreover, let $\bar{C}_P$, $\bar{D}_P$, $\bar{E}_Q$, and $\bar{D}_Q$ be defined as in (6.49). Then, $J_p^*(x_0, w_0, \Pi, \Gamma)$ and $J_{sp}^*(x_0, w_0, \Pi, \Gamma)$ exist, and are given by

$$\begin{aligned} J_p^*(x_0, w_0, \Pi, \Gamma) &= J_{sp}^*(x_0, w_0, \Pi, \Gamma) \\ &= \bar{E}^T \bar{P} \bar{E} + \operatorname{trace}[\bar{C}_P \bar{Q} \bar{C}_P^T] \\ &= \bar{E}^T \bar{P} \bar{E} + \operatorname{trace}[(\bar{A}^T \bar{P} + \bar{P} \bar{A} + \bar{C}_z^T \bar{C}_z) \bar{Q}] \\ &= \operatorname{trace}[\bar{C}_z \bar{Q} \bar{C}_z^T] + \operatorname{trace}[(\bar{Q} \bar{A}^T + \bar{A} \bar{Q} + \bar{E} \bar{E}^T) \bar{P}]. \end{aligned} \tag{6.50}$$

Proof : It follows from Lemma 6.5.1 when appropriate results from H_2 optimal control theory are utilized, see [61]. In general, the infimum over the class of controllers of the form (6.43) with $\bar{D}_{c,2} = 0$ (which have a $1-1$ relationship to strictly proper regulators for Σ) might obviously be larger than or equal to the infimum over proper controllers and might be smaller than or equal to the infimum over strictly proper controllers. However, we note that the infimum of performance measure for proper and strictly proper controllers for the auxiliary system are the same in general. This proves the above theorem. ■

For continuous-time systems, the following theorem presents the conditions under which the infimum for the average cost $\tilde{J}_p^*(\Pi, \Gamma)$ (or $\tilde{J}_{sp}^*(\Pi, \Gamma)$) exists and gives an expression for it.

Theorem 6.5.2 *For continuous-time systems, consider the system Σ as in (6.1) and the exosystem Σ_E as in (6.2). Let Assumptions A.1, A.2, A.3, and A.4 of Section 6.2 hold. Also, let a solution (Π, Γ) of the regulator equation (6.4) be given. Finally, let the basis, with respect to which the average cost is defined, be given by $e_i = Q_1\tilde{e}_i$ $(i = 1, \ldots, n)$ and $f_i = Q_2\tilde{f}_i$ $(i = 1, \ldots, s)$ with $\tilde{e}_i$ $(i = 1, \ldots, n)$ and $\tilde{f}_i$ $(i = 1, \ldots, s)$ as orthonormal bases. Moreover, let $\bar{A}$, $\bar{B}$, $\bar{C}_z$, $\bar{D}_{zu}$, and $\bar{C}_y$ be given by (6.42), but define $\bar{E}$ by*

$$\bar{E} = \begin{pmatrix} Q_1 - \Pi Q_2 \\ Q_2 \end{pmatrix}. \tag{6.51}$$

Let $\bar{P}$ and $\bar{Q}$ be the unique symmetric semi-stabilizing solutions of (6.45) and (6.46) respectively. Moreover, let $\bar{C}_P$, $\bar{D}_P$, $\bar{E}_Q$, and $\bar{D}_Q$ be defined as in (6.49). Then, $\tilde{J}_p^*(\Pi, \Gamma)$ and $\tilde{J}_{sp}^*(\Pi, \Gamma)$ exist, and are given by

$$\begin{aligned} \tilde{J}_p^*(\Pi, \Gamma) &= \tilde{J}_{sp}^*(\Pi, \Gamma) \\ &= \bar{E}^T\bar{P}\bar{E} + \text{trace}[\bar{C}_P\bar{Q}\bar{C}_P^T] \\ &= \bar{E}^T\bar{P}\bar{E} + \text{trace}[(\bar{A}^T\bar{P} + \bar{P}\bar{A} + \bar{C}_z^T\bar{C}_z)\bar{Q}] \\ &= \text{trace}[\bar{C}_z\bar{Q}\bar{C}_z^T] + \text{trace}[(\bar{Q}\bar{A}^T + \bar{A}\bar{Q} + \bar{E}\bar{E}^T)\bar{P}]. \end{aligned} \tag{6.52}$$

Proof : It follows from Lemma 6.5.1 when appropriate results from H_2 optimal control theory are utilized, see [61]. For the strictly proper case, the arguments from the proof of Theorem 6.5.1 also apply in this case. ■

In order to present a result similar to the above one for discrete-time systems, we need to define a matrix $\bar{R}^*$ by

$$\bar{R}^* := (\bar{D}_{\mathrm{P}}^{\mathrm{T}})^{\dagger}(\bar{D}_{\mathrm{P}}^{\mathrm{T}}\bar{C}_{\mathrm{P}}\bar{Q}\bar{C}_y^{\mathrm{T}})(\bar{D}_{\mathrm{Q}}^{\mathrm{T}})^{\dagger} \tag{6.53}$$

where the generalized inverse (or Moore-Penrose inverse) of a matrix M is denoted by $M^{\dagger}$. We can now present, for discrete-time systems, the following theorem which first gives the conditions under which the infimum $J_p^*(x_0, w_0, \Pi, \Gamma)$ (or $J_{sp}^*(x_0, w_0, \Pi, \Gamma)$) exists, and then gives an expression for it.

Theorem 6.5.3 *For discrete-time systems, consider the system Σ as in (6.1) and the exosystem Σ_{E} as in (6.2). Let Assumptions A.1, A.2, A.3, and A.4 of Section 6.2 hold. Also, let a solution (Π, Γ) of the regulator equation (6.4) be given. Moreover, let $\bar{A}$, $\bar{B}$, $\bar{E}$, $\bar{C}_z$, $\bar{D}_{zu}$, and $\bar{C}_y$ be given by (6.42).*

Let $\bar{P}$ and $\bar{Q}$ be the unique semi-stabilizing and strongly rank minimizing solutions of (6.47) and (6.48) respectively. Moreover, let $\bar{C}_{\mathrm{P}}$, $\bar{D}_{\mathrm{P}}$, $\bar{E}_{\mathrm{Q}}$, and $\bar{D}_{\mathrm{Q}}$ be given by (6.49). Also, let $\bar{R}^*$ be as in (6.53). Then, $J_p^*(x_0, w_0, \Pi, \Gamma)$ and $J_{sp}^*(x_0, w_0, \Pi, \Gamma)$ exist, and are given by

$$\begin{aligned} J_p^*(x_0, w_0, \Pi, \Gamma) &= \bar{E}^T\bar{P}\bar{E} + \operatorname{trace}[\bar{C}_P\bar{Q}\bar{C}_P^T] - \operatorname{trace}[\bar{R}^*(\bar{R}^*)^T] \\ &= \bar{E}^T\bar{P}\bar{E} + \operatorname{trace}[(\bar{A}^T\bar{P}\bar{A} - \bar{P} + \bar{C}_z^T\bar{C}_z)\bar{Q}] \\ &\qquad - \operatorname{trace} \bar{R}^*(\bar{R}^*)^T \\ &= \operatorname{trace}[\bar{C}_z\bar{Q}\bar{C}_z^T] + \operatorname{trace}[(\bar{A}\bar{Q}\bar{A}^T - \bar{Q} + \bar{E}\bar{E}^T)\bar{P}] \\ &\qquad - \operatorname{trace} \bar{R}^*(\bar{R}^*)^T, \end{aligned} \tag{6.54}$$

$$\begin{aligned} J_{sp}^*(x_0, w_0, \Pi, \Gamma) &= \bar{E}^T\bar{P}\bar{E} + \operatorname{trace} \bar{C}_P\bar{Q}\bar{C}_P^T \\ &= \bar{E}^T\bar{P}\bar{E} + \operatorname{trace}[(\bar{A}^T\bar{P}\bar{A} - \bar{P} + \bar{C}_z^T\bar{C}_z)\bar{Q}] \\ &= \operatorname{trace} \bar{C}_z\bar{Q}\bar{C}_z^T + \operatorname{trace}[(\bar{A}\bar{Q}\bar{A}^T - \bar{Q} + \bar{E}\bar{E}^T)\bar{P}]. \end{aligned} \tag{6.55}$$

Proof : It follows from Lemma 6.5.1 when appropriate results from H_2 optimal control theory are utilized, see [61]. For the proper case this is immediate. For the strictly proper case we need to find the infimum over all stabilizing controllers of the form (6.43) with

$$\bar{D}_{\mathrm{c}} = \begin{pmatrix} \bar{D}_{\mathrm{c},1} \\ \bar{D}_{\mathrm{c},2} \end{pmatrix} = \begin{pmatrix} \bar{D}_{\mathrm{c},1} \\ 0 \end{pmatrix}.$$

Note that Lemma 6.5.4 of [61] gives a lower bound for the achievable H_2 norm for a fixed direct feedthrough matrix. For the auxiliary system we need to look for direct feedthrough matrices with a specific structure but the lower bound of Lemma 6.5.4 can be shown to be **independent** of the specific direct feedthrough matrix (this makes actually use of Lemma 8.4.1 which is presented later in this book). Using this specific structure the result then follows directly from the arguments in [61]. ■

Remark 6.5.2 *Let us emphasize that, in contrast with the case of continuous-time systems, for discrete-time systems the optimal performance depends on whether proper or strictly proper controllers are used.*

We note that the expressions for $J_p^*(x_0, w_0, \Pi, \Gamma)$ and $J_{sp}^*(x_0, w_0, \Pi, \Gamma)$ given in (6.50), (6.54), and (6.55) indeed point out the minimal possible transient energy one could possibly attain under proper or strictly proper measurement feedback controllers as the case may be. As such they point out a fundamental limitation or characteristic of the given system Σ of (6.1) and the exosystem Σ_E of (6.2).

For discrete-time systems, the following theorem presents the conditions under which the infimum for the average cost $\tilde{J}_p^*(\Pi, \Gamma)$ (or $\tilde{J}_{sp}^*(\Pi, \Gamma)$) exists and gives an expression for it.

Theorem 6.5.4 *For discrete-time systems, consider the system Σ as in (6.1) and the exosystem Σ_E as in (6.2). Let Assumptions A.1, A.2, A.3, and A.4 of Section 6.2 hold. Also, let a solution (Π, Γ) of the regulator equation (6.4) be given. Moreover, let $\bar{A}$, $\bar{B}$, $\bar{C}_z$, $\bar{D}_{zu}$, and $\bar{C}_y$ be given by (6.42), but define $\bar{E}$ by (6.51).*

Let $\bar{P}$ and $\bar{Q}$ be the unique semi-stabilizing and strongly rank minimizing solutions of (6.47) and (6.48) respectively. Moreover, let $\bar{C}_P$, $\bar{D}_P$, $\bar{E}_Q$, and $\bar{D}_Q$ be defined as in (6.49). Also, let $\bar{R}^*$ be as in (6.53). Then, $\tilde{J}_p^*(\Pi, \Gamma)$ and $\tilde{J}_{sp}^*(\Pi, \Gamma)$ exist, and are given by

$$\begin{aligned}
\tilde{J}_p^*(\Pi, \Gamma) &= \bar{E}^T \bar{P} \bar{E} + \operatorname{trace}[\bar{C}_P \bar{Q} \bar{C}_P^T] - \operatorname{trace}[\bar{R}^* (\bar{R}^*)^T] \\
&= \bar{E}^T \bar{P} \bar{E} + \operatorname{trace}[(\bar{A}^T \bar{P} \bar{A} - \bar{P} + \bar{C}_z^T \bar{C}_z) \bar{Q}] \\
&\qquad - \operatorname{trace} \bar{R}^* (\bar{R}^*)^T \\
&= \operatorname{trace}[\bar{C}_z \bar{Q} \bar{C}_z^T] + \operatorname{trace}[(\bar{A} \bar{Q} \bar{A}^T - \bar{Q} + \bar{E}\bar{E}^T) \bar{P}] \\
&\qquad - \operatorname{trace} \bar{R}^* (\bar{R}^*)^T,
\end{aligned} \tag{6.56}$$

$$\begin{aligned}\tilde{J}^*_{sp}(\Pi,\Gamma) &= \bar{E}^T\bar{P}\bar{E} + \operatorname{trace}\bar{C}_P\bar{Q}\bar{C}_P^T \\ &= \bar{E}^T\bar{P}\bar{E} + \operatorname{trace}[(\bar{A}^T\bar{P}\bar{A} - \bar{P} + \bar{C}_z^T\bar{C}_z)\bar{Q}] \\ &= \operatorname{trace}\bar{C}_z\bar{Q}\bar{C}_z^T + \operatorname{trace}[(\bar{A}\bar{Q}\bar{A}^T - \bar{Q} + \bar{E}\bar{E}^T)\bar{P}]. \qquad (6.57)\end{aligned}$$

Proof : It follows from Lemma 6.5.1 when appropriate results from H_2 optimal control theory are utilized, see [61]. For the strictly proper case we need the same arguments as given in the proof of Theorem 6.5.3. ■

6.5.2 Solvability conditions

In this subsection we develop the solvability conditions for the optimal and suboptimal output regulation problems posed in Section 6.2 via measurement feedback. The proofs in this section for the proper controllers are a consequence of Lemma 6.5.1 when the necessary and sufficient conditions for the existence of a proper H_2 optimal controller for the system $\bar{\Sigma}$ are taken into consideration (see [61]). Hence we omit the proofs. For the strictly proper case we have to be a bit more careful because of imposing an additional structure on its direct feedthrough matrix in finding an H_2 optimal controller for the auxiliary system.

For continuous-time systems, we have the following theorem that deals with using proper measurement feedback regulators.

Theorem 6.5.5 *For continuous-time systems, consider Problem 6.2.2 via proper measurement feedback for Σ as in (6.1) and the exosystem Σ_E as in (6.2). Let Assumptions A.1, A.2 , A.3 , and A.4 of Section 6.2 hold. Also, let (Π,Γ) be any solution of the regulator equation (6.4). Then, the optimal output regulation problem for a given x_0 and w_0 (i.e. Problem 6.2.2) is solvable via proper measurement feedback for Σ if and only if the following holds:*

The H_2 optimal control problem for $\bar{\Sigma}$ given in (6.41) is solvable via proper measurement feedback. That is, the following conditions hold:

(*i*) $\operatorname{im}\bar{E}_Q \subseteq \mathcal{V}^-(\bar{A},\bar{B},\bar{C}_P,\bar{D}_P) + B\ker\bar{D}_P$,

(*ii*) $\ker\bar{C}_P \supseteq \mathcal{S}^-(\bar{A},\bar{E}_Q,\bar{C}_y,\bar{D}_Q)\cap\bar{C}_y^{-1}\{\operatorname{im}\bar{D}_Q\}$,

(*iii*) $\mathcal{S}^-(\bar{A},\bar{E}_Q,\bar{C}_y,\bar{D}_Q) \subseteq \mathcal{V}^-(\bar{A},\bar{B},\bar{C}_P,\bar{D}_P)$, *where $\bar{P}$ and $\bar{Q}$ are respectively the unique semi-stabilizing solutions of the linear matrix inequalities (6.45) and (6.46), and the matrices $\bar{C}_P$, $\bar{D}_P$, $\bar{E}_Q$, and $\bar{D}_Q$ are as in (6.49).*

The following theorem is the analog of Theorem 6.5.5, however it considers the use of strictly proper measurement feedback regulators.

Theorem 6.5.6 *For continuous-time systems, consider Problem 6.2.2 via strictly proper measurement feedback for Σ as in (6.1) and the exosystem Σ_E as in (6.2). Let Assumptions A.1, A.2 , A.3 , and A.4 of Section 6.2 hold. Also, let (Π, Γ) be any solution of the regulator equation (6.4). Then, the optimal output regulation problem for a given x_0 and w_0 (i.e. Problem 6.2.2) is solvable via strictly proper measurement feedback for Σ if and only if the following holds:*

The H_2 optimal control problem for $\bar{\Sigma}$ given in (6.41) is solvable via strictly proper measurement feedback. That is, the following conditions hold:

(*i*) $\operatorname{im} \bar{E}_Q \subseteq \mathcal{V}^-(\bar{A}, \bar{B}, \bar{C}_P, \bar{D}_P)$,

(*ii*) $\ker \bar{C}_P \supseteq \mathcal{S}^-(\bar{A}, \bar{E}_Q, \bar{C}_y, \bar{D}_Q)$,

(*iii*) $A\mathcal{S}^-(\bar{A}, \bar{E}_Q, \bar{C}_y, \bar{D}_Q) \subseteq \mathcal{V}^-(\bar{A}, \bar{B}, \bar{C}_P, \bar{D}_P)$,

(*iv*) $\mathcal{S}^-(\bar{A}, \bar{E}_Q, \bar{C}_y, \bar{D}_Q) \subseteq \mathcal{V}^-(\bar{A}, \bar{B}, \bar{C}_P, \bar{D}_P)$,

where $\bar{P}$ and $\bar{Q}$ are respectively the unique semi-stabilizing solutions of linear matrix inequalities (6.45) and (6.46), and the matrices $\bar{C}_P$, $\bar{D}_P$, $\bar{E}_Q$, and $\bar{D}_Q$ are as in (6.49).

Proof : We know that we are looking for an H_2 optimal controller for the auxiliary system with the additional structure that $\bar{D}_{c,2} = 0$. Assume we fix the direct feedthrough matrix $\bar{D}_c$ of the controller. Then after a preliminary feedback $\bar{u} = \bar{D}_c\bar{y} + \tilde{u}$ we are looking for an H_2 optimal strictly proper controller from $\bar{y}$ to $\tilde{u}$. The resulting existence conditions are known from [61]. It is then straightforward to check that due to the structure of the direct feedthrough matrix we have $\bar{D}_P\bar{D}_c = 0$. But then by a direct verification we obtain that the existence of an optimal controller is independent of the specific direct feedthrough matrix we have started with. Therefore, we can equally well choose $\bar{D}_c = 0$ and find a strictly proper controller for the auxiliary system. ■

Remark 6.5.3 *The conditions for the existence of a proper optimal regulator for a given x_0 and w_0 as given in Theorem 6.5.5 are weaker than those for the existence of a strictly proper optimal regulator for a given x_0 and w_0 as given in Theorem 6.5.6. That is, the conditions in Theorem 6.5.5 are implied by those in Theorem 6.5.6. Thus, it is possible for a system to have a proper optimal regulator for a given x_0 and w_0 while not having a strictly proper optimal regulator.*

For discrete-time systems, the following theorem considers the use of proper measurement feedback regulators.

Theorem 6.5.7 *For discrete-time systems, consider Problem 6.2.2 via proper measurement feedback for Σ as in (6.1) and the exosystem Σ_E as in (6.2). Let Assumptions A.1, A.2 , A.3 , and A.4 of Section 6.2 hold. Also, let (Π, Γ) be any solution of the regulator equation (6.4). Then, the optimal output regulation problem for a given x_0 and w_0 (i.e. Problem 6.2.2) is solvable via proper measurement feedback for Σ if and only if the following holds:*

> *The H_2 optimal control problem for $\bar{\Sigma}$ given in (6.41) is solvable via proper measurement feedback. That is, the following conditions hold:*
>
> (*i*) $\operatorname{im}[\bar{E}_Q + \bar{B}\bar{D}_P^\dagger \bar{R}^*] \subseteq \mathcal{V}^\ominus(\bar{A}, \bar{B}, \bar{C}_P, \bar{D}_P)$,
>
> (*ii*) $\ker[\bar{C}_P - \bar{R}^*\bar{D}_Q^\dagger \bar{C}_y] \supseteq \mathcal{S}^\ominus(\bar{A}, \bar{E}_Q, \bar{C}_y, \bar{D}_Q)$,
>
> (*iii*) $\mathcal{S}^\ominus(\bar{A}, \bar{E}_Q, \bar{C}_y, \bar{D}_Q) \subseteq \mathcal{V}^\ominus(\bar{A}, \bar{B}, \bar{C}_P, \bar{D}_P)$,
>
> (*iv*) $(\bar{A} - \bar{B}\bar{D}_P^\dagger \bar{R}^*\bar{D}_Q^\dagger \bar{C}_y)\mathcal{S}^\ominus(\bar{A}, \bar{E}_Q, \bar{C}_y, \bar{D}_Q) \subseteq \mathcal{V}^\ominus(\bar{A}, \bar{B}, \bar{C}_P, \bar{D}_P)$,
>
> *where $\bar{P}$ and $\bar{Q}$ are respectively the unique semi-stabilizing and strongly rank minimizing solutions of linear matrix inequalities (6.47) and (6.48), and the matrices $\bar{C}_P$, $\bar{D}_P$, $\bar{E}_Q$, and $\bar{D}_Q$ are as in (6.49).*

Similarly, for discrete-time systems, the following theorem considers using strictly proper measurement feedback regulators.

Theorem 6.5.8 *For discrete-time systems, consider Problem 6.2.2 via strictly proper measurement feedback for Σ as in (6.1) and the exosystem Σ_E as in (6.2). Let Assumptions A.1, A.2 , A.3 , and A.4 of Section 6.2 hold. Also, let (Π, Γ) be any solution of the regulator equation (6.4). Then, the optimal output regulation problem for a given x_0 and w_0 (i.e. Problem 6.2.2) is solvable* *via strictly proper measurement feedback for* Σ *if and only if the following holds:*

> *The* H_2 *optimal control problem for* $\bar{\Sigma}$ *given in (6.41) is solvable via strictly proper measurement feedback. That is, the following conditions hold:*
>
> (*i*) $\operatorname{im} \bar{E}_{\mathrm{Q}} \subseteq \mathcal{V}^{\ominus}(\bar{A}, \bar{B}, \bar{C}_{\mathrm{P}}, \bar{D}_{\mathrm{P}})$,
>
> (*ii*) $\ker \bar{C}_{\mathrm{P}} \supseteq \mathcal{S}^{\ominus}(\bar{A}, \bar{E}_{\mathrm{Q}}, \bar{C}_y, \bar{D}_{\mathrm{Q}})$,
>
> (*iii*) $A\mathcal{S}^{\ominus}(\bar{A}, \bar{E}_{\mathrm{Q}}, \bar{C}_y, \bar{D}_{\mathrm{Q}}) \subseteq \mathcal{V}^{\ominus}(\bar{A}, \bar{B}, \bar{C}_{\mathrm{P}}, \bar{D}_{\mathrm{P}})$,
>
> (*iv*) $\mathcal{S}^{\ominus}(\bar{A}, \bar{E}_{\mathrm{Q}}, \bar{C}_y, \bar{D}_{\mathrm{Q}}) \subseteq \mathcal{V}^{\ominus}(\bar{A}, \bar{B}, \bar{C}_{\mathrm{P}}, \bar{D}_{\mathrm{P}})$,
>
> *where* $\bar{P}$ *and* $\bar{Q}$ *are respectively the unique semi-stabilizing and strongly rank minimizing solutions of linear matrix inequalities (6.47) and (6.48), and the matrices* $\bar{C}_{\mathrm{P}}$, $\bar{D}_{\mathrm{P}}$, $\bar{E}_{\mathrm{Q}}$, *and* $\bar{D}_{\mathrm{Q}}$ *are as in (6.49).*

Proof : This follows via the same arguments as in the proof of Theorem 6.5.6 and [61]. Note that the direct feedthrough matrix of an H_2 optimal controller in discrete-time is normally determined via an optimization as given in Lemma 6.5.5 of [61] but due to the imposed structure of the direct feedthrough matrix it can be shown that any feedthrough matrix is optimal with respect to this optimization. ■

As studied by various theorems above, optimal output regulation problem for a given x_0 and w_0 (i.e. Problem 6.2.2) is solvable provided certain conditions are satisfied by the given system and by the given initial conditions. We emphasize that in general the optimal output regulation problem (i.e. Problem 6.2.3) is not solvable via proper or strictly proper measurement feedback regulators. In other words, **in general there does not exists a fixed proper or strictly proper measurement feedback regulator** that attains the optimal performance $J_p^*(x_0, w_0, \Pi, \Gamma)$ (or $J_{sp}^*(x_0, w_0, \Pi, \Gamma)$) for all x_0 and w_0 for which Problem 6.2.2 is solvable.

Looking from a different point of view, optimal output regulation that seeks to achieve exactly the best possible transient performance is perhaps a rigid requirement. It is natural then to enquire whether a compromise is possible by relaxing the requirement on the transient performance that is sought. This leads to the study of suboptimal output regulation problems.

The following theorem shows that, under some natural assumptions, the suboptimal output regulation problem for a given x_0 and w_0 is solvable via proper or strictly proper measurement feedback for both continuous- and discrete-time systems.

Theorem 6.5.9 *For both continuous- and discrete-time systems, consider the Problem 6.2.4 via proper or strictly proper measurement feedback for* Σ *as in (6.1) and the exosystem* Σ_E *as in (6.2). Let Assumptions A.1, A.2, A.3, and A.4 hold. Also, let* (Π, Γ) *be any solution of the regulator equation (6.4). Moreover, let* $J_p^*(x_0, w_0, \Pi, \Gamma)$ *and* $J_{sp}^*(x_0, w_0, \Pi, \Gamma)$ *be the corresponding optimal performance for a given* x_0 *and* w_0*. Then, the suboptimal regulation problem as formulated in Problem 6.2.4 is solvable via proper or strictly proper measurement feedback for any given* $x_0 \in \mathbb{R}^n$ *and* $w_0 \in \mathbb{R}^s$*.*

Obviously the conditions for solvability of controllers which are optimal with respect to the average cost are also of interest. These results are available but only amount to changing $\bar{E}$ in the above theorems.

The above Theorems 6.5.5 (continuous-time, proper), 6.5.6 (continuous-time, strictly proper), 6.5.7 (discrete-time proper), 6.5.8 (discrete-time, strictly proper), and 6.5.9 (suboptimal) give conditions for the existence of an optimal output regulator with respect to the initial conditions x_0, w_0 where $\bar{E}$ is defined by (6.42). The existence of optimal and suboptimal measurement feedback regulators with respect to the average cost yields exactly the same solvability conditions except for the fact that $\bar{E}$ is no longer defined by (6.42) but is given by

$$\bar{E} = \begin{pmatrix} Q_1 - \Pi Q_2 \\ Q_2 \end{pmatrix}$$

where the bases, with respect to which the average cost is defined, are given by $e_i = Q_1 \tilde{e}_i$ $(i = 1, \ldots, n)$ and $f_i = Q_2 \tilde{f}_i$ $(i = 1, \ldots, s)$ with $\tilde{e}_i$ $(i = 1, \ldots, n)$ and $\tilde{f}_i$ $(i = 1, \ldots, s)$ as orthonormal bases.

6.5.3 Perfect output regulation

As discussed earlier in Subsection 6.4.3, the notion of perfect output regulation consists of two parts, (1) $J_p^*(x_0, w_0, \Pi, \Gamma) = 0$ (or $J_{sp}^*(x_0, w_0, \Pi, \Gamma) = 0$) for all $x_0 \in \mathbb{R}^n$ and $w_0 \in \mathbb{R}^s$, and (2) the existence of a fixed sequence $\{\Sigma_C(\varepsilon) \mid \varepsilon > 0\}$ of regulators by selecting a member of which one can render the attained performance $J(x_0, w_0, \Sigma_C(\varepsilon))$ as small as desired uniformly for all $x_0 \in \mathbb{R}^n$ and $w_0 \in \mathbb{R}^s$. In other words, the problem of perfect output regulation is tantamount to the suboptimal output regulation problem (Problem 6.2.5) with the additional property that $J_p^*(x_0, w_0, \Pi, \Gamma) = 0$ (or

$J_{sp}^*(x_0, w_0, \Pi, \Gamma) = 0$). In Subsection 6.4.3 we formulated structural conditions on the given system Σ under which perfect output regulation can be achieved via state feedback. In this subsection, we revisit the topic of perfect output regulation while using measurement feedback regulators instead of state feedback regulators. Unlike in the case of state feedback, a close examination of the topic reveals that for a given system Σ one can almost never achieve perfect output regulation via measurement feedback regulators.

On the other hand, instead of seeking conditions on the given system Σ under which the transient performance measure can be achieved as close to zero as desired uniformly for all initial conditions $x_0 \in \mathbb{R}^n$ and $w_0 \in \mathbb{R}^s$, one can seek a limited formulation. Namely, for a given system Σ, one can seek a set of initial conditions x_0 and w_0, say $\mathcal{S}_{x_0 w_0}$, such that the transient performance measure $J(x_0, w_0, \Sigma_c)$ can be attained as close to zero as desired uniformly for all initial conditions x_0 and w_0 in $\mathcal{S}_{x_0 w_0}$, i.e. $J_p^*(x_0, w_0, \Pi, \Gamma)$ (or $J_{sp}^*(x_0, w_0, \Pi, \Gamma)$) is equal to zero uniformly for all initial conditions x_0 and w_0 in $\mathcal{S}_{x_0 w_0}$.

Lemma 6.5.1 comes to our aid in characterizing the set $\mathcal{S}_{x_0 w_0}$ as well. In view of Lemma 6.5.1, the problem of characterizing the set of all initial conditions x_0 and w_0 for which $J_p^*(x_0, w_0, \Pi, \Gamma)$ (or $J_{sp}^*(x_0, w_0, \Pi, \Gamma)$) is equal to zero is equivalent to characterizing all x_0 and w_0 for which we can make the H_2 norm of the transfer function from $\bar{r}$ to $\bar{z}$ of the auxiliary system $\bar{\Sigma}$ arbitrarily close to zero.

The problem of making the H_2 norm of the transfer function from $\bar{r}$ to $\bar{z}$ of the auxiliary system $\bar{\Sigma}$ arbitrary small is nothing else than the so called H_2-ADDPMS (H_2 Almost Disturbance Decoupling Problem with Measurement feedback and Stability) for $\bar{\Sigma}$. In view of (6.42), since one of the coefficient matrices, namely $\bar{E}$, that describes $\bar{\Sigma}$ depends on x_0 and w_0, by studying the structural conditions under which H_2-ADDPMS can be achieved for $\bar{\Sigma}$, one can characterize the set $\mathcal{S}_{x_0 w_0}$.

We have the following result for continuous-time systems when either proper or strictly proper measurement feedback regulators are used.

Theorem 6.5.10 *For continuous-time systems, consider the system Σ as in (6.1) and the exosystem Σ_E as in (6.2). Let Assumptions A.1, A.2, A.3, and A.4 of Section 6.2 hold. Also, let (Π, Γ) be a given solution of the regulator equation (6.4). Then, $J_p^*(x_0, w_0, \Pi, \Gamma)$ and $J_{sp}^*(x_0, w_0, \Pi, \Gamma)$ are equal to zero uniformly for all initial conditions x_0 and w_0 that satisfy the following conditions:*

$$(i)\ \operatorname{im}\begin{pmatrix} x_0 - \Pi w_0 \\ w_0 \end{pmatrix} \subseteq \mathcal{S}^*(\bar{A}, \bar{B}, \bar{C}_z, \bar{D}_{zu}) + \mathcal{V}^{-0}(\bar{A}, \bar{B}, \bar{C}_z, \bar{D}_{zu}).$$

(ii) $\mathcal{V}^*(\bar{A}, \begin{pmatrix} x_0 - \Pi w_0 \\ w_0 \end{pmatrix}, \bar{C}_y, 0) \cap \mathcal{S}^{-0}(\bar{A}, \begin{pmatrix} x_0 - \Pi w_0 \\ w_0 \end{pmatrix}, \bar{C}_y, 0) \subseteq \ker \bar{C}_z.$

(iii) $\mathcal{V}^*(\bar{A}, \begin{pmatrix} x_0 - \Pi w_0 \\ w_0 \end{pmatrix}, \bar{C}_y, 0) \cap \mathcal{S}^{-0}(\bar{A}, \begin{pmatrix} x_0 - \Pi w_0 \\ w_0 \end{pmatrix}, \bar{C}_y, 0)$
$\subseteq \mathcal{S}^*(\bar{A}, \bar{B}, \bar{C}_z, \bar{D}_{zu}) + \mathcal{V}^{-0}(\bar{A}, \bar{B}, \bar{C}_z, \bar{D}_{zu}).$

The set $\mathcal{S}_{x_0 w_0}$ is indeed the set of all x_0 and w_0 that satisfy the above conditions.

Proof : It follows from Lemma 6.5.1 when appropriate conditions to solve H_2-ADDPMS for $\bar{\Sigma}$ via proper or strictly proper measurement feedback controllers are taken into account, see Theorem 5.9.1 of [61]. ■

An important question arises at this point as to whether there exists a fixed sequence $\{\Sigma_C(\varepsilon) \mid \varepsilon > 0\}$ of regulators by selecting a member of which one can render the attained performance measure $J(x_0, w_0, \Sigma_C(\varepsilon))$ as arbitrarily small as desired uniformly for all initial conditions x_0 and w_0 that belong to the set $\mathcal{S}_{x_0 w_0}$ characterized in Theorem 6.5.10.

In the following theorem we provide sufficient conditions for the existence of such a sequence of regulators. To do so, let $\check{E}$ be such that

$$\operatorname{im} \check{E} = \mathcal{S}^*(\bar{A}, \bar{B}, \bar{C}_z, \bar{D}_{zu}) + \mathcal{V}^{-0}(\bar{A}, \bar{B}, \bar{C}_z, \bar{D}_{zu}). \tag{6.58}$$

Theorem 6.5.11 *For continuous-time systems, consider the system Σ as in (6.1) and the exosystem Σ_E as in (6.2). Let Assumptions A.1, A.2, A.3, and A.4 of Section 6.2 hold. Also, let (Π, Γ) be a given solution of the regulator equation (6.4). Moreover, consider the set $\mathcal{S}_{x_0 w_0}$ that is characterized in Theorem 6.5.10. Then, there exists a fixed sequence of proper (or strictly proper) measurement feedback regulators $\{\Sigma_C(\varepsilon) \mid \varepsilon > 0\}$ such that the attained performance measure $J(x_0, w_0, \Sigma_C(\varepsilon))$ tends to zero as ε tends to zero uniformly for all initial conditions x_0 and w_0 that belong to the set $\mathcal{S}_{x_0 w_0}$ if the following conditions hold:*

(i) $\mathcal{V}^*(\bar{A}, \check{E}, \bar{C}_y, 0) \cap \mathcal{S}^{-0}(\bar{A}, \check{E}, \bar{C}_y, 0) \subseteq \ker \bar{C}_z,$

(ii) $\mathcal{V}^*(\bar{A}, \check{E}, \bar{C}_y, 0) \cap \mathcal{S}^{-0}(\bar{A}, \check{E}, \bar{C}_y, 0)$
$\subseteq \mathcal{S}^*(\bar{A}, \bar{B}, \bar{C}_z, \bar{D}_{zu}) + \mathcal{V}^{-0}(\bar{A}, \bar{B}, \bar{C}_z, \bar{D}_{zu}),$

where $\check{E}$ is as defined in (6.58).

Proof : It is obvious since the conditions given in this theorem (which are independent of x_0 and w_0) imply that the conditions given in Theorem 6.5.10 hold for all x_0 and w_0 that belong to the set $\mathcal{S}_{x_0 w_0}$. ■

The above theorem gives only sufficient conditions; providing conditions that are both necessary and sufficient is still an open problem.

We now proceed with discrete-time systems. Unlike in continuous-time systems, we need to consider proper and strictly proper measurement feedback regulators separately for discrete-time systems. The following theorem characterizes the set $\mathcal{S}_{x_0 w_0}$ when strictly proper measurement feedback regulators are used.

Theorem 6.5.12 *For discrete-time systems, consider the system Σ as in (6.1) and the exosystem Σ_{E} as in (6.2). Let Assumptions A.1, A.2, A.3, and A.4 of Section 6.2 hold. Also, let (Π, Γ) be a given solution of the regulator equation (6.4). Then, $J^*_{sp}(x_0, w_0, \Pi, \Gamma)$ is equal to zero uniformly for all initial conditions x_0 and w_0 that satisfy the following conditions:*

$$(i)\ \operatorname{im}\begin{pmatrix} x_0 - \Pi w_0 \\ w_0 \end{pmatrix} \subseteq \mathcal{V}^{\otimes}(\bar{A}, \bar{B}, \bar{C}_z, \bar{D}_{zu}).$$

$$(ii)\ \mathcal{S}^{\otimes}(\bar{A}, \begin{pmatrix} x_0 - \Pi w_0 \\ w_0 \end{pmatrix}, \bar{C}_y, 0) \subseteq \ker \bar{C}_z.$$

$$(iii)\ \bar{A}\mathcal{S}^{\otimes}(\bar{A}, \begin{pmatrix} x_0 - \Pi w_0 \\ w_0 \end{pmatrix}, \bar{C}_y, 0) \subseteq \mathcal{V}^{\otimes}(\bar{A}, \bar{B}, \bar{C}_z, \bar{D}_{zu}).$$

The set $\mathcal{S}_{x_0 w_0}$ is indeed the set of all x_0 and w_0 that satisfy the above conditions.

Proof : It follows from Lemma 6.5.1 when appropriate conditions to solve H_2-ADDPMS for $\bar{\Sigma}$ via strictly proper measurement feedback controllers are taken into account, see Theorem 6.9.1 of [61]. ■

In the following theorem we provide sufficient conditions for the existence of such a strictly proper measurement feedback regulator sequence. To do so, let $\check{E}$ be defined by

$$\operatorname{im} \check{E} = \mathcal{V}^{\otimes}(\bar{A}, \bar{B}, \bar{C}_z, \bar{D}_{zu}). \tag{6.59}$$

Theorem 6.5.13 *For discrete-time systems, consider the system Σ as in (6.1) and the exosystem Σ_{E} as in (6.2). Let Assumptions A.1, A.2, A.3, and A.4 of Section 6.2 hold. Also, let (Π, Γ) be a given solution of the regulator equation (6.4). Moreover, consider the set $\mathcal{S}_{x_0 w_0}$ that is characterized in Theorem 6.5.12. Then, there exists a fixed sequence of strictly proper measurement feedback regulators $\{\Sigma_{\mathrm{C}}(\varepsilon) \mid \varepsilon > 0\}$ such that the attained performance measure $J(x_0, w_0, \Sigma_{\mathrm{C}}(\varepsilon))$ tends to zero as ε tends to zero uniformly for all initial conditions x_0 and w_0 that belong to the set $\mathcal{S}_{x_0 w_0}$ if the following conditions hold:*

(*i*) $\mathcal{S}^{\otimes}(\bar{A}, \check{E}, \bar{C}_y, 0) \subseteq \ker \bar{C}_z$,

(*ii*) $\bar{A}\mathcal{S}^{\otimes}(\bar{A}, \check{E}, \bar{C}_y, 0) \subseteq \mathcal{V}^{\otimes}(\bar{A}, \bar{B}, \bar{C}_z, \bar{D}_{zu})$,

where $\check{E}$ is as defined in (6.59).

Proof : It is obvious since the conditions given in this theorem (which are independent of x_0 and w_0) imply that the conditions given in Theorem 6.5.12 hold for all x_0 and w_0 that belong to the set $\mathcal{S}_{x_0 w_0}$. ■

The above theorem gives only sufficient conditions; providing necessary and sufficient conditions is still an open problem.

For discrete-time systems, the following theorem characterizes the set $\mathcal{S}_{x_0 w_0}$ when proper measurement feedback regulators are used.

Theorem 6.5.14 *For discrete-time systems, consider the system Σ as in (6.1) and the exosystem Σ_{E} as in (6.2). Let Assumptions A.1, A.2, A.3, and A.4 of Section 6.2 hold. Also, let (Π, Γ) be a given solution of the regulator equation (6.4). Then, $J_p^*(x_0, w_0, \Pi, \Gamma)$ is equal to zero uniformly for all initial conditions x_0 and w_0 that satisfy the following conditions:*

(*i*) $\operatorname{im} \begin{pmatrix} x_0 - \Pi w_0 \\ w_0 \end{pmatrix} \subseteq \mathcal{V}^{\otimes}(\bar{A}, \bar{B}, \bar{C}_z, \bar{D}_{zu}) + \bar{B} \ker \bar{D}_{zu}$.

(*ii*) $\mathcal{S}^{\otimes}(\bar{A}, \begin{pmatrix} x_0 - \Pi w_0 \\ w_0 \end{pmatrix}, \bar{C}_y, 0) \cap \ker \bar{C}_y \subseteq \ker \bar{C}_z$.

(*iii*) $\mathcal{S}^{\otimes}(\bar{A}, \begin{pmatrix} x_0 - \Pi w_0 \\ w_0 \end{pmatrix}, \bar{C}_y, 0) \subseteq \mathcal{V}^{\otimes}(\bar{A}, \bar{B}, \bar{C}_z, \bar{D}_{zu})$.

The set $\mathcal{S}_{x_0 w_0}$ is indeed the set of all x_0 and w_0 that satisfy the above conditions.

Proof : It follows from Lemma 6.5.1 when appropriate conditions to solve H_2-ADDPMS for $\bar{\Sigma}$ via proper measurement feedback controllers are taken into account, see Theorem 6.9.2 of [61]. ■

In the following theorem we provide sufficient conditions for the existence of such a strictly proper measurement feedback regulator sequence. To do so, let $\check{E}$ be defined by

$$\operatorname{im} \check{E} = \mathcal{V}^{\otimes}(\bar{A}, \bar{B}, \bar{C}_z, \bar{D}_{zu}) + \bar{B} \ker \bar{D}_{zu}. \tag{6.60}$$

Theorem 6.5.15 *For discrete-time systems, consider the system Σ as in (6.1) and the exosystem Σ_E as in (6.2). Let Assumptions A.1, A.2, A.3, and A.4 of Section 6.2 hold. Also, let (Π, Γ) be a given solution of the regulator equation (6.4). Moreover, consider the set $\mathcal{S}_{x_0 w_0}$ that is characterized in Theorem 6.5.14. Then, there exists a fixed sequence of proper measurement feedback regulators $\{\Sigma_C(\varepsilon) \mid \varepsilon > 0\}$ such that the attained performance measure $J(x_0, w_0, \Sigma_C(\varepsilon))$ tends to zero as ε tends to zero uniformly for all initial conditions x_0 and w_0 that belong to the set $\mathcal{S}_{x_0 w_0}$ if the following conditions hold:*

(i) $\mathcal{S}^{\otimes}(\bar{A}, \check{E}, \bar{C}_y, 0) \cap \ker \bar{C}_y \subseteq \ker \bar{C}_z$,

(ii) $\mathcal{S}^{\otimes}(\bar{A}, \check{E}, \bar{C}_y, 0) \subseteq \mathcal{V}^{\otimes}(\bar{A}, \bar{B}, \bar{C}_z, \bar{D}_{zu})$,

where $\check{E}$ is as defined in (6.60).

Proof : It is obvious since the conditions given in this theorem (which are independent of x_0 and w_0) imply that the conditions given in Theorem 6.5.14 hold for all x_0 and w_0 that belong to the set $\mathcal{S}_{x_0 w_0}$. ■

The above theorem gives only sufficient conditions; providing necessary and sufficient conditions is still an open problem.

6.5.4 Construction of a sequence of measurement feedback regulators that achieve perfect output regulation

We now proceed to construct a fixed sequence of proper or strictly proper measurement feedback regulators for all initial conditions x_0 and w_0 that belong to the set $\mathcal{S}_{x_0 w_0}$.

Step 1: Construct (Π, Γ), a solution of the regulator equation (6.4).

Step 2: For continuous-time systems construct a matrix $\check{E}$ satisfying the equation (6.58). Similarly, for discrete-time systems construct a matrix $\check{E}$ satisfying the equation (6.59) or (6.60) depending upon whether strictly proper or proper measurement feedback controllers are used.

Consider the system,

$$\bar{\Sigma} : \begin{cases} \rho\bar{x} = \bar{A}\bar{x} + \bar{B}\bar{u} + \check{E}\bar{r} \\ \bar{y} = \bar{C}_y\bar{x} \\ \bar{z} = \bar{C}_z\bar{x} + \bar{D}_{zu}\bar{u}. \end{cases} \tag{6.61}$$

Construct a parameterized sequence of H_2 suboptimal proper (or strictly proper) measurement feedback controllers for the above system where in the H_2 norm of the transfer function from $\bar{r}$ to $\bar{z}$ is the performance index. Let such a sequence be characterized by the sequence of quadruples given by $\{\bar{A}_c(\varepsilon)$, $\bar{B}_c(\varepsilon)$, $\bar{C}_c(\varepsilon)$, $\bar{D}_c(\varepsilon)$, $\varepsilon > 0\}$.

Step 3: Utilizing the sequence of quadruples $\{\bar{A}_c(\varepsilon)$, $\bar{B}_c(\varepsilon)$, $\bar{C}_c(\varepsilon)$, $\bar{D}_c(\varepsilon)$, $\varepsilon > 0\}$, construct a sequence of controllers $\{\Sigma_C(\varepsilon), \varepsilon > 0\}$ each member of which is as given in (6.44).

The sequence of proper or strictly proper measurement feedback controllers $\{\Sigma_C(\varepsilon), \varepsilon > 0\}$ constructed above achieves perfect output regulation for all initial conditions x_0 and w_0 that belong to the set $\mathcal{S}_{x_0 w_0}$. □

6.6 Transient performance in structurally stable output regulation

In this section, we would like to study structurally stable output regulation while taking into account transient performance requirements. It can be seen from Section 2.8 that such a study is equivalent to achieving output regulation for an auxiliary system given in (2.42) while taking into account transient performance requirements.

Consider the case where we additionally want to ensure optimal transient performance for a particular set of initial conditions for the nominal system. A problem we encounter here is that the auxiliary system has a totally different exosystem (2.41) and there is not a transparent way to connect initial conditions of the original exosystem to initial conditions of the exosystem for the auxiliary system.

A solution for this problem exists and we will briefly outline its main idea. We first follow the design methodology as described in Section 2.8 and construct the auxiliary system (2.42) and the corresponding exosystem (2.41). Next we consider the following system,

$$\tilde{\Sigma}: \begin{cases} \rho\tilde{x} = \bar{A}_0\tilde{x} + \bar{B}_0\tilde{u} + \bar{E}_{w,0}w \\ y = \bar{C}_{y,0}\tilde{x} + \bar{D}_{yu,0}\tilde{u} + \tilde{D}_{yw}\tilde{w} + \bar{D}_{yw,0}w \\ e = \bar{C}_{e,0}\tilde{x} + \bar{D}_{eu,0}\tilde{u} + \tilde{D}_{ew}\tilde{w} + \bar{D}_{ew,0}w \\ z = \bar{C}_{z,0}\tilde{x} + \bar{D}_{zu,0}\tilde{u} + \tilde{D}_{zw}\tilde{w} + \bar{D}_{zw,0}w \end{cases} \tag{6.62}$$

which has two exosystems

$$\rho\tilde{w} = \tilde{S}_p\tilde{w}$$

and the original exosystem

$$\rho w = Sw$$

and where

$$\bar{C}_{z,0} = Q^{1/2}\bar{C}_{e,0}, \quad \bar{D}_{zu,0} = Q^{1/2}\bar{D}_{eu,0}, \quad \bar{D}_{zw,0} = Q^{1/2}\bar{D}_{ew,0},$$
$$\tilde{D}_{zw} = Q^{1/2}\tilde{D}_{ew}.$$

Note that, for $\tilde{w} = 0$, we obtain the original nominal system after the preliminary feedback $u = Ny + \tilde{u}$. Clearly we must therefore achieve output regulation with respect to w. For structural stability it is necessary that we obtain output regulation with respect to $\tilde{w}$. Linearity then guarantees that we must also achieve output regulation with respect to both $\tilde{w}$ and w.

But now it is easy to see that achieving optimal transient performance for initial conditions $x(0) = x_0$, and $w(0) = w_0$ with a structurally stable output regulation requirement for the original system is equivalent to achieving optimal transient performance for initial conditions $x(0) = x_0$, $\tilde{w}(0) = 0$ and $w(0) = w_0$ with a standard output regulation requirement for this auxiliary system. The latter problem can of course be solved by the tools introduced in this chapter.

6.A Continuous-time linear matrix inequalities

Our goal here is to introduce briefly continuous-time linear matrix inequalities and their properties to the extent needed for our purpose. Our presentation here is an extract of [61] where linear matrix inequalities are dealt with in detail along with the required proofs for their properties.

We introduce a continuous-time linear matrix inequality in the following definition.

Definition 6.A.1 ***Continuous-time linear matrix inequality*** *Let $A \in \mathbb{R}^{n\times n}$, $B \in \mathbb{R}^{n\times m}$, $Q \in \mathbb{R}^{n\times n}$, $R \in \mathbb{R}^{m\times m}$, and $S \in \mathbb{R}^{n\times m}$ with Q and R being symmetric. The matrix inequality of the form*

$$L(X) \geq 0, \tag{6.63}$$

with unknown $X \in \mathbb{R}^{n\times n}$ where

$$L(X) := \begin{pmatrix} Q + A^T X + XA & XB + S \\ B^T X + S^T & R \end{pmatrix},$$

is called a continuous-time linear matrix inequality (CLMI). Moreover, when X satisfies (6.63), it is referred to as a solution of the linear matrix inequality. Note that $L(X) \geq 0$ implies that $R \geq 0$, i.e. R cannot be indefinite.

We often encounter a linear matrix inequality in which the matrices Q, R, and S satisfy the positive semi-definite condition, namely

$$\begin{pmatrix} Q & S \\ S^T & R \end{pmatrix} \geqslant 0. \tag{6.64}$$

Under the above condition, it follows that there exists matrices $C \in \mathbb{R}^{p\times n}$ and $D \in \mathbb{R}^{p\times m}$ with $\begin{pmatrix} C & D \end{pmatrix}$ of full rank such that

$$\begin{pmatrix} Q & S \\ S^T & R \end{pmatrix} = \begin{pmatrix} C & D \end{pmatrix}^T \begin{pmatrix} C & D \end{pmatrix}.$$

As shown in [61], an important property of a solution X of a linear matrix inequality with matrices Q, R, and S satisfying the positive semi-definite condition (6.64) is that $X\mathcal{S}^*(A, B, C, D) = 0$ where the detectable strongly controllable subspace $\mathcal{S}^*(A, B, C, D)$ is as in Definition 1.2.1.

We denote the set of all real symmetric solutions of the linear matrix inequality in (6.63) as $\mathbf{\Gamma}$, i.e.

$$\mathbf{\Gamma} := \left\{ X \in \mathbb{R}^{n\times n} \mid X = X^{\mathrm{T}} \text{ and } L(X) \geq 0 \right\}. \tag{6.65}$$

A relevant set of solutions of a linear matrix inequality in the context of H_2 optimal control theory is a set of what are called rank minimizing solutions. In order to develop a definition for such solutions, we first need to state some properties of linear matrix inequalities.

To start with we observe that for every $X \in \mathbf{\Gamma}$, there exists real matrices C_{X} and D_{X} such that

$$L(X) = \begin{pmatrix} C_{\mathrm{X}} & D_{\mathrm{X}} \end{pmatrix}^{\mathrm{T}} \begin{pmatrix} C_{\mathrm{X}} & D_{\mathrm{X}} \end{pmatrix},$$

and such that $\begin{pmatrix} C_{\mathrm{X}} & D_{\mathrm{X}} \end{pmatrix}$ is of full rank. Then we can define a system Σ_{X} characterized by the quadruple $(A, B, C_{\mathrm{X}}, D_{\mathrm{X}})$. The transfer function of Σ_{X} is then given by

$$H_{\mathrm{X}}(s) := C_{\mathrm{X}}(sI - A)^{-1}B + D_{\mathrm{X}}.$$

We have the following lemma.

Lemma 6.A.1 *Let $X \in \mathbf{\Gamma}$. Then we have*

$$\text{normrank}\, H_{\mathrm{X}} := \rho(X) = \rho := \text{normrank}\, \hat{H},$$

where $\hat{H}(s)$ is defined as

$$\hat{H}(s) := \begin{pmatrix} B^T(-sI - A^T)^{-1} & I \end{pmatrix} \begin{pmatrix} Q & S \\ S^T & R \end{pmatrix} \begin{pmatrix} (sI - A)^{-1}B \\ I \end{pmatrix}.$$

Remark 6.A.1 *It follows from Lemma 6.A.1 that the normal rank of $H_{\mathrm{X}}(s)$ is independent of X for any $X \in \mathbf{\Gamma}$.*

The following lemma provides a lower bound on the rank of $L(X)$ for any $X \in \mathbf{\Gamma}$.

Lemma 6.A.2 *For any $X \in \mathbf{\Gamma}$, we have*

$$\text{rank}\, L(X) \geq \rho$$

and, moreover, the equality holds if and only if Σ_{X} is right invertible.

Now we are ready to define the set of rank minimizing solutions of a linear matrix inequalities.

Definition 6.A.2 *A solution $X \in \mathbf{\Gamma}$ is said to be rank minimizing if*

$$\operatorname{rank} L(X) = \rho.$$

Moreover, we denote the set of all rank minimizing solutions of the linear matrix inequality (6.63) by $\mathbf{\Gamma}_{\text{min}}$, i.e.

$$\mathbf{\Gamma}_{\text{min}} := \left\{ X \in \mathbf{\Gamma} \mid \operatorname{rank} L(X) = \rho \right\}. \tag{6.66}$$

Also, the set of all positive semi-definite rank minimizing solutions of the linear matrix inequality in (6.63) is defined as $\mathbf{\Gamma}^{\text{psd}}_{\text{min}}$,

$$\mathbf{\Gamma}^{\text{psd}}_{\text{min}} := \left\{ X \in \mathbf{\Gamma}_{\text{min}} \mid X \geq 0 \right\}.$$

Similarly, the set of all positive definite rank minimizing solutions of the linear matrix inequality (6.63) is denoted by $\mathbf{\Gamma}^{\text{pd}}_{\text{min}}$,

$$\mathbf{\Gamma}^{\text{pd}}_{\text{min}} := \left\{ X \in \mathbf{\Gamma}_{\text{min}} \mid X \geq 0 \right\}.$$

The following lemma says that under a very weak condition that the pair (A, B) is stabilizable, the set $\mathbf{\Gamma}_{\text{min}}$ is non-empty.

Lemma 6.A.3 *Consider a linear matrix inequality as in (6.63) with the matrices Q, R, and S satisfying the positive semi-definite condition (6.64), and the pair (A, B) being stabilizable. Then there exists at least one rank minimizing solution to the linear matrix inequality (6.63), i.e. $\mathbf{\Gamma}_{\text{min}}$ is non-empty.*

Remark 6.A.2 *The condition of Lemma 6.A.3 can be weakened further. In fact, in [74] it is shown that the result of Lemma 6.A.3 holds under the condition that the uncontrollable eigenvalues of (A, B), say $\lambda_1, \lambda_2, \cdots, \lambda_k$, are such that $\lambda_i + \lambda_j \neq 0$ for any i and j. This condition is obviously weaker than the requirement of (A, B) being stabilizable.*

We next have the following additional definition regarding semi-stabilizing and stabilizing solutions.

Definition 6.A.3 *Consider a matrix pencil*

$$N(s, X) := \begin{pmatrix} M(s) \\ L(X) \end{pmatrix}$$

where $M(s) := \begin{pmatrix} sI - A & -B \end{pmatrix}$. *Then a solution* $X \in \mathbf{\Gamma}_{\min}$ *is said to be a semi-stabilizing solution if all the finite zeros of the matrix pencil* $N(s, X)$ *are in the closed left half plane, i.e. in* $\mathbb{C}^0 \cup \mathbb{C}^-$. *Similarly, a solution* $X \in \mathbf{\Gamma}_{\min}$ *is said to be a stabilizing solution if all the finite zeros of* $N(s, X)$ *are in the open left half plane, i.e. in* $\mathbb{C}^-$. *The set of all semi-stabilizing solutions of a linear matrix inequality is denoted by* $\mathbf{\Gamma}^{ss}_{\min}$. *Similarly, the set of all stabilizing solutions of a linear matrix inequality is denoted by* $\mathbf{\Gamma}^{s}_{\min}$.

Remark 6.A.3 *For the case when the matrices* Q, R, *and* S *satisfy the positive semi-definite condition (6.64), and when the matrix* D *is injective, Definition 6.A.3 for a semi-stabilizing or a stabilizing solution can be simplified. That is,* $X \in \mathbf{\Gamma}$ *is a semi-stabilizing solution if the matrix*

$$A - B(D^TD)^{-1}(B^TX + D^TC) \tag{6.67}$$

has all its eigenvalues in the closed left-half complex plane $\mathbb{C}^0 \cup \mathbb{C}^-$. *Similarly,* $X \in \mathbf{\Gamma}$ *is a stabilizing solution if the matrix (6.67) has all its eigenvalues in the open left-half complex plane* $\mathbb{C}^-$.

We now begin to judiciously compile some important properties of the linear matrix inequality in (6.63).

Property 6.A.1 *Consider a linear matrix inequality as in (6.63) with the matrices* Q, R, *and* S *satisfying the positive semi-definite condition (6.64), and the pair* (A, B) *being stabilizable. Then the following hold:*

(i) *A real symmetric semi-stabilizing solution, say* X_{ss}, *if it exists, is larger than any solution* X *of the given linear matrix inequality, i.e.* $X_{ss} \geq X$ *for all* $X \in \mathbf{\Gamma}$.

(ii) *A real symmetric semi-stabilizing solution* X_{ss}, *if it exists, is unique, i.e.* $\mathbf{\Gamma}^{ss}_{\min}$ *is at most a singleton set.*

(iii) *A real symmetric semi-stabilizing solution of it, if it exists, is also positive semi-definite.*

We would like to study next the existence conditions for some specific types of solutions to a linear matrix inequality. We have the following theorem.

Theorem 6.A.1 *Consider a linear matrix inequality as in (6.63) with the matrices* Q, R, *and* S *satisfying the positive semi-definite condition (6.64). Then* *there exists a positive semi-definite rank minimizing solution, i.e.* $\Gamma_{\min}^{\text{psd}}$ *is non-empty, if and only if*

$$X_{-}(A) + \langle A \mid \operatorname{im} B \rangle + \mathcal{V}^*(A, B, C, D) = \mathbb{R}^n,$$

where, $X_{-}(A)$ *is the stable modal subspace of* $\mathbb{R}^n$ *related to* A, $\langle A \mid \operatorname{im} B \rangle$ *is the controllable subspace of the pair* (A, B)*, and* $\mathcal{V}^*(A, B, C, D)$ *as given in Definition 1.2.1 represents the weakly unobservable subspace of the system* Σ *characterized by the quadruple* (A, B, C, D).

Theorem 6.A.2 *Consider a linear matrix inequality as in (6.63) with the matrices* Q, R, *and* S *satisfying the positive semi-definite condition (6.64), and the pair* (A, B) *being stabilizable. Then, there exists a unique semi-stabilizing solution of it. Moreover, this solution is positive semi-definite and is larger than any other symmetric solution of it.*

Theorem 6.A.3 *Consider a linear matrix inequality as in (6.63) with the matrices* Q, R, *and* S *satisfying the positive semi-definite condition (6.64), and the pair* (A, B) *being stabilizable. Then* $X = 0$ *is a unique positive semi-definite semi-stabilizing solution of it if and only if the system represented by* (A, B, C, D) *is right invertible and has no invariant zeros in the open right-half plane* $\mathbb{C}^+$.

Theorem 6.A.2 guarantees that there exists a unique positive semi-definite semi-stabilizing solution of a linear matrix inequality whenever the matrices Q, R, and S satisfy the positive semi-definite condition (6.64), and when the pair (A, B) is stabilizable. We would like to examine next the continuity of such a solution with respect to the parameters Q, R, and S. To start with, we parameterize the matrices Q, R, and S with a scalar parameter ε, and rewrite the linear matrix inequality (6.63) as

$$L^{\varepsilon}(X^{\varepsilon}) := \begin{pmatrix} Q^{\varepsilon} + A^{\mathrm{T}} X^{\varepsilon} + X^{\varepsilon} A & X^{\varepsilon} B + S^{\varepsilon} \\ B^{\mathrm{T}} X^{\varepsilon} + (S^{\varepsilon})^{\mathrm{T}} & R^{\varepsilon} \end{pmatrix} \geqslant 0, \tag{6.68}$$

where the matrices Q^{ε}, R^{ε}, and S^{ε} satisfy the positive semi-definite condition (6.64), and where the pair (A, B) is stabilizable. It follows from Theorem 6.A.2 that for each ε there exists a unique positive semi-definite semi-stabilizing solution of the linear matrix inequality (6.68). Let this solution be denoted by X_{ss}^{ε}. The following theorem establishes the continuity of the mapping $\varepsilon \to X_{ss}^{\varepsilon}$.

Theorem 6.A.4 *Consider a linear matrix inequality as in (6.68). Let the following hold:*

A1. (A, B) *is stabilizable.*

A2. The matrices Q^ε, R^ε, *and* S^ε *satisfy the positive semi-definite condition (6.64) for each* $\varepsilon \in [0,\ \delta]$ *for some* $\delta > 0$.

A3. The mapping $\varepsilon \to \begin{pmatrix} Q^\varepsilon & S^\varepsilon \\ (S^\varepsilon)^T & R^\varepsilon \end{pmatrix}$ *is continuous at* $\varepsilon = 0$.

A4. $\begin{pmatrix} Q^{\varepsilon_1} & S^{\varepsilon_1} \\ (S^{\varepsilon_1})^T & R^{\varepsilon_1} \end{pmatrix} \leqslant \begin{pmatrix} Q^{\varepsilon_2} & S^{\varepsilon_2} \\ (S^{\varepsilon_2})^T & R^{\varepsilon_2} \end{pmatrix}$ *for all* $0 \leqslant \varepsilon_1 \leqslant \varepsilon_2 \leqslant \delta$.

Then the unique positive semi-definite semi-stabilizing solution X_{ss}^ε *of the linear matrix inequality (6.68) is continuous at* $\varepsilon = 0$. *That is,* $X_{ss}^\varepsilon \to X_{ss}^0$ *as* $\varepsilon \to 0$.

6.B Discrete-time linear matrix inequalities

Our goal here is to introduce briefly discrete-time linear matrix inequalities and their properties to the extent needed for our purpose. As in the case of continuous-time linear matrix inequality, only what are called rank minimizing solutions of linear matrix inequality are of interest to us. However, a subset of all rank minimizing solutions which we shall refer to as a set of strongly rank minimizing solutions is the one that is pertinent to H_2 optimal control theory. As such, our focus here is on strongly rank minimizing solutions.

To preserve conceptual thought process, the notations used for several objects here are the same as those in Section 6.A which discusses continuous-time linear matrix inequality. However, whenever we refer to a particular result, we shall distinguish between continuous- and discrete-time linear matrix inequalities by quoting appropriate definitions or equations.

Our presentation here is again an extract of [61] where linear matrix inequalities are dealt with in detail along with the required proofs for their properties.

We introduce a discrete-time linear matrix inequality in the following definition.

Definition 6.B.1 ***Discrete-time linear matrix inequality*** *Let* $A \in \mathbb{R}^{n\times n}$, $B \in \mathbb{R}^{n\times m}$, $Q \in \mathbb{R}^{n\times n}$, $R \in \mathbb{R}^{m\times m}$, *and* $S \in \mathbb{R}^{n\times m}$ *with* Q *and* R *being symmetric.*

The matrix inequality of the form

$$L(X) \geq 0, \tag{6.69}$$

with unknown $X \in \mathbb{R}^{n\times n}$ where

$$L(X) := \begin{pmatrix} Q + A^T X A - X & A^T X B + S \\ B^T X A + S^T & B^T X B + R \end{pmatrix},$$

is called a discrete-time linear matrix inequality (DLMI). Moreover, when X satisfies (6.69), it is referred to as a solution of the linear matrix inequality.

As in Appendix 6.A, we require often that the matrices Q, R, and S in (6.69) to satisfy the positive semi-definite condition, namely

$$\begin{pmatrix} Q & S \\ S^{\mathrm{T}} & R \end{pmatrix} \geqslant 0. \tag{6.70}$$

Under the above condition, it follows that there exists matrices $C \in \mathbb{R}^{p\times n}$ and $D \in \mathbb{R}^{p\times m}$ with $\begin{pmatrix} C & D \end{pmatrix}$ of full rank such that

$$\begin{pmatrix} Q & S \\ S^{\mathrm{T}} & R \end{pmatrix} = \begin{pmatrix} C & D \end{pmatrix}^{\mathrm{T}} \begin{pmatrix} C & D \end{pmatrix}.$$

We denote the set of real symmetric solutions of the linear matrix inequality in (6.69) as $\boldsymbol{\Gamma}$, i.e.

$$\boldsymbol{\Gamma} := \left\{ X \in \mathbb{R}^{n\times n} \mid X = X^{\mathrm{T}} \text{ and } L(X) \geq 0 \right\}. \tag{6.71}$$

As shown in [61], an important property of a solution X of a linear matrix inequality with matrices Q, R, and S satisfying the positive semi-definite condition (6.70) is that

$$\mathcal{R}^*(A, B, C, D) \subseteq \ker X$$

where the subspace $\mathcal{R}^*(A, B, C, D)$ is as in Definition 1.2.3.

We next introduce a certain transfer function $\hat{H}(z)$ which plays a strong role in the study of linear matrix inequalities,

$$\hat{H}(z) := \begin{pmatrix} B^{\mathrm{T}}(-zI - A^{\mathrm{T}})^{-1} & I \end{pmatrix} \begin{pmatrix} Q & S \\ S^{\mathrm{T}} & R \end{pmatrix} \begin{pmatrix} (zI - A)^{-1}B \\ I \end{pmatrix}. \tag{6.72}$$

It turns out that the set of solutions of a linear matrix inequality coincides with that of an appropriate general discrete-time algebraic Riccati inequality. This is explored in the following lemma.

Lemma 6.B.1 *Consider a linear matrix inequality as in (6.69) and its solution set* $\mathbf{\Gamma}$ *as in (6.71). Then the solution set* $\mathbf{\Gamma}$ *coincides with the set of all real symmetric solutions of the general discrete-time algebraic Riccati inequality defined by*

$$B^T X B + R \geq 0, \tag{6.73a}$$

$$\ker(B^T X B + R) \subseteq \ker(A^T X B + S), \tag{6.73b}$$

and

$$A^T X A - X - (A^T X B + S)(R + B^T X B)^{\dagger}(B^T X A + S^T) + Q \geq 0. \tag{6.73c}$$

As in the case of continuous-time, it turns out that a relevant set of solutions of a discrete-time linear matrix inequality in the context of H_2 optimal control theory is a set of what are called rank minimizing solutions. In order to develop a definition for such solutions, as in the case of continuous-time, we first need to state some properties of discrete-time linear matrix inequalities.

To start with we observe that for every $X \in \mathbf{\Gamma}$, there exists real matrices C_{X} and D_{X} such that

$$L(X) = \begin{pmatrix} C_{\mathrm{X}} & D_{\mathrm{X}} \end{pmatrix}^{\mathrm{T}} \begin{pmatrix} C_{\mathrm{X}} & D_{\mathrm{X}} \end{pmatrix}, \tag{6.74}$$

and such that $\begin{pmatrix} C_{\mathrm{X}} & D_{\mathrm{X}} \end{pmatrix}$ is of full rank. Then we can define a linear system Σ_{X} characterized by the quadruple $(A, B, C_{\mathrm{X}}, D_{\mathrm{X}})$. The transfer function of Σ_{X} is then given by

$$H_{\mathrm{X}}(z) := C_{\mathrm{X}}(zI - A)^{-1} B + D_{\mathrm{X}}.$$

We have the following lemma.

Lemma 6.B.2 *Let* $X \in \mathbf{\Gamma}$. *Then we have*

$$\text{normrank } H_{\mathrm{X}} := \rho(X) = \rho := \text{normrank } \hat{H},$$

where $\hat{H}(z)$ *is as defined in (6.72).*

The following lemma provides a lower bound on the rank of $\mathbf{L}(X)$ for any $X \in \mathbf{\Gamma}$.

Lemma 6.B.3 *For any $X \in \mathbf{\Gamma}$, we have*

$$\operatorname{rank} \boldsymbol{L}(X) \geq \rho$$

and, moreover, the equality holds if and only if Σ_X is right invertible.

Now we are ready to define the set of rank minimizing solutions of a linear matrix inequality.

Definition 6.B.2 *A solution $X \in \mathbf{\Gamma}$ is said to be rank minimizing if*

$$\operatorname{rank} \boldsymbol{L}(X) = \rho.$$

Moreover, we denote the set of all rank minimizing solutions of the linear matrix inequality (6.69) by $\mathbf{\Gamma}_{\text{min}}$, i.e.

$$\mathbf{\Gamma}_{\text{min}} := \left\{ X \in \mathbf{\Gamma} \mid \operatorname{rank} \boldsymbol{L}(X) = \rho \right\}. \tag{6.75}$$

We introduce next the notions of semi-stabilizing and stabilizing solutions of linear matrix inequalities.

Definition 6.B.3 *Consider a matrix*

$$N(z, X) := \begin{pmatrix} M(z) \\ \boldsymbol{L}(X) \end{pmatrix}$$

where $M(z) := \begin{pmatrix} zI - A & -B \end{pmatrix}$. Then a solution $X \in \mathbf{\Gamma}_{\text{min}}$ is said to be a semi-stabilizing solution if all the finite zeros of the matrix pencil $N(z, X)$ are inside or on the unit circle, i.e. in $\mathbb{C}^{\circ} \cup \mathbb{C}^{\ominus}$, and if the number of zeros at infinity of $N(z, X)$ is equal to the rank of $\boldsymbol{L}(X)$. Similarly, a solution $X \in \mathbf{\Gamma}_{\text{min}}$ is said to be a stabilizing solution if all the finite zeros of $N(z, X)$ are inside the unit circle, i.e. in $\mathbb{C}^{\ominus}$, and if the number of zeros at infinity of $N(z, X)$ is equal to the rank of $\boldsymbol{L}(X)$. The set of all semi-stabilizing solutions of a linear matrix inequality is denoted by $\mathbf{\Gamma}^{\text{ss}}_{\text{min}}$. Similarly, the set of all stabilizing solutions of a linear matrix inequality is denoted by $\mathbf{\Gamma}^{\text{s}}_{\text{min}}$.

A subset of all rank minimizing solutions, which we shall refer to as strongly rank minimizing solutions, are pertinent to H_2 optimal control theory. We recall below their formal definition.

Definition 6.B.4 *A solution $X \in \mathbf{\Gamma}$ is said to be a strongly rank minimizing solution if*

$$\operatorname{rank} L(X) = \operatorname{rank}(B^T X B + R). \tag{6.76}$$

Moreover, we denote the set of all strongly rank minimizing solutions of the linear matrix inequality (6.69) as

$$\mathcal{L}_{\min} := \left\{ \; X \in \mathbf{\Gamma} \mid \operatorname{rank} L(X) = \operatorname{rank}(B^T X B + R) \; \right\}.$$

Also, the set of all positive semi-definite strongly rank minimizing solutions of the linear matrix inequality (6.69) is defined as $\mathcal{L}_{\min}^{\text{psd}}$,

$$\mathcal{L}_{\min}^{\text{psd}} := \{ \; X \in \mathcal{L}_{\min} \mid X \geq 0 \} \, .$$

Similarly, the set of all positive definite strongly rank minimizing solutions of the linear matrix inequality (6.69) is defined as $\mathcal{L}_{\min}^{\text{pd}}$,

$$\mathcal{L}_{\min}^{\text{pd}} := \{ \; X \in \mathcal{L}_{\min} \mid X > 0 \} \, .$$

The following theorem discusses the conditions under which $\mathcal{L}_{\min}$ is non-empty.

Theorem 6.B.1 *Consider a linear matrix inequality as in (6.69) with the matrices Q, R, and S satisfying the positive semi-definite condition (6.70), and the pair (A, B) being stabilizable. Then there exists at least one strongly rank minimizing solution, i.e. $\mathcal{L}_{\min}$ is non-empty.*

Now we are ready to say that $\mathcal{L}_{\min} \subseteq \mathbf{\Gamma}_{\min}$. This is formulated as the following lemma.

Lemma 6.B.4

$$\mathcal{L}_{\min} \subseteq \mathbf{\Gamma}_{\min}.$$

A natural question that arises next is under what conditions $\mathcal{L}_{\min}$ coincides with $\mathbf{\Gamma}_{\min}$. This is answered in the next lemma.

Lemma 6.B.5

$$\mathcal{L}_{\min} = \Gamma_{\min}$$

if the matrix pencil

$$\begin{pmatrix} Q & S & z^{-1}I - A^T \\ S^T & R & -B^T \\ zI - A & -B & 0 \end{pmatrix} \tag{6.77}$$

has no zeros at infinity.

It turns out that the stabilizing and semi-stabilizing solutions of a linear matrix inequality are in fact its strongly rank minimizing solutions. The following lemma formalizes this.

Lemma 6.B.6 *A semi-stabilizing or a stabilizing solution of a linear matrix inequalities indeed a strongly rank minimizing solution of it. That is,*

$$\Gamma^{ss}_{\min} \subseteq \mathcal{L}_{\min}.$$

Remark 6.B.1 *Definition 6.B.3 can be rewritten as follows. A solution X of a linear matrix inequality is said to be a stabilizing (respectively, semi-stabilizing) solution if all the eigenvalues of the matrix*

$$A - B(B^TXB + R)^{\dagger}(B^TXA + S^T) \\ - B(I - (B^TXB + R)^{\dagger}(B^TXB + R))F$$

are inside the unit circle (respectively, inside and/or on the unit circle) for some suitably chosen matrix F.

As in the case of continuous-time linear matrix inequalities, one can easily obtain the following property of a discrete-time linear matrix inequality.

Property 6.B.1 *Consider a linear matrix inequality as in (6.69) with the matrices Q, R, and S satisfying the positive semi-definite condition (6.70), and the pair (A, B) being stabilizable. Then the following hold:*

(*i*) *A real symmetric semi-stabilizing solution, say X_{ss}, if it exists, is larger than any solution X of the given, linear matrix inequalities i.e. $X_{ss} \geq X$ for all $X \in \Gamma$.*

(ii) *A real symmetric semi-stabilizing solution* X_{ss}, *if it exists, is unique, i.e.* Γ^{ss}_{min} *is at most a singleton set.*

(iii) *A real symmetric semi-stabilizing solution of it, if it exists, is also positive semi-definite.*

Remark 6.B.2 *It is shown in [74] that a semi-stabilizing solution of a linear matrix inequality (6.69), if it exists, is a stabilizing solution if and only if the matrix pencil*

$$\begin{pmatrix} Q & S & z^{-1}I - A^T \\ S^T & R & -B^T \\ zI - A & -B & 0 \end{pmatrix}$$

has the full rank for all z *on the unit circle* $\mathbb{C}^{\circ}$.

We study next the existence conditions for some specific types of solutions to a linear matrix inequality. We have the following theorems.

Theorem 6.B.2 *Consider a linear matrix inequality as in (6.69) with the matrices* Q, R, *and* S *satisfying the positive semi-definite condition (6.70), and the pair* (A, B) *being stabilizable. Also, assume that the system represented by* (A, B, C, D) *has no invariant zeros on the unit circle* $\mathbb{C}^{\circ}$. *Then the given linear matrix inequality has a unique stabilizing solution. Moreover, this solution is positive semi-definite and is the largest among all the real symmetric solutions of it.*

Theorem 6.B.3 *Consider a linear matrix inequality as in (6.69) with the matrices* Q, R, *and* S *satisfying the positive semi-definite condition (6.70), and the pair* (A, B) *being stabilizable. Then there exists a unique semi-stabilizing solution of it. Moreover, this solution is positive semi-definite and is the largest among all strongly rank minimizing solutions.*

Theorem 6.B.4 *Consider a linear matrix inequality as in (6.69) with the matrices* Q, R, *and* S *satisfying the positive semi-definite condition (6.70), and the pair* (A, B) *being stabilizable. Then* $X = 0$ *is a unique positive semi-definite semi-stabilizing solution of it if and only if the system represented by* (A, B, C, D) *is right invertible, has no invariant zeros in* $\mathbb{C}^{\oplus}$, *and has no infinite zeros of order greater than or equal to one.*

Theorem 6.B.3 guarantees that there exists a unique semi-stabilizing solution of a linear matrix inequality whenever the matrices Q, R, and S satisfy the positive semi-definite condition (6.70), and when the pair (A, B) is stabilizable. We now examine the continuity of such a solution with respect to the parameters Q, R, and S. To start with, we parameterize the matrices Q, R, and S with a scalar parameter ε, and rewrite the linear matrix inequality (6.69) as

$$L^{\varepsilon}(X^{\varepsilon}) := \begin{pmatrix} Q^{\varepsilon} + A^{\mathrm{T}}X^{\varepsilon}A - X^{\varepsilon} & A^{\mathrm{T}}X^{\varepsilon}B + S^{\varepsilon} \\ B^{\mathrm{T}}X^{\varepsilon}A + (S^{\varepsilon})^{\mathrm{T}} & B^{\mathrm{T}}X^{\varepsilon}B + R^{\varepsilon} \end{pmatrix} \geq 0, \tag{6.78}$$

where the matrices Q^{ε}, R^{ε}, and S^{ε} satisfy the positive semi-definite condition (6.70), and where the pair (A, B) is stabilizable. It follows from Theorem 6.B.3 that for each ε there exists a unique semi-stabilizing solution of the linear matrix inequality (6.78). Let this solution be denoted by X_{ss}^{ε}. The following theorem establishes the continuity of the mapping $\varepsilon \to X_{ss}^{\varepsilon}$.

Theorem 6.B.5 *Consider a linear matrix inequality as in (6.78). Let the following hold:*

- **A1.** *(A, B) is stabilizable.*
- **A2.** *The matrices Q^{ε}, R^{ε}, and S^{ε} satisfy the positive semi-definite condition (6.70) for each $\varepsilon \in [0, \ \delta]$ for some $\delta > 0$.*
- **A3.** *The mapping $\varepsilon \to \begin{pmatrix} Q^{\varepsilon} & S^{\varepsilon} \\ (S^{\varepsilon})^{T} & R^{\varepsilon} \end{pmatrix}$ is continuous at $\varepsilon = 0$.*
- **A4.** $\begin{pmatrix} Q^{\varepsilon_1} & S^{\varepsilon_1} \\ (S^{\varepsilon_1})^{T} & R^{\varepsilon_1} \end{pmatrix} \leq \begin{pmatrix} Q^{\varepsilon_2} & S^{\varepsilon_2} \\ (S^{\varepsilon_2})^{T} & R^{\varepsilon_2} \end{pmatrix}$ *for all $0 \leq \varepsilon_1 \leq \varepsilon_2 \leq \delta$.*

Then the unique semi-stabilizing solution X_{ss}^{ε} of the linear matrix inequality (6.78) is continuous at $\varepsilon = 0$. That is, $X_{ss}^{\varepsilon} \to X_{ss}^{0}$ as $\varepsilon \to 0$.

Chapter 7

Achieving a desired performance with an output regulation constraint

7.1 Introduction

So far we have studied the output regulation problem which seeks to find a controller that internally stabilizes a given plant or system while asymptotically tracking a reference signal even in the presence of persistent disturbances. In the last chapter, we considered an additional performance requirement of optimizing the transient performance. In this chapter we explore output regulation with a more general performance constraint.

It is common in modern control theory to formulate the design of appropriate controllers that achieve a desired performance as optimization problems. In an analytical design of control systems, certain given specifications are at first transformed into a performance criterion, and then control laws are sought which would achieve a desired performance criterion. Typically, performance is measured by the H_2 or H_∞ norm of a chosen transfer function matrix although any other norm such as the L_1 norm could be used. When the performance is measured by, say the H_2 or H_∞ norm, one casts the problem of finding an appropriate controller as an H_2 or H_∞ optimal control problem.

Our focus in this chapter is to study here a multi-objective problem where we seek to find a controller that results in a desired performance as indicated by a certain closed-loop transfer function matrix (or its characteristics such as H_2, H_∞ or L_1 norm) while simultaneously achieving output regulation. Our goal is to establish the necessary and sufficient conditions under which the multi-objective problem we pose admits a solution, and then provide a vehicle for designing suitable controllers that solve such a problem. Since the multi-objective problem we pose has a performance requirement as well as the output regulation constraint, such a problem when looked at it directly turns out to be a constrained optimization problem, and as such it is difficult to solve it in a straightforward manner. Our technique to solve such a problem is somewhat indirect. Our method basically amounts to transforming the *constrained* optimization problem for a given system to an *unconstrained* optimization problem, however, for a certain auxiliary system which is to be explicitly constructed. The auxiliary system is obtained by adding to the plant dynamics certain other dynamics which are directly related to the internal model principle.

In this chapter we do not consider explicitly any specific performance measure. Our development here, for both continuous- and discrete-time systems, is general and it simply concentrates on transforming the *constrained* optimization problem for a given system to an *unconstrained* optimization problem for the auxiliary system in such a way that the closed-loop transfer function matrix is the same for both the *constrained* and the *unconstrained* optimization problems. In order to measure performance, one can use any norm of the closed-loop transfer function matrix. In the subsequent chapters, we revisit the problem while using the H_2 or H_∞ norm measures.

We would like to say next that the multi-objective problem of achieving a desired performance along with the output regulation constraint can be strengthened further by requiring structurally stable output regulation. As we discussed in Section 2.8, structurally stable output regulation is a well known classical concept that requires output regulation to be maintained even in the presence of arbitrarily small perturbation of the data from its nominal value of the given plant (but not of the exosystem). Requiring structurally stable output regulation does not add any complexity as it can be reduced to the classical problem of output regulation. As such our main focus in this chapter is only on the problem of achieving a desired closed-loop transfer function along with output regulation.

This chapter is based on the recent research work of authors [77].

7.2 Problem formulation and assumptions

We start with a linear system with state space realization,

$$\Sigma : \begin{cases} \rho x = Ax + Bu + E_w w + E_d d \\ e = C_e x + D_{eu} u + D_{ew} w \\ z = C_z x + D_{zu} u + D_{zd} d \\ y = C_y x + D_{yu} u + D_{yw} w + D_{yd} d, \end{cases} \tag{7.1}$$

where, as usual, ρ denotes the time derivative, $\rho x(t) = \frac{dx}{dt}(t)$, for continuous-time systems, and the shift, $(\rho x)(k) = x(k+1)$, for discrete-time systems. In the above representation, Σ describes the plant with state $x \in \mathbb{R}^n$, control input $u \in \mathbb{R}^m$, measured output $y \in \mathbb{R}^p$, tracking error $e \in \mathbb{R}^q$, and the controlled variable (for performance requirement) $z \in \mathbb{R}^\ell$. The exogenous disturbance input $w \in \mathbb{R}^s$ is generated by an exosystem Σ_E with state space realization,

$$\Sigma_E \; : \; \rho w = Sw. \tag{7.2}$$

Finally, the variable d denotes an external disturbance. Graphically, the given plant and the exosystem are depicted in Figure 7.1.

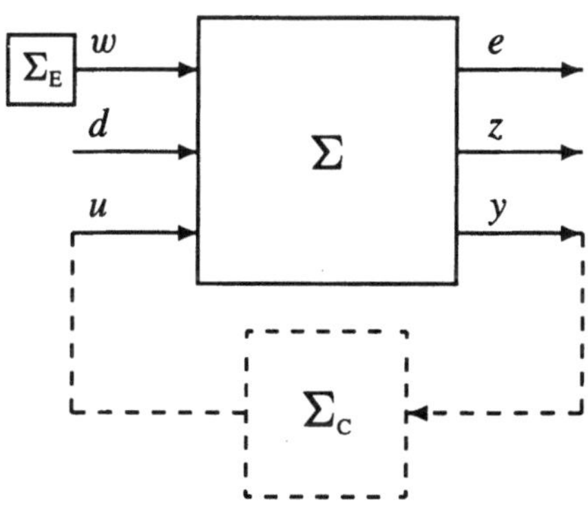

Figure 7.1: Performance with output regulation

As in previous chapters, in general, we seek measurement feedback controllers which are of the form,

$$\Sigma_C : \begin{cases} \rho v = A_c v + B_c y, \\ u = C_c v + D_c y. \end{cases} \tag{7.3}$$

The closed-loop system consisting of the given system Σ and the controller Σ_C is denoted by $\Sigma \times \Sigma_C$. Also, the transfer matrix from d to z of $\Sigma \times \Sigma_C$ is denoted by $T_{d,z}(\Sigma \times \Sigma_C)$.

The specific multi-objective problem which we shall refer to as the *problem of performance with output regulation constraint* is stated below.

Problem 7.2.1 ***(Problem of achieving a desired performance with output regulation constraint)*** *Consider the system Σ and the exosystem Σ_E as given in (7.1) and (7.2). Find, if possible, a controller Σ_C of the form (7.3) such that the following conditions hold:*

(*i*) ***(Internal Stability)*** *In the absence of the disturbances w and d, the closed-loop system $\Sigma \times \Sigma_C$ is internally stable.*

(*ii*) ***(Performance Measure)*** *A desired performance measure based on the transfer matrix $T_{d,z}(\Sigma \times \Sigma_C)$ is obtained.*

(*iii*) ***(Output Regulation)*** *For any $d \in L_2$ (continuous-time) or $d \in \ell_2$ (discrete-time), and for all $x(0) \in \mathbb{R}^n$ and $w(0) \in \mathbb{R}^s$, the solution of the closed-loop system $\Sigma \times \Sigma_C$ satisfies*

$$\lim_{t \to \infty} e(t) = 0.$$

Remark 7.2.1 *The part (iii) of the above problem formulation is equivalent to the following: In the absence of external signal d, i.e. when $d = 0$, Σ_C achieves output regulation for Σ.*

The above statement of the *problem of achieving a desired performance with output regulation constraint* is intentionally general. No specific performance measure is prescribed. Any performance measure based on the transfer matrix $T_{d,z}(\Sigma \times \Sigma_C)$ can be used. As we said earlier, in this chapter, we focus our attention in getting a desired matrix for $T_{d,z}(\Sigma \times \Sigma_C)$. However, in subsequent chapters, our focus will be mainly on H_2 or H_∞ performance measures. That is, one of the desired performance objectives could be to attain $\|T_{d,z}(\Sigma \times \Sigma_C)\|_2$ equal to the infimum of such norms. This leads to the H_2 performance measure. Another desired performance objective could be to attain $\|T_{d,z}(\Sigma \times \Sigma_C)\|_\infty$ at most a specified value γ, i.e. to achieve an H_∞ γ-suboptimal performance. This leads to the H_∞ performance measure. We note that when the performance measure is removed, the above *problem of achieving a desired performance with output regulation constraint* is simply the *output regulation problem* which was our main focus in previous chapters.

Note that the direct feedthrough matrix D_{ed} from d to e and the direct feedthrough matrix D_{zw} from w to z are zero because they are irrelevant in this problem. We only consider the effect of w on e and not on z. Similarly, we only consider the effect of d on z and not on e.

Our objectives in this chapter are two fold, the first is to find the necessary and sufficient conditions under which the above posed problem can be solved. The second and our main objective in this chapter is to transform the above posed problem to another problem of achieving a desired performance without having any output regulation constraint.

With regards to the first objective, we make the following assumptions. Later on we will show how the posed problem can be solved under these assumptions.

A.1. (A, B) is stabilizable.

A.2. The matrix S is anti-Hurwitz-stable for continuous-time systems and anti-Schur-stable for discrete-time systems.

A.3. $\left(\begin{pmatrix} C_y & D_{yw} \end{pmatrix}, \begin{pmatrix} A & E_w \\ 0 & S \end{pmatrix}\right)$ is detectable.

A.4. There exist Π and Γ solving the regulator equation,

$$\begin{aligned} \Pi S &= A\Pi + B\Gamma + E_w, \\ 0 &= C_e\Pi + D_{eu}\Gamma + D_{ew}. \end{aligned} \tag{7.4}$$

Note that these assumptions have been defined earlier, for instance on page 215.

Let us briefly comment why the above assumptions are necessary to solve the problem we posed. From Chapter 2, we know that Assumptions A.1, A.3a, and A.3b (defined on page 19) are necessary for achieving output regulation. If Assumption A.3 is not satisfied, but only the necessary condition Assumptions A.3a, A.3b are satisfied, then a reduction technique, discussed in Chapter 2, can be used to reduce a given output regulation problem that does not satisfy A.3 to another output regulation problem that does satisfy A.3. Note that this applies to both continuous- and discrete-time systems having detectability and stabilizability with respect to the negative half plane or the unit disc. As explained in Chapter 2, Assumption A.2 is often made because, in the absence of it, output regulation is a mere consequence of internal stabilization. Also, from Chapter 2, we know that when Assumptions A.1, A.2, and A.3 hold, Assumption A.4 is a necessary and sufficient condition for the existence of a solution to the output regulation problem. Thus, each one of the Assumptions A.1, A.2, A.3, and A.4 is obviously necessary to solve any output regulation problem, and thus the *problem of achieving a desired performance with output regulation constraint.*

Note that in Chapter 6, where we optimized transient performance, we basically considered a configuration which is very close to the configuration as illustrated in Figure 7.1 on page 273. The main difference is that we had an external signal (which was called r) which effects both the system and the exosystem as illustrated in Figure 7.2. In Chapter 6 it was then shown that minimizing the H_2 norm from r to z is equivalent to optimal output regulation. This shows that the basic setup of this chapter is closely related to a more

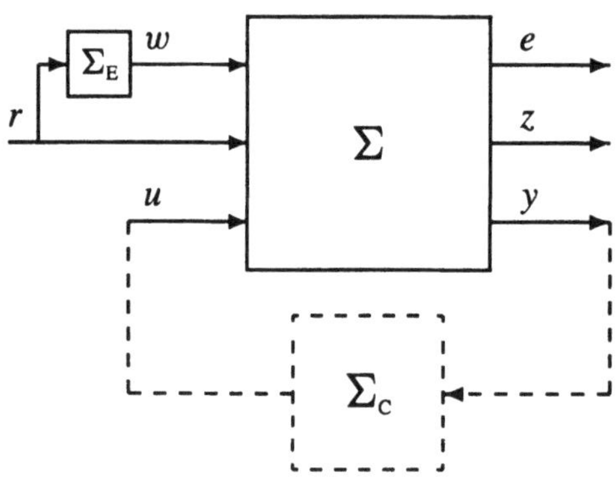

Figure 7.2: Performance with transient performance

specific problem of output regulation with optimal transient performance.

In the case when the regulator equation (7.4) has a non-unique solution for (Π, Γ), for simplicity of presentation, we assume throughout the chapter that a solution (Π, Γ) of (7.4) has been chosen, and all our development here builds on such a solution.

7.3 Reduction of a constrained optimal control problem to an unconstrained optimal control problem

As we stated earlier, basically our goal in this chapter is to study how to achieve performance under the output regulation constraint. One normally solves an optimization problem in order to achieve as much performance as possible. Thus, the problem at hand is obviously a constrained optimization problem which is known to be complex and hard to deal with directly. Our objective in this section is to solve such a *constrained optimization* problem for the given system by transforming it to an *unconstrained optimization* problem for a certain auxiliary system.

To do so, we first formulate the auxiliary system denoted by $\bar{\Sigma}$. For each controller $\bar{\Sigma}_{\mathrm{C}}$ for the auxiliary system, we formulate a corresponding controller Σ_{C} for the given system Σ, and in so doing we generate a class of controllers for the given system. We note that a controller for $\bar{\Sigma}$ can be related to a controller for Σ and vice versa. Then, we develop our main result. It states that a certain transfer function (say, from $\bar{d}$ to $\bar{z}$) of the auxiliary system with an internally stabilizing controller $\bar{\Sigma}_{\mathrm{C}}$, is exactly the same as the transfer function from d to z in the given system with a controller Σ_{C} that corresponds to $\bar{\Sigma}_{\mathrm{C}}$. Also, such a Σ_{C} internally stabilizes Σ, and is such that the error e tends to zero asymptotically. Thus, we basically transform the *problem of achieving a desired performance with output regulation constraint* for the given system to a similar one for the auxiliary system however without any output regulation constraint. This transformation is valid whatever may be the chosen performance measure as long as it is based on the transfer matrix $T_{d,z}(\Sigma \times \Sigma_{\mathrm{C}})$ from d to z.

Before we proceed further, let us recall from Section 2.6 and Theorem 2.6.1 what the so-called internal model principle says: a controller that achieves output regulation (for $d = 0$) must contain a (partial) copy of the exosystem. A careful examination of the internal model principle guides us to construct the following auxiliary system $\bar{\Sigma}$, its controller $\bar{\Sigma}_{\mathrm{C}}$, and a corresponding controller Σ_{C} for the given system Σ. Consider the auxiliary system $\bar{\Sigma}$ defined by

$$\bar{\Sigma}: \begin{cases} \rho\bar{x} = \bar{A}\bar{x} + \bar{B}\bar{u} + \bar{E}_d\bar{d} \\ \bar{z} \ = \bar{C}_z\bar{x} + \bar{D}_{zu}\bar{u} + D_{zd}\bar{d} \\ \bar{y} \ = \bar{C}_y\bar{x} + \bar{D}_{yu}\bar{u} + D_{yd}\bar{d}, \end{cases} \tag{7.5}$$

where

$$\bar{A} = \begin{pmatrix} A & -B\Gamma \\ 0 & S \end{pmatrix}, \quad \bar{B} = \begin{pmatrix} 0 & B \\ I & 0 \end{pmatrix}, \quad \bar{E}_d = \begin{pmatrix} E_d \\ 0 \end{pmatrix}, \tag{7.6a}$$

$$\bar{C}_z = \begin{pmatrix} C_z & -D_{zu}\Gamma \end{pmatrix}, \quad \bar{D}_{zu} = \begin{pmatrix} 0 & D_{zu} \end{pmatrix}, \tag{7.6b}$$

$$\bar{C}_y = \begin{pmatrix} C_y & (D_{yw} + C_y\Pi) \end{pmatrix}, \quad \bar{D}_{yu} = \begin{pmatrix} 0 & D_{yu} \end{pmatrix}. \tag{7.6c}$$

We note that the auxiliary system $\bar{\Sigma}$ is constructed from the data of the given system Σ and the exosystem Σ_{E} respectively as in (7.1) and (7.2), and the matrices Π and Γ that solve the regulator equation (7.4).

It turns out that two subsystems of the above auxiliary system $\bar{\Sigma}$ have the same zeros as two other subsystems of the given system Σ have. This property is needed in the subsequent chapters.

We have the following lemma.

Lemma 7.3.1

- *The zeros of the system characterized by* $(\bar{A}, \bar{B}, \bar{C}_z, \bar{D}_{zu})$ *are those of the system characterized by* (A, B, C_z, D_{zu})*. This does not depend on the eigenvalues of* S*.*

- *The zeros of the system characterized by* $(\bar{A}, \bar{E}_d, \bar{C}_y, D_{yd})$ *are those of the system characterized by* (A, E_d, C_y, D_{yd}) *plus a subset of the eigenvalues of* S*.*

Proof : This follows directly from the definition of invariant zeros as points were Rosenbrock's system matrix loses rank. ∎

We will next consider for $\bar{\Sigma}$ a general class of measurement feedback controllers with a state space representation $(\bar{A}_c, \bar{B}_c, \bar{C}_c, \bar{D}_c)$. Consider a controller $\bar{\Sigma}_C$ given by

$$\bar{\Sigma}_C : \begin{cases} \rho\bar{v} = \bar{A}_c\bar{v} + \bar{B}_c\bar{y} \\ \bar{u} = \bar{C}_c\bar{v} + \bar{D}_c\bar{y}. \end{cases} \tag{7.7}$$

Next, for any measurement feedback controller $\bar{\Sigma}_C$ that is constructed for the auxiliary system $\bar{\Sigma}$, we can define a corresponding controller Σ_C for the given system Σ. Besides being parameterized in the parameters $\bar{A}_c$, $\bar{B}_c$, $\bar{C}_c$, and $\bar{D}_c$ as in the case of $\bar{\Sigma}_C$, the controller Σ_C depends on the data of the given system Σ and the exosystem Σ_E as well as the matrices Π and Γ that solve the regulator equation (7.4). The controller Σ_C is given by

$$\Sigma_C : \begin{cases} \rho v_1 = \quad Sv_1 + \bar{C}_{c,1}v_2 + \bar{D}_{c,1}(y + (D_{yw} + D_{yu}\Gamma + C_y\Pi)v_1) \\ \rho v_2 = \qquad\qquad \bar{A}_c v_2 + \bar{B}_c(y + (D_{yw} + D_{yu}\Gamma + C_y\Pi)v_1) \\ u \quad = -\Gamma v_1 + \bar{C}_{c,2}v_2 + \bar{D}_{c,2}(y + (D_{yw} + D_{yu}\Gamma + C_y\Pi)v_1), \end{cases} \tag{7.8}$$

where $\bar{C}_{c,1}$, $\bar{C}_{c,2}$, $\bar{D}_{c,1}$, and $\bar{D}_{c,2}$ are obtained by partitioning $\bar{C}_c$ and $\bar{D}_c$ in conformity with the partitioning of $\bar{A}$,

$$\bar{C}_c = \begin{pmatrix} \bar{C}_{c,1} \\ \bar{C}_{c,2} \end{pmatrix}, \quad \text{and } \bar{D}_c = \begin{pmatrix} \bar{D}_{c,1} \\ \bar{D}_{c,2} \end{pmatrix}.$$

In order to understand the motivations in introducing the auxiliary system (7.5), consider the following intermediate auxiliary system,

$$\bar{\Sigma}_a : \begin{cases} \rho\bar{x} = \bar{A}\bar{x} & + \quad \bar{B}\bar{u} + \bar{E}_d d \\ \rho w = & Sw \\ z = \bar{C}_z\bar{x} + \bar{D}_{zw}w + \bar{D}_{zu}\bar{u} + D_{zd}d \\ e = \bar{C}_e\bar{x} & + \bar{D}_{eu}\bar{u} \\ \bar{y} = \bar{C}_y\bar{x} & + \bar{D}_{yu}\bar{u} + D_{yd}d, \end{cases} \tag{7.9}$$

where

$$\bar{C}_e = \begin{pmatrix} C_e & -D_{eu}\Gamma \end{pmatrix}, \quad \bar{D}_{zw} = C_z\Pi + D_{zu}\Gamma, \quad \bar{D}_{eu} = \begin{pmatrix} 0 & D_{eu} \end{pmatrix}. \tag{7.10}$$

It is then immediate to check that the interconnection of Σ and Σ_C is then equal to the interconnection of $\bar{\Sigma}_C$ and $\bar{\Sigma}_a$. This can be done by observing the following relationships between different variables,

$$\bar{x}_1 = x - \Pi w, \quad \bar{x}_2 = v_1 + w, \quad \bar{u}_2 = u + \Gamma v_1, \quad \text{and} \quad \bar{v} = v_2 \tag{7.11}$$

where $\bar{x}_1$, $\bar{x}_2$, $\bar{u}_1$ and $\bar{u}_2$ are obtained by partitioning $\bar{x}$ and $\bar{u}$ in conformity with the partitioning of $\bar{A}$, $\bar{B}$, and $\bar{E}_d$,

$$\bar{x} = \begin{pmatrix} \bar{x}_1 \\ \bar{x}_2 \end{pmatrix}, \quad \text{and} \quad \bar{u} = \begin{pmatrix} \bar{u}_1 \\ \bar{u}_2 \end{pmatrix}.$$

Next, if we look at the structure of $\bar{\Sigma}_a$ then it is obvious that w does not effect either $\bar{y}$ or e. Therefore, no matter what controller from $\bar{y}$ to $\bar{u}$ we apply to this system, w does not effect the error signal e. Obviously w does effect the signal z. However, our objective is to minimize the effect of w on e (which is automatically achieved) while our performance objective is not related to the effect of w on z but only the effect of d on z. Therefore, we can equally well omit w and its dynamics. We then obtain the system $\bar{\Sigma}$. Hence our objective becomes to find a controller $\bar{\Sigma}_C$ for the system $\bar{\Sigma}$ which internally stabilizes and achieves a desired performance for the transfer matrix from $\bar{d}$ to $\bar{z}$. This is an unconstrained performance objective where we do not have to worry about the output regulation constraint for the original system.

Next, we consider the converse question. We have seen that every stabilizing controller $\bar{\Sigma}_C$ for the auxiliary system $\bar{\Sigma}$ which yields a closed loop transfer matrix G from $\bar{d}$ to $\bar{z}$, generates a controller Σ_C for the original system which internally stabilizes Σ (omitting as usual the unstable dynamics of w), achieves output regulation, and achieves the transfer matrix G from d

to z. This class of controllers that achieves output regulation for the original system might be limited. Perhaps, there are other controllers which achieve output regulation and a better performance from d to z but which cannot be generated via a stabilizing controller for the auxiliary system. The following theorem is a key to our development and shows the converse. Namely, any controller Σ_C which achieves output regulation for the system and a transfer matrix G from d to z generates a stabilizing controller $\bar{\Sigma}_C$ for the auxiliary system $\bar{\Sigma}$ with the transfer matrix G from $\bar{d}$ to $\bar{z}$. This result strongly depends on the internal model principle as derived in Section 2.6.

Theorem 7.3.1 *Consider the given system Σ as in (7.1), and the exosystem Σ_E as in (7.2). Let Assumptions A.1, A.2, and A.3 hold. Also, let Π and Γ be a solution of the regulator equation (7.4). Then, there exists a controller of the form*

$$\begin{aligned} \rho v &= A_c v + B_c y \\ u &= C_c v + D_c y \end{aligned} \tag{7.12}$$

which, when applied to Σ and Σ_E of (7.1) and (7.2), solves the output regulation problem, and achieves a closed-loop transfer matrix G from d to z if and only if the following condition holds:

> *There exists a finite dimensional controller $\bar{\Sigma}_C$ of the form (7.7) which internally stabilizes the auxiliary system $\bar{\Sigma}$ given in (7.5). Moreover, when such an internally stabilizing controller $\bar{\Sigma}_C$ is applied to $\bar{\Sigma}$, the achieved closed-loop transfer matrix from $\bar{d}$ to $\bar{z}$ equals G.*

Furthermore, given any such controller $\bar{\Sigma}_C$ for the auxiliary system $\bar{\Sigma}$, the correspondingly constructed controller Σ_C given in (7.8) internally stabilizes the given system Σ, achieves output regulation when applied to (7.1) and (7.2), and also yields a closed-loop transfer matrix from d to z equal to G.

Remark 7.3.1 *Note that there does not exist a $1-1$ relationship between strictly proper stabilizing controllers for the auxiliary system $\bar{\Sigma}$ and strictly proper controllers that achieve output regulation for the system Σ. Instead there exists a $1-1$ relationship between strictly proper controllers that achieve output regulation for the system Σ and proper stabilizing controllers for the auxiliary system $\bar{\Sigma}$ whose direct feedthrough matrix satisfies the structural constraint,*

$$\bar{D}_c = \begin{pmatrix} \bar{D}_{c,1} \\ 0 \end{pmatrix}.$$

Proof : For ease of exposition we prove this result only for the case $D_{yu} = 0$.

Suppose that we have a controller of the form (7.12) which when applied to Σ achieves output regulation as well as a transfer matrix from d to z equal to certain G. Also, as already noted in the proof of Theorem 2.6.1, there exist Π and Θ satisfying (2.17), that is,

$$\Pi S = A\Pi + E_w + BD_c(C_y\Pi + D_{yw}) + BC_c\Theta, \tag{7.13}$$
$$\Theta S = A_c\Theta + B_c(C_y\Pi + D_{yw}), \tag{7.14}$$
$$0 = C_e\Pi + D_{ew} + D_{eu}D_c(C_y\Pi + D_{yw}) + D_{eu}C_c\Theta. \tag{7.15}$$

In what follows, we first claim that there always exists a controller which when applied to Σ solves the output regulation problem, and achieves the transfer matrix from d to z equal to G, and furthermore the corresponding Θ is injective. To see this, given the results of Theorem 2.6.1, we know that in a suitable basis we have

$$S = \begin{pmatrix} S_1 & S_{12} \\ 0 & S_{22} \end{pmatrix}, \qquad \mathcal{V} = \operatorname{im}\begin{pmatrix} I \\ 0 \end{pmatrix}, \qquad \Theta = \begin{pmatrix} -I & \Theta_{12} \\ 0 & \Theta_{22} \end{pmatrix},$$
$$A_c = \begin{pmatrix} S_1 & A_{c,12} \\ 0 & A_{c,22} \end{pmatrix}, \qquad B_c = \begin{pmatrix} B_{c,1} \\ B_{c,2} \end{pmatrix}, \qquad C_c = \begin{pmatrix} C_{c,1} & C_{c,2} \end{pmatrix},$$
$$D_{yw} = \begin{pmatrix} -C_y\Pi_1 & D_{yw2} \end{pmatrix}, \quad \Gamma = \begin{pmatrix} \Gamma_1 & \Gamma_2 \end{pmatrix}, \quad \Pi = \begin{pmatrix} \Pi_1 & \Pi_2 \end{pmatrix},$$

such that the pair $(C_y\Pi_2 + D_{y2},\ S_{22})$ is observable. Choose $B_{c,3}$ such that

$$A_{c,33} = S_{22} - B_{c,3}(C_y\Pi_2 + D_{y2})$$

is Hurwitz-stable for continuous-time systems and Schur-stable for discrete-time systems. Then the controller characterized by the matrices,

$$\tilde{A}_c = \begin{pmatrix} S_1 & A_{c,12} & 0 \\ 0 & A_{c,22} & 0 \\ 0 & 0 & A_{c,33} \end{pmatrix}, \qquad \tilde{B}_c = \begin{pmatrix} B_{c,1} \\ B_{c,2} \\ B_{c,3} \end{pmatrix},$$
$$\tilde{C}_c = \begin{pmatrix} C_{c,1} & C_{c,2} & 0 \end{pmatrix}, \qquad \tilde{D}_c = D_c,$$

yields the same closed-loop system from d to z, and solves the output regulation problem. On the other hand for this controller, (7.13), (7.14), and (7.15) are satisfied with the same Π but the following Θ,

$$\Theta = \begin{pmatrix} -I & \Theta_{12} \\ 0 & \Theta_{22} \\ 0 & I \end{pmatrix}.$$

Obviously this new Θ is injective.

If Θ is injective then in a suitable basis we can guarantee that $\Theta = \begin{pmatrix} -I & 0 \end{pmatrix}^{\mathrm{T}}$. In this new basis the controller takes a special form,

$$A_{\mathrm{c}} = \begin{pmatrix} S + B_{\mathrm{c},1}(C_y\Pi + D_{yw}) & A_{\mathrm{c},12} \\ B_{\mathrm{c},2}(C_y\Pi + D_{yw}) & A_{\mathrm{c},22} \end{pmatrix}, \qquad B_{\mathrm{c}} = \begin{pmatrix} B_{\mathrm{c},1} \\ B_{\mathrm{c},2} \end{pmatrix},$$
$$C_{\mathrm{c}} = \begin{pmatrix} -\Gamma + D_{\mathrm{c}}(C_y\Pi + D_{yw}) & C_{\mathrm{c},2} \end{pmatrix},$$

while D_{c} remains the same as before. It is then obvious that this controller applied to (7.1) yields the same closed loop dynamics as the controller $\bar{\Sigma}_{\mathrm{C}}$ of (7.7) when applied to the auxiliary system $\bar{\Sigma}_a$ of (7.5) provided $\bar{A}_{\mathrm{c}} = A_{\mathrm{c},22}$, $\bar{B}_{\mathrm{c}} = B_{\mathrm{c},2}$, $\bar{C}_{\mathrm{c},1} = A_{\mathrm{c},12}$, $\bar{C}_{\mathrm{c},2} = C_{\mathrm{c},2}$, $\bar{D}_{\mathrm{c},1} = B_{\mathrm{c},1}$, and $\bar{D}_{\mathrm{c},2} = D_{\mathrm{c}}$ and we substitute the relations given in (7.11).

The controller Σ_{C} internally stabilizes Σ and hence the only unstable dynamics in the interconnection of Σ_{C} and Σ is the dynamics of the exosystem. But then also the only unstable dynamics in the interconnection of $\bar{\Sigma}_{\mathrm{C}}$ and $\bar{\Sigma}_a$ is the dynamics of the exosystem. Obviously, the interconnection of $\bar{\Sigma}_{\mathrm{C}}$ and $\bar{\Sigma}_a$ achieves output regulation since for $d = 0$ we have that $e(t) \to 0$ as $t \to 0$ since the unstable dynamics do not affect the error signal e and the stable dynamics obviously converges to 0. Suppose we omit this unstable dynamics of the exosystem from the interconnection of $\bar{\Sigma}_{\mathrm{C}}$ and $\bar{\Sigma}_a$. We then obtain the same dynamics as the interconnection of $\bar{\Sigma}_{\mathrm{C}}$ and $\bar{\Sigma}_a$ which must therefore be internally stable. Finally if we look at the transfer matrix from d to z in the interconnection of Σ_{C} and Σ then this is the same as the transfer matrix from d to z in the interconnection of $\bar{\Sigma}_{\mathrm{C}}$ and $\bar{\Sigma}_a$. But this transfer matrix obviously is independent of any initial conditions and therefore we can without loss of generality set $w(0) = 0$ and hence $w = 0$. But then the the transfer matrix from d to z in the interconnection of $\bar{\Sigma}_{\mathrm{C}}$ and $\bar{\Sigma}_a$ is clearly equal to the transfer matrix from $\bar{d}$ to $\bar{z}$ in the interconnection of $\bar{\Sigma}_{\mathrm{C}}$ and $\bar{\Sigma}$. ■

We will now consider two special cases of measurement that is available for feedback. The first case corresponds to *state feedback* where the state x of Σ and the state w of Σ_{E} are both available for feedback, and the second corresponds to *full information feedback* where states x, w as well as the disturbance d are available for feedback. For these cases, the auxiliary system $\bar{\Sigma}$ simplifies and has the same dimension n as that of the given system Σ. Also, one can use only static feedback to solve the posed problems. We will consider each of them separately.

State feedback case:

As said above, in this case, we assume that both the states x and w are available for feedback, i.e. in the notation of equations (7.1) and (7.2), we assume that,

$$y = \begin{pmatrix} x \\ w \end{pmatrix}, \quad C_y = \begin{pmatrix} I \\ 0 \end{pmatrix}, \quad D_{yw} = \begin{pmatrix} 0 \\ I \end{pmatrix}, \quad \text{and } D_{yd} = 0.$$

For this case, the construction of the auxiliary system $\bar{\Sigma}$ can be simplified. Consider the simplified auxiliary system $\bar{\Sigma}$ given by

$$\bar{\Sigma} : \begin{cases} \rho\bar{x} = A\bar{x} + E_d d + B\bar{u} \\ \bar{z} = C_z\bar{x} + D_{zd}d + D_{zu}\bar{u} \\ \bar{y} = \bar{x}. \end{cases} \tag{7.16}$$

For $\bar{\Sigma}$ given in (7.16), we consider a general class of static state feedback controllers given by

$$\bar{\Sigma}_C : \bar{u} = \bar{F}\bar{x}. \tag{7.17}$$

For any static state feedback controller $\bar{\Sigma}_C$ of (7.17) that is constructed for the auxiliary system $\bar{\Sigma}$ of (7.16), we can define a corresponding controller Σ_C for the given system Σ. Besides being parameterized in $\bar{F}$, the controller Σ_C depends on the matrices Π and Γ that solve the regulator equation (7.4), and it is given by

$$\Sigma_C : u = \bar{F}(x - \Pi w) + \Gamma w. \tag{7.18}$$

We have the following theorem.

Theorem 7.3.2 *Consider the given system Σ as in (7.1), and the exosystem Σ_E as in (7.2). Let Assumptions A.1 and A.2 hold. Also, let Π and Γ be a solution of the regulator equation (7.4). Then, there exists a static feedback controller of the form $u = F_x x + G_w w$, which, when applied to Σ and Σ_E of (7.1) and (7.2), solves the output regulation problem, and achieves a closed-loop transfer matrix G from d to z if and only if the following condition holds:*

> *There exists a static feedback controller $\bar{\Sigma}_C$ of the form (7.17) which internally stabilizes the auxiliary system $\bar{\Sigma}$ given in (7.16). Moreover, when such an internally stabilizing controller $\bar{\Sigma}_C$ is applied to $\bar{\Sigma}$, the achieved closed-loop transfer matrix from $\bar{d}$ to $\bar{z}$ equals G.*

Furthermore, given any such controller $\bar{\Sigma}_C$ for $\bar{\Sigma}$, the correspondingly constructed controller Σ_C given in (7.18) internally stabilizes the given system Σ, achieves output regulation when applied to (7.1) and (7.2), and also yields a closed-loop transfer matrix from d to z equal to G.

Proof : It follows along the same lines as the proof of Theorem 7.3.1. ■

We will now consider the other special case.

Full information feedback:

In this case we assume not only both the states x and w are available for feedback, but also the disturbance d is available for feedback. That is, we assume that

$$y = \begin{pmatrix} x \\ w \\ d \end{pmatrix}, \quad C_y = \begin{pmatrix} I \\ 0 \\ 0 \end{pmatrix}, \quad D_{yw} = \begin{pmatrix} 0 \\ I \\ 0 \end{pmatrix}, \quad \text{and } D_{yd} = \begin{pmatrix} 0 \\ 0 \\ I \end{pmatrix}.$$

For this case, the construction of the auxiliary system $\bar{\Sigma}$ can be simplified as

$$\bar{\Sigma} : \begin{cases} \rho\bar{x} = A\bar{x} + E_d d + B\bar{u} \\ \bar{z} = C_z\bar{x} + D_{zd}d + D_{zu}\bar{u} \\ \bar{y} = \begin{pmatrix} \bar{x} \\ d \end{pmatrix}. \end{cases} \tag{7.19}$$

For $\bar{\Sigma}$ given in (7.19), we consider a general class of static state feedback controllers given by

$$\bar{\Sigma}_C \;:\; \left\{ \bar{u} = \bar{F}\bar{x} + \bar{G}d. \right. \tag{7.20}$$

For any static state feedback controller $\bar{\Sigma}_C$ of (7.20) that is constructed for the auxiliary system $\bar{\Sigma}$ of (7.19), we can define a corresponding controller Σ_C for the given system Σ. Besides being parameterized in $\bar{F}$ and $\bar{G}$, the controller Σ_C depends on the matrices Π and Γ that solve the regulator equation (7.4), and it is given by

$$\Sigma_C \;:\; u = \bar{F}(x - \Pi w) + \Gamma w + \bar{G}d. \tag{7.21}$$

We have the following theorem.

Theorem 7.3.3 *Consider the given system Σ as in (7.1), and the exosystem Σ_E as in (7.2). Let Assumptions A.1 and A.2 hold. Also, let Π and Γ be a* *solution of the regulator equation (7.4). Then, there exists a static feedback controller of the form* $u = F_x x + G_w w + G_d d$ *which, when applied to* Σ *and* Σ_E *of (7.1) and (7.2), solves the regulator problem, and achieves a closed-loop transfer matrix* G *from* d *to* z *if and only if the following condition holds:*

> *There exists a static feedback controller* $\bar{\Sigma}_C$ *of the form (7.20) which internally stabilizes the auxiliary system* $\bar{\Sigma}$ *given in (7.19). Moreover, when such an internally stabilizing controller* $\bar{\Sigma}_C$ *is applied to* $\bar{\Sigma}$*, the achieved closed-loop transfer matrix from* $\bar{d}$ *to* $\bar{z}$ *equals* G*.*

Furthermore, given any such controller $\bar{\Sigma}_C$ *for* $\bar{\Sigma}$*, the correspondingly constructed controller* Σ_C *given in (7.21) internally stabilizes the given system* Σ*, achieves output regulation when applied to (7.1) and (7.2), and also yields a closed-loop transfer matrix from* d *to* z *equal to* G*.*

Proof : It follows along the same lines as the proof of Theorem 7.3.1. ■

It is obvious that, with the help of Theorems 7.3.1, 7.3.2, and 7.3.3, we can transform the *problem of achieving a desired performance with output regulation constraint* for the given system to a similar one for the auxiliary system however without any output regulation constraint. This transformation is valid whatever may be the chosen performance measure as long as it is based on $T_{d,z}(\Sigma \times \Sigma_C)$. There are a number of multivariable design techniques that can make use of the above theorems. In particular, we can name H_2 optimal control, H_∞ optimal control, closed-loop transfer recovery [57] among many others. Thus, Theorems 7.3.1, 7.3.2, and 7.3.3 enable us to invoke the standard optimization theory without any constraints to find a suitable controller for the auxiliary system, and thus for the given system. There is only a problem if a solution to the regulator equation is not unique since then there are many different auxiliary systems which one has to check to be guaranteed of optimal performance. For H_2 and H_∞ optimal control problems this turns out to be rather straightforward as can be seen in the subsequent chapters.

7.4 Achieving a desired performance with structurally stable output regulation constraint

The output regulation requirement in our *problem of achieving a desired performance with output regulation constraint* as defined in Section 7.2 can be further strengthened by requiring structurally stable output regulation. Let us briefly recall the concept of *structurally stable output regulation* which strengthens the concept of output regulation and thus renders it practically useful. In any given practical situation, the data of the plant cannot always be determined exactly. The data usually lies in a given neighborhood. With this in background, the *structurally stable output regulation* seeks a *fixed* controller that solves the output regulation problem not only for the nominal plant but also for plants obtained by arbitrarily small perturbations of their parameters from their nominal values. Thus, even if the values of the plant parameters drift but are confined to the given neighborhood, the same controller always achieves output regulation. In this sense, one fixed controller solves the output regulation problem for a *family* of plants, i.e. the family of all plants whose parameters range in a given neighborhood of the nominal point in a parameter space. The concept of *structurally stable output regulation* is a classical concept as discussed in Section 2.8 and Appendix 2.B, and it is known that to achieve *structurally stable output regulation* the controller should contain multiple copies of the exosystem.

Now, coming back to our problem, instead of simply considering the *problem of achieving a desired performance with output regulation constraint*, we can consider in an obvious way a strengthened *problem of achieving a desired performance with structurally stable output regulation constraint.* However, it turns out that the *problem of achieving a desired performance with structurally stable output regulation constraint* can be reduced to a *problem of achieving a desired performance with output regulation constraint.* It is feasible to do so since, as pointed out in Section 2.8, a *structurally stable output regulation* problem for a given system can be solved by transforming it to the *output regulation* problem for another auxiliary system, provided that the matrix

$$\begin{pmatrix} A_0 - \lambda I & B_0 \\ C_{e,0} & 0 \end{pmatrix} \tag{7.22}$$

has full row-rank for each λ which is an eigenvalue of S_0, where A_0, B_0, and $C_{e,0}$ are the nominal values of A, B, and C_e respectively. Let a system

$$\Sigma : \begin{cases} \rho x = Ax + Bu + E_w w + E_d d \\ e = C_e x + D_{eu} u + D_{ew} w + D_{ed} d \\ z = C_z x + D_{zu} u + D_{zd} d \\ y = C_y x + D_{yu} u + D_{yw} w + D_{yd} d, \end{cases} \tag{7.23}$$

and an exosystem Σ_E as in (7.2) be given. Obviously we need that the error signal as a component of the measurement signal, that is

$$C_y = \begin{pmatrix} C_e \\ C_{y2} \end{pmatrix}, \quad D_{yu} = \begin{pmatrix} D_{eu} \\ D_{yu2} \end{pmatrix}, \quad D_{yw} = \begin{pmatrix} D_{ew} \\ D_{yw2} \end{pmatrix}, \quad D_{yd} = \begin{pmatrix} D_{ed} \\ D_{yd2} \end{pmatrix}. \tag{7.24}$$

We included here the term $D_{ed}d$ in e, although it has no effect on the output regulation, it might be needed in order to have the error signal part of the measurements. But we can set this matrix D_{ed} to an arbitrary value since it does not effect our performance requirements.

We provide next the steps to transform a *problem of achieving a desired performance with structurally stable output regulation constraint* into a *problem of performance with the output regulation constraint.*

(*i*) We apply a preliminary feedback $u = Ne + \tilde{u}$ such that

$$A_0 + B_0 N(I - D_{eu,0}N)^{-1} C_{e,0}$$

and S have no eigenvalues in common and we obtain the system,

$$\bar{\Sigma} : \begin{cases} \rho x = \bar{A}_0 x + \bar{B}_0 \tilde{u} + \bar{E}_{w,0} w + \bar{E}_{d,0} d \\ y = \bar{C}_{y,0} x + \bar{D}_{yu,0} \tilde{u} + \bar{D}_{yw,0} w + \bar{D}_{yd,0} d \\ e = \bar{C}_{e,0} x + \bar{D}_{eu,0} \tilde{u} + \bar{D}_{ew,0} w + \bar{D}_{ed,0} d \\ z = \bar{C}_{z,0} x + \bar{D}_{zu,0} \tilde{u} + \bar{D}_{zw,0} w + \bar{D}_{zd,0} d, \end{cases} \tag{7.25}$$

where

$$\begin{aligned}
\bar{A}_0 &= A_0 + B_0 N(I - D_{eu,0}N)^{-1} C_{e,0}, \\
\bar{B}_0 &= B_0 (I - N D_{eu,0})^{-1}, \\
\bar{E}_{w,0} &= E_{w,0} + B_0 N(I - D_{eu,0}N)^{-1} D_{ew,0}, \\
\bar{E}_{d,0} &= E_{d,0} + B_0 N(I - D_{eu,0}N)^{-1} D_{ed,0}, \\
\bar{C}_{e,0} &= (I - D_{eu,0}N)^{-1} C_{e,0}, \\
\bar{C}_{y,0} &= C_{y,0} + D_{yu,0} N(I - D_{eu,0}N)^{-1} C_{e,0}, \\
\bar{C}_{z,0} &= C_{z,0} + D_{zu,0} N(I - D_{eu,0}N)^{-1} C_{e,0}, \\
\bar{D}_{eu,0} &= (I - D_{eu,0}N)^{-1} D_{eu,0}, \\
\bar{D}_{ew,0} &= (I - D_{eu,0}N)^{-1} D_{ew,0}, \\
\bar{D}_{ed,0} &= (I - D_{eu,0}N)^{-1} D_{ed,0}, \\
\bar{D}_{yu,0} &= D_{yu,0} (I - N D_{eu,0})^{-1}, \\
\bar{D}_{yw,0} &= D_{yw,0} + D_{yu,0} N(I - D_{eu,0}N)^{-1} D_{ew,0}.
\end{aligned}$$

$$\bar{D}_{yd,0} = D_{yd,0} + D_{yu,0}N(I - D_{eu,0}N)^{-1}D_{ed,0}.$$
$$\bar{D}_{zu,0} = D_{zu,0}(I - ND_{eu,0})^{-1},$$
$$\bar{D}_{zw,0} = D_{zu,0}N(I - D_{eu,0}N)^{-1}D_{ew,0},$$
$$\bar{D}_{zd,0} = D_{zd,0} + D_{zu,0}N(I - D_{eu,0}N)^{-1}D_{ed,0}.$$

(*ii*) Given a system Σ and an exosystem Σ_{E} as in (7.23) and (7.2), we formulate an auxiliary system $\tilde{\Sigma}$ and another auxiliary exosystem $\tilde{\Sigma}_{\mathrm{E}}$. Without loss of generality assume that the matrix S of Σ_{E} is of block diagonal form,

$$S = \begin{pmatrix} S^* & 0 \\ 0 & S_{\min} \end{pmatrix},$$

in which S^* is a certain matrix not of much concern to us, and $S_{\min}$ is a matrix whose characteristic polynomial coincides with the minimal polynomial of S (thus, $S_{\min}$ is a cyclic matrix, i.e. its characteristic and minimal polynomials coincide). Now define the auxiliary exosystem $\tilde{\Sigma}_{\mathrm{E}}$ as

$$\tilde{\Sigma}_{\mathrm{E}} : \ \rho\tilde{w} = \tilde{S}_q\tilde{w} \tag{7.26}$$

where $\tilde{w} \in \mathbb{R}^{q\tilde{s}}$ and $\tilde{S}_q$ is a block diagonal matrix given by

$$\tilde{S}_q = \begin{pmatrix} S_{\min} & 0 & \cdots & 0 \\ 0 & S_{\min} & \ddots & \vdots \\ \vdots & \ddots & \ddots & 0 \\ 0 & \cdots & 0 & S_{\min} \end{pmatrix}.$$

We note that the auxiliary exosystem given in (7.26) is composed of q identical copies of a constituent exosystem where q is the dimension of the measurement vector y. Next we define the auxiliary system $\tilde{\Sigma}$ as

$$\tilde{\Sigma} : \begin{cases} \rho\tilde{x} = \bar{A}_0\tilde{x} + \bar{B}_0\tilde{u} + \bar{E}_{d,0}d \\ y = \bar{C}_{y,0}\tilde{x} + \bar{D}_{yu,0}\tilde{u} + \tilde{D}_{yw}\tilde{w} + \bar{D}_{yd,0}d \\ e = \bar{C}_{e,0}\tilde{x} + \bar{D}_{eu,0}\tilde{u} + \tilde{D}_{ew}\tilde{w} + \bar{D}_{ed,0}d \\ z = \bar{C}_{z,0}\tilde{x} + \bar{D}_{zu,0}\tilde{u} + \bar{D}_{zd,0}d \end{cases} \tag{7.27}$$

and the matrix $\tilde{D}_{ew}$ and $\tilde{D}_{yw}$ are partitioned as,

$$\tilde{D}_{yw} = \begin{pmatrix} \tilde{D}_{ew} \\ \tilde{D}_{yw2} \end{pmatrix} = \begin{pmatrix} \tilde{D}_{ew1} & 0 \\ 0 & \tilde{D}_{yw22} \end{pmatrix},$$

where

$$\tilde{D}_{ew1} = \left(\tilde{D}_{ew1,1} \quad \tilde{D}_{ew1,2} \quad \cdots \quad \tilde{D}_{ew1,q}\right),$$
$$\tilde{D}_{yw22} = \left(\tilde{D}_{yw22,1} \quad \tilde{D}_{yw22,2} \quad \cdots \quad \tilde{D}_{yw22,p-q}\right).$$

Here the matrices $\tilde{D}_{ew1}$ and $\tilde{D}_{yw22}$ are selected so that the pairs of matrices $(\tilde{D}_{ew1}, \tilde{S}_q)$ and $(\tilde{D}_{yw22}, \tilde{S}_{p-q})$ are detectable. Note that we can set $\bar{D}_{ed}$ equal to zero without loss of generality since $\bar{D}_{ed}$ does not effect any of our performance objectives.

(*iii*) Solve the *problem of achieving a desired performance with the output regulation constraint* for the auxiliary systems $\tilde{\Sigma}$ and $\tilde{\Sigma}_{\mathrm{E}}$ utilizing Theorem 7.3.1.

(*iv*) The above solution is the solution to the *problem of performance with structurally stable output regulation constraint* for the original system Σ and the exosystem Σ_{E}.

Chapter 8

H_2 optimal control with an output regulation constraint – continuous-time systems

8.1 Introduction

In Chapter 7, we dealt with the problem of achieving a desired performance subject to an output regulation constraint. We showed how one can solve such a problem without solving directly a constrained optimization problem. The method we developed there basically amounts to transforming the constrained optimization problem for a given system to an *unconstrained* optimization problem, however, for a certain auxiliary system. The development in Chapter 7 is general i.e. it does not consider explicitly any specific performance measure; typically any performance measure can be used. In this chapter, we will use the H_2 norm of the closed-loop transfer function matrix as the performance measure.

To be explicit, the multi-objective problem we consider in this chapter is an H_2 optimal (or suboptimal) control problem along with the added output regulation constraint. That is, we seek a controller that achieves output regulation and results in the smallest possible (or arbitrarily close to the minimum possible) H_2 norm of a transfer function from a certain input signal to a desired output signal of the system. Such a problem clearly takes into account certain performance requirements. This is obvious when we recognize that the H_2 norm, being an integral square operation, takes into account performance measures such as mean square error, total energy consumed etc.

Utilizing the results of Chapter 7, the H_2 optimal (or suboptimal) control problem with the added output regulation constraint for the given system is solved by transforming it to an H_2 optimal (or suboptimal) control problem without any output regulation constraint for the auxiliary system. It is interesting to point out that even if one starts with an H_2 optimal control problem for the originally given system (ignoring the output regulation constraint) that satisfies the regularity conditions, the corresponding H_2 optimal control problem for the auxiliary system inherently *does not* satisfy the regularity conditions [61], i.e. in general it turns out to be a singular H_2 optimal control problem. We can then obtain a solution to such a singular H_2 optimal control problem for the auxiliary system by using the recently developed methods discussed in [61,71].

It is clear that the H_2 optimal (or suboptimal) control problem with the regulation constraint for the given system can be solved via solving an unconstrained H_2 optimal (or suboptimal) control problem for the auxiliary system. However, a fundamental and significant issue that still remains to be clarified is this; does the added output regulation constraint compromise the achievable performance? In this regard, for continuous-time systems, we will show that in the present case where we measure performance by the H_2 norm of a transfer function matrix, there is no loss at all in the achievable performance.

The H_2 optimal control problem with the added output regulation constraint has been studied earlier in [3,4,69]. The methods pursued there to solve the posed problem restrict attention to certain special cases, and the general problem is not considered. Our objective here is to present an elegant and to the point derivation based on combining results from a number of recent papers to solve the general problem without unnecessary restrictions.

The mathematical aspects of our development depends on whether we consider continuous- or discrete-time systems. As such, in this chapter, we focus on continuous-time systems while the next chapter considers discrete-time systems. This chapter is based on the recent research work of authors [77].

This chapter is organized as follows. Section 8.2 formulates H_2 optimal and suboptimal control problems with the output regulation constraint, while Section 8.3 presents solvability conditions for such problems and also develops methods of constructing regulators that solve such problems whenever they are solvable. Finally Section 8.4 shows that the added output regulation constraint in the given H_2 optimal control problem does not compromise the achievable performance.

8.2 Problem formulations

As in Chapter 7, we start with a linear system with state space realization,

$$\Sigma : \begin{cases} \dot{x} = Ax + Bu + E_w w + E_d d \\ e = C_e x + D_{eu} u + D_{ew} w \\ z = C_z x + D_{zu} u \\ y = C_y x + D_{yw} w + D_{yd} d, \end{cases} \tag{8.1}$$

where, as usual, $x \in \mathbb{R}^n$, $u \in \mathbb{R}^m$, $y \in \mathbb{R}^p$, $e \in \mathbb{R}^q$, and $z \in \mathbb{R}^\ell$. The exogenous disturbance input $w \in \mathbb{R}^s$ is generated by an exosystem Σ_E with state space realization,

$$\Sigma_E : \dot{w} = Sw. \tag{8.2}$$

As before, the variable d denotes an external disturbance. We seek measurement feedback controllers which are of the form,

$$\Sigma_C : \begin{cases} \dot{v} = A_c v + B_c y, \\ u = C_c v + D_c y. \end{cases} \tag{8.3}$$

The closed-loop system consisting of the given system Σ and the controller Σ_C is denoted by $\Sigma \times \Sigma_C$. Also, the transfer matrix from d to z of $\Sigma \times \Sigma_C$ is denoted by $T_{d,z}(\Sigma \times \Sigma_C)$. As usual, we define,

$$\|T_{d,z}(\Sigma \times \Sigma_C)\|_2 := \left(\frac{1}{2\pi} \int_{-\infty}^{\infty} \text{trace}\, T_{d,z}(\Sigma \times \Sigma_C)^{\mathrm{T}}(-i\omega) T_{d,z}(\Sigma \times \Sigma_C)(i\omega) d\omega \right)^{1/2}.$$

Before we state formally the specific multi-objective problems of interest here, we define the following notations:

$$\gamma_{2p,r}^* = \inf \{ \|T_{d,z}(\Sigma \times \Sigma_C)\|_2 \mid \Sigma_C \text{ is a proper controller that achieves output regulation for } \Sigma \},$$

$$\gamma_{2p}^* = \inf \{ \|T_{d,z}(\Sigma \times \Sigma_C)\|_2 \mid \Sigma_C \text{ is a proper controller that internally stabilizes } \Sigma \},$$

$$\gamma_{2sp,r}^* = \inf \{ \|T_{d,z}(\Sigma \times \Sigma_C)\|_2 \mid \Sigma_C \text{ is a strictly proper controller that achieves output regulation for } \Sigma \},$$

$$\gamma_{2sp}^* = \inf \{ \|T_{d,z}(\Sigma \times \Sigma_C)\|_2 \mid \Sigma_C \text{ is a strictly proper controller that internally stabilizes } \Sigma \}.$$

We can now state formally the H_2 optimal control problem with the output regulation constraint.

Problem 8.2.1 ***(H_2 optimal control problem with the output regulation constraint)*** *Consider the system Σ and the exosystem Σ_E as given in (8.1) and (8.2). Find, if possible, a proper (or a strictly proper) controller Σ_C such that the following conditions hold:*

(i) ***(Internal Stability)*** *In the absence of the disturbances w and d, the closed-loop system $\Sigma \times \Sigma_C$ is internally stable.*

(ii) ***(Performance Measure)*** $\|T_{d,z}(\Sigma \times \Sigma_C)\|_2$ *is equal to the infimum $\gamma^*_{2p,r}$ (or $\gamma^*_{2sp,r}$).*

(iii) ***(Output Regulation)*** *For any $d \in L_2$, and for all $x(0) \in \mathbb{R}^n$ and $w(0) \in \mathbb{R}^s$, the solution of the closed-loop system $\Sigma \times \Sigma_C$ satisfies*

$$\lim_{t\to\infty} e(t) = 0.$$

Remark 8.2.1 *The part (iii) of the above problem formulation is equivalent to the following: In the absence of external signal d, i.e. when $d = 0$, Σ_C achieves output regulation for Σ.*

We can also define the H_2 suboptimal control problem with the output regulation constraint.

Problem 8.2.2 ***(H_2 suboptimal control problem with the output regulation constraint)*** *Consider the system Σ and the exosystem Σ_E as given in (8.1) and (8.2). Find, if possible, a parameterized family $\{\Sigma_C(\varepsilon) \mid \varepsilon > 0\}$ of proper controllers of the form (8.3) (or a parameterized family $\{\Sigma_C(\varepsilon) \mid \varepsilon > 0\}$ of strictly proper controllers of the form (8.3) with $D_c = 0$) such that the following conditions hold:*

(i) ***(Internal Stability)*** *In the absence of the disturbances w and d, the closed-loop system $\Sigma \times \Sigma_C(\varepsilon)$ is internally stable.*

(ii) ***(Performance Measure)*** *As $\varepsilon \to 0$, we have $\|T_{d,z}(\Sigma \times \Sigma_C(\varepsilon))\|_2$ tending to the infimum $\gamma^*_{2p,r}$ (or $\gamma^*_{2sp,r}$).*

(iii) ***(Output Regulation)*** *For any $d \in L_2$, and for all $x(0) \in \mathbb{R}^n$ and $w(0) \in \mathbb{R}^s$, the solution of the closed-loop system $\Sigma \times \Sigma_C(\varepsilon)$ satisfies*

$$\lim_{t\to\infty} e(t) = 0.$$

We note that when the output regulation constraint is removed from the above problem formulations, the resulting problems are simply *unconstrained H_2 optimal and suboptimal control problems.*

8.3 Solvability conditions

As we discussed in Chapter 7, the following assumptions are reasonable and almost necessary to solve the problems defined above:

A.1. (A, B) is stabilizable.

A.2. The matrix S is anti-Hurwitz-stable, i.e. it has all its eigenvalues in the closed right-half plane.

A.3. $\left(\begin{pmatrix} C_y & D_{yw} \end{pmatrix}, \begin{pmatrix} A & E_w \\ 0 & S \end{pmatrix}\right)$ is detectable.

A.4. There exist Π and Γ solving the regulator equation,

$$\begin{aligned} \Pi S &= A\Pi + B\Gamma + E_w, \\ 0 &= C_e\Pi + D_{eu}\Gamma + D_{ew}. \end{aligned} \tag{8.4}$$

Note that these assumptions have been defined earlier, say for instance on page 215.

In the case when the regulator equation (8.4) has a non-unique solution for (Π, Γ), for simplicity of presentation, we assume throughout the chapter that a solution (Π, Γ) of (8.4) has been chosen, and all our development here builds on such a solution.

Following the results of Chapter 7, a solution to the H_2 *optimal or suboptimal control problems with output regulation constraint* for Σ is given in terms of a solution to the H_2 *optimal or suboptimal control problem without any output regulation constraint* for the auxiliary system $\bar{\Sigma}$ which is as in (7.5), and is repeated below for convenience:

$$\bar{\Sigma}: \begin{cases} \dot{\bar{x}} = \bar{A}\bar{x} + \bar{B}\bar{u} + \bar{E}_d\bar{d} \\ \bar{z} = \bar{C}_z\bar{x} + \bar{D}_{zu}\bar{u} \\ \bar{y} = \bar{C}_y\bar{x} + D_{yd}\bar{d}, \end{cases} \tag{8.5}$$

where

$$\bar{A} = \begin{pmatrix} A & -B\Gamma \\ 0 & S \end{pmatrix}, \quad \bar{B} = \begin{pmatrix} 0 & B \\ I & 0 \end{pmatrix}, \quad \bar{E}_d = \begin{pmatrix} E_d \\ 0 \end{pmatrix}, \tag{8.6a}$$

$$\bar{C}_z = \begin{pmatrix} C_z & -D_{zu}\Gamma \end{pmatrix}, \quad \bar{D}_{zu} = \begin{pmatrix} 0 & D_{zu} \end{pmatrix}, \tag{8.6b}$$

$$\bar{C}_y = \begin{pmatrix} C_y & (D_{yw} + C_y\Pi) \end{pmatrix}. \tag{8.6c}$$

For $\bar{\Sigma}$, a controller $\bar{\Sigma}_C$ with state space representation $(\bar{A}_c, \bar{B}_c, \bar{C}_c, \bar{D}_c)$ is given by

$$\bar{\Sigma}_C : \begin{cases} \dot{\bar{v}} = \bar{A}_c \bar{v} + \bar{B}_c \bar{y} \\ \bar{u} = \bar{C}_c \bar{v} + \bar{D}_c \bar{y}. \end{cases} \tag{8.7}$$

For each controller $\bar{\Sigma}_C$ for the auxiliary system $\bar{\Sigma}$, we formulate a corresponding controller Σ_C for the given system Σ. It is given by

$$\Sigma_C : \begin{cases} \dot{v}_1 = \quad S v_1 + \bar{C}_{c,1} v_2 + \bar{D}_{c,1}(y + (D_{yw} + C_y \Pi) v_1) \\ \dot{v}_2 = \qquad\qquad \bar{A}_c v_2 + \bar{B}_c (y + (D_{yw} + C_y \Pi) v_1) \\ u \;= -\Gamma v_1 + \bar{C}_{c,2} v_2 + \bar{D}_{c,2}(y + (D_{yw} + C_y \Pi) v_1), \end{cases} \tag{8.8}$$

where $\bar{C}_{c,1}$, $\bar{C}_{c,2}$, $\bar{D}_{c,1}$, and $\bar{D}_{c,2}$ are obtained by partitioning $\bar{C}_c$ and $\bar{D}_c$ in conformity with the partitioning of $\bar{A}$,

$$\bar{C}_c = \begin{pmatrix} \bar{C}_{c,1} \\ \bar{C}_{c,2} \end{pmatrix}, \quad \text{and} \quad \bar{D}_c = \begin{pmatrix} \bar{D}_{c,1} \\ \bar{D}_{c,2} \end{pmatrix}.$$

We note that there is a $1-1$ relationship between the proper controller $\bar{\Sigma}_C$ as given in (8.7) and the proper controller Σ_C as given in (8.8). Furthermore, we observe that with the constraint that $\bar{D}_{c,2} = 0$, the controller Σ_C as given in (8.8) is strictly proper. In other words, by imposing the constraint that $\bar{D}_{c,2} = 0$ in $\bar{\Sigma}_C$, we can generate the class of strictly proper controllers Σ_C for the given system Σ. We denote below the class of controllers $\bar{\Sigma}_C$ with the constraint $\bar{D}_{c,2} = 0$ by $\bar{\Sigma}_C^s$,

$$\bar{\Sigma}_C^s : \begin{cases} \dot{\bar{v}} = \qquad \bar{A}_c \bar{v} + \qquad \bar{B}_c \bar{y} \\ \bar{u} = \begin{pmatrix} \bar{C}_{c,1} \\ \bar{C}_{c,2} \end{pmatrix} \bar{v} + \begin{pmatrix} \bar{D}_{c,1} \\ 0 \end{pmatrix} \bar{y}. \end{cases} \tag{8.9}$$

The following theorems provide the conditions under which the H_2 optimal control problem with output regulation constraint can be solved via proper or via strictly proper controllers. Also, they provide a procedure of constructing an appropriate controller that solves the posed problem whenever it is solvable.

Theorem 8.3.1 *Let Assumptions A.1, A.2, A.3, and A.4 hold. Consider the given system Σ as in (8.1), and the exosystem Σ_E as in (8.2). Also, consider the auxiliary system $\bar{\Sigma}$ as in (8.5). Then, the following statements hold:*

> *The H_2 optimal control problem with output regulation constraint for Σ is solvable via a proper controller if and only if the H_2 optimal control problem for $\bar{\Sigma}$ is solvable via a proper controller. Moreover, a proper controller $\bar{\Sigma}_C$ of the form given in (8.7) is a proper H_2 optimal controller for $\bar{\Sigma}$ if and only if the corresponding proper controller Σ_C of the form given in (8.8) solves the H_2 optimal control problem with the output regulation constraint for Σ.*

Proof : The proof follows from Theorem 7.3.1. ■

Theorem 8.3.2 *Let Assumptions A.1, A.2, A.3, and A.4 hold. Consider the given system Σ as in (8.1), and the exosystem Σ_E as in (8.2). Also, consider the auxiliary system $\bar{\Sigma}$ as in (8.5). Then, the following statements hold:*

> *The H_2 optimal control problem with output regulation constraint for Σ is solvable via a strictly proper controller if and only if the H_2 optimal control problem for $\bar{\Sigma}$ is solvable via a strictly proper controller. Moreover, a strictly proper controller of the form given in (8.7) with $\bar{D}_c = 0$ is an H_2 optimal controller for $\bar{\Sigma}$ if and only if the corresponding strictly proper controller Σ_C of the form given in (8.8) solves the H_2 optimal control problem with the output regulation constraint for Σ.*

Remark 8.3.1 *Note that by restricting controllers of the auxiliary system to be strictly proper instead of the more general form $\bar{\Sigma}_C^s$, we do not lose performance but we might lose some flexibility in the sense that we do not capture all optimal controllers.*

Proof : We apply Theorem 7.3.1. It is then clear that H_2 optimal control problem with output regulation constraint for Σ is solvable via a strictly proper controller if and only if the H_2 optimal control problem for $\bar{\Sigma}$ is solvable via a controller of the form $\bar{\Sigma}_C^s$. However, for the auxiliary system $\bar{\Sigma}$ there exists an H_2 optimal controller of the form $\bar{\Sigma}_C^s$ if and only if there exists a strictly proper H_2 optimal controller. The latter follows directly from the conditions in [61], the special structure of the parameters of the auxiliary system and the special structure of the semi-stabilizing solution of the linear matrix inequality as we will establish in Lemma 8.4.1. ■

We can now state the following theorems which provide the conditions under which the H_2 suboptimal control problem with output regulation constraint can be solved. Also, they provide a procedure of constructing an appropriate sequence of controllers that solves the posed problem.

Theorem 8.3.3 *Let Assumptions A.1, A.2, A.3, and A.4 hold. Consider the given system Σ as in (8.1), and the exosystem Σ_E as in (8.2). Also, consider the auxiliary system $\bar{\Sigma}$ as in (8.5). Then, the following statements hold:*

The H_2 suboptimal control problem with output regulation constraint for Σ is solvable via a family $\{\bar{\Sigma}_C(\varepsilon) \mid \varepsilon > 0\}$ of proper (or strictly proper) controllers.

Moreover, a family $\{\bar{\Sigma}_C(\varepsilon) \mid \varepsilon > 0\}$ of proper controllers of the form (8.7) is H_2 suboptimal for $\bar{\Sigma}$ if and only if the corresponding family $\{\Sigma_C(\varepsilon) \mid \varepsilon > 0\}$ of proper controllers of the form (8.8) solves the H_2 suboptimal control problem with the output regulation constraint for Σ.

Similarly, a family $\{\bar{\Sigma}_C^s(\varepsilon) \mid \varepsilon > 0\}$ of controllers of the form (8.9) is H_2 suboptimal for $\bar{\Sigma}$ if and only if the corresponding family $\{\Sigma_C(\varepsilon) \mid \varepsilon > 0\}$ of strictly proper controllers of the form (8.8) solves the H_2 suboptimal control problem with the output regulation constraint for Σ.

Remark 8.3.2 *By exploiting the structure of the problem it can be shown that there exists a family $\{\bar{\Sigma}_C^s(\varepsilon) \mid \varepsilon > 0\}$ of controllers of the form (8.9) which is H_2 suboptimal for $\bar{\Sigma}$ if and only if there exists a family $\{\bar{\Sigma}_C(\varepsilon) \mid \varepsilon > 0\}$ of controllers of the form (8.7) with $\bar{D}_c = 0$ which is H_2 suboptimal for $\bar{\Sigma}$. In other words, it is sufficient to look for strictly proper controllers for the auxiliary system $\bar{\Sigma}$.*

Proof : The proof is obvious from Theorem 7.3.1 when one takes into account the fact that the H_2 suboptimal control problem for $\bar{\Sigma}$ is solvable under the Assumptions A.1 and A.3. ■

Design of a measurement feedback regulator that solves the H_2 optimal control problem with output regulation constraint:

Theorems 8.3.1 and 8.3.2 suggest the following two step procedure:

Step 1: Construct a controller $\bar{\Sigma}_C$ of the form (8.7) (or a controller $\bar{\Sigma}_C$ of the form (8.7)) with $\bar{D}_c = 0$ so that it solves the H_2 optimal control problem for $\bar{\Sigma}$.

Step 2: Knowing the parameters $\bar{A}_c$, $\bar{B}_c$, $\bar{C}_c$, and $\bar{D}_c$ of the controller $\bar{\Sigma}_C$ (or $\bar{\Sigma}_C^s$) obtained in Step 1, construct a corresponding controller Σ_C as given in (8.8).

Clearly Σ_C constructed in Step 2 solves the H_2 optimal control problem with output regulation constraint for Σ via a proper (or a strictly proper) controller. □

The above procedure can be modified in an obvious way to obtain a sequence of proper (or strictly proper) controllers that solves the H_2 suboptimal control problem with output regulation constraint for Σ. Also, as will be shown shortly in Theorem 8.4.2 of Section 8.4, we have $\gamma_{2p,r}^* = \gamma_{2p}^* = \gamma_{2sp,r}^* = \gamma_{2sp}^*$. Hence one can design without any penalty a sequence of either proper or strictly proper controllers to solve the H_2 suboptimal control problem with output regulation constraint. In what follows, we simply give the design of a sequence of proper controllers.

Design of a sequence of measurement feedback regulators that solves the H_2 suboptimal control problem with output regulation constraint:

Step 1: Construct a family $\{\bar{\Sigma}_C(\varepsilon) \mid \varepsilon > 0\}$ of proper or strictly proper measurement feedback regulators that solves the H_2 suboptimal control problem for $\bar{\Sigma}$.

Step 2: Knowing the parameterized matrix quadruples $\{(\bar{A}_c(\varepsilon), \bar{B}_c(\varepsilon), \bar{C}_c(\varepsilon), \bar{D}_c(\varepsilon)) \mid \varepsilon > 0\}$ that characterize $\{\bar{\Sigma}_C(\varepsilon) \mid \varepsilon > 0\}$ in Step 1, construct a corresponding sequence of proper or strictly proper measurement feedback regulators $\{\Sigma_C(\varepsilon) \mid \varepsilon > 0\}$ each element of which is as given in (8.8).

Clearly $\{\Sigma_C(\varepsilon) \mid \varepsilon > 0\}$ constructed in Step 2 solves the H_2 suboptimal control problem with output regulation constraint for Σ via a sequence of proper controllers. □

8.4 Relationship between $\gamma_{2p,r}^*$, $\gamma_{2sp,r}^*$, γ_{2p}^*, and γ_{2sp}^*

Our primary objective in this section is to study how the achievable performance is affected by having the output regulation constraint. It turns out that there is no loss at all in the achievable performance, i.e. $\gamma_{2p,r}^*$, $\gamma_{2sp,r}^*$, γ_{2p}^*, and γ_{2sp}^* are all equal to each other.

To proceed, we need to recall certain results from [61, 71]. Consider the system Σ described by (8.1), and for any matrix $P \in \mathbb{R}^{n\times n}$, define a continuous-time linear matrix inequality as

$$F(P) \geq 0, \tag{8.10}$$

where

$$F(P) := \begin{pmatrix} A^{\mathrm{T}}P + PA + C_z^{\mathrm{T}}C_z & PB + C_z^{\mathrm{T}}D_{zu} \\ B^{\mathrm{T}}P + D_{zu}^{\mathrm{T}}C_z & D_{zu}^{\mathrm{T}}D_{zu} \end{pmatrix}.$$

As shown in [61] and as explained further in Appendix 6.A, whenever the pair (A, B) is stabilizable, there exists a unique semi-stabilizing solution P of the linear matrix inequality (8.10). Moreover, such a solution P is positive semi-definite, rank minimizing, and is the largest among all symmetric solutions.

Let us also define a dual version of the above linear matrix inequality. For any matrix $Q \in \mathbb{R}^{n\times n}$, let

$$G(Q) \geq 0, \tag{8.11}$$

where

$$G(Q) := \begin{pmatrix} AQ + QA^{\mathrm{T}} + E_d E_d^{\mathrm{T}} & QC_y^{\mathrm{T}} + E_d D_{yd}^{\mathrm{T}} \\ C_y Q + D_{yd} E_d^{\mathrm{T}} & D_{yd} D_{yd}^{\mathrm{T}} \end{pmatrix}.$$

Again, whenever the pair (C_y, A) is detectable, there exists a unique semi-stabilizing solution Q of the linear matrix inequality (8.11). Moreover, such a solution Q is positive semi-definite, rank minimizing, and is the largest among all symmetric solutions.

We will next recall a theorem from [61, 71] that gives an expression for the minimal achievable H_2 norm of $T_{d,z}(\Sigma \times \Sigma_{\mathrm{C}})$ for any arbitrary system, and in particular for the given system Σ.

Theorem 8.4.1 *Consider the system (8.1) with $w = 0$. Assume that (A, B) is stabilizable and (C_y, A) is detectable. Then, the infimum of the H_2 norm of $T_{d,z}(\Sigma \times \Sigma_{\mathrm{C}})$ over all proper or strictly proper controllers of the form (8.3) which internally stabilize Σ is equal to*

$$\gamma_{2p}^* = \gamma_{2sp}^* = \left(\operatorname{trace} E_d^T P E_d + \operatorname{trace}(A^T P + PA + C_z^T C_z) Q\right)^{1/2},$$

where P and Q are respectively the unique semi-stabilizing, and thus positive semi-definite and rank minimizing solutions of (8.10) and (8.11).

In order to study the possibility that $\gamma_{2p,r}^*$ equals γ_{2p}^* or $\gamma_{2sp,r}^*$ equals γ_{2sp}^*, we need to compare the achievable H_2 norms for the systems Σ and $\bar{\Sigma}$. We study this by relating the associated semi-stabilizing, rank-minimizing solutions of the linear matrix inequalities for the two systems. We have the following lemma.

Lemma 8.4.1 *Let P and Q be respectively the unique semi-stabilizing solutions of the linear matrix inequalities (8.10) and (8.11) for the system Σ. Similarly, let $\bar{P}$ and $\bar{Q}$ be respectively the unique semi-stabilizing solutions of the linear matrix inequalities corresponding to (8.10) and (8.11) defined for the auxiliary system $\bar{\Sigma}$. Then we have,*

$$\bar{P} = \begin{pmatrix} P & 0 \\ 0 & 0 \end{pmatrix}, \quad \textit{and} \quad \bar{Q} = \begin{pmatrix} Q & 0 \\ 0 & 0 \end{pmatrix}.$$

Conversely, let $\bar{P}$ and $\bar{Q}$ be respectively the unique semi-stabilizing solutions of the linear matrix inequalities corresponding to (8.10) and (8.11) defined for the auxiliary system $\bar{\Sigma}$. Decompose $\bar{P}$ and $\bar{Q}$ to be compatible with the decompositions in (8.6),

$$\bar{P} = \begin{pmatrix} P_{11} & P_{12} \\ P_{21} & P_{22} \end{pmatrix} \quad \textit{and} \quad \bar{Q} = \begin{pmatrix} Q_{11} & Q_{12} \\ Q_{21} & Q_{22} \end{pmatrix}.$$

Then P_{11} and Q_{11} are the unique semi-stabilizing solutions of the linear matrix inequalities (8.10) and (8.11) for the system Σ.

Proof : Given P and Q are respectively the unique semi-stabilizing solutions of the linear matrix inequalities (8.10) and (8.11) for the system Σ, it is easy to check that

$$\begin{pmatrix} P & 0 \\ 0 & 0 \end{pmatrix} \text{ and } \begin{pmatrix} Q & 0 \\ 0 & 0 \end{pmatrix}$$

are the semi-stabilizing solutions of the linear matrix inequalities corresponding to (8.10) and (8.11) defined for the auxiliary system $\bar{\Sigma}$. Then, the result follows right away since such solutions are unique.

Conversely, if $\bar{P}$ and $\bar{Q}$ are respectively the unique semi-stabilizing solutions of the linear matrix inequalities corresponding to (8.10) and (8.11) defined for the auxiliary system $\bar{\Sigma}$, then it is straightforward to check that P_{11} and Q_{11} are indeed respectively the unique semi-stabilizing solutions of the linear matrix inequalities (8.10) and (8.11) defined for Σ. ■

The above lemma enables us to present the main result of this section given in the next theorem.

Theorem 8.4.2 *Let the system Σ with realization (8.1) be given. Also, let Assumptions A.1, A.2, A.3, and A.4 hold. Then, we have $\gamma^*_{2p,r} = \gamma^*_{2p} = \gamma^*_{2sp,r} = \gamma^*_{2sp}$.*

Proof : By Theorem 8.4.1 we have,

$$\gamma_{2p}^* = \gamma_{2sp}^* = \left(\text{trace } E_d^{\mathrm{T}} P E_d + \text{trace}(A^{\mathrm{T}} P + PA + C_z^{\mathrm{T}} C_z) Q\right)^{1/2}.$$

On the other hand, if we restrict attention to the special class of controllers of the form (8.8), then we know that we achieve output regulation. Moreover, obviously we have $\gamma_{2p}^* \leqslant \gamma_{2p,r}^*$, and $\gamma_{2sp}^* \leqslant \gamma_{2sp,r}^*$, where

$$\gamma_{2p,r}^* = \gamma_{2sp,r}^* = \left(\text{trace}\begin{pmatrix} E_d^{\mathrm{T}} & 0 \end{pmatrix} \bar{P} \begin{pmatrix} E_d \\ 0 \end{pmatrix} + \text{trace}\left[\begin{pmatrix} A^{\mathrm{T}} & 0 \\ -\Gamma^{\mathrm{T}} B^{\mathrm{T}} & S^{\mathrm{T}} \end{pmatrix} \bar{P} \right.\right.$$
$$\left.\left. + \bar{P} \begin{pmatrix} A & -B\Gamma \\ 0 & S \end{pmatrix} + \begin{pmatrix} C_z^{\mathrm{T}} \\ -\Gamma^{\mathrm{T}} D_{zu}^{\mathrm{T}} \end{pmatrix} \begin{pmatrix} C_z & -D_{zu}\Gamma \end{pmatrix} \right] \bar{Q} \right)^{1/2}.$$

Using Lemma 8.4.1, it is then immediate that $\gamma_{2p,r}^* = \gamma_{2p}^* = \gamma_{2sp,r}^* = \gamma_{2sp}^*$. We note from [61] that the imposition of a constraint on a feedthrough matrix of a class of controllers for H_2 optimal control of continuous-time systems does not affect the infimum. ■

Remark 8.4.1 *It turns out that the achievable performance of the auxiliary system is independent of the specific solution of the regulator equation used in the construction of the auxiliary system. Hence even in the case of multiple solutions of the regulator equation, it is easy to find a suitable controller.*

Chapter 9

H_2 optimal control with an output regulation constraint – discrete-time systems

9.1 Introduction

This chapter is a discrete-time analog of Chapter 8. That is we formulate here an output regulation problem which seeks to achieve simultaneously the infimum (or arbitrarily close to the infimum) H_2 norm of a closed-loop transfer function. Such a problem can equivalently be viewed as an H_2 optimal (or suboptimal) control problem with the output regulation constraint. As we discussed in the previous chapter, although a suitable controller which solves the posed problem for the given system can be constructed via the construction of a controller that solves an H_2 optimal (or suboptimal) control problem without the output regulation constraint for a certain auxiliary system, one fundamental question still needs to be answered. Namely, whether the added output regulation constraint in a problem compromises the achievable performance. In this regard, we will show again, as in the previous chapter, that there is no loss at all in the achievable performance because of the added output regulation constraint whenever proper (or strictly proper) controllers are used. However, although the achievable performance is not compromised because of the added output regulation constraint, as well known in H_2 optimal control theory, the achievable performance for discrete-time systems is different over the class of proper controllers compared to that over the class of strictly proper controllers. This chapter is based on the recent research work of authors [77].

This chapter is organized as follows. Section 9.2 formulates H_2 optimal and suboptimal control problems with the output regulation constraint, while Section 9.3 presents solvability conditions for such problems and also develops methods of constructing regulators that solve such problems whenever they are solvable. Section 9.4 shows that the added output regulation constraint in the given H_2 optimal control problem does not compromise the achievable performance whenever optimization is done over the class of proper (or strictly proper) controllers.

9.2 Problem formulations

As in Chapter 8, we start with a linear system with state space realization,

$$\Sigma : \begin{cases} x(k+1) = Ax(k) + Bu(k) + E_w w(k) + E_d d(k) \\ e(k) = C_e x(k) + D_{eu} u(k) + D_{ew} w(k) \\ z(k) = C_z x(k) + D_{zu} u(k) \\ y(k) = C_y x(k) + D_{yw} w(k) + D_{yd} d(k), \end{cases} \tag{9.1}$$

where, as usual, $x \in \mathbb{R}^n$, $u \in \mathbb{R}^m$, $y \in \mathbb{R}^p$, $e \in \mathbb{R}^q$, and $z \in \mathbb{R}^\ell$. The exogenous disturbance input $w \in \mathbb{R}^s$ is generated by an exosystem Σ_E with state space realization,

$$\Sigma_E : w(k+1) = Sw(k). \tag{9.2}$$

As before, the variable d denotes an external disturbance. We seek measurement feedback controllers which are of the form,

$$\Sigma_C : \begin{cases} v(k+1) = A_c v(k) + B_c y(k), \\ u(k) = C_c v(k) + D_c y(k). \end{cases} \tag{9.3}$$

The closed-loop system consisting of the given system Σ and the controller Σ_C is denoted by $\Sigma \times \Sigma_C$. Also, the transfer matrix from d to z of $\Sigma \times \Sigma_C$ is denoted by $T_{d,z}(\Sigma \times \Sigma_C)$. As usual, we define,

$$\|T_{d,z}(\Sigma \times \Sigma_C)\|_2 := \left(\frac{1}{2\pi} \int_{-\pi}^{\pi} \text{trace}\, T_{d,z}(\Sigma \times \Sigma_C)^{\mathrm{T}}(e^{-i\omega}) T_{d,z}(\Sigma \times \Sigma_C)(e^{i\omega}) d\omega \right)^{1/2}.$$

Before we state formally the specific multi-objective problems of interest here, we define the following notations:

$$\begin{aligned}\gamma_{2p,r}^* &= \inf \{ \ \|T_{d,z}(\Sigma \times \Sigma_C)\|_2 \mid \Sigma_C \text{ is a proper controller that} \\ &\qquad\qquad \text{achieves output regulation for } \Sigma \ \}, \\ \gamma_{2p}^* &= \inf \{ \ \|T_{d,z}(\Sigma \times \Sigma_C)\|_2 \mid \Sigma_C \text{ is a proper controller that} \\ &\qquad\qquad \text{internally stabilizes } \Sigma \ \}, \\ \gamma_{2sp,r}^* &= \inf \{ \ \|T_{d,z}(\Sigma \times \Sigma_C)\|_2 \mid \Sigma_C \text{ is a strictly proper controller} \\ &\qquad\qquad \text{that achieves output regulation for } \Sigma \ \}, \\ \gamma_{2sp}^* &= \inf \{ \ \|T_{d,z}(\Sigma \times \Sigma_C)\|_2 \mid \Sigma_C \text{ is a strictly proper controller} \\ &\qquad\qquad \text{that internally stabilizes } \Sigma \ \}.\end{aligned}$$

We can now state formally the H_2 optimal control problem with the output regulation constraint.

Problem 9.2.1 ***(H_2 optimal control problem with the output regulation constraint)*** *Consider the system Σ and the exosystem Σ_E as given in (9.1) and (9.2). Find, if possible, a proper (or a strictly proper) controller Σ_C of the form (9.3) such that the following conditions hold:*

(i) ***(Internal Stability)*** *In the absence of the disturbances w and d, the closed-loop system $\Sigma \times \Sigma_C$ is internally stable.*

(ii) ***(Performance Measure)*** *$\|T_{d,z}(\Sigma \times \Sigma_C)\|_2$ is equal to the infimum $\gamma_{2p,r}^*$ (or $\gamma_{2sp,r}^*$).*

(iii) ***(Output Regulation)*** *For any $d \in \ell_2$, and for all $x(0) \in \mathbb{R}^n$ and $w(0) \in \mathbb{R}^s$, the solution of the closed-loop system $\Sigma \times \Sigma_C$ satisfies*

$$\lim_{k \to \infty} e(k) = 0.$$

Remark 9.2.1 *The part (iii) of the above problem formulation is equivalent to the following: In the absence of external signal d, i.e. when $d = 0$, Σ_C achieves output regulation for Σ.*

We can also define the H_2 suboptimal control problem with the output regulation constraint.

Problem 9.2.2 ***(H_2 suboptimal problem with the output regulation constraint)*** *Consider the system Σ and the exosystem Σ_E as given in (9.1) and (9.2).*

Find, if possible, a parameterized family $\{\Sigma_C(\varepsilon) \mid \varepsilon > 0\}$ of proper controllers of the form (9.3) (or a parameterized family $\{\Sigma_C(\varepsilon) \mid \varepsilon > 0\}$ of strictly proper controllers of the form (9.3) with $D_c = 0$) such that the following conditions hold:

(i) ***(Internal Stability)*** *In the absence of the disturbances w and d, the closed-loop system $\Sigma \times \Sigma_C(\varepsilon)$ is internally stable.*

(ii) ***(Performance Measure)*** *As $\varepsilon \to 0$, we have $\|T_{d,z}(\Sigma \times \Sigma_C(\varepsilon))\|_2$ tending to the infimum $\gamma^*_{2p,r}$ (or $\gamma^*_{2sp,r}$).*

(iii) ***(Output Regulation)*** *For any $d \in l_2$, and for all $x(0) \in \mathbb{R}^n$ and $w(0) \in \mathbb{R}^s$, the solution of the closed-loop system $\Sigma \times \Sigma_C(\varepsilon)$ satisfies*

$$\lim_{k\to\infty} e(k) = 0.$$

We note that when the output regulation constraint is removed from the above problem formulations, the resulting problems are simply *unconstrained H_2 optimal and suboptimal control problems.*

9.3 Solvability conditions

As we discussed in Chapter 7, the following assumptions are reasonable and almost necessary to solve the problems defined above:

A.1. (A, B) is stabilizable.

A.2. The matrix S is anti-Schur-stable, i.e. all its eigenvalues are on the unit circle or outside the unit circle.

A.3. $\left(\begin{pmatrix} C_y & D_{yw} \end{pmatrix}, \begin{pmatrix} A & E_w \\ 0 & S \end{pmatrix}\right)$ is detectable.

A.4. There exist Π and Γ solving the regulator equation,

$$\begin{aligned} \Pi S &= A\Pi + B\Gamma + E_w, \\ 0 &= C_e\Pi + D_{eu}\Gamma + D_{ew}. \end{aligned} \tag{9.4}$$

Note that these assumptions have been defined before, for instance on page 215.

In the case when the regulator equation (9.4) has a non-unique solution for (Π, Γ), for simplicity of presentation, we assume throughout the chapter that a solution $(\Pi,\ \Gamma)$ of (8.4) has been chosen, and all our development here builds on such a solution.

Following the results of Chapter 7, a solution to the H_2 *optimal or suboptimal control problem with output regulation constraint* for Σ is given in terms of a solution to the H_2 *optimal or suboptimal control problem without any output regulation constraint* for the auxiliary system $\bar{\Sigma}$ which is as in (7.5), and is repeated below for convenience:

$$\bar{\Sigma}:\begin{cases} \bar{x}(k+1) = \bar{A}\bar{x}(k) + \bar{B}\bar{u}(k) + \bar{E}_d\bar{d}(k) \\ \bar{z}(k) \quad = \bar{C}_z\bar{x}(k) + \bar{D}_{zu}\bar{u}(k) \\ \bar{y}(k) \quad = \bar{C}_y\bar{x}(k) \qquad\qquad + D_{yd}\bar{d}(k), \end{cases} \tag{9.5}$$

where

$$\bar{A} = \begin{pmatrix} A & -B\Gamma \\ 0 & S \end{pmatrix}, \quad \bar{B} = \begin{pmatrix} 0 & B \\ I & 0 \end{pmatrix}, \quad \bar{E}_d = \begin{pmatrix} E_d \\ 0 \end{pmatrix}, \tag{9.6a}$$

$$\bar{C}_z = \begin{pmatrix} C_z & -D_{zu}\Gamma \end{pmatrix}, \quad \bar{D}_{zu} = \begin{pmatrix} 0 & D_{zu} \end{pmatrix}, \tag{9.6b}$$

$$\bar{C}_y = \begin{pmatrix} C_y & (D_{yw} + C_y\Pi) \end{pmatrix}. \tag{9.6c}$$

For $\bar{\Sigma}$, a controller $\bar{\Sigma}_C$ with state space representation $(\bar{A}_c,\ \bar{B}_c,\ \bar{C}_c,\ \bar{D}_c)$ is given by

$$\bar{\Sigma}_C:\begin{cases} \bar{v}(k+1) = \bar{A}_c\bar{v}(k) + \bar{B}_c\bar{y}(k) \\ \bar{u}(k) \quad = \bar{C}_c\bar{v}(k) + \bar{D}_c\bar{y}(k). \end{cases} \tag{9.7}$$

For each controller $\bar{\Sigma}_C$ for the auxiliary system $\bar{\Sigma}$, we formulate a corresponding controller Σ_C for the given system Σ. It is given by

$$\Sigma_C:\begin{cases} \rho v_1 = \quad Sv_1 + \bar{C}_{c,1}v_2 + \bar{D}_{c,1}(y + (D_{yw} + C_y\Pi)v_1) \\ \rho v_2 = \qquad\qquad \bar{A}_c v_2 + \bar{B}_c(y + (D_{yw} + C_y\Pi)v_1) \\ u \quad = -\Gamma v_1 + \bar{C}_{c,2}v_2 + \bar{D}_{c,2}(y + (D_{yw} + C_y\Pi)v_1), \end{cases} \tag{9.8}$$

where $\bar{C}_{c,1}$, $\bar{C}_{c,2}$, $\bar{D}_{c,1}$, and $\bar{D}_{c,2}$ are obtained by partitioning $\bar{C}_c$ and $\bar{D}_c$ in conformity with the partitioning of $\bar{A}$,

$$\bar{C}_c = \begin{pmatrix} \bar{C}_{c,1} \\ \bar{C}_{c,2} \end{pmatrix}, \quad \text{and } \bar{D}_c = \begin{pmatrix} \bar{D}_{c,1} \\ \bar{D}_{c,2} \end{pmatrix}.$$

We note that there is a $1-1$ relationship between the proper controller $\bar{\Sigma}_C$ as given in (9.7) and the proper controller Σ_C as given in (9.8). Furthermore, we observe that with the constraint that $\bar{D}_{c,2} = 0$, the controller Σ_C as given in (9.8) is strictly proper. In other words, by imposing the constraint that

$\bar{D}_{c,2} = 0$ in $\bar{\Sigma}_C$, we can generate the class of strictly proper controllers Σ_C for the given system Σ. We denote below the class of controllers $\bar{\Sigma}_C$ with the constraint $\bar{D}_{c,2} = 0$ by $\bar{\Sigma}_C^s$,

$$\bar{\Sigma}_C^s : \begin{cases} \bar{v}(k+1) = & \bar{A}_c \bar{v}(k) + \quad \bar{B}_c \bar{y}(k) \\ \bar{u}(k) = & \begin{pmatrix} \bar{C}_{c,1} \\ \bar{C}_{c,2} \end{pmatrix} \bar{v}(k) + \begin{pmatrix} \bar{D}_{c,1} \\ 0 \end{pmatrix} \bar{y}(k). \end{cases} \tag{9.9}$$

The following theorems provide the conditions under which the H_2 optimal control problem with output regulation constraint can be solved via proper or via strictly proper controllers. Also, they provide a procedure of constructing an appropriate controller that solves the posed problem whenever it is solvable.

Theorem 9.3.1 *Let Assumptions A.1, A.2, A.3, and A.4 hold. Consider the given system Σ as in (9.1), and the exosystem Σ_E as in (9.2). Also, consider the auxiliary system $\bar{\Sigma}$ as in (9.5). Then, the following statements hold:*

> *The H_2 optimal control problem with output regulation constraint for Σ is solvable via a proper controller if and only if the H_2 optimal control problem for $\bar{\Sigma}$ is solvable via a proper controller. Moreover, a proper controller $\bar{\Sigma}_C$ of the form given in (9.7) is a proper H_2 optimal controller for $\bar{\Sigma}$ if and only if the corresponding proper controller Σ_C of the form given in (9.8) solves the H_2 optimal control problem with the output regulation constraint for Σ.*

Proof : The proof follows from Theorem 7.3.1. ■

Theorem 9.3.2 *Let Assumptions A.1, A.2, A.3, and A.4 hold. Consider the given system Σ as in (9.1), and the exosystem Σ_E as in (9.2). Also, consider the auxiliary system $\bar{\Sigma}$ as in (9.5). Then, the following statements hold:*

> *The H_2 optimal control problem with output regulation constraint for Σ is solvable via a strictly proper controller if and only if the H_2 optimal control problem for $\bar{\Sigma}$ is solvable via a strictly proper controller. Moreover, a controller $\bar{\Sigma}_C$ of the form given in (9.7) is an H_2 optimal controller for $\bar{\Sigma}$ if and only if the corresponding strictly proper controller Σ_C of the form given in (9.8) solves the H_2 optimal control problem with the output regulation constraint for Σ.*

Remark 9.3.1 *Note that by restricting controllers of the auxiliary system to be strictly proper instead of the more general form $\bar{\Sigma}_{\mathrm{C}}^{s}$, we do not lose performance but we might lose some flexibility in the sense that we do not capture all optimal controllers.*

Proof : From Theorem 7.3.1, it is clear that the H_2 optimal control problem with output regulation constraint for Σ is solvable via a strictly proper controller if and only if the H_2 optimal control problem for $\bar{\Sigma}$ is solvable over the class of controllers of the form $\bar{\Sigma}_{\mathrm{C}}^{s}$ as given in (9.9). Next, as will be evident from the proof of Theorem 9.4.3 of Subsection 9.4.2, whenever the H_2 optimal control problem for $\bar{\Sigma}$ is solvable over the class of controllers of the form $\bar{\Sigma}_{\mathrm{C}}^{s}$ as given in (9.9), there exists an H_2 optimal controller with a direct feedthrough term equal to zero, i.e. $\bar{D}_{c,1} = 0$. Hence the result follows. ■

We can now state the following theorems which provide the conditions under which the H_2 suboptimal control problem with output regulation constraint can be solved. Also, they provide a procedure of constructing an appropriate sequence of controllers that solves the posed problem.

Theorem 9.3.3 *Let Assumptions A.1, A.2, A.3, and A.4 hold. Consider the given system Σ as in (9.1), and the exosystem Σ_{E} as in (9.2). Also, consider the auxiliary system $\bar{\Sigma}$ as in (9.5). Then, the following statements hold:*

> *The H_2 suboptimal control problem with output regulation constraint for Σ is solvable via a family $\{\bar{\Sigma}_{\mathrm{C}}(\varepsilon) \mid \varepsilon > 0\}$ of proper (or strictly proper) controllers.*
>
> *Moreover, a family $\{\bar{\Sigma}_{\mathrm{C}}(\varepsilon) \mid \varepsilon > 0\}$ of proper controllers of the form (9.7) is H_2 suboptimal for $\bar{\Sigma}$ if and only if the corresponding family $\{\Sigma_{\mathrm{C}}(\varepsilon) \mid \varepsilon > 0\}$ of proper controllers of the form (9.8) solves the H_2 suboptimal control problem with the output regulation constraint for Σ.*
>
> *Similarly, a family $\{\bar{\Sigma}_{\mathrm{C}}^{s}(\varepsilon) \mid \varepsilon > 0\}$ of controllers of the form (9.9) is H_2 suboptimal for $\bar{\Sigma}$ if and only if the corresponding family $\{\Sigma_{\mathrm{C}}(\varepsilon) \mid \varepsilon > 0\}$ of strictly proper controllers of the form (9.8) solves the H_2 suboptimal control problem with the output regulation constraint for Σ.*

Proof : In view of Theorem 7.3.1 and the proof of Theorem 9.3.2, the proof of this theorem is obvious when one takes into account the fact that the H_2

suboptimal control problem for $\bar{\Sigma}$ is solvable under the Assumptions A.1 and A.3. ■

Design of a measurement feedback regulator that solves the H_2 optimal control problem with output regulation constraint:

Theorems 9.3.1 and 9.3.2 suggest the following two step procedure:

Step 1: Construct a proper or strictly proper controller $\bar{\Sigma}_C$ of the form (9.7) which solves the H_2 optimal control problem for $\bar{\Sigma}$.

Step 2: Knowing the parameters $\bar{A}_c$, $\bar{B}_c$, $\bar{C}_c$, and $\bar{D}_c$ of the controller $\bar{\Sigma}_C$ obtained in Step 1, construct a corresponding controller Σ_C as given in (9.8).

Clearly Σ_C constructed in Step 2 solves the H_2 optimal control problem with output regulation constraint for Σ via a proper (or a strictly proper) controller. □

The above procedure can be modified in an obvious way to obtain a sequence of proper (or strictly proper) controllers that solves the H_2 suboptimal control problem with output regulation constraint for Σ.

Design of a sequence of measurement feedback regulators that solves the H_2 suboptimal control problem with output regulation constraint:

Theorem 9.3.3 suggests the following two step procedure.

Step 1: Construct a family $\{\bar{\Sigma}_C(\varepsilon) \mid \varepsilon > 0\}$ (or $\{\bar{\Sigma}_C(\varepsilon) \mid \bar{D}_c(\varepsilon) = 0,\ \varepsilon > 0\}$) of measurement feedback regulators that solves the H_2 suboptimal control problem for $\bar{\Sigma}$.

Step 2: Knowing the parameterized matrix quadruples $\{(\bar{A}_c(\varepsilon), \bar{B}_c(\varepsilon), \bar{C}_c(\varepsilon), \bar{D}_c(\varepsilon)) \mid \varepsilon > 0\}$ that characterize $\{\bar{\Sigma}_C(\varepsilon) \mid \varepsilon > 0\}$ in Step 1, construct a corresponding family $\{\Sigma_C(\varepsilon) \mid \varepsilon > 0\}$ of proper (or strictly proper) measurement feedback regulators of the form (9.8).

Clearly $\{\Sigma_C(\varepsilon) \mid \varepsilon > 0\}$ constructed in Step 2 solves the H_2 suboptimal control problem with output regulation constraint for Σ via a sequence of proper (or strictly proper) controllers. □

9.4 Relationship between $\gamma^*_{2p,r}$, $\gamma^*_{2sp,r}$, γ^*_{2p}, and γ^*_{2sp}

Our primary objective in this section is to study how the achievable performance is affected by having the output regulation constraint. As in Chapter 8, it turns out that there is no loss at all in the achievable performance if optimization is done over the class of proper (or strictly proper) controllers, i.e. $\gamma^*_{2p,r} = \gamma^*_{2p}$, and $\gamma^*_{2sp,r} = \gamma^*_{2sp}$.

To proceed with, we need to recall first some results from [61, 86]. Consider the system Σ described by (9.1), and for any matrix $P \in \mathbb{R}^{n\times n}$, define a discrete-time linear matrix inequality as

$$F(P) \geq 0, \tag{9.10}$$

where

$$F(P) := \begin{pmatrix} A^{\mathrm{T}}PA - P + C_z^{\mathrm{T}}C_z & C_z^{\mathrm{T}}D_{zu} + A^{\mathrm{T}}PB \\ D_{zu}^{\mathrm{T}}C_z + B^{\mathrm{T}}PA & D_{zu}^{\mathrm{T}}D_{zu} + B^{\mathrm{T}}PB \end{pmatrix}.$$

As shown in [61] and as explained further in Appendix 6.B, whenever the pair (A, B) is stabilizable, there exists a unique semi-stabilizing solution P of the linear matrix inequality (9.10). Moreover, such a solution P is positive semi-definite, strongly rank minimizing, and is the largest among all strongly rank minimizing solutions.

Let us also define a dual version of the above linear matrix inequality. For any matrix $Q \in \mathbb{R}^{n\times n}$, let

$$G(Q) \geq 0, \tag{9.11}$$

where

$$G(Q) := \begin{pmatrix} AQA^{\mathrm{T}} - Q + E_dE_d^{\mathrm{T}} & E_dD_{yd}^{\mathrm{T}} + AQC_y^{\mathrm{T}} \\ D_{yd}E_d^{\mathrm{T}} + C_yQA^{\mathrm{T}} & D_{yd}D_{yd}^{\mathrm{T}} + C_yQC_y^{\mathrm{T}} \end{pmatrix}.$$

Again, whenever the pair (C_y, A) is detectable, there exists a unique semi-stabilizing solution Q of the linear matrix inequality (9.11). Moreover, such a solution Q is positive semi-definite, strongly rank minimizing, and is the largest among all strongly rank minimizing solutions.

We will next recall a theorem from [61, 86] that gives an expression for the minimal achievable H_2 norm of $T_{d,z}(\Sigma \times \Sigma_{\mathrm{C}})$ for any arbitrary system, and in particular for the given system Σ.

Theorem 9.4.1 *Consider the system (9.1) with $w = 0$. Assume that (A, B) is stabilizable and (C_y, A) is detectable. Then, the infimum of the H_2 norm of $T_{d,z}(\Sigma \times \Sigma_{\mathrm{C}})$ over all the proper controllers of the form (9.3) which internally stabilize Σ is equal to*

$$\gamma_{2p}^* = \Big(\mathrm{trace}\, E_d^T P E_d + \mathrm{trace}(A^T PA - P + C_z^T C_z)Q \\ - \mathrm{trace}\, R^*(R^*)^T\Big)^{1/2},$$

where P and Q are respectively the unique semi-stabilizing, and thus positive semi-definite and strongly rank minimizing, solutions of the linear matrix inequalities (9.10) and (9.11), and R^* is defined by

$$R^* := [(D_{zu}^T D_{zu} + B^T P B)^{1/2}]^\dagger [(D_{zu}^T C_z + B^T P A) Q C_y^T + B^T P E_d D_{yd}^T] \times [(D_{yd} D_{yd}^T + C_y Q C_y^T)^{1/2}]^\dagger. \tag{9.12}$$

Here for a given matrix M, we denote by $M^\dagger$ its Moore-Penrose inverse.

Also, the infimum of the H_2 norm of $T_{d,z}(\Sigma \times \Sigma_C)$ over all the strictly proper controllers of the form (9.3) with $D_c = 0$ which internally stabilize Σ is equal to

$$\gamma_{2sp}^* = \left(\text{trace } E_d^T P E_d + \text{trace}(A^T P A - P + C_z^T C_z) Q\right)^{1/2},$$

where again P and Q are respectively the unique semi-stabilizing, and thus positive semi-definite and strongly rank minimizing, solutions of the linear matrix inequalities (9.10) and (9.11).

In order to study the relationship between $\gamma_{2p,r}^*$ and γ_{2p}^*, as well as the relationship between $\gamma_{2sp,r}^*$ and γ_{2sp}^*, we need to relate the associated semi-stabilizing, strongly rank-minimizing solutions of the linear matrix inequalities for the two systems Σ and $\bar{\Sigma}$. We have the following lemma.

Lemma 9.4.1 *Let P and Q be respectively the unique semi-stabilizing and strongly rank minimizing solutions of the linear matrix inequalities (9.10) and (9.11) for the system Σ. Similarly, let $\bar{P}$ and $\bar{Q}$ be respectively the unique semi-stabilizing and strongly rank minimizing solutions of the linear matrix inequalities corresponding to (9.10) and (9.11) defined for the auxiliary system $\bar{\Sigma}$. Then we have,*

$$\bar{P} = \begin{pmatrix} P & 0 \\ 0 & 0 \end{pmatrix}, \quad \text{and} \quad \bar{Q} = \begin{pmatrix} Q & 0 \\ 0 & 0 \end{pmatrix}.$$

Conversely, let $\bar{P}$ and $\bar{Q}$ be the unique semi-stabilizing and strongly rank minimizing solutions of the linear matrix inequalities corresponding to (9.10) and (9.11) respectively defined for the auxiliary system $\bar{\Sigma}$. Decompose $\bar{P}$ and $\bar{Q}$ to be compatible with the decompositions in (9.6),

$$\bar{P} = \begin{pmatrix} P_{11} & P_{12} \\ P_{21} & P_{22} \end{pmatrix} \quad \text{and} \quad \bar{Q} = \begin{pmatrix} Q_{11} & Q_{12} \\ Q_{21} & Q_{22} \end{pmatrix}.$$

Then P_{11} and Q_{11} are respectively the unique semi-stabilizing and strongly rank minimizing solutions of the linear matrix inequalities (9.10) and (9.11) for the system Σ.

Proof : Let P and Q be the unique semi-stabilizing and strongly rank minimizing solutions of the linear matrix inequalities (9.10) and (9.11) respectively for the system Σ. Then it is easy to check that

$$\begin{pmatrix} P & 0 \\ 0 & 0 \end{pmatrix} \text{ and } \begin{pmatrix} Q & 0 \\ 0 & 0 \end{pmatrix}$$

are the semi-stabilizing and strongly rank minimizing solutions of the linear matrix inequalities corresponding to (9.10) and (9.11) defined for the auxiliary system $\bar{\Sigma}$. Then, the result follows right away since such solutions are unique.

Conversely, if $\bar{P}$ and $\bar{Q}$ are respectively the unique semi-stabilizing and strongly rank minimizing solutions of the linear matrix inequalities corresponding to (9.10) and (9.11) defined for the auxiliary system $\bar{\Sigma}$, then it is straightforward to check that $\bar{P}_{11}$ and $\bar{Q}_{11}$ are indeed respectively the unique semi-stabilizing and strongly rank minimizing solutions of the linear matrix inequalities (9.10) and (9.11) defined for Σ. ■

9.4.1 Relationship between $\gamma^*_{2p,r}$ and γ^*_{2p}

In this subsection, we study $\gamma^*_{2p,r}$ and γ^*_{2p}, and show that $\gamma^*_{2p,r} = \gamma^*_{2p}$.

Theorem 9.4.2 *Let the system Σ with realization (9.1) be given. Also, let Assumptions A.1, A.2, A.3, and A.4 hold. Then, we have $\gamma^*_{2p,r} = \gamma^*_{2p}$.*

Proof : By Theorem 9.4.1 we have,

$$\gamma^*_{2p} = \left(\text{trace } E_d^{\mathrm{T}} P E_d + \text{trace}(A^{\mathrm{T}} P A - P + C_z^{\mathrm{T}} C_z) Q - \text{trace } R^*(R^*)^{\mathrm{T}}\right)^{1/2}.$$

On the other hand, if we restrict attention to the special class of controllers of the form (9.8), then we know that we achieve output regulation. Moreover, obviously we have $\gamma^*_{2p} \leqslant \gamma^*_{2p,r}$, where

$$\gamma^*_{2p,r} = \left(\text{trace}\begin{pmatrix} E_d^{\mathrm{T}} & 0 \end{pmatrix} \bar{P} \begin{pmatrix} E_d \\ 0 \end{pmatrix} + \text{trace}\left[\begin{pmatrix} A^{\mathrm{T}} & 0 \\ -\Gamma^{\mathrm{T}} B^{\mathrm{T}} & S^{\mathrm{T}} \end{pmatrix} \bar{P} \begin{pmatrix} A & -B\Gamma \\ 0 & S \end{pmatrix} \right.\right.$$
$$\left.\left. - \bar{P} + \begin{pmatrix} C_z^{\mathrm{T}} \\ -\Gamma^{\mathrm{T}} D_{zu}^{\mathrm{T}} \end{pmatrix} \begin{pmatrix} C_z & -D_{zu}\Gamma \end{pmatrix}\right] \bar{Q} - \text{trace } \bar{R}^*(\bar{R}^*)^{\mathrm{T}}\right)^{1/2},$$

where $\bar{R}^*$ is equal to R^* as defined in (9.12) for the system $\bar{\Sigma}$. Using Lemma 9.4.1, it is then immediate that $\gamma^*_{2p} = \gamma^*_{2p,r}$. ■

Remark 9.4.1 *It turns out that the achievable performance of the auxiliary system is independent of the specific solution of the regulator equation used in the construction of the auxiliary system. Hence even in the case of multiple solutions of the regulator equation, it is easy to find a suitable proper controller.*

9.4.2 Relationship between $\gamma_{2sp,r}^*$ and γ_{2sp}^*

In this subsection, we study $\gamma_{2sp,r}^*$ and γ_{2sp}^*, and as in the previous subsection show that they are equal to one another.

Theorem 9.4.3 *Let the system Σ with realization (9.1) be given. Also, let Assumptions A.1, A.2, A.3, and A.4 hold. Then, we have $\gamma_{2sp,r}^* = \gamma_{2sp}^*$.*

Proof : We first explicitly calculate γ_{2sp}^* and $\gamma_{2sp,r}^*$. In fact, by Theorem 9.4.1 we have,

$$\gamma_{2sp}^* = \left(\text{trace } E_d^{\mathrm{T}} P E_d + \text{trace}(A^{\mathrm{T}} P A - P + C_z^{\mathrm{T}} C_z) Q\right)^{1/2}.$$

On the other hand, if we restrict attention to the special class of strictly proper controllers which are of the form (9.8) with the constraint $\bar{D}_{c,2} = 0$, then we know that we achieve output regulation. Moreover, utilizing Theorem 9.3.2 and using Lemma 6.5.5 of [61], we can calculate $\gamma_{2sp,r}^*$. To do so, let us define

$$\Phi(\bar{D}_c) = 2\,\text{trace}[D_{yd}^{\mathrm{T}} \bar{D}_c^{\mathrm{T}} \bar{B}^{\mathrm{T}} \bar{P} \bar{E}_d] + 2\,\text{trace}[\bar{D}_c \bar{C}_y \bar{Q} (\bar{C}_{zu} \bar{D}_{zu} + \bar{A}^{\mathrm{T}} \bar{P} \bar{B})] \\ + \text{trace}[(\bar{D}_P \bar{D}_c \bar{D}_Q)(\bar{D}_P \bar{D}_c \bar{D}_Q)^{\mathrm{T}}],$$

where

$$\bar{D}_P = (\bar{D}_{zu}^{\mathrm{T}} \bar{D}_{zu} + \bar{B}^{\mathrm{T}} \bar{P} \bar{B})^{1/2}, \quad \bar{D}_Q = (D_{yd} D_{yd}^{\mathrm{T}} + \bar{C}_y \bar{Q} \bar{C}_y^{\mathrm{T}})^{1/2}.$$

Also, let

$$\Phi^* := \min\left\{ \Phi(\bar{D}_c) \mid \bar{D}_c = \begin{pmatrix} \bar{D}_{c,1} \\ 0 \end{pmatrix} \in \mathbb{R}^{(m+n)\times p} \right\}.$$

It is easy to see that $\bar{D}_{c,1} = 0$ is a solution to the above minimization, and this leads to $\Phi^* = 0$. Then, from Lemma 6.5.5 of [61], we have

$$\gamma_{2sp,r}^* = \left(\text{trace} \begin{pmatrix} E_d \\ 0 \end{pmatrix}^{\mathrm{T}} \bar{P} \begin{pmatrix} E_d \\ 0 \end{pmatrix} + \text{trace}\left[\begin{pmatrix} A^{\mathrm{T}} & 0 \\ -\Gamma^{\mathrm{T}} B^{\mathrm{T}} & S^{\mathrm{T}} \end{pmatrix} \bar{P} \begin{pmatrix} A & -B\Gamma \\ 0 & S \end{pmatrix} \right.\right. \\ \left.\left. - \bar{P} + \begin{pmatrix} C_z^{\mathrm{T}} \\ -\Gamma^{\mathrm{T}} D_{zu}^{\mathrm{T}} \end{pmatrix} \begin{pmatrix} C_z & -D_{zu}\Gamma \end{pmatrix} \right] \bar{Q} \right)^{1/2}.$$

Using Lemma 9.4.1, it is then immediate that $\gamma^*_{2sp,r} = \gamma^*_{2sp}$. ■

Remark 9.4.2 *A remark similar to Remark 9.4.1 can be made. That is, the achievable performance of the auxiliary system is independent of the specific solution of the regulator equation used in the construction of the auxiliary system. Hence even in the case of multiple solutions of the regulator equation, it is easy to find a suitable strictly proper controller.*

Chapter 10

H_∞ optimal control with an output regulation constraint – continuous-time systems

10.1 Introduction

For continuous-time systems, we consider here an output regulation problem which seeks to achieve simultaneously a desired H_∞ norm of a closed-loop transfer function. That is, we seek here a controller that achieves output regulation while resulting in at most a specified value for the H_∞ norm of a transfer function from a certain input signal to a desired output signal of the system. Such a multi-objective problem clearly takes into account certain performance requirements. This is obvious when we recognize that the H_∞ norm can identify robustness to unstructured plant uncertainties.

The above posed problem can be viewed as an H_∞ optimal control problem with an added output regulation constraint. As discussed in the previous chapters, utilizing the results of Chapter 7, such a problem for the given system is solved by transforming it to an H_∞ optimal control problem without any output regulation constraint for a certain auxiliary system. As in Chapters 8 and 9, it can be seen that for the originally given system even if one starts with an H_∞ optimal control problem (ignoring the output regulation constraint) that satisfies the regularity conditions, the corresponding H_∞ optimal control problem for the auxiliary system inherently *does not* satisfy the regularity conditions, i.e. in general it turns out to be a singular H_∞ optimal control problem. We can then obtain a solution to such a singular H_∞ optimal control problem for the auxiliary system by using the recently developed methods discussed in [73].

Also, as in Chapters 8 and 9, one fundamental issue arises. Does the added output regulation constraint compromise the achievable performance? In this regard, we will show that in the present case where we measure performance by the H_∞ norm of a transfer function matrix, there is a certain loss or decay in the achievable performance due to the added output regulation constraint. We will present a very explicit expression for such a decay in performance. This decay will be expressed in terms of a static optimization problem.

The H_∞ optimal control problem with an added output regulation constraint has been studied earlier in [1,92]. The methods pursued there to solve the posed problem restrict attention only to certain special cases, and a general problem is not considered. To avoid the unnecessary restrictions and to solve a general problem, our objective as in previous chapters is to present an elegant and to the point derivation based on combining results from a number of recent papers.

The mathematical aspects of our development depends on whether we consider continuous- or discrete-time systems. As such, in this chapter, we focus on continuous-time systems while the next chapter considers discrete-time systems. This chapter is based on the recent research work of authors [77].

This chapter is organized as follows. Section 10.2 formulates the H_∞ γ-suboptimal control problem with the output regulation constraint, while Section 10.3 presents solvability conditions for such a problem and also develops methods of constructing a regulator that solves such a problem whenever it is solvable. Section 10.4 shows that the added output regulation constraint in the given H_∞ γ-suboptimal control problem indeed compromises the achievable performance, and an explicit expression is derived showing such a compromise.

10.2 Problem formulation

As in previous chapters, we start with a linear system with state space realization,

$$\Sigma : \begin{cases} \dot{x} = Ax + Bu + E_w w + E_d d \\ e = C_e x + D_{eu} u + D_{ew} w \\ z = C_z x + D_{zu} u \\ y = C_y x + D_{yw} w + D_{yd} d, \end{cases} \tag{10.1}$$

where, as usual, $x \in \mathbb{R}^n$, $u \in \mathbb{R}^m$, $y \in \mathbb{R}^p$, $e \in \mathbb{R}^q$, and $z \in \mathbb{R}^\ell$. The exogenous disturbance input $w \in \mathbb{R}^s$ is generated by an exosystem Σ_E with state space realization,

$$\Sigma_E \; : \; \dot{w} = Sw. \tag{10.2}$$

As before, the variable d denotes an external disturbance. We seek measurement feedback controllers which are of the form,

$$\Sigma_C : \begin{cases} \dot{v} = A_c v + B_c y, \\ u = C_c v + D_c y. \end{cases} \tag{10.3}$$

The closed-loop system consisting of the given system Σ and the controller Σ_C is denoted by $\Sigma \times \Sigma_C$. Also, the transfer matrix from d to z of $\Sigma \times \Sigma_C$ is denoted by $T_{d,z}(\Sigma \times \Sigma_C)$. As usual, we define,

$$\|T_{d,z}(\Sigma \times \Sigma_C)\|_\infty := \sup_{\omega \in [-\infty,\infty)} \sigma_{\max}\left[T_{d,z}(\Sigma \times \Sigma_C)(i\omega)\right],$$

where $\sigma_{\max}$ denotes the largest singular value.

Before we state formally the specific multi-objective problems of interest here, we define the following notations:

$$\begin{aligned}
\gamma^*_{\infty\, p,r} &= \inf \{ \; \|T_{d,z}(\Sigma \times \Sigma_C)\|_\infty \mid \Sigma_C \text{ is a proper controller that} \\
&\qquad\qquad \text{achieves output regulation for } \Sigma \; \}, \\
\gamma^*_{\infty\, p} &= \inf \{ \; \|T_{d,z}(\Sigma \times \Sigma_C)\|_\infty \mid \Sigma_C \text{ is a proper controller that} \\
&\qquad\qquad \text{internally stabilizes } \Sigma \; \}, \\
\gamma^*_{\infty\, sp,r} &= \inf \{ \; \|T_{d,z}(\Sigma \times \Sigma_C)\|_\infty \mid \Sigma_C \text{ is a strictly proper controller} \\
&\qquad\qquad \text{that achieves output regulation for } \Sigma \; \}, \\
\gamma^*_{\infty\, sp} &= \inf \{ \; \|T_{d,z}(\Sigma \times \Sigma_C)\|_\infty \mid \Sigma_C \text{ is a strictly proper controller} \\
&\qquad\qquad \text{that internally stabilizes } \Sigma \; \}.
\end{aligned}$$

Note that it is known that for a system with a direct feedthrough matrix from d to z with norm less than γ (in our case it is even zero) then we have in continuous-time that $\gamma^*_{\infty\, sp} = \gamma^*_{\infty\, p}$. If the norm of this direct feedthrough matrix is not less than γ than we must have a preliminary static output injection which makes this direct feedthrough matrix less than γ. In the resulting system we can make the closed loop H_∞ norm strictly less than γ by strictly proper controllers if and only if we can achieve this objective via proper controllers. For details we refer to [70].

We can now state the specific multi-objective problem of interest here, namely the *H_∞ γ-suboptimal control problem with output regulation constraint.*

Problem 10.2.1 ***(H_∞ γ-suboptimal control problem with the output regulation constraint)*** *Consider the system Σ and the exosystem Σ_E as given in (10.1) and (10.2). Find, if possible, a controller Σ_C of the form (10.3) such that the following conditions hold:*

(i) ***(Internal Stability)*** *In the absence of the disturbances w and d, the closed-loop system $\Sigma \times \Sigma_C$ is internally stable.*

(ii) ***(Performance Measure)*** *$\|T_{d,z}(\Sigma \times \Sigma_C)\|_\infty$ is strictly less than a specified value γ, i.e. an H_∞ γ-suboptimal performance is obtained.*

(iii) ***(Output Regulation)*** *For any $d \in L_2$, and for all $x(0) \in \mathbb{R}^n$ and $w(0) \in \mathbb{R}^s$, the solution of the closed-loop system $\Sigma \times \Sigma_C$ satisfies*

$$\lim_{t\to\infty} e(t) = 0.$$

Remark 10.2.1 *The part (iii) of the above problem formulation is equivalent to the following: In the absence of external signal d, i.e. when $d = 0$, Σ_C achieves output regulation for Σ.*

We note that when the output regulation constraint is removed from the above problem formulation, the resulting problem is simply an *unconstrained H_∞ γ-suboptimal control problem.*

10.3 Solvability conditions

As we discussed in Chapter 7, the following assumptions are reasonable and almost necessary to solve the problem defined above:

A.1. (A, B) is stabilizable.

A.2. The matrix S is anti-Hurwitz-stable, i.e. it has all its eigenvalues in the closed right-half plane.

A.3. $\left(\begin{pmatrix} C_y & D_{yw} \end{pmatrix}, \begin{pmatrix} A & E_w \\ 0 & S \end{pmatrix} \right)$ is detectable.

A.4. There exist Π and Γ solving the regulator equation,

$$\begin{cases} \Pi S = A\Pi + B\Gamma + E_w, \\ 0 = C_e\Pi + D_{eu}\Gamma + D_{ew}. \end{cases} \tag{10.4}$$

Note that these assumptions have been defined before, for instance on page 215.

In the case when the regulator equation (10.4) has a non-unique solution for (Π, Γ), for simplicity of presentation, we assume throughout the chapter that a solution (Π, Γ) of (10.4) has been chosen, and all our development here builds on such a solution.

Following the results of Chapter 7, a solution to the H_∞ *γ-suboptimal control problem with output regulation constraint* for Σ is given in terms of a solution to the H_∞ *γ-suboptimal control problem without any output regulation constraint* for the auxiliary system $\bar{\Sigma}$ which is as in (7.5), and is repeated below for convenience:

$$\bar{\Sigma} : \begin{cases} \dot{\bar{x}} = \bar{A}\bar{x} + \bar{B}\bar{u} + \bar{E}_d\bar{d} \\ \bar{z} = \bar{C}_z\bar{x} + \bar{D}_{zu}\bar{u} \\ \bar{y} = \bar{C}_y\bar{x} + D_{yd}\bar{d}, \end{cases} \tag{10.5}$$

where

$$\bar{A} = \begin{pmatrix} A & -B\Gamma \\ 0 & S \end{pmatrix}, \quad \bar{B} = \begin{pmatrix} 0 & B \\ I & 0 \end{pmatrix}, \quad \bar{E}_d = \begin{pmatrix} E_d \\ 0 \end{pmatrix}, \tag{10.6a}$$

$$\bar{C}_z = \begin{pmatrix} C_z & -D_{zu}\Gamma \end{pmatrix}, \quad \bar{D}_{zu} = \begin{pmatrix} 0 & D_{zu} \end{pmatrix}, \tag{10.6b}$$

$$\bar{C}_y = \begin{pmatrix} C_y & (D_{yw} + C_y\Pi) \end{pmatrix}. \tag{10.6c}$$

For $\bar{\Sigma}$, a controller $\bar{\Sigma}_C$ with state space representation $(\bar{A}_c, \bar{B}_c, \bar{C}_c, \bar{D}_c)$ is given by

$$\bar{\Sigma}_C : \begin{cases} \dot{\bar{v}} = \bar{A}_c\bar{v} + \bar{B}_c\bar{y} \\ \bar{u} = \bar{C}_c\bar{v} + \bar{D}_c\bar{y}. \end{cases} \tag{10.7}$$

For each controller $\bar{\Sigma}_C$ for the auxiliary system $\bar{\Sigma}$, we formulate a corresponding controller Σ_C for the given system Σ. It is given by

$$\Sigma_C : \begin{cases} \dot{v}_1 = Sv_1 + \bar{C}_{c,1}v_2 + \bar{D}_{c,1}(y + (D_{yw} + C_y\Pi)v_1) \\ \dot{v}_2 = \bar{A}_c v_2 + \bar{B}_c(y + (D_{yw} + C_y\Pi)v_1) \\ u = -\Gamma v_1 + \bar{C}_{c2}v_2 + \bar{D}_{c,2}(y + (D_{yw} + C_y\Pi)v_1), \end{cases} \tag{10.8}$$

where $\bar{C}_{c,1}$, $\bar{C}_{c,2}$, $\bar{D}_{c,1}$, and $\bar{D}_{c,2}$ are obtained by partitioning $\bar{C}_c$ and $\bar{D}_c$ in conformity with the partitioning of $\bar{A}$,

$$\bar{C}_c = \begin{pmatrix} \bar{C}_{c,1} \\ \bar{C}_{c,2} \end{pmatrix}, \quad \text{and} \quad \bar{D}_c = \begin{pmatrix} \bar{D}_{c,1} \\ \bar{D}_{c,2} \end{pmatrix}.$$

We note that there is a $1-1$ relationship between the proper controller $\bar{\Sigma}_C$ as given in (10.7) and the proper controller Σ_C as given in (10.8). Furthermore, we observe that with the constraint that $\bar{D}_{c,2} = 0$, the controller Σ_C as given in (10.8) is strictly proper. In other words, by imposing the constraint that $\bar{D}_{c,2} = 0$ in $\bar{\Sigma}_C$, we can generate the class of strictly proper controllers Σ_C for the given system Σ. We denote below the class of controllers $\bar{\Sigma}_C$ with the constraint $\bar{D}_{c,2} = 0$ by $\bar{\Sigma}_C^s$,

$$\bar{\Sigma}_C^s : \begin{cases} \dot{\bar{v}} = \bar{A}_c \bar{v} + \bar{B}_c \bar{y} \\ \bar{u} = \begin{pmatrix} \bar{C}_{c,1} \\ \bar{C}_{c,2} \end{pmatrix} \bar{v} + \begin{pmatrix} \bar{D}_{c,1} \\ 0 \end{pmatrix} \bar{y}. \end{cases} \tag{10.9}$$

The following theorems provide the conditions under which the H_∞ γ-suboptimal control problem with output regulation constraint can be solved via proper or via strictly proper controllers. Also, they provide a procedure of constructing an appropriate controller that solves the posed problem whenever it is solvable.

Theorem 10.3.1 *Let Assumptions A.1, A.2, A.3, and A.4 hold. Consider the given system Σ as in (10.1), and the exosystem Σ_E as in (10.2). Also, consider the auxiliary system $\bar{\Sigma}$ as in (10.5). Then, the following statements hold:*

> *The H_∞ γ-suboptimal control problem with output regulation constraint for Σ is solvable via a proper controller if and only if the H_∞ γ-suboptimal control problem for $\bar{\Sigma}$ is solvable via a proper controller.*
>
> *Moreover, a proper controller $\bar{\Sigma}_C$ of the form given in (10.7) is a proper H_∞ γ-suboptimal controller for $\bar{\Sigma}$ if and only if the corresponding proper controller Σ_C of the form given in (10.8) solves the H_∞ γ-suboptimal control problem with the output regulation constraint for Σ.*

Proof : The proof follows from Theorem 7.3.1. ∎

Theorem 10.3.2 *Let Assumptions A.1, A.2, A.3, and A.4 hold. Consider the given system Σ as in (10.1), and the exosystem Σ_E as in (10.2). Also, consider the auxiliary system $\bar{\Sigma}$ as in (10.5). Then, the following statements hold:*

> *The H_∞ γ-suboptimal control problem with output regulation constraint for Σ is solvable via a strictly proper controller if and only if the H_∞ γ-suboptimal control problem for $\bar{\Sigma}$ is solvable via a strictly proper controller.*
>
> *Moreover, a strictly proper controller is an H_∞ γ-suboptimal controller for $\bar{\Sigma}$ if and only if the corresponding strictly proper controller Σ_C of the form given in (10.8) solves the H_∞ γ-suboptimal control problem with the output regulation constraint for Σ.*

Remark 10.3.1 *To have a complete $1-1$ relationship between an H_∞ γ-suboptimal controller for $\bar{\Sigma}$ and a strictly proper controller Σ_C of the form given in (10.8) that solves the H_∞ γ-suboptimal control problem with the output regulation constraint for Σ, we need to consider a controller $\bar{\Sigma}_C^s$ of the form given in (10.9) for $\bar{\Sigma}$. However, by restricting attention to strictly proper controllers for $\bar{\Sigma}$, we know that we do not lose performance because the achievable H_∞ norm for $\bar{\Sigma}$ is less than γ over the class of proper controllers if and only if the achievable H_∞ norm is less than γ over the class of strictly proper controllers.*

Proof : The proof follows from Theorem 7.3.1 combined with the fact that we can make the H_∞ norm less than γ by controllers of the form $\bar{\Sigma}_C^s$ as given in (10.9) if and only if we can make the H_∞ norm less than γ by strictly proper controllers. ■

Design of a measurement feedback regulator that solves the H_∞ γ-suboptimal control problem with output regulation constraint:

Theorems 10.3.1 and 10.3.2 suggest the following two step procedure:

Step 1: Construct a proper or strictly proper controller $\bar{\Sigma}_C$ so that it solves the H_∞ γ-suboptimal control problem for $\bar{\Sigma}$.

Step 2: Knowing the parameters $\bar{A}_c$, $\bar{B}_c$, $\bar{C}_c$, and $\bar{D}_c$ of the controller $\bar{\Sigma}_C$ obtained in Step 1, construct a corresponding controller Σ_C as given in (10.8).

Clearly Σ_C constructed in Step 2 solves the H_∞ γ-suboptimal control problem with output regulation constraint for Σ via a proper (or a strictly proper) controller. □

10.4 The relationship between $\gamma^*_{\infty\, p,r}$, $\gamma^*_{\infty\, p}$, $\gamma^*_{\infty\, sp,r}$ and $\gamma^*_{\infty\, sp}$

Our primary objective in this section is to study how the achievable performance is affected by having the output regulation constraint. From the previous section we know that it does not matter whether we consider proper or strictly proper controllers since we know that $\gamma^*_{\infty\, sp,r} = \gamma^*_{\infty\, p,r}$ and $\gamma^*_{\infty\, p} = \gamma^*_{\infty\, sp}$.

It turns out that there exists in general a certain loss in the achievable performance because of the output regulation constraint; i.e. $\gamma^*_{\infty\, p,r} \geqslant \gamma^*_{\infty\, p}$ and $\gamma^*_{\infty\, sp,r} \geqslant \gamma^*_{\infty\, sp}$. We will find here a precise relationship between them.

Before we relate them, we need to recall first some results from [61, 73].

Consider the system Σ described by (10.1), and for matrices $P \in \mathbb{R}^{n\times n}$ and $Q \in \mathbb{R}^{n\times n}$, we define two continuous-time quadratic matrix inequalities,

$$F_\gamma(P) \geq 0, \qquad G_\gamma(Q) \geqslant 0, \tag{10.10}$$

where

$$F_\gamma(P) := \begin{pmatrix} A^{\mathrm T}P + PA + C_z^{\mathrm T}C_z + \gamma^{-2}PE_dE_d^{\mathrm T}P & PB + C_z^{\mathrm T}D_{zu} \\ B^{\mathrm T}P + D_{zu}^{\mathrm T}C_z & D_{zu}^{\mathrm T}D_{zu} \end{pmatrix},$$

$$G_\gamma(Q) := \begin{pmatrix} AQ + QA^{\mathrm T} + E_dE_d^{\mathrm T} + \gamma^{-2}QC_z^{\mathrm T}C_zQ & QC_y^{\mathrm T} + E_dD_{yd}^{\mathrm T} \\ C_yQ + D_{yd}E_d^{\mathrm T} & D_{yd}D_{yd}^{\mathrm T} \end{pmatrix}.$$

If $F_\gamma(P) \geqslant 0$, we say that P is a solution of the quadratic matrix inequality and if $G_\gamma(Q) \geqslant 0$, we say that Q is a solution of the dual quadratic matrix inequality. We also note that in Appendix 10.A it is shown that each quadratic matrix inequality is related to a matrix pencil. Hence we define the following two matrix pencils which are related to the two quadratic matrix inequalities defined before,

$$N_{F_\gamma}(s, P) := \begin{pmatrix} \begin{pmatrix} sI - A - \gamma^{-2}E_dE_d^{\mathrm T}P & -B \end{pmatrix} \\ F_\gamma(P) \end{pmatrix}, \tag{10.11}$$

$$N_{G_\gamma}(s, Q) := \begin{pmatrix} \begin{pmatrix} sI - A - \gamma^{-2}QC_z^{\mathrm T}C_z \\ -C_y \end{pmatrix} & G_\gamma(Q) \end{pmatrix}. \tag{10.12}$$

We now recall a theorem from [61, 73].

Theorem 10.4.1 *Consider the system (10.1) with $w = 0$, and let $\gamma > 0$ be given. Then the following two statements are equivalent:*

(*i*) *For the system (10.1), there exists a controller* Σ_C *such that the resulting closed-loop system is internally stable, and the closed-loop transfer matrix from d to z, namely* $T_{d,z}(\Sigma \times \Sigma_C)$ *has an* H_∞ *norm less than* γ, *i.e.* $\|T_{d,z}(\Sigma \times \Sigma_C)\|_\infty < \gamma$.

(*ii*) *The following hold:*

(*a*) *There exist semi-stabilizing, and rank-minimizing solutions P and Q of the quadratic matrix inequalities* $F_\gamma(P) \geqslant 0$ *and* $G_\gamma(Q) \geqslant 0$ *which satisfy* $r(PQ) < \gamma^2$.

(*b*) *The number of zeros on the imaginary axis, counting multiplicities, of the matrix pencil* $N_{F_\gamma}(s, P)$ *is equal to the number of zeros on the imaginary axis of the system characterized by* (A, B, C_z, D_{zu}).

(*c*) *The number of zeros on the imaginary axis, counting multiplicities, of the matrix pencil* $N_{G_\gamma}(s, Q)$ *is equal to the number of zeros on the imaginary axis of the system characterized by* (A, E_d, C_y, D_{yd}).

(*d*) *For any zero* λ *on the imaginary axis of the system characterized by* (A, B, C_z, D_{zu}) *or of the system characterized by* (A, E_d, C_y, D_{yd}), *there exists a matrix* Θ *such that* $\lambda I - A - B\Theta C_y$ *is invertible and*

$$\|(C_z + D_{zu}\Theta C_y)(\lambda I - A - B\Theta C_y)^{-1}(E_d + B\Theta D_{yd}) + D_{zu}\Theta D_{yd}\|_\infty < \gamma. \quad (10.13)$$

Remark 10.4.1 *We note that in contrast to the* H_2 *case of Chapter 8, the semi-stabilizing, and rank-minimizing solutions P and Q of the quadratic matrix inequalities (10.10) are not uniquely determined. However, later on we work with a particular semi-stabilizing, and rank-minimizing solution which has a special structure and turns out to be the smallest positive semi-definite, semi-stabilizing, and rank-minimizing solution.*

Remark 10.4.2 *The existence of a suitable* Θ *can be most easily checked via a variation on the Youla parameterization. Let F and H be such that* $A + BF$ *and* $A + HC_y$ *are Hurwitz-stable. Define*

$$T_1(s) = \begin{pmatrix} C_z + D_{zu}F & -D_{zu}F \end{pmatrix} \times \begin{pmatrix} sI - A - BF & BF \\ 0 & sI - A - HC_y \end{pmatrix}^{-1} \begin{pmatrix} E_d \\ E_d + HD_{yd} \end{pmatrix},$$
$$T_2(s) = (C_z + D_{zu}F)(sI - A - BF)^{-1}B + D_{zu},$$
$$T_3(s) = C_y(sI - A - HC_y)^{-1}(E_d + HD_{yd}) + D_{yd}.$$

Then the last condition of the theorem is satisfied if for any zero λ on the imaginary axis of the system (A, B, C_z, D_{zu}) or of the system (A, E_d, C_y, D_{yd}) there exists a matrix Q_0 such that

$$\|T_1(\lambda) + T_2(\lambda)Q_0T_3(\lambda)\|_\infty < \gamma.$$

This is obviously a straightforward convex feasibility problem.

Theorem 10.4.1 gives conditions to check whether we can make the H_∞ norm of the closed-loop transfer function matrix less than γ for any arbitrary system. As in Chapter 8, in order to study the relationship between $\gamma^*_{\infty\,p,r}$ and $\gamma^*_{\infty\,p}$ we need to compare the achievable H_∞ norms of the closed-loop transfer function matrices for the systems Σ, given by (10.1), and $\bar{\Sigma}$, given by (10.5). We do so by relating the associated semi-stabilizing, and rank-minimizing solutions of the quadratic matrix inequalities for the two systems. We have the following lemma.

Lemma 10.4.1 *Let P and Q be the smallest positive semi-definite rank-minimizing semi-stabilizing solutions of the quadratic matrix inequalities associated to the system Σ which satisfy $r(PQ) < \gamma^2$. We define,*

$$\bar{P} = \begin{pmatrix} P & 0 \\ 0 & 0 \end{pmatrix}, \quad \text{and} \quad \bar{Q} = \begin{pmatrix} Q & 0 \\ 0 & 0 \end{pmatrix}.$$

Then $\bar{P}$ and $\bar{Q}$ are the smallest positive semi-definite rank-minimizing semi-stabilizing solutions of the quadratic matrix inequalities associated to the system $\bar{\Sigma}$ which satisfy $r(\bar{P}\bar{Q}) < \gamma^2$.

Conversely, let $\bar{P}$ and $\bar{Q}$ be positive semi-definite rank-minimizing semi-stabilizing solutions of the quadratic matrix inequalities associated to the system $\bar{\Sigma}$ which satisfy $r(\bar{P}\bar{Q}) < \gamma^2$. Decompose $\bar{P}$ and $\bar{Q}$ to be compatible with the decompositions in (10.6a),

$$\bar{P} = \begin{pmatrix} P_{11} & P_{12} \\ P_{21} & P_{22} \end{pmatrix} \quad \text{and} \quad \bar{Q} = \begin{pmatrix} Q_{11} & Q_{12} \\ Q_{21} & Q_{22} \end{pmatrix}.$$

Then P_{11} and Q_{11} are positive semi-definite matrices that satisfy the conditions of Theorem 10.4.1 for the system Σ which satisfy $r(P_{11}Q_{11}) < \gamma^2$.

Proof : It is easy to check that

$$\begin{pmatrix} P & 0 \\ 0 & 0 \end{pmatrix}, \text{ and } \begin{pmatrix} Q & 0 \\ 0 & 0 \end{pmatrix},$$

are positive semi-definite, semi-stabilizing, and rank-minimizing solutions of the quadratic matrix inequalities associated with the system $\bar{\Sigma}$. Therefore, the smallest positive semi-definite, semi-stabilizing, and rank-minimizing solutions of the quadratic matrix inequalities will have the same structure, i.e. we have

$$\bar{P} = \begin{pmatrix} \tilde{P} & 0 \\ 0 & 0 \end{pmatrix}, \text{ and } \bar{Q} = \begin{pmatrix} \tilde{Q} & 0 \\ 0 & 0 \end{pmatrix},$$

with $\tilde{P} \leqslant P$ and $\tilde{Q} \leqslant Q$ because of the fact that $\bar{P}$ and $\bar{Q}$ are the smallest positive semi-definite, semi-stabilizing, and rank-minimizing solutions of the quadratic matrix inequalities associated with the system $\bar{\Sigma}$. Conversely, it is easy to check that $\tilde{P}$ and $\tilde{Q}$ are positive semi-definite, semi-stabilizing, and rank-minimizing solutions of the quadratic matrix inequalities associated with the system Σ. Since P and Q were the smallest matrices with this property we find $\tilde{P} \geqslant P$ and $\tilde{Q} \geqslant Q$. Combining the above yields $\tilde{P} = P$ and $\tilde{Q} = Q$ and the result follows.

Finally, if $\bar{P}$ and $\bar{Q}$ are positive semi-definite matrices that satisfy the conditions of Theorem 10.4.1, then it is straightforward to check that P_{11} and Q_{11} are positive semi-definite matrices that satisfy the conditions of Theorem 10.4.1 for the system Σ. ■

If we forget condition (10.13), then the above lemma implies the following; if part (2) of Theorem 10.4.1 is satisfied for Σ, then it is also satisfied for the system $\bar{\Sigma}$. In order to study the condition (10.13), we note that Lemma 7.3.1 yields an explicit relationship between the zeros of subsystems of the systems Σ and $\bar{\Sigma}$. It is easy to check then that the condition (10.13) for $\bar{\Sigma}$ is satisfied for the zeros of systems characterized by (A, B, C_z, D_{zu}) and (A, E_d, C_y, D_{yd}) (except for possibly eigenvalues of S) assuming that the condition (10.13) is satisfied for the system Σ. With respect to the eigenvalues of S we define the following:

Definition 10.4.1 *Denote by γ_s the smallest γ such that there exist Γ and Π satisfying (10.4) such that for any eigenvalue λ of S, there exists a matrix Θ*

satisfying the property,

$$\Big\| \big[\begin{pmatrix} C_z & -D_{zu}\Gamma\end{pmatrix} + \begin{pmatrix} 0 & D_{zu}\end{pmatrix}\Theta\begin{pmatrix} C_y & (D_{yw}+C_y\Pi)\end{pmatrix}\big]$$
$$\left[\lambda I - \begin{pmatrix} A & -B\Gamma \\ 0 & S\end{pmatrix} - \begin{pmatrix} 0 & B \\ I & 0\end{pmatrix}\Theta\begin{pmatrix} C_y & (D_{yw}+C_y\Pi)\end{pmatrix}\right]^{-1}$$
$$\left[\begin{pmatrix} E_d \\ 0\end{pmatrix} + \begin{pmatrix} 0 & B \\ I & 0\end{pmatrix}\Theta D_{yd}\right] + \begin{pmatrix} 0 & D_{zu}\end{pmatrix}\Theta D_{yd}\Big\|_\infty < \gamma.$$

Finding a matrix Θ such that the norm given above is less than γ can be reduced to finding a matrix Q_0 such that

$$\|T_1(\lambda) + T_2(\lambda)Q_0T_3(\lambda)\|_\infty < \gamma$$

for suitably chosen transfer matrices T_1, T_2, and T_3 in the same way as described after Theorem 10.4.1.

Note that, in the special case that the system characterized by (A, B, C_e, D_{eu}) is left-invertible, the matrices Π and Γ are uniquely determined by (10.4). Hence, for this special case, the determination of γ_S is greatly simplified.

We can now present a relationship between $\gamma^*_{\infty p,r}$ and $\gamma^*_{\infty p}$, as well as between $\gamma^*_{\infty sp,r}$ and $\gamma^*_{\infty sp}$, and it is given in the following theorem.

Theorem 10.4.2 *Let the system Σ with realization (10.1) be given, and also let Assumptions A.1, A.2, A.3, and A.4 hold. Then, we have*

$$\gamma^*_{\infty sp,r} = \gamma^*_{\infty p,r} = \max\{\gamma_S, \gamma^*_{\infty p}\}.$$

Proof : According to Lemma 10.4.1 there exist matrices P and Q that satisfy the conditions of Theorem 10.4.1 for the system Σ if and only if there exist matrices $\bar{P}$ and $\bar{Q}$ that satisfy the conditions of Theorem 10.4.1 for the system $\bar{\Sigma}$. Therefore the only difference in achievable H_∞ norm has to be caused by the additional assumption (10.13). Then, given the definition of γ_S, the result follows immediately. ■

Remark 10.4.3 *The intuitive reason why H_2 performance does not change when you impose output regulation constraint but H_∞ does is related to the fact that in the frequency where we have a pole we get constraints on the amount with which we can reduce the peak. This increases unavoidably the*

H_∞ norm. However, for H_2 a high peak which is very narrow has a small norm and therefore even though the peak cannot be lowered by making the peak more narrow its affect on the H_2 norm can be reduced enough to avoid an increase in H_2 norm.

10.A Continuous-time quadratic matrix inequalities

As it is well known, algebraic Riccati equations are primary tools used in solving what is known as the *regular* H_2 and H_∞ control problems. However, as Riccati equations do in the case of regular problems, quadratic matrix inequalities introduced earlier by Stoorvogel [70] play a central role in solving the singular H_∞ control problems. It turns out that quadratic matrix inequalities have properties which are somewhat similar to those of linear matrix inequalities discussed earlier in Appendix 6.A. We introduce now a continuous-time quadratic matrix inequality (CQMI) in the following definition.

Definition 10.A.1 *Let $A \in \mathbb{R}^{n\times n}$, $B \in \mathbb{R}^{n\times m}$, $Q \in \mathbb{R}^{n\times n}$, $R \in \mathbb{R}^{m\times m}$, $S \in \mathbb{R}^{n\times m}$, and $E \in \mathbb{R}^{n\times \ell}$ with Q and R being symmetric. The matrix inequality of the form*

$$\mathcal{Q}(X) \geq 0, \tag{10.14}$$

with $X \in \mathbb{R}^{n\times n}$ where

$$\mathcal{Q}(X) := \begin{pmatrix} Q + A^{T}X + XA + XEE^{T}X & XB + S \\ B^{T}X + S^{T} & R \end{pmatrix},$$

is called a continuous-time quadratic matrix inequality (CQMI). Moreover, when X satisfies (10.14), it is referred to as a solution of the quadratic matrix inequality.

Remark 10.A.1 *Note that the quadratic matrix inequality in (10.14) when the quadratic term $XEE^{T}X$ is dropped is precisely the same as the linear matrix inequality given in (6.63).*

We restrict ourselves here to the case when the matrices Q, R, and S satisfy the positive semi-definite condition

$$\begin{pmatrix} Q & S \\ S^{T} & R \end{pmatrix} \geqslant 0. \tag{10.15}$$

In particular, we assume that there exists matrices $C \in \mathbb{R}^{p\times n}$ and $D \in \mathbb{R}^{p\times m}$ with $\begin{pmatrix} C & D \end{pmatrix}$ of full rank such that

$$Q = C^{T}C, \quad R = D^{T}D, \quad \text{and} \quad S = C^{T}D.$$

This restriction is not necessary; however, we have chosen it here merely for the sake of simplicity. Moreover, this special class of quadratic matrix inequalities is relevant to H_∞ control theory. With the above restriction, the quadratic matrix inequality in (10.14) is rewritten as

$$Q(X) = \begin{pmatrix} C^{\mathrm{T}}C + A^{\mathrm{T}}X + XA + XEE^{\mathrm{T}}X & XB + C^{\mathrm{T}}D \\ B^{\mathrm{T}}X + D^{\mathrm{T}}C & D^{\mathrm{T}}D \end{pmatrix} \geq 0. \quad (10.16)$$

We denote the set of real symmetric solutions of the quadratic matrix inequality in (10.14) as $\mathbf{\Gamma}$, i.e.

$$\mathbf{\Gamma} := \left\{ X \in \mathbb{R}^{n \times n} \mid X = X^{\mathrm{T}} \text{ and } Q(X) \geq 0 \right\}. \quad (10.17)$$

As discussed in [61], whenever X is a solution of the quadratic matrix inequality (10.16), we have $X\mathcal{S}^*(A, B, C, D) = 0$ where the detectable strongly controllable subspace $\mathcal{S}^*(A, B, C, D)$ is as in Definition 1.2.1.

As in the case of linear matrix inequalities, a relevant set of solutions of a quadratic matrix inequalities in the context of H_∞ theory is the set of rank minimizing solutions. Before we develop the definition of rank minimizing solutions, we need to examine the following property.

Lemma 10.A.1 *Consider the quadratic matrix inequality in (10.16). Let the normal rank of $D + C(sI - A)^{-1}B$ be denoted by $\bar{\rho}$. Then, for all $X \in \mathbf{\Gamma}$, the rank of $Q(X)$ is always greater than or equal to $\bar{\rho}$, i.e.*

$$\operatorname{rank}\{Q(X)\} \geq \bar{\rho} \quad \forall\, X \in \mathbf{\Gamma}.$$

Now, we are ready to define the set of all rank minimizing solutions of a quadratic matrix inequality

Definition 10.A.2 *A solution $X \in \mathbf{\Gamma}$ is said to be rank minimizing if*

$$\operatorname{rank} Q(X) = \bar{\rho}.$$

Also, let us denote the set of all rank minimizing solutions of the quadratic matrix inequality in (10.16) as $\mathbf{\Gamma}_{\min}$, i.e.

$$\mathbf{\Gamma}_{\min} := \left\{ X \in \mathbf{\Gamma} \mid \operatorname{rank} Q(X) = \bar{\rho} \right\}. \quad (10.18)$$

A rank minimizing solution X such that $X \geq 0$ is said to be a positive semi-definite rank minimizing solution, and similarly a rank minimizing solution X such that $X > 0$ is said to be a positive definite rank minimizing solution.

We next have the following definition regarding semi-stabilizing and stabilizing solutions of a quadratic matrix inequality.

Definition 10.A.3 *Consider a matrix pencil*

$$N(s, X) := \begin{pmatrix} M(s, X) \\ Q(X) \end{pmatrix}$$

where $M(s, X) := \begin{pmatrix} sI - A - EE^{T}X & -B \end{pmatrix}$. *Then a solution* $X \in \boldsymbol{\Gamma}_{\min}$ *is said to be a semi-stabilizing solution if all the finite zeros of the matrix pencil* $N(s, X)$ *are in the closed left half plane, i.e. in* $\mathbb{C}^0 \cup \mathbb{C}^-$. *Similarly, a solution* $X \in \boldsymbol{\Gamma}_{\min}$ *is said to be a stabilizing solution if all the finite zeros of* $N(s, X)$ *are in the open left half plane, i.e. in* $\mathbb{C}^-$.

It turns out that the problem of obtaining the existence conditions for semi-stabilizing, or stabilizing, or positive semi-definite, or positive definite solutions of a quadratic matrix inequality reduces to the one of obtaining the existence conditions for similar solutions of an associated continuous-time algebraic Riccati equation. Also, in the H_∞ literature, a pertinent solution of a quadratic matrix inequality that has been used often is a stabilizing solution. As shown in [61], a stabilizing solution of a quadratic matrix inequality, if it exists, is also unique. Methods of constructing semi-stabilizing or stabilizing solutions as well as other solutions of a quadratic matrix inequality are given in [61]. Such methods are based on transforming a given quadratic matrix inequality to an associated Riccati equation.

Finally, we would like to make a comment on discrete-time quadratic matrix inequalities. Although discrete-time quadratic matrix inequalities can be discussed following the concepts introduced in this section for continuous-time quadratic matrix inequalities, we note that nothing has yet been done in this regard in the existing literature. This may be because of the lack of a clear understanding as to the potential of such discrete-time quadratic matrix inequalities.

Chapter 11

H_∞ optimal control with an output regulation constraint – discrete-time systems

11.1 Introduction

This chapter is a discrete-time analog of Chapter 10. That is we formulate here an output regulation problem with an H_∞ optimal control constraint. Such a problem can equivalently be viewed as an H_∞ optimal control problem with the output regulation constraint. As in the previous chapters, although a suitable controller which solves the posed problem for a given system can be constructed via the construction of a controller that solves an H_∞ optimal control problem without the output regulation constraint for a certain auxiliary system, a fundamental and significant issue still needs to be answered. Namely, whether the added output regulation constraint in a problem compromises the achievable performance. In this regard, as in Chapter 10, there is a certain loss or decay in the achievable performance due to the added output regulation constraint, and this decay will be explicitly expressed in terms of a static optimization problem.

This chapter is based on the recent research work of the authors [77], and is organized as follows. Section 11.2 formulates the H_∞ γ-suboptimal control problem with the output regulation constraint, while Section 11.3 presents solvability conditions for such a problem and also develops methods of constructing a regulator that solves such a problem whenever it is solvable. Finally, Section 11.4 shows that the added output regulation constraint in the given H_∞ γ-suboptimal control problem indeed compromises the achievable performance, and an explicit expression is derived showing such a compromise.

11.2 Problem formulation

As in previous chapters, we start with a linear system with state space realization,

$$\Sigma : \begin{cases} x(k+1) = Ax(k) + Bu(k) + E_w w(k) + E_d d(k) \\ e(k) \quad = C_e x(k) + D_{eu} u(k) + D_{ew} w(k) \\ z(k) \quad = C_z x(k) + D_{zu} u(k) \\ y(k) \quad = C_y x(k) \qquad\qquad + D_{yw} w(k) + D_{yd} d(k), \end{cases} \tag{11.1}$$

where, as usual, $x \in \mathbb{R}^n$, $u \in \mathbb{R}^m$, $y \in \mathbb{R}^p$, $e \in \mathbb{R}^q$, and $z \in \mathbb{R}^\ell$. The exogenous disturbance input $w \in \mathbb{R}^s$ is generated by an exosystem Σ_{E} with state space realization,

$$\Sigma_{\mathrm{E}} \; : \; w(k+1) = Sw(k). \tag{11.2}$$

As before, the variable d denotes an external disturbance. We seek measurement feedback controllers which are of the form,

$$\Sigma_{\mathrm{C}} : \begin{cases} v(k+1) = A_c v(k) + B_c y(k), \\ u(k) \quad = C_c v(k) + D_c y(k). \end{cases} \tag{11.3}$$

The closed-loop system consisting of the given system Σ and the controller Σ_{C} is denoted by $\Sigma \times \Sigma_{\mathrm{C}}$. Also, the transfer matrix from d to z of $\Sigma \times \Sigma_{\mathrm{C}}$ is denoted by $T_{d,z}(\Sigma \times \Sigma_{\mathrm{C}})$. As usual, we define,

$$\|T_{d,z}(\Sigma \times \Sigma_{\mathrm{C}})\|_\infty := \sup_{\omega \in [-\pi,\pi)} \sigma_{\max}\left[T_{d,z}(\Sigma \times \Sigma_{\mathrm{C}})(e^{i\omega})\right],$$

where $\sigma_{\max}$ denotes the largest singular value.

Before we state formally the specific multi-objective problems of interest here, we define the following notations:

$$\gamma^*_{\infty p,r} = \inf \{ \; \|T_{d,z}(\Sigma \times \Sigma_{\mathrm{C}})\|_\infty \mid \Sigma_{\mathrm{C}} \text{ is a proper controller that achieves output regulation for } \Sigma \; \},$$

$$\gamma^*_{\infty p} = \inf \{ \; \|T_{d,z}(\Sigma \times \Sigma_{\mathrm{C}})\|_\infty \mid \Sigma_{\mathrm{C}} \text{ is a proper controller that internally stabilizes } \Sigma \; \},$$

$$\gamma^*_{\infty sp,r} = \inf \{ \; \|T_{d,z}(\Sigma \times \Sigma_C)\|_\infty \mid \Sigma_C \text{ is a strictly proper controller that achieves output regulation for } \Sigma \; \},$$

$$\gamma^*_{\infty sp} = \inf \{ \; \|T_{d,z}(\Sigma \times \Sigma_C)\|_\infty \mid \Sigma_C \text{ is a strictly proper controller that internally stabilizes } \Sigma \; \}.$$

We can now state the specific multi-objective problem of interest here, namely the *H_∞ γ-suboptimal control problem with output regulation constraint.*

Problem 11.2.1 ***(H_∞ γ-suboptimal control problem with the output regulation constraint)*** *Consider the system Σ and the exosystem Σ_E as given in (11.1) and (11.2). Find, if possible, a controller Σ_C of the form (11.3) such that the following conditions hold:*

(i) ***(Internal Stability)*** *In the absence of the disturbances w and d, the closed-loop system $\Sigma \times \Sigma_C$ is internally stable.*

(ii) ***(Performance Measure)*** *$\|T_{d,z}(\Sigma \times \Sigma_C)\|_\infty$ is at most a specified value γ, i.e. an H_∞ γ-suboptimal performance is obtained.*

(iii) ***(Output Regulation)*** *For any $d \in \ell_2$, and for all $x(0) \in \mathbb{R}^n$ and $w(0) \in \mathbb{R}^s$, the solution of the closed-loop system $\Sigma \times \Sigma_C$ satisfies*

$$\lim_{k \to \infty} e(k) = 0.$$

Remark 11.2.1 *The part (iii) of the above problem formulation is equivalent to the following: In the absence of external signal d, i.e. when $d = 0$, Σ_C achieves output regulation for Σ.*

We note that when the output regulation constraint is removed from the above problem formulation, the resulting problem is simply an *unconstrained H_∞ γ-suboptimal control problem.*

11.3 Solvability conditions

As we discussed in Chapter 7, the following assumptions are reasonable and almost necessary to solve the problem defined above:

A.1. (A, B) is stabilizable.

A.2. The matrix S is anti-Schur-stable, i.e. all the eigenvalues of S are on the unit circle or outside the unit circle.

A.3. $\left(\begin{pmatrix} C_y & D_{yw} \end{pmatrix}, \begin{pmatrix} A & E_w \\ 0 & S \end{pmatrix}\right)$ is detectable.

A.4. There exist Π and Γ solving the regulator equation,

$$\begin{cases} \Pi S = A\Pi + B\Gamma + E_w, \\ 0 = C_e\Pi + D_{eu}\Gamma + D_{ew}. \end{cases} \tag{11.4}$$

Note that these assumptions have been defined before, for instance on page 215.

In the case when the regulator equation (11.4) has a non-unique solution for (Π, Γ), for simplicity of presentation, we assume throughout the chapter that a solution (Π, Γ) of (11.4) has been chosen, and all our development here builds on such a solution.

Following the results of Chapter 7, a solution to the H_∞ *γ-suboptimal control problem with output regulation constraint* for Σ is given in terms of a solution to the H_∞ *γ-suboptimal control problem without any output regulation constraint* for the auxiliary system $\bar{\Sigma}$ which is as in (7.5), and is repeated below for convenience:

$$\bar{\Sigma}: \begin{cases} \bar{x}(k+1) = \bar{A}\bar{x}(k) + \bar{B}\bar{u}(k) + \bar{E}_d\bar{d}(k) \\ \bar{z}(k) = \bar{C}_z\bar{x}(k) + \bar{D}_{zu}\bar{u}(k) \\ \bar{y}(k) = \bar{C}_y\bar{x}(k) + D_{yd}\bar{d}(k), \end{cases} \tag{11.5}$$

where

$$\bar{A} = \begin{pmatrix} A & -B\Gamma \\ 0 & S \end{pmatrix}, \quad \bar{B} = \begin{pmatrix} 0 & B \\ I & 0 \end{pmatrix}, \quad \bar{E}_d = \begin{pmatrix} E_d \\ 0 \end{pmatrix}, \tag{11.6a}$$

$$\bar{C}_z = \begin{pmatrix} C_z & -D_{zu}\Gamma \end{pmatrix}, \quad \bar{D}_{zu} = \begin{pmatrix} 0 & D_{zu} \end{pmatrix}, \tag{11.6b}$$

$$\bar{C}_y = \begin{pmatrix} C_y & (D_{yw} + C_y\Pi) \end{pmatrix}. \tag{11.6c}$$

For $\bar{\Sigma}$, a controller $\bar{\Sigma}_C$ with state space representation $(\bar{A}_c, \bar{B}_c, \bar{C}_c, \bar{D}_c)$ is given by

$$\bar{\Sigma}_C: \begin{cases} \bar{v}(k+1) = \bar{A}_c\bar{v}(k) + \bar{B}_c\bar{y}(k) \\ \bar{u}(k) = \bar{C}_c\bar{v}(k) + \bar{D}_c\bar{y}(k). \end{cases} \tag{11.7}$$

For each controller $\bar{\Sigma}_C$ for the auxiliary system $\bar{\Sigma}$, we formulate a corresponding controller Σ_C for the given system Σ. It is given by

$$\Sigma_C: \begin{cases} \rho v_1 = S v_1 + \bar{C}_{c,1}v_2 + \bar{D}_{c,1}(y + (D_{yw} + C_y\Pi)v_1) \\ \rho v_2 = \bar{A}_c v_2 + \bar{B}_c(y + (D_{yw} + C_y\Pi)v_1) \\ u = -\Gamma v_1 + \bar{C}_{c,2}v_2 + \bar{D}_{c,2}(y + (D_{yw} + C_y\Pi)v_1), \end{cases} \tag{11.8}$$

where $\bar{C}_{c,1}$, $\bar{C}_{c,2}$, $\bar{D}_{c,1}$, and $\bar{D}_{c,2}$ are obtained by partitioning $\bar{C}_c$ and $\bar{D}_c$ in conformity with the partitioning of $\bar{A}$,

$$\bar{C}_c = \begin{pmatrix} \bar{C}_{c,1} \\ \bar{C}_{c,2} \end{pmatrix}, \quad \text{and} \quad \bar{D}_c = \begin{pmatrix} \bar{D}_{c,1} \\ \bar{D}_{c,2} \end{pmatrix}.$$

We note that there is a $1-1$ relationship between the proper controller $\bar{\Sigma}_C$ as given in (11.7) and the proper controller Σ_C as given in (11.8). Furthermore, we observe that with the constraint that $\bar{D}_{c,2} = 0$, the controller Σ_C as given in (11.8) is strictly proper. In other words, by imposing the constraint that $\bar{D}_{c,2} = 0$ in $\bar{\Sigma}_C$, we can generate the class of strictly proper controllers Σ_C for the given system Σ. We denote below the class of controllers $\bar{\Sigma}_C$ with the constraint $\bar{D}_{c,2} = 0$ by $\bar{\Sigma}_C^s$,

$$\bar{\Sigma}_C^s : \begin{cases} \bar{v}(k+1) = \quad \bar{A}_c \bar{v}(k) + \quad \bar{B}_c \bar{y}(k) \\ \bar{u}(k) \quad = \begin{pmatrix} \bar{C}_{c,1} \\ \bar{C}_{c,2} \end{pmatrix} \bar{v}(k) + \begin{pmatrix} \bar{D}_{c,1} \\ 0 \end{pmatrix} \bar{y}(k). \end{cases} \tag{11.9}$$

The following theorems provide the conditions under which the H_∞ γ-suboptimal control problem with output regulation constraint can be solved via proper or via strictly proper controllers. Also, they provide a procedure of constructing an appropriate controller that solves the posed problem whenever it is solvable.

Theorem 11.3.1 *Let Assumptions A.1, A.2, A.3, and A.4 hold. Consider the given system Σ as in (11.1), and the exosystem Σ_E as in (11.2). Also, consider the auxiliary system $\bar{\Sigma}$ as in (11.5). Then, the following statements hold:*

> *The H_∞ γ-suboptimal control problem with output regulation constraint for Σ is solvable via a proper controller if and only if the H_∞ γ-suboptimal control problem for $\bar{\Sigma}$ is solvable via a proper controller.*
>
> *Moreover, a proper controller $\bar{\Sigma}_C$ of the form given in (11.7) is a proper H_∞ γ-suboptimal controller for $\bar{\Sigma}$ if and only if the corresponding proper controller Σ_C of the form given in (11.8) solves the H_∞ γ-suboptimal control problem with the output regulation constraint for Σ.*

Proof : The proof follows from Theorem 7.3.1. ■

Theorem 11.3.2 *Let Assumptions A.1, A.2, A.3, and A.4 hold. Consider the given system Σ as in (11.1), and the exosystem Σ_E as in (11.2). Also, consider the auxiliary system $\bar{\Sigma}$ as in (11.5). Then, the following statements hold:*

> *The H_∞ γ-suboptimal control problem with output regulation constraint for Σ is solvable via a strictly proper controller if and only if the H_∞ γ-suboptimal control problem for $\bar{\Sigma}$ is solvable via a strictly proper controller.*
>
> *Moreover, a strictly proper controller is an H_∞ γ-suboptimal controller for $\bar{\Sigma}$ if and only if the corresponding strictly proper controller Σ_C of the form given in (10.8) solves the H_∞ γ-suboptimal control problem with the output regulation constraint for Σ.*

Remark 11.3.1 *To have a complete $1-1$ relationship between an H_∞ γ-suboptimal controller for $\bar{\Sigma}$ and a strictly proper controller Σ_C of the form given in (10.8) that solves the H_∞ γ-suboptimal control problem with the output regulation constraint for Σ, we need to consider a controller $\bar{\Sigma}_C^s$ of the form given in (10.9) for $\bar{\Sigma}$. However, by restricting attention to strictly proper controllers for $\bar{\Sigma}$, we know that we do not lose performance due to the specific structure as illustrated in the proof.*

Proof : We need to establish that the H_∞ γ-suboptimal control problem for $\bar{\Sigma}$, is solvable via a controller $\bar{\Sigma}_C^s$ of the form given in (11.9) if and only if H_∞ γ-suboptimal control problem for $\bar{\Sigma}$, is solvable via a strictly proper controller. The rest of the proof follows from Theorem 7.3.1.

In [76] the discrete-time H_∞ control problem with a structural constraint on the direct feedthrough matrix has been considered. Using the structure we will establish in Lemma 11.4.1, the above result then follows intrinsically directly from [76]. However, [76] assumes that the system has no zeros on the unit circle; an assumption which is never satisfied in our case. Nevertheless, combining the paper [76] with the results from [73], the result can be obtained straightforwardly. ■

Design of a measurement feedback regulator that solves the H_∞ γ-suboptimal control problem with output regulation constraint:

Theorems 11.3.1 and 11.3.2 suggest the following two step procedure:

Step 1: Construct a controller $\bar{\Sigma}_C$ of the form (11.7) (or a controller $\bar{\Sigma}_C^s$ of the form (11.9)) so that it solves the H_∞ γ-suboptimal control problem for $\bar{\Sigma}$.

Step 2: Knowing the parameters $\bar{A}_c$, $\bar{B}_c$, $\bar{C}_c$, and $\bar{D}_c$ of the controller $\bar{\Sigma}_C$ (or $\bar{\Sigma}_C^s$) obtained in Step 1, construct a corresponding controller Σ_C as given in (11.8).

Clearly Σ_C constructed in Step 2 solves the H_∞ γ-suboptimal control problem with output regulation constraint for Σ via a proper (or a strictly proper) controller. □

11.4 The relationship between $\gamma^*_{\infty p,r}$, $\gamma^*_{\infty p}$, $\gamma^*_{\infty sp,r}$ and $\gamma^*_{\infty sp}$

Our primary objective in this section is to study how the achievable performance is affected by having the output regulation constraint. As in Chapter 10, it turns out that there exists in general a certain loss in the achievable performance because of the output regulation constraint; i.e. $\gamma^*_{\infty p,r} \geqslant \gamma^*_{\infty p}$ and $\gamma^*_{\infty sp,r} \geqslant \gamma^*_{\infty sp}$. We will find here a precise relationship between them.

Before we relate them, we need to recall first some results from [73].

Theorem 11.4.1 *Consider the system (11.1) with $w = 0$, and let $\gamma > 0$ be given. The following statements are equivalent:*

(i) There exists a controller Σ_C for the system (11.1), such that the resulting closed-loop system is internally stable, and the closed-loop transfer matrix from d to z, namely $T_{d,z}(\Sigma \times \Sigma_C)$, has an H_∞ norm less than γ, i.e. $\|T_{d,z}(\Sigma \times \Sigma_C)\|_\infty < \gamma$.

(ii) There exist symmetric matrices $P \geqslant 0$ and $Q \geqslant 0$ such that the following hold:

(a) We have $R > 0$ where

$$V := B^T P B + D_{zu}^T D_{zu},$$
$$R := \gamma^2 I - E_d^T P E_d + E_d^T P B V^\dagger B^T P E_d.$$

(b) There exists a semi-stabilizing solution P of the generalized discrete-time algebraic Riccati equation[1],

[1]In Appendix 11.A, we develop briefly certain definitions regarding generalized Riccati equations as well as certain properties of them as needed for our purposes here.

$$P = A^T P A + C_z^T C_z - \begin{pmatrix} B^T P A + D_{zu}^T C_z \\ E_d^T P A \end{pmatrix}^T G(P)^\dagger \begin{pmatrix} B^T P A + D_{zu}^T C_z \\ E_d^T P A \end{pmatrix}, \tag{11.10}$$

where

$$G(P) := \begin{pmatrix} D_{zu}^T D_{zu} & 0 \\ 0 & -\gamma^2 I \end{pmatrix} + \begin{pmatrix} B^T \\ E_d^T \end{pmatrix} P \begin{pmatrix} B & E_d \end{pmatrix}. \tag{11.11}$$

(c) *We have* $T > 0$ *where*

$$W := D_{yd} D_{yd}^T + C_y Q C_y^T,$$
$$T := \gamma^2 I - C_z Q C_z^T + C_z Q C_y^T W^\dagger C_y Q C_z^T.$$

(d) *There exists a semi-stabilizing solution* Q *of the dual generalized Riccati equation,*

$$Q = A Q A^T + E_d E_d^T - \begin{pmatrix} C_y Q A^T + D_{yd} E_d^T \\ C_z Q A^T \end{pmatrix}^T H(Q)^\dagger \begin{pmatrix} C_y Q A^T + D_{yd} E_d^T \\ C_z Q A^T \end{pmatrix}, \tag{11.12}$$

where

$$H(Q) := \begin{pmatrix} D_{yd} D_{yd}^T & 0 \\ 0 & -\gamma^2 I \end{pmatrix} + \begin{pmatrix} C_y \\ C_z \end{pmatrix} Q \begin{pmatrix} C_y^T & C_z^T \end{pmatrix}. \tag{11.13}$$

(e) $r(PQ) < \gamma^2$ *where* $r(\cdot)$ *denotes the spectral radius.*

Finally, for any zero λ *on the unit circle of the system* (A, B, C_z, D_{zu}) *or of the system* (A, E_d, C_y, D_{yd})*, there exists a matrix* Θ *such that* $\lambda I - A - B\Theta C_y$ *is invertible, and*

$$\|(C_z + D_{zu}\Theta C_y)(\lambda I - A - B\Theta C_y)^{-1}(E_d + B\Theta D_{yd}) + D_{zu}\Theta D_d\|_\infty < \gamma. \tag{11.14}$$

Remark 11.4.1 *We note that in contrast to the* H_2 *case of Chapter 9, the matrices* P *and* Q *are not uniquely determined. However, later on we work with a particular semi-stabilizing solution which has a special structure and turns out to be the smallest positive semi-definite, semi-stabilizing solution.*

Remark 11.4.2 *The definition of a semi-stabilizing or stabilizing solution as given in [73] looks different from our Definition 11.A.6 given in Appendix 11.A. For instance, in order to guarantee P is a stabilizing solution of (11.10) and Q is a stabilizing solution of (11.12), in [73] the following conditions are imposed:*

(*i*) *For all $z \in \mathbb{C}^{\circ} \cup \mathbb{C}^{\oplus}$, the solution P of the generalized Riccati equation (11.10) must be such that its associated matrix pencil, namely*

$$\begin{pmatrix} zI - A & -B & -E_d \\ B^T P A + D_{zu}^T C_z & B^T P B + D_{zu}^T D_{zu} & B^T P E_d \\ E_d^T P A & E_d^T P B & E_d^T P E_d - \gamma^2 I \end{pmatrix}, \tag{11.15}$$

has normal rank equal to

$$n + \ell_d + \text{normrank}\{C_z(zI - A)^{-1}B + D_{zu}\}$$

where ℓ_d is the dimension of the disturbance vector d.

(*ii*) *Similarly, for all $z \in \mathbb{C}^{\circ} \cup \mathbb{C}^{\oplus}$, the solution Q of the generalized Riccati equation (11.12) must be such that its associated matrix pencil, namely*

$$\begin{pmatrix} zI - A & AQC_y^T + E_d D_{yd}^T & AQC_z^T \\ -C_y & C_y Q C_y^T + D_{yd} D_{yd}^T & C_y Q C_z^T \\ -C_z & C_z Q C_y^T & C_z Q C_z^T - \gamma^2 I \end{pmatrix} \tag{11.16}$$

has normal rank equal to

$$n + \ell + \text{normrank}\{C_y(zI - A)^{-1}E_d + D_{yd}\}$$

where ℓ is the dimension of the vector z.

However, with some simple straightforward manipulations, one can show that the above conditions are equivalent to those imposed by Definition 11.A.6.

Theorem 11.4.1 gives conditions to check whether we can make the H_∞ norm of the closed-loop transfer function matrix less than γ for any arbitrary system. As in Chapter 10, in order to study the relationship between $\gamma^*_{\infty\,p,r}$ and $\gamma^*_{\infty\,p}$, or between $\gamma^*_{\infty\,sp,r}$ and $\gamma^*_{\infty\,sp}$, we need to compare the achievable H_∞ norms of the closed-loop transfer function matrices for the systems Σ, given by (11.1), and $\bar{\Sigma}$, given by (11.5). We do so by relating the associated semi-stabilizing, rank-minimizing solutions of the quadratic matrix inequalities for the two systems. We have the following lemma the proof of which is based on an identical argument as used in the proof of Lemma 10.4.1. Note that solutions of two generalized Riccati equations (using the notation of Theorem 11.4.1) need to satisfy $R > 0$ and $T > 0$. We denote by R and T these two matrices associated to the system Σ and by $\bar{R}$ and $\bar{T}$ these two matrices associated to the system $\bar{\Sigma}$.

Lemma 11.4.1 *Let P and Q be the smallest positive semi-definite semi-stabilizing solutions of the generalized Riccati equations associated to the system Σ which satisfy $r(PQ) < \gamma^2$ and yield $R > 0$ and $T > 0$. We define,*

$$\bar{P} = \begin{pmatrix} P & 0 \\ 0 & 0 \end{pmatrix}, \quad \text{and } \bar{Q} = \begin{pmatrix} Q & 0 \\ 0 & 0 \end{pmatrix}.$$

Then, $\bar{P}$ and $\bar{Q}$ are the smallest positive semi-definite semi-stabilizing solutions of the generalized Riccati equations associated to the system $\bar{\Sigma}$ which satisfy $r(\bar{P}\bar{Q}) < \gamma^2$ and yields $\bar{R} > 0$ and $\bar{T} > 0$.

Conversely, let $\bar{P}$ and $\bar{Q}$ be positive semi-definite semi-stabilizing solutions of the generalized Riccati equations associated to the system $\bar{\Sigma}$ which satisfy $r(\bar{P}\bar{Q}) < \gamma^2$ and yield $\bar{R} > 0$ and $\bar{T} > 0$. Decompose $\bar{P}$ and $\bar{Q}$ to be compatible with the decompositions in (11.6),

$$\bar{P} = \begin{pmatrix} P_{11} & P_{12} \\ P_{21} & P_{22} \end{pmatrix} \quad \text{and } \bar{Q} = \begin{pmatrix} Q_{11} & Q_{12} \\ Q_{21} & Q_{22} \end{pmatrix}.$$

Then P_{11} and Q_{11} are positive semi-definite positive semi-definite semi-stabilizing solutions of the generalized Riccati equations associated to the system Σ which satisfy $r(P_{11}Q_{11}) < \gamma^2$ and yields $R > 0$ and $T > 0$.

If we forget condition (11.14), then the above lemma implies that if part (ii) of Theorem 11.4.1 is satisfied for Σ, then it is also satisfied for the system $\bar{\Sigma}$. In order to study this last condition we note that Lemma 7.3.1 points out an explicit relationship between the zeros of subsystems of the systems Σ and $\bar{\Sigma}$. It is easy to check then that the final condition of part (ii) for $\bar{\Sigma}$ is satisfied for the zeros of systems characterized by (A, B, C_z, D_{zu}) and (A, E_d, C_y, D_{yd}) (except for possibly eigenvalues of S) assuming that part (ii) is satisfied for the system Σ. With respect to the eigenvalues of S we have the following definition.

Definition 11.4.1 *Denote by γ_S the smallest γ such that there exist Γ and Π satisfying (11.4) such that for any eigenvalue λ of S there exists a matrix Θ*

satisfying the property,

$$\Big\| \big[\begin{pmatrix} C_z & -D_{zu}\Gamma \end{pmatrix} + \begin{pmatrix} 0 & D_{zu} \end{pmatrix} \Theta \begin{pmatrix} C_y & (D_{yw} + C_y\Pi) \end{pmatrix} \big]$$

$$\left[\lambda I - \begin{pmatrix} A & -B\Gamma \\ 0 & S \end{pmatrix} - \begin{pmatrix} 0 & B \\ I & 0 \end{pmatrix} \Theta \begin{pmatrix} C_y & (D_{yw} + C_y\Pi) \end{pmatrix} \right]^{-1}$$

$$\times \left[\begin{pmatrix} E_d \\ 0 \end{pmatrix} + \begin{pmatrix} 0 & B \\ I & 0 \end{pmatrix} \Theta D_{yd} \right] + \begin{pmatrix} 0 & D_{zu} \end{pmatrix} \Theta D_{yd} \Big\|_\infty < \gamma.$$

Note that in the special case that the system (A, B, C_e, D_{eu}) is left-invertible the matrices Π and Γ are uniquely determined by (11.4). Hence, for this special case the determination of γ_S is greatly simplified. However, in general it can always be transformed into a convex optimization using the same arguments as in Remark 10.4.2.

We can now present a relationship between $\gamma^*_{\infty p,r}$ and $\gamma^*_{\infty p}$, as well as between $\gamma^*_{\infty sp,r}$ and $\gamma^*_{\infty sp}$, and it is given in the following theorem.

Theorem 11.4.2 *Let the system Σ with a realization (11.1) be given. Also, let Assumptions A.1, A.2, A.3, and A.4 hold. Then, we have*

$$\gamma^*_{\infty p,r} = \max\{\gamma_S, \gamma^*_{\infty p}\} \quad \textit{and} \quad \gamma^*_{\infty sp,r} = \max\{\gamma_S, \gamma^*_{\infty sp}\}.$$

Proof : The proof follows along the same arguments used in the proof of Theorem 10.4.2. ∎

11.A Discrete-time algebraic Riccati equations

As the continuous-time algebraic Riccati equations do, the discrete-time algebraic Riccati equations play important roles. The intent of this section is to study briefly Riccati equation as well as a generalization called the generalized Riccati equation.

Our presentation here is again an extract of [61] where Riccati equations are dealt with in detail along with the required proofs for their properties.

We introduce a discrete-time Riccati equation in the following definition.

Definition 11.A.1 ***Discrete-time Riccati equation*** *Let $A \in \mathbb{R}^{n\times n}$, $B \in \mathbb{R}^{n\times m}$, $Q \in \mathbb{R}^{n\times n}$, $R \in \mathbb{R}^{m\times m}$, and $S \in \mathbb{R}^{n\times m}$ with Q and R being symmetric. Then the constrained quadratic matrix equation for an unknown $n \times n$ matrix X given by*

$$X = A^T XA - (A^T XB + S)(R + B^T XB)^{-1}(B^T XA + S^T) + Q, \tag{11.17}$$

is called a discrete-time algebraic Riccati equation.

The following is a special Riccati equation that satisfies $R + B^T XB > 0$, and is called an H_2 Riccati equation because of its occurrence in some H_2 optimal control problems.

Definition 11.A.2 ***H_2 Riccati equation*** *Let $A \in \mathbb{R}^{n\times n}$, $B \in \mathbb{R}^{n\times m}$, $Q \in \mathbb{R}^{n\times n}$, $R \in \mathbb{R}^{m\times m}$, and $S \in \mathbb{R}^{n\times m}$ with Q and R being symmetric. Then the constrained quadratic matrix equation for an unknown $n \times n$ matrix X given by*

$$R + B^T XB > 0, \tag{11.18a}$$

and

$$X = A^T XA - (A^T XB + S)(R + B^T XB)^{-1}(B^T XA + S^T) + Q, \tag{11.18b}$$

is called an H_2 discrete-time algebraic Riccati equation.

Definition 11.A.3 *A solution of the Riccati equation (11.17) (or the H_2 Riccati equation in (11.18)) is said to be a semi-stabilizing solution if $A - B(R + B^T XB)^{-1}(B^T XA + S^T)$ has each of its eigenvalues either on the unit circle* *or within the unit circle of the complex plane, i.e. in $\mathbb{C}^{\otimes}$. Similarly, a solution of the Riccati equation is said to be a stabilizing solution if $A - B(R + B^T X B)^{-1}(B^T X A + S^T)$ has all its eigenvalues entirely within the unit circle of the complex plane, i.e. in $\mathbb{C}^{\ominus}$.*

A more general discrete-time algebraic Riccati equation that has been used in the optimal control literature is defined next.

Definition 11.A.4 ***(Generalized discrete-time Riccati equation)*** *Given $A \in \mathbb{R}^{n\times n}$, $B \in \mathbb{R}^{n\times m}$, $Q \in \mathbb{R}^{n\times n}$, $R \in \mathbb{R}^{m\times m}$, and $S \in \mathbb{R}^{n\times m}$ with Q and R being symmetric. Then the matrix equation for an unknown matrix $X \in R^{n\times n}$ given by*

$$X = A^T X A - (A^T X B + S)(R + B^T X B)^{\dagger}(B^T X A + S^T) + Q, \quad (11.19a)$$

and

$$\ker[R + B^T X B] \subseteq \ker[A^T X B + S], \quad (11.19b)$$

is called a generalized discrete-time algebraic Riccati equation where, as usual, $A^{\dagger}$ denotes the generalized or Moore-Penrose inverse of A.

Remark 11.A.1 *Whenever $R + B^T X B$ is invertible, obviously the subspace inclusion in (11.19b) is automatically satisfied, and thus the generalized Riccati equation in (11.19) reduces to the Riccati equation in (11.17).*

We introduce next a special generalized Riccati equation that satisfies $R + B^T X B \geq 0$. It is called an H_2 generalized Riccati equation.

Definition 11.A.5 ***H_2 generalized Riccati equation*** *Let $A \in \mathbb{R}^{n\times n}$, $B \in \mathbb{R}^{n\times m}$, $Q \in \mathbb{R}^{n\times n}$, $R \in \mathbb{R}^{m\times m}$, and $S \in \mathbb{R}^{n\times m}$ with Q and R being symmetric. Then the constrained quadratic matrix equation for an unknown $n \times n$ matrix X given by*

$$X = A^T X A - (A^T X B + S)(R + B^T X B)^{\dagger}(B^T X A + S^T) + Q, \quad (11.20a)$$

$$\ker R + B^T X B \subseteq \ker A^T X B + S, \quad (11.20b)$$

and

$$R + B^T X B \geq 0, \quad (11.20c)$$

is called an H_2 generalized Riccati equation.

Definition 11.A.6 *Consider a matrix*

$$N(z, X) := \begin{pmatrix} M(z) \\ L(X) \end{pmatrix}$$

where

$$M(z) := \begin{pmatrix} zI - A & -B \end{pmatrix}$$

and

$$L(X) := \begin{pmatrix} Q + A^T X A - X & A^T X B + S \\ B^T X A + S^T & B^T X B + R \end{pmatrix}.$$

Then a solution X of the generalized Riccati equation in (11.19) (or the H_2 generalized Riccati equation in (11.20)) is said to be a semi-stabilizing solution if the rank of $N(z, X)$ is equal to its normal rank for all z outside the unit circle of the complex plane, i.e. for all $z \in \mathbb{C}^{\oplus}$. Similarly, a solution X of the generalized Riccati equation is said to be a stabilizing solution if the rank of $N(z, X)$ is equal to its normal rank for all z on the unit circle or outside the unit circle, i.e. for all $z \in \mathbb{C}^{\circ} \cup \mathbb{C}^{\oplus}$.

Remark 11.A.2 *Obviously, Definition 11.A.6 reduces to Definition 11.A.3 whenever the matrix $(R + B^T X B)$ is non-singular.*

Remark 11.A.3 *Definition 11.A.6 can be rewritten as follows. A solution X of a generalized Riccati equation is said to be a stabilizing (respectively, semi-stabilizing) solution if all the eigenvalues of the matrix*

$$A - B(B^T X B + R)^{\dagger}(B^T X A + S^T) \\ - B(I - (B^T X B + R)^{\dagger}(B^T X B + R))F$$

are inside the unit circle (respectively, inside or on the unit circle) for some suitably chosen matrix F.

We often require that the matrices Q, R, and S in the above definitions of (generalized) Riccati equations to satisfy the positive semi-definite condition, namely

$$\begin{pmatrix} Q & S \\ S^T & R \end{pmatrix} \geqslant 0. \tag{11.21}$$

Under the above condition, it follows that there exists matrices $C \in \mathbb{R}^{p\times n}$ and $D \in \mathbb{R}^{p\times m}$ with $\begin{pmatrix} C & D \end{pmatrix}$ of full rank such that

$$\begin{pmatrix} Q & S \\ S^{\mathrm{T}} & R \end{pmatrix} = \begin{pmatrix} C & D \end{pmatrix}^{\mathrm{T}} \begin{pmatrix} C & D \end{pmatrix}.$$

Let us next define a rational matrix $\hat{H}(z)$,

$$\hat{H}(z) := \begin{pmatrix} B^{\mathrm{T}}(z^{-1}I - A^{\mathrm{T}})^{-1} & I \end{pmatrix} \begin{pmatrix} Q & S \\ S^{\mathrm{T}} & R \end{pmatrix} \begin{pmatrix} (zI - A)^{-1}B \\ I \end{pmatrix}. \tag{11.22}$$

We note that, whenever the matrices Q, R, and S satisfy the positive semi-definite condition (11.21), we can rewrite $\hat{H}(z)$ as

$$\hat{H}(z) = G^{\mathrm{T}}(z^{-1})G(z)$$

where $G(z)$ is the transfer function of the system characterized by the matrix quadruple (A, B, C, D).

We observe that the only difference between a standard and a generalized Riccati equation, as formulated in our Definitions 11.A.1 and 11.A.4, concerns with whether $(R + B^{\mathrm{T}}XB)$ is non-singular or not. A revealing fundamental property of a generalized Riccati equation is that either *all* or *none* of the solutions have the property that the corresponding $(R + B^{\mathrm{T}}XB)$ is non-singular. That is, a given generalized Riccati equation cannot have one solution, say X_1, for which $(R + B^{\mathrm{T}}X_1B)$ is non-singular, and another solution, say X_2, for which $(R + B^{\mathrm{T}}X_2B)$ is singular. Thus, in one case the study of a generalized Riccati equation is exactly the same as that of a standard Riccati equation, while in another case it is not. The distinction between the two cases depends on the normal rank of the rational matrix $\hat{H}(z)$ being full or not. Similarly, the difference between an H_2 generalized and standard Riccati equation, as formulated in our definitions, is the additional assumption that $R + B^{\mathrm{T}}XB \geq 0$ in the case of a generalized Riccati equation. Again, a revealing fact is that either *all* or *none* of the solutions of the generalized Riccati equation satisfy the condition $R + B^{\mathrm{T}}XB \geq 0$ depending on the normal rank of $\hat{H}(z)$.

We have the following lemma [74].

Lemma 11.A.1 *Consider a generalized Riccati equation as in (11.19). Let X be any symmetric solution of it. Then the following hold:*

(i) $\hat{H}(z)$ has full normal rank if and only if $B^{\mathrm{T}}XB + R$ is invertible.

(ii) *The inertia*[2] *of* $B^T X B + R$ *is equal to the inertia of* $\hat{H}(z)$ *for all but finitely many* z *on the unit circle.*

(iii) $B^T X B + R \geqslant 0$ *if and only if* $\hat{H}(z) \geqslant 0$ *for any point* z *on the unit circle.*

The last point in the above lemma is basically a special case of the second point but listed separately since it plays an important role. Note that the above lemma implies that a necessary condition for the existence of a solution to a generalized Riccati equation is that the inertia of $\hat{H}(z)$ is independent of z except for some possible singularities. A necessary condition for the existence of a solution with $B^T X B + R \geqslant 0$ (i.e. a solution to an H_2 generalized Riccati equation) is that $\hat{H}(z) \geqslant 0$ for all z on the unit circle. Finally, note that $\hat{H}(z)$, being of full normal rank, guarantees that the generalized inverse in (11.19a) is a normal inverse and that (11.19b) is automatically satisfied, and thus then a generalized Riccati equation reverts back to a standard Riccati equation.

There are certain strong relationships between the solutions of generalized and standard discrete-time Riccati equations and certain appropriately defined continuous-time Riccati equations. Details can be found in [61]. From such interconnections, one can study the conditions under which semi-stabilizing solutions of a generalized discrete-time Riccati equation exist. In general, when such solutions exist, they are not necessarily unique.

[2] The inertia of a matrix is defined as the triple of the number of eigenvalues outside the unit circle, the number of eigenvalues on the unit circle, and the number of eigenvalues inside the unit circle.

Chapter 12

Robust output regulation

12.1 Introduction, problem formulation and some discussions

In Section 2.8 we considered the problem of structural stability. There are a few issues in this respect we would like to discuss in this chapter:

- For structural stability we needed to restrict perturbations in the system parameters since otherwise structural stability is never achievable. We cannot perturb the exosystem and, in the case of saturation, we have to take care that the eigenvalues of A remain in the closed left half plane for continuous-time systems and within or on the unit circle for discrete-time systems.

- By definition it only considers arbitrarily small perturbations of the closed-loop system and does not guarantee that output regulation is preserved for larger perturbations.

- In robust control, it has been an established fact that there is often a lot of a priori information about model uncertainty, and using this a priori information gives us much stronger results. In some cases only certain parameters are subject to perturbation instead of all the parameters. Also, sometimes there is uncertain dynamics which requires dynamic uncertainty instead of parametric uncertainty.

- In view of Corollary 2.8.1 we know that any controller that achieves structural stability must contain q copies of the exosystem where q is the dimension of the error signal. This is a serious limitation since this might make the order of the controller too large to apply it in many applications.

Note that the restrictions in structural stability such as no uncertainty in the exosystem are not related to a priori information but are fundamental limitations of the concept of structural stability. Also note that Theorem 2.8.2 tells us that we actually guarantee output regulation for all perturbations which preserve stability in the closed-loop system. This partially addresses the second point. However, it still leaves the question on how to find a controller which achieves output regulation and guarantees internal stability for all possible perturbations.

We already briefly noted on page 49 that perhaps structural stability is not the right question to ask. To ask for perfect tracking for all possible perturbations is a lot to ask for. Perhaps, keeping the tracking error small over all possible perturbations is more natural. We first restate the concept as already formulated on page 49:

Problem 12.1.1 ***(Local practical structurally stable output regulation problem)*** *Given any nominal values for the system parameters* $(A_0, B_0, E_{w,0}, C_{e,0}, D_{eu,0}, D_{ew,0}, C_{y,0}, D_{yu,0}, D_{yw,0}, S_0)$ *for which the output regulation problem is solvable, for any* $\varepsilon > 0$*, find a controller, if it exists, such that there exists a neighborhood* $\mathcal{P}_0$ *of* $(A_0, B_0, E_{w,0}, C_{e,0}, D_{eu,0}, D_{ew,0}, C_{y,0}, D_{yu,0}, D_{yw,0}, S_0)$ *in the parameter space* $\mathcal{P}$ *such that the interconnection of the given controller with any system with parameters* $(A, B, E_w, C_e, D_{eu}, D_{ew}, C_y, D_{yu}, D_{yw}, S)$ *in* $\mathcal{P}_0$ *yields internal stability and*

$$\lim_{t\to\infty} \|e(t)\| < \varepsilon.$$

It is not hard to see that any controller that achieves internal stability and output regulation also achieves practical structural stability as defined in the above problem. We define below another related problem.

Problem 12.1.2 ***(Practical structurally stable output regulation problem)*** *Given any nominal values for the system parameters* $(A_0, B_0, E_{w,0}, C_{e,0}, D_{eu,0}, D_{ew,0}, C_{y,0}, D_{yu,0}, D_{yw,0}, S_0)$ *for which the output regulation problem is solvable, and given a neighborhood* $\mathcal{P}_0$ *of* $(A_0, B_0, E_{w,0}, C_{e,0}, D_{eu,0}, D_{ew,0}, C_{y,0}, D_{yu,0}, D_{yw,0}, S_0)$ *in the parameter space* $\mathcal{P}$*, for any* $\varepsilon > 0$*, find a controller, if it exists, such that the interconnection of the given controller with any system with parameters* $(A, B, E_w, C_e, D_{eu}, D_{ew}, C_y, D_{yu}, D_{yw}, S)$ *in* $\mathcal{P}_0$ *yields internal stability and*

$$\lim_{t\to\infty} \|e(t)\| < \varepsilon.$$

This problem is much more difficult than Problem 12.1.1. In particular, the question is how the solvability depends on the set $\mathcal{P}_0$. We obviously need to be able to achieve robust stability when the parameters vary over the set $\mathcal{P}_0$. The big question is how far almost output regulation imposes additional constraints on $\mathcal{P}_0$ and whether we still need the multiple copies of the exosystem in the controller. Unfortunately these problems are still open.

To address the problem regarding structure in the model uncertainty and increasing the size of allowed perturbations we consider the model as given in Figure 12.1. where Σ is given by:

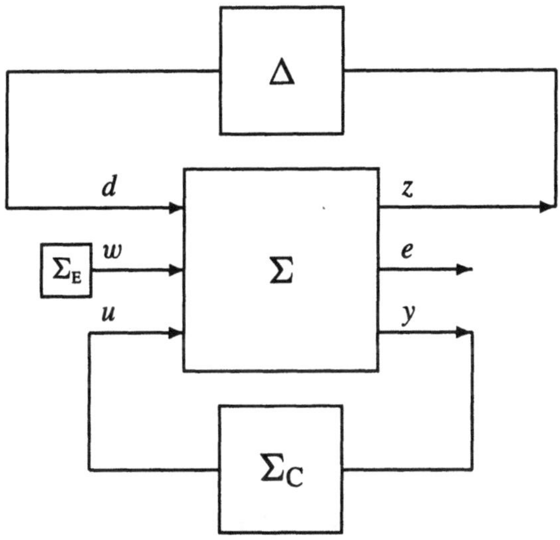

Figure 12.1: Output regulation with model uncertainty

$$\Sigma : \begin{cases} \rho x = Ax + Bu + E_w w + E_d d \\ e = C_e x + D_{eu} u + D_{ew} w \\ z = C_z x + D_{zu} u \\ y = C_y x + D_{yw} w + D_{yd} d. \end{cases} \tag{12.1}$$

As is common in H_∞ optimal control (see any standard textbook such as [44, 95]), the Δ indicates the model uncertainty. It is less than γ in H_∞ norm and can have all kind of additional structure such as square or (block) diagonal, dynamic or static, linear or nonlinear, time-invariant or time-varying. For ease of exposition we denote by $\mathbf{\Delta}$ the set of possible Δ's. We can then formulate the following two problems which are in our view the crucial questions in robust output regulation.

Problem 12.1.3 ***(Robust exact output regulation problem)*** *Consider the interconnection in Figure 12.1 on the page before. Find a controller which yields internal stability and*

$$\lim_{t\to\infty} e(t) = 0$$

for all $\Delta \in \boldsymbol{\Delta}$.

However, as we have seen in structural stability, exact output regulation after perturbation often requires multiple copies of the exosystem in the controller. Therefore by weakening exact output regulation to almost output regulation we can hope that these multiple copies are no longer needed. The robust almost output regulation problem is formulated in the following problem.

Problem 12.1.4 ***(Robust almost output regulation problem)*** *Consider the interconnection in Figure 12.1 on the preceding page. For each* $\varepsilon > 0$ *find a controller which yields internal stability and*

$$\lim_{t\to\infty} \|e(t)\| < \varepsilon$$

for all $\Delta \in \boldsymbol{\Delta}$.

12.2 A result (sufficient condition)

In [2], the robust exact output regulation problem was studied with the additional requirement that the system achieves structural stability. We feel that this paper offers an interesting sufficient condition to the robust exact output regulation problem but we also feel that imposing the structural stability requirement is badly motivated but is crucial in this paper to obtain a solution. Actually, also the results from Chapters 10 (continuous-time) and 11 (discrete-time) allow us to solve this problem with this additional requirement. We describe this solution ihere.

Theorem 12.2.1 *Consider the interconnection in Figure 12.1 on the page before. Let the system* Σ *be given by (7.23). Let Assumptions A.1, A.2, A.3, and A.4 of Chapter 7 hold for the system (7.23). The robust exact output regulation problem is solvable if there exists a controller which achieves regulation and robust stability for the auxiliary system (7.27).*

This result is an immediate consequence of the property we showed in section 2.8 that if we achieve structural stability (which is formulated on the basis of arbitrary small perturbations) then we are guaranteed that we also achieve regulation for any perturbation which preserves internal stability. We know that we achieve structural stability for the system (7.23) if and only if we achieve regulation for the system (7.27). Moreover, for a particular uncertainty Δ, i.e. $d = \Delta z$, the interconnection of the controller and (7.23) is internally stable if and only if the interconnection of the controller and (7.27) is internally stable since for $w = 0$ both systems are the same.

The next question is how to achieve regulation and robust stability for the auxiliary system (7.27). Here we can actually use the arguments as illustrated in the design of a general measurement feedback regulator on page 30. We basically design another auxiliary system such that stabilizing controllers for this second auxiliary system have a $1 - 1$ relationship with stabilizing controllers that achieve regulation for the first auxiliary system. Then the problem of achieving robust stability for the second auxiliary system is a classical robust control problem. The controller we then obtain is connected with a controller which achieves robust stability and regulation for the first auxiliary system. Moreover, this same controller achieves structural stability and robust stability for the original system and therefore the problem is solved.

We emphasize the following aspects:

- In Chapters 10 and 11, we analyzed the loss in performance due to the output regulation constraint. This can obviously also be used to analyze the loss of performance due to a structural stability constraint. Note that structural stability is not necessary to solve the problems of exact or almost robust output regulation and using this analysis of the loss of performance we can check how conservative our results are.

- The main problem of this sufficient condition is the need of multiple copies of the exosystem in the controller which might in many cases not be needed for exact robust output regulation. At the moment we have no method available to obtain controllers which do not contain multiple copies of the exosystem. Also robust almost output regulation might help us to reduce the complexity of the controller but this problem remains open.

Chapter 13

Generalized output regulation

13.1 Introduction

The classical output regulation problem we dealt with so far, although it occupies a marked status in modern control literature, has certain shortcomings. To motivate our work in this chapter, we mention below some of the prominent shortcomings.

(*i*) In the classical output regulation problem, both the reference signals to be tracked and the external disturbances that act on the plant are modeled by an autonomous linear dynamic system called the exosystem. Since the exosystem is autonomous, the reference signal as well as the external disturbances could have only known frequency components. This implies that the class of signals to be tracked as well as the class of external disturbances that act on the plant are rather narrow.

(*ii*) In some classes of problems the system is affected by exogenous signals which are basically unknown but for which information is available such as being differentiable with bounded derivative which cannot be handled via a classical weighting function because this weighting function would be intrinsically unstable.

Our motivation here to reconsider the classical output regulation problem arises from a desire to stamp out the above shortcomings among others. By generalizing the modeling of exosystem, we have the following advantages:

- Any arbitrary reference signal can be treated.
- The derivative or feedforward information of reference signals whenever it is available can be utilized.

- It opens up new avenues to pursue whenever exact output regulation problem is not solvable; for instance, one can study almost output regulation in which we quantify the asymptotic tracking error via one of several possible performance criteria, e.g. the supremum of asymptotic tracking error, is less than a specified fraction of a specified norm of the reference signal.

The modeling method we follow here can easily be introduced via the example illustrated by the master/slave block diagram of Figure 13.1. The master is a non-autonomous dynamic system that produces a desirable behavior for the slave while also modeling external disturbances. The controller or regulator has access to two sets of information, measured outputs of the slave (plant) as well as a certain output of the master, and it generates an input u for the slave. The slave (plant) controlled by this input u produces an output which tries to track the desired behavior dictated by the master. The job of the controller is to generate u so that the tracking error e has certain desirable and specified characteristics. The master/slave configuration of Figure 13.1 has applications in other areas of engineering, e.g. synchronization in communication systems.

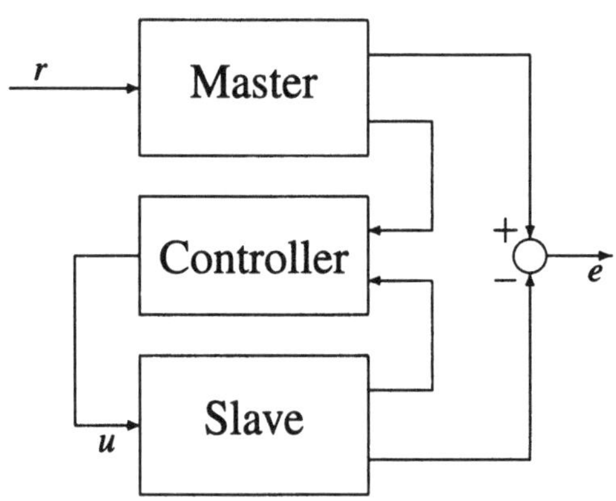

Figure 13.1: Master/Slave topology.

The generalized output regulation as developed here depends critically on the classical notions of exact and almost disturbance decoupling. However, these notions of exact and almost disturbance decoupling have to be examined under a very broad category of controllers beyond the classical state and measurement feedback. This aspect is developed in Section 13.2. Section 13.3 re-formulates the output regulation problem under a generalized model.

Steady state tracking performance is examined in Section 13.4. Here, we formulate and study three hierarchically ordered layers of problems all dealing with steady state tracking performance. The first layer of problems is concerned with the strongest requirement on the steady state tracking error. As in the case of classical output regulation problems, we seek here to render the steady state tracking error exactly zero. The second layer concerns with a weaker requirement of rendering the steady state tracking error arbitrarily close to zero. The third layer imposes the weakest requirement. Here we seek the norm of the amplitude of steady state tracking error be in a specified acceptable range.

This chapter is based mostly on the recent research work of authors [65] and [62,63].

13.2 Exact and almost disturbance decoupling problems

Our purpose in this section is to examine the classical concepts of exact and almost disturbance decoupling under a variety of controllers. As the results available in the literature are not stated in the form necessary for our development in subsequent sections, we need to carefully examine the notions of exact and almost disturbance decoupling and state the new results as needed. The results we present here are indeed generalizations of the existing classical results (see, for instance, [49,58,78,85,88–91]). As in the literature, the solvability conditions of the exact and almost disturbance decoupling problems are expressed in terms of some geometric subspaces defined in Section 1.2.

We proceed now with our development of exact or almost disturbance decoupling. As usual, ρ will denote an operator indicating the time derivative $\frac{d}{dt}$ for continuous-time systems and a forward unit time shift for discrete-time systems. We consider the following linear system,

$$\begin{aligned} \rho x &= Ax + Bu + E_r r \\ y &= C_y x + D_y r \\ z &= C_z x + D_{zu} u, \end{aligned} \tag{13.1}$$

where $x \in \mathbb{R}^n$ is the state, $u \in \mathbb{R}^m$ is the control input, $z \in \mathbb{R}^\ell$ is the to be controlled output, and r stands for a disturbance input.

Our goal in this section is to develop conditions under which the controlled output variable z is insensitive (to be made precise soon) to the disturbance r by an appropriate choice of a controller among any prescribed class of controllers. Traditionally, as well as earlier in this book, only two classes of controllers have been used, (1) state feedback controllers, and (2) measurement feedback controllers. In addition to these classes of controllers, in this chapter, we will also use enhanced versions of these controllers in which certain information regarding the signal $r(t)$ is available for feedback, such as information about $r(t)$, its derivatives or its future values on an interval $[t, t+T]$ for continuous-time systems; information about $r(k)$ or future values on an interval $[k, k+j]$ for discrete-time systems. Information about r is often only available for part of the signal r, e.g. command signals are often available for feedback while disturbances are generally unknown. We assume that information about

$$\bar{r} = Rr$$

is known for some matrix R and the rest of the signal r is unknown and not available for feedback.

We enumerate below different classes of controllers which we will use:

(*i*) **State Feedback Controllers:** In this case, the control law for both continuous- and discrete-time systems is of the form,

$$\Sigma_{C1} : u = Fx. \tag{13.2}$$

(*ii*) **State Feedback + PD Controllers:** In this case, for both continuous- and discrete-time systems the control law is of the form,

$$\Sigma_{C2} : u = Fx + \sum_{i=0}^{j} H_i \rho^i \bar{r}, \tag{13.3}$$

for some integer j where for continuous-time systems $\rho^i \bar{r}$ denotes $\frac{d^i \bar{r}}{dt^i}$ and $\rho^0 \bar{r} = \bar{r}$ while, for discrete-time systems, $(\rho^i \bar{r})(k) = \bar{r}(k+i)$.

(*iii*) **State Feedback + j-th order PD Controllers:** In this case the control law is of the form (13.3) but j is no longer a design parameter but a prescribed integer.

(*iv*) **Measurement Feedback Controllers:** In this case, the control law for both continuous- and discrete-time systems is of the form,

$$\Sigma_{C3} : \begin{cases} \rho v = A_c v + B_c y \\ u \;= C_c v + D_c y. \end{cases} \tag{13.4}$$

(*v*) **Measurement Feedback + PD Controllers:** In this case, for both continuous- and discrete-time systems, the control law is of the form,

$$\Sigma_{C4}: \begin{cases} \rho v = A_c v + B_c y + \sum_{i=0}^{j} G_i \rho^i \bar{r} \\ u \;= C_c v + D_c y + \sum_{i=0}^{j} H_i \rho^i \bar{r} \end{cases} \tag{13.5}$$

for some integer j.

(*vi*) **Measurement Feedback + j-th order PD Controllers:** In this case, the control law is of the form (13.5) but j is no longer a design parameter but a prescribed integer.

(*vii*) **Measurement Feedback + PD + Feedforward Controllers:** In this case, for continuous-time systems, the control law is of the form,

$$\Sigma_{C5}: \begin{cases} \dot{v} = A_c v + B_c y + \sum_{i=0}^{j} G_i \rho^i \bar{r} + \mathcal{G}\bar{r} \\ u = C_c v + D_c y + \sum_{i=0}^{j} H_i \rho^i \bar{r} + \mathcal{H}\bar{r} \end{cases} \tag{13.6}$$

where $\mathcal{G}$ and $\mathcal{H}$ are the operators,

$$(\mathcal{G}\bar{r})(t) = \int_t^{t+T} g(t)\bar{r}(t)dt, \qquad (\mathcal{H}\bar{r})(t) = \int_t^{t+T} h(t)\bar{r}(t)dt$$

for a fixed T and for arbitrary continuous functions g and h.

Remark 13.2.1 *We observe that the state feedback + 0-th order PD controllers with $R = I$, i.e. controllers of the form $u = Fx + H_0 r$, are traditionally called full information feedback controllers.*

We would like to make an important comment on various PD controllers described by Σ_{C2}, Σ_{C4} and Σ_{C5}. For discrete-time systems, the controllers described by Σ_{C2}, Σ_{C4} utilize at every time step k the disturbance $\bar{r}(k)$ as well as its forward shifted values. This implies that these PD controllers anticipate the disturbance that would act on the system in the future. An analogous situation holds for continuous-time systems. In this case, the controllers described by Σ_{C2}, Σ_{C4} utilize at every time instant t the disturbance $\bar{r}(t)$ as well as the values of its time derivatives while the controller described by Σ_{C5} utilizes at time t the disturbance r on a time interval $[t, t+T]$.

Our basic objectives in utilizing any and each of the above classes of controllers are stated precisely in the following two definitions.

Problem 13.2.1 ***Exact disturbance decoupling problem:*** *For the linear system given by (13.1), the exact disturbance decoupling problem with internal stability via any prescribed class of controllers is to find a feedback controller in that class such that the following properties hold:*

(i) *The poles of the closed-loop system consisting of the given plant (13.1) and the controller are in $\mathbb{C}^g$ with $\mathbb{C}^g \subseteq \mathbb{C}^-$ for continuous-time systems, and $\mathbb{C}^g \subseteq \mathbb{C}^{\ominus}$ for discrete-time systems.*

(ii) *For all initial conditions and all r, we have $z(t) \to 0$ as $t \to \infty$.*

Remark 13.2.2 *Note that the above defined problem is termed as the exact disturbance decoupling problem since, whenever a controller exists that solves the above problem, the closed-loop transfer matrix from r to z is equal to* 0.

Problem 13.2.2 ***Almost Disturbance Decoupling Problem:*** *For the linear system given by (13.1), the almost disturbance decoupling problem with internal stability via any prescribed class of controllers is to find a family of feedback controllers parameterized by $\varepsilon > 0$ (i.e. every parameter in the prescribed family of controllers is considered to be a function of a scalar parameter $\varepsilon > 0$) such that the following properties hold:*

(i) *The poles of the closed-loop system consisting of the given plant (13.1) and any controller in the family of feedback controllers are in $\mathbb{C}^g$ with $\mathbb{C}^g \subseteq \mathbb{C}^-$ for continuous-time systems, and $\mathbb{C}^g \subseteq \mathbb{C}^{\ominus}$ for discrete-time systems.*

(ii) *For all initial conditions, and all $r \in L_\infty$ (or $r \in \ell_\infty$), the solution of the closed-loop system satisfies*

$$\limsup_{t\to\infty} \|z(t)\| \leqslant \varepsilon \|r\|_\infty.$$

Remark 13.2.3 *We would like to note that part (ii) is equivalent to the condition that the feedback is such that for all initial conditions there exists a $T > 0$ such that for all $r \in L_\infty$ (or $r \in \ell_\infty$), the solution of the closed-loop system satisfies for all $t > T$,*

$$\|z(t)\| \leqslant \varepsilon\,(\|r\|_\infty + 1)\,. \tag{13.7}$$

It is also equivalent to internal stability combined with the property that for all zero *initial conditions we have*

$$\|z\|_\infty \leqslant \varepsilon \|r\|_\infty.$$

Therefore the above defined problem is termed as the almost disturbance decoupling problem since, whenever a family of controllers exists that solves the above problem, the closed-loop input-output operator from r to z has an L_∞-induced (or ℓ_∞-induced) operator norm less than ε.

Note that due to the relationship between the H_∞ norm and the L_∞-induced (or ℓ_∞-induced) operator norm we know that the almost disturbance decoupling problem is solvable with an L_∞-induced (or ℓ_∞-induced) operator norm if and only if it is solvable with respect to an L_2-induced (or ℓ_2-induced) operator norm. Solvability with respect to an H_2 norm yields slightly weaker solvability conditions.

We categorized earlier different classes of controllers any one of which can be utilized for the purpose of achieving either exact or almost disturbance decoupling. We proceed now to study the solvability conditions for exact and almost disturbance decoupling problems via each of the classes of controllers categorized earlier. At first we like to make certain general observations. Our first observation is well known and concerns exact disturbance decoupling. For any specified type of controller and for a specified stability set $\mathbb{C}^g \subseteq \mathbb{C}^-$ for continuous-time systems and $\mathbb{C}^g \subseteq \mathbb{C}^\ominus$ for discrete-time systems, the solvability conditions for exact disturbance decoupling are the same for both continuous- and discrete-time systems except for the obvious difference that the controlled and conditioned invariant subspaces are taken with respect to different stability sets.

An important question that arises next is this. For a given class of controllers, by relaxing the requirement, say from exact disturbance decoupling to almost disturbance decoupling, can the solvability conditions be weakened? In this regard, it is known that, when continuous-time systems are considered, for any specified type of controller, the solvability conditions required to achieve almost disturbance decoupling are significantly weaker than those required to achieve exact disturbance decoupling. This observation is due to the wise use of asymptotic theory of low-gain and high-gain which plays a crucial role in the realm of continuous-time systems. However, this observation does not quite hold for discrete-time systems since high-gain asymptotic theory does not help us for discrete-time systems due to the bounded stability set. As such, for discrete-time systems, if $\mathbb{C}^g$ is closed, one cannot weaken the solvability conditions by relaxing the requirement from exact to almost disturbance decoupling. On the other hand, if $\mathbb{C}^g$ is not closed, the solvability conditions can be slightly weakened for almost disturbance decoupling owing to the possible use of low-gain asymptotic theory.

Another fundamental query that arises is this. In the context of disturbance decoupling, what is the impact of the availability of $r(t)$ and its derivatives or $r(k)$ and its feedforward values on the solvability conditions required to achieve a given type of (exact or almost) disturbance decoupling? The following observation answers this query for continuous-time systems when state feedback controllers are utilized.

Observation 13.2.1 *For continuous-time systems, if there exists a state feedback + PD controller that renders the L_p induced norm of the closed-loop input-output operator from the disturbance r to the controlled output z exactly equal to a certain value γ, then for any arbitrarily specified $\varepsilon > 0$, there exists as well a state feedback controller that renders the L_p induced norm of the closed-loop input-output operator from r to z exactly equal to $\gamma + \varepsilon$.*

The above fundamental observation leads to an important corollary of it, namely for continuous-time systems whenever a state feedback + PD controller exists that achieves *exact* disturbance decoupling, there exists as well a state feedback controller of the type Σ_{CI} that achieves *almost* disturbance decoupling.

A second corollary of the above observation is also possible, namely for continuous-time systems whenever a family of state feedback + PD controllers exists that achieves *almost* disturbance decoupling, there exists as well a family of state feedback controllers that achieve *almost* disturbance decoupling.

The basic reason why Observation 13.2.1 and its corollaries hold for continuous-time systems is that one can use for them wisely the notion of high-gain asymptotic theory. Observation 13.2.1 and its corollaries do not hold for discrete-time systems because a state feedback + PD controller can actually use future values of the reference signal. That is, for discrete-time systems, the existence of a state feedback + PD controller that renders the ℓ_p induced norm of the closed-loop input-output operator from the disturbance r to the controlled output z exactly equal to a certain value γ, does not necessarily imply the existence of a state feedback controller that renders the ℓ_p induced norm of the closed-loop input-output operator from r to z less than $\gamma + \varepsilon$ for any arbitrarily specified $\varepsilon > 0$. In fact, it turns out that, for discrete-time systems, when we are interested in achieving almost disturbance decoupling by utilizing state feedback, the availability of $r(k)$ at each step k already significantly weakens the solvability conditions compared to the case of state feedback controllers.

The above discussion concerns with state feedback controllers. On the other hand, when we utilize measurement feedback controllers, the following observation shows the impact of the availability of r on the required solvability conditions to achieve either exact or almost disturbance decoupling.

Observation 13.2.2 *For both continuous- and discrete-time systems, when we utilize measurement feedback controllers, the availability of the signal $r(t)$ at each instant t alleviates the conditions needed for the existence of an appropriate observer.*

Throughout this section, the pair (A, B) is said to be $\mathbb{C}^g$-stabilizable if the uncontrollable eigenvalues of A are in the set $\mathbb{C}^g$. Similarly, the pair (C_y, A) is said to be $\mathbb{C}^g$-detectable if the unobservable eigenvalues of A are in the set $\mathbb{C}^g$. Also, a given matrix is said to be $\mathbb{C}^g$-stable if all its eigenvalues are in the set $\mathbb{C}^g$.

We have the following specific results regarding exact disturbance decoupling.

Theorem 13.2.1 *Consider the system given in (13.1) for both continuous- and discrete-time systems. Let $\mathbb{C}^g \subseteq \mathbb{C}^-$ for continuous-time systems, and $\mathbb{C}^g \subseteq \mathbb{C}^\ominus$ for discrete-time systems. Then, the necessary and sufficient conditions under which the exact disturbance decoupling problem with internal stability via any specified class of controllers is solvable are categorized below under each class of controllers:*

(*i*) ***State feedback controllers of the type Σ_{C1} as given in (13.2):***

(*a*) (A, B) *is $\mathbb{C}^g$-stabilizable.*

(*b*) $\operatorname{im} E_r \subseteq \mathcal{V}^g(A, B, C_z, D_{zu})$.

(*ii*) ***State feedback + PD controllers of the type Σ_{C2} as given in (13.3) with j a design parameter:***

(*a*) (A, B) *is $\mathbb{C}^g$-stabilizable.*

(*b*) $E_r \ker R \subseteq \mathcal{V}^g(A, B, C_z, D_{zu})$.

(*c*) $\operatorname{im} E_r \subseteq \mathcal{V}^g(A, B, C_z, D_{zu}) + \mathcal{S}^*(A, B, C_z, D_{zu})$.

(*iii*) ***State feedback + j-th order PD controllers of the type Σ_{C2} as given in (13.3) with j fixed:***

(*a*) (A, B) *is $\mathbb{C}^g$-stabilizable.*

(*b*) $E_r \ker R \subseteq \mathcal{V}^g(A, B, C_z, D_{zu})$.

(*c*) $\operatorname{im} E_r \subseteq \mathcal{V}^g(A, B, C_z, D_{zu}) + \mathcal{S}_j^*(A, B, C_z, D_{zu})$.

(*iv*) ***Measurement feedback controllers of the type Σ_{C3} as given in (13.4):***

(*a*) (A, B) *is* $\mathbb{C}^g$*-stabilizable and* (C_y, A) *is* $\mathbb{C}^g$*-detectable.*

(*b*) $\operatorname{im} E_r \subseteq B \ker D_{zu} + \mathcal{V}^g(A, B, C_z, D_{zu})$.

(*c*) $\ker C_z \supseteq C_y^{-1} \operatorname{im} D_y \cap \mathcal{S}^g(A, E_r, C_y, D_y)$.

(*d*) $\mathcal{S}^g(A, E_r, C_y, D_y) \subseteq \mathcal{V}^g(A, B, C_z, D_{zu})$.

(*v*) ***Measurement feedback + PD controllers of the type Σ_{C4} as given in (13.5) with j a design parameter:***

(*a*) (A, B) *is* $\mathbb{C}^g$*-stabilizable and* (C_y, A) *is* $\mathbb{C}^g$*-detectable.*

(*b*) $E_r \ker R \subseteq B \ker D_{zu} + \mathcal{V}^g(A, B, C_z, D_{zu})$.

(*c*) $\operatorname{im} E_r \subseteq \mathcal{V}^g(A, B, C_z, D_{zu}) + \mathcal{S}^*(A, B, C_z, D_{zu})$.

(*d*) $\ker C_z \supseteq C_y^{-1} \operatorname{im} D_y \cap \mathcal{S}^g(A, E_r Q, C_y, D_y Q)$.

(*e*) $\mathcal{S}^g(A, E_r Q, C_y, D_y Q) \subseteq \mathcal{V}^g(A, B, C_z, D_{zu})$.

Here and in what follows Q is such that $\ker R = \operatorname{im} Q$.

(*vi*) ***Measurement feedback + j-th order PD controllers of the type Σ_{C4} as given in (13.5) with j fixed:***

(*a*) (A, B) *is* $\mathbb{C}^g$*-stabilizable and* (C_y, A) *is* $\mathbb{C}^g$*-detectable.*

(*b*) $E_r \ker R \subseteq B \ker D_{zu} + \mathcal{V}^g(A, B, C_z, D_{zu})$.

(*c*) $\operatorname{im} E_r \subseteq \mathcal{V}^g(A, B, C_z, D_{zu}) + \mathcal{S}_j^*(A, B, C_z, D_{zu})$.

(*d*) $\ker C_z \supseteq C_y^{-1} \operatorname{im} D_y \cap \mathcal{S}^g(A, E_r Q, C_y, D_y Q)$.

(*e*) $\mathcal{S}^g(A, E_r Q, C_y, D_y Q) \subseteq \mathcal{V}^g(A, B, C_z, D_{zu})$.

(*vii*) ***Measurement feedback + PD + feedforward controllers of the type Σ_{C5} as given in (13.6):*** *The conditions are the same as in item* (*v*).

Remark 13.2.4 *We would like to emphasize that, unlike the case of almost disturbance decoupling to be discussed shortly, Theorem 13.2.1 which concerns with exact disturbance decoupling holds for any stability set* $\mathbb{C}^g$*, no matter whether it is closed or open, for instance* $\mathbb{C}^g$ *equal to* $\mathbb{C}^-$.

For continuous-time systems we can of course also have derivative information and information of future values with either j fixed or a design parameter. But also in that case the information on future values does not weaken solvability conditions.

Remark 13.2.5 *For state feedback + 0-th order PD controllers with $R = I$, i.e. for full information feedback controllers of the type $u = Fx + H_0 r$, the solvability conditions given in item (iii) of the above theorem still hold, however in this case $\mathcal{S}_0^*(A, B, C_z, D_{zu})$ can equivalently be replaced by $B \ker D_{zu}$.*

Proof : Part (i) and (iv) are standard results which can for instance be found in [78]. (ii) is a consequence of (iii) and (v) is a consequence of (vi) when we note that from Definition 1.2.2 that $\mathcal{S}_i^*$ is increasing and equal to $\mathcal{S}^*$ after a finite number of steps.

It remains to show (iii), (vi) and (vii). It is easy to see that conditions in (vi) are necessary. Conditions (vi)a),(vi)d) and (vi)e) are necessary to solve the disturbance decoupling problem with measurement feedback for the subset of disturbances r satisfying $Rr = 0$. We can parameterize this subset of disturbances by $\tilde{r}$ where $r = Q\tilde{r}$ and it is obviously necessary for our problem that the disturbance decoupling problem with measurement feedback is solvable for this subset of disturbances since in this case $Rr = 0$.

Conditions (vi)b) and (vi)c) are necessary to solve disturbance decoupling with state feedback + PD information and are hence obviously also necessary to solve the measurement feedback + PD information disturbance decoupling problem. Finally for (iii) it is obviously necessary that disturbance decoupling with state feedback is solvable for $r = Q\tilde{r}$ which yields conditions (iii)a) and (iii)b). The necessity of (iii)c) follows almost directly from the definition of $\mathcal{V}^g(A, B, C_z, D_{zu})$ and $\mathcal{S}_i^*(A, B, C_z, D_{zu})$.

Finally we show that the conditions in (iii) and (vi) are also sufficient. Note that from the definition of $\mathcal{S}_i^*(A, B, C_z, D_{zu})$ it follows almost directly that there exist $F_0, F_1, \ldots, F_i$ such that for any $\tilde{E}$ such that $\operatorname{im} \tilde{E} \subseteq \mathcal{S}_i^*$ we obtain for

$$u_r = \sum_{i=0}^{j} B F_i \tilde{E} \rho^i r,$$

and for zero initial conditions, $z = 0$ for the system

$$\begin{aligned} \rho x &= Ax + Bu_r + \tilde{E}r, \\ z &= C_z x + D_{zu} u_r. \end{aligned}$$

Similarly there exists a F such that $A + BF$ is Hurwitz-stable for continuous-time systems and Schur-stable for discrete-time systems and for any $\operatorname{im} \tilde{E} \subseteq \mathcal{V}^g(A, B, C_z, D_{zu})$ with $u = Fx$ and for zero initial conditions we get $z = 0$

for the system

$$\begin{aligned} \rho x &= Ax + Bu + \tilde{E}r, \\ z &= C_z x + D_{zu} u. \end{aligned}$$

Combining the above it is easy to see that there exist $F, F_0, F_1, \ldots, F_i$ such that for any $\tilde{E}$ such that

$$\operatorname{im} \tilde{E} \subseteq \mathcal{V}^g(A, B, C_z, D_{zu}) + \mathcal{S}_i^*(A, B, C_z, D_{zu}),$$

and for

$$u_r = Fx + \sum_{i=0}^{j} BF_i \tilde{E} \rho^i r,$$

with zero initial conditions, we obtain $z = 0$ for the system

$$\begin{aligned} \rho x &= Ax + Bu_r + \tilde{E}r \\ z &= C_z x + D_{zu} u_r. \end{aligned}$$

Note that this matrix F must be such that $\mathcal{V}^g(A, B, C_z, D_{zu})$ is $(A + BF)$-invariant and contained in $\ker(C + DF)$. Our claim is that the controller

$$u_r = Fx + \sum_{i=0}^{j} BF_i E_r R^\dagger \rho^i \bar{r},$$

will then solve our disturbance decoupling problem where $R^\dagger$ denotes the Moore-Penrose generalized inverse of R. We note that we can always decompose $r = r_1 + r_2 = R^\dagger R r + (I - R^\dagger R) r$. If $r = r_1$ then $E_r R^\dagger \bar{r} = E_r r_1$ and condition (iii)c) guarantees $z = 0$. On the other hand if $r = r_2$ then $\bar{r} = 0$ and hence $u = Fx$. But then condition (iii)b) guarantees that $z = 0$. Due to linearity of the system and controller, any linear combination of r_1 and r_2 then must also yield an output $z = 0$.

Next we consider the measurement feedback problem. According to [61] we can solve the disturbance decoupling problem for the following system

$$\begin{aligned} \rho x &= Ax + Bu + E_r Q r \\ y &= C_y x + D_y Q r \\ z &= C_z x + D_{zu} u \end{aligned} \tag{13.8}$$

by a controller of the form

$$\begin{aligned} \rho v &= Av + Bu + K(C_y v - y) \\ u &= \tilde{F} v + N(C_y v - y), \end{aligned} \tag{13.9}$$

and because the F we constructed before in the state feedback case has the property that $A + BF$ is Hurwitz-stable for continuous-time systems and Schur-stable for discrete-time systems, and $\mathcal{V}^g$ is $(A + BF)$-invariant and contained in $\ker(C + DF)$, we know that we can choose $\tilde{F} = F$.

We claim that the following controller (13.10) then solves the disturbance decoupling problem with internal stability for the original system (13.1),

$$\begin{aligned} \rho v &= Av + Bu + K(C_y v + D_y R^\dagger \bar{r} - y) + E_r R^\dagger \bar{r} \\ u &= \tilde{F} v + N(C_y v + D_y R^\dagger \bar{r} - y) + \textstyle\sum_{i=0}^{j} BF_i E_r R^\dagger \rho^i \bar{r}. \end{aligned} \tag{13.10}$$

That this controller achieves internal stability is obvious. In order to show that it achieves disturbance decoupling we again decompose $r = r_1 + r_2 = R^\dagger Rr + (I - R^\dagger R)r$. If $r = r_2$ then $r = Q\bar{r}$ and $\bar{r} = 0$. Then from the fact that the controller (13.9) achieves $z = 0$ for the system (13.8), we immediately find that the controller (13.10) achieves $z = 0$ for the original system (13.1). On the other hand if $r = r_1$ then in the interconnection of (13.10) and (13.1) it is immediate that for zero initial conditions we obtain $x = v$ and $y = C_y v + D_y r$. But then

$$u = Fx + \sum_{i=0}^{j} BF_i E_r \rho^i r$$

and we already know that in that case we have $z = 0$. Since both r_1 and r_2 yield $z = 0$ and the system and controller are linear, we find that also for arbitrary linear combinations of r_1 and r_2 we get $z = 0$.

A proof in full detail of (vii) requires quite some work. However, intuitively it is quite obvious. Using the Youla parameterization we can see that we have to find stable Q_s such that $T_1 - T_2 Q_s T_3 = 0$ where T_i are all stable rational transfer matrices but then it is obvious that if there exists a Q_s which solves this equation then there also exists a rational Q_s that solves the equation. But obviously a rational Q_s corresponds to a controller of the form (13.5). Obviously it is restrictive if we impose that Q_s must be proper but nonrational solutions for instance controllers of the form (13.6) will not help us. ■

Theorem 13.2.1 deals with the exact disturbance decoupling. We now proceed to study the solvability conditions for the almost disturbance decoupling. Unlike in the case of exact disturbance decoupling, the solvability conditions for almost disturbance decoupling depend on whether we consider continuous- or discrete-time systems. Also, they depend on the nature of the stability set $\mathbb{C}^g$.

We have the following results regarding almost disturbance decoupling for continuous-time systems.

Theorem 13.2.2 *Consider the system given in (13.1) for continuous-time systems. Assume that $\mathbb{C}^g = \mathbb{C}^-$. Then, the necessary and sufficient conditions under which the almost disturbance decoupling problem with internal stability via any specified class of controllers is solvable are categorized below under each class of controllers:*

(i) ***State feedback controllers of the type Σ_{C1} as given in (13.2):***

(a) (A, B) *is* $\mathbb{C}^-$*-stabilizable.*

(b) $\operatorname{im} E_r \subseteq \mathcal{V}^{-0}(A, B, C_z, D_{zu}) + \mathcal{S}^*(A, B, C_z, D_{zu})$.

(c) *Let F be such that $A + BF$ is Hurwitz-stable. For any invariant zero λ on the imaginary axis of the system characterized by the quadruple (A, B, C_z, D_{zu}), we have*

$$\operatorname{im} C_z(\lambda I - A - BF)^{-1}E_r \subseteq \operatorname{im}[C_z(\lambda I - A - BF)^{-1}B + D_{zu}].$$

(ii) ***State feedback + PD controllers of the type Σ_{C2} as given in (13.3) with j a design parameter:*** *The conditions are the same as in item (i).*

(iii) ***State feedback + j-th order PD controllers of the type Σ_{C2} as given in (13.3) with j fixed:*** *The conditions are the same as in item (i).*

(iv) ***Measurement feedback controllers of the type Σ_{C3} as given in (13.4):***

(a) (A, B) *is* $\mathbb{C}^-$*-stabilizable and* (C_y, A) *is* $\mathbb{C}^-$*-detectable.*

(b) $\operatorname{im} E_r \subseteq \mathcal{V}^{-0}(A, B, C_z, D_{zu}) + \mathcal{S}^*(A, B, C_z, D_{zu})$.

(c) $\ker C_z \supseteq \mathcal{S}^{-0}(A, E_r, C_y, D_y) \cap \mathcal{V}^*(A, E_r, C_y, D_y)$.

(d) $$\mathcal{S}^{-0}(A, E_r, C_y, D_y) \cap \mathcal{V}^*(A, E_r, C_y, D_y) \subseteq \mathcal{V}^{-0}(A, B, C_z, D_{zu}) + \mathcal{S}^*(A, B, C_z, D_{zu}).$$

(e) *Let F and K be such that $A + BF$ and $A + KC_y$ are Hurwitz-stable. For any invariant zero λ on the imaginary axis of the system characterized by the quadruple (A, B, C_z, D_{zu}),*

$$\operatorname{im} C_z(\lambda I - A - BF)^{-1}E_r \subseteq \operatorname{im}[C_z(\lambda I - A - BF)^{-1}B + D_{zu}],$$

while for any invariant zero λ *on the imaginary axis of the system characterized by the quadruple* (A, E_r, C_y, D_y),

$$\ker[C_y(\lambda I - A - KC_y)^{-1}E_r + D_y] \subseteq \ker C_z(\lambda I - A - C_yK)^{-1}E_r.$$

(*v*) ***Measurement feedback + PD controllers of the type*** Σ_{C4} ***as given in (13.5) with*** j ***a design parameter:***

(*a*) (A, B) *is* $\mathbb{C}^-$*-stabilizable and* (C_y, A) *is* $\mathbb{C}^-$*-detectable.*

(*b*) $\operatorname{im} E_r \subseteq \mathcal{V}^{-0}(A, B, C_z, D_{zu}) + \mathcal{S}^*(A, B, C_z, D_{zu})$.

(*c*) $\ker C_z \supseteq \mathcal{S}^{-0}(A, E_rQ, C_y, D_yQ) \cap \mathcal{V}^*(A, E_rQ, C_y, D_yQ)$.

(*d*) $\mathcal{S}^{-0}(A, E_rQ, C_y, D_yQ) \cap \mathcal{V}^*(A, E_rQ, C_y, D_yQ) \subseteq \mathcal{V}^{-0}(A, B, C_z, D_{zu}) + \mathcal{S}^*(A, B, C_z, D_{zu})$.

(*e*) *Let* F *and* K *be such that* $A + BF$ *and* $A + KC_y$ *are Hurwitz-stable. For any invariant zero* λ *on the imaginary axis of the system characterized by the quadruple* (A, B, C_z, D_{zu}),

$$\operatorname{im} C_z(\lambda I - A - BF)^{-1}E_r \subseteq \operatorname{im}[C_z(\lambda I - A - BF)^{-1}B + D_{zu}],$$

while for any invariant zero λ *on the imaginary axis of the system characterized by the quadruple* (A, E_r, C_y, D_y),

$$\ker R \cap \ker[C_y(\lambda I - A - KC_y)^{-1}E_r + D_y] \subseteq \ker C_z(\lambda I - A - C_yK)^{-1}E_r.$$

In all the above conditions, again Q *is such that* $\ker R = \operatorname{im} Q$.

(*vi*) ***Measurement feedback +*** j***-th order PD controllers of the type*** Σ_{C4} ***as given in (13.5) with*** j ***fixed:*** *The conditions are the same as in item* (*v*).

(*vii*) ***Measurement feedback with feedforward controllers of the type*** Σ_{C5} ***as given in (13.6):*** *The conditions are the same as in item* (*v*).

Remark 13.2.6 *Theorem 13.2.2 considers the case when* $\mathbb{C}^g$ *is* $\mathbb{C}^-$. *If* $\mathbb{C}^g$ *is not* $\mathbb{C}^-$ *but it is any closed set in* $\mathbb{C}^-$, *the results of Theorem 13.2.2 would still hold provided two modifications are made, (1) the super-scripts* $^{-0}$ *in the conditions given in Theorem 13.2.2 is replaced by* ^{g}, *and (2) the conditions associated with the invariant zeros on the imaginary axis of a system characterized by either the quadruple* (A, B, C_z, D_{zu}) *or* (A, E_r, C_y, D_y) *are removed.*

Proof : Part (i) and (iv) are standard results which can for instance be found in [61]. Note that in that book, the authors minimize the H_2 or H_∞ norm of the closed-loop transfer matrix from r to z while we use the L_∞ induced operator norm. However, we have the following relationship between L_∞ induced operator norm and H_∞ norm as can be found in [8],

$$\|G\|_{H_\infty} \leqslant \|G\|_{L_\infty, L_\infty} \leqslant (2n+1)\|G\|_{H_\infty}$$

where n is the McMillan degree of G. Since the controllers that solve the H_∞ almost disturbance decoupling problem have bounded McMillan degree, this shows that H_∞ almost disturbance decoupling and almost disturbance decoupling in the L_∞ induced operator norm are equivalent. It is also known that the derivative information does not help in almost disturbance decoupling. Therefore the solvability conditions of (iii) and (ii) are the same as in (i).

Although in the measurement feedback case the derivative information does not help us, the fact that Rr is known yields additional information. Therefore the result in part (v) can be obtained from the standard almost disturbance decoupling literature by using the following modified measurement equation,

$$\tilde{y} = \begin{pmatrix} C_y \\ 0 \end{pmatrix} x + \begin{pmatrix} D_y \\ R \end{pmatrix} r.$$

Parts (vi) and (v) have the same solvability conditions because again the derivative information does not weaken the solvability conditions.

Finally, (vii) is a consequence of the known fact that in the H_∞ control problem using feedforward information does not help us in reducing the closed-loop H_∞ norm. The state feedback version of this result is explicitly given in [70] while the measurement feedback case follows from a detailed analysis of the proof in [70] of the measurement feedback H_∞ control problem. ■

The following theorem deals with almost disturbance decoupling for discrete-time systems.

Theorem 13.2.3 *Consider the system given in (13.1) for discrete-time systems. Assume that $\mathbb{C}^g = \mathbb{C}^{\ominus}$. Then, the necessary and sufficient conditions under which the almost disturbance decoupling problem with internal stability via any specified class of controllers is solvable are categorized below under each class of controllers:*

(i) ***State feedback controllers of the type Σ_{CI} as given in (13.2):***

(*a*) (A, B) *is* $\mathbb{C}^{\ominus}$*-stabilizable.*

(*b*) $\operatorname{im} E_r \subseteq \mathcal{V}^{\otimes}(A, B, C_z, D_{zu})$.

(*c*) *Let F be such that* $A + BF$ *is Schur-stable. For any invariant zero* λ *on the unit circle of the system characterized by the quadruple* (A, B, C_z, D_{zu}),

$$\operatorname{im} C_z(\lambda I - A - BF)^{-1} E_r \subseteq \operatorname{im}[C_z(\lambda I - A - BF)^{-1} B + D_{zu}].$$

(*ii*) ***State feedback + PD controllers of the type*** Σ_{C2} ***as given in (13.3) with*** j ***a design parameter:***

(*a*) (A, B) *is* $\mathbb{C}^{\ominus}$*-stabilizable.*

(*b*) $E_r \ker R \subseteq \mathcal{V}^{\otimes}(A, B, C_z, D_{zu})$.

(*c*) $\operatorname{im} E_r \subseteq \mathcal{V}^{\otimes}(A, B, C_z, D_{zu}) + \mathcal{S}^*(A, B, C_z, D_{zu})$.

(*d*) *Let F be such that* $A + BF$ *is Schur-stable. For any invariant zero* λ *on the unit circle of the system characterized by the quadruple* (A, B, C_z, D_{zu}),

$$\operatorname{im} C_z(\lambda I - A - BF)^{-1} E_r \subseteq \operatorname{im}[C_z(\lambda I - A - BF)^{-1} B + D_{zu}].$$

(*iii*) ***State feedback +*** j***-th order PD controllers of the type*** Σ_{C3} ***as given in (13.3) with*** j ***fixed:***

(*a*) (A, B) *is* $\mathbb{C}^{\ominus}$*-stabilizable.*

(*b*) $E_r \ker R \subseteq \mathcal{V}^{\otimes}(A, B, C_z, D_{zu})$.

(*c*) $\operatorname{im} E_r \subseteq \mathcal{V}^{\otimes}(A, B, C_z, D_{zu}) + \mathcal{S}_j^*(A, B, C_z, D_{zu})$.

(*d*) *Let F be such that* $A + BF$ *is Schur-stable. For any invariant zero* λ *on the unit circle of the system characterized by the quadruple* (A, B, C_z, D_{zu}),

$$\operatorname{im} C_z(\lambda I - A - BF)^{-1} E_r \subseteq \operatorname{im}[C_z(\lambda I - A - BF)^{-1} B + D_{zu}].$$

(*iv*) ***Measurement feedback controllers of the type*** Σ_{C3} ***as given in (13.4):***

(*a*) (A, B) *is* $\mathbb{C}^{\ominus}$*-stabilizable and* (C_y, A) *is* $\mathbb{C}^{\ominus}$*-detectable.*

(*b*) $\operatorname{im} E_r \subseteq B \ker D_{zu} + \mathcal{V}^{\otimes}(A, B, C_z, D_{zu})$.

(*c*) $\ker C_z \supseteq C_y^{-1} \operatorname{im} D_y \cap \mathcal{S}^{\otimes}(A, E_r, C_y, D_y)$.

(*d*) $\mathcal{S}^{\otimes}(A, E_r, C_y, D_y) \subseteq \mathcal{V}^{\otimes}(A, B, C_z, D_{zu})$.

(*e*) *Let F and K be such that $A + BF$ and $A + KC_y$ are Schur-stable. For any invariant zero λ on the unit circle of the system characterized by the quadruple (A, B, C_z, D_{zu}),*

$$\operatorname{im} C_z(\lambda I - A - BF)^{-1} E_r \subseteq \operatorname{im}[C_z(\lambda I - A - BF)^{-1} B + D_{zu}],$$

while for any invariant zero λ on the unit circle of the system characterized by the quadruple (A, E_r, C_y, D_y),

$$\ker[C_y(\lambda I - A - KC_y)^{-1} E_r + D_y] \subseteq \ker C_z(\lambda I - A - C_y K)^{-1} E_r.$$

(*v*) ***Measurement feedback + PD controllers of the type Σ_{C4} as given in (13.5) with j a design parameter:***

(*a*) (A, B) *is* $\mathbb{C}^{\ominus}$*-stabilizable and* (C_y, A) *is* $\mathbb{C}^{\ominus}$*-detectable.*

(*b*) $E_r \ker R \subseteq B \ker D_{zu} + \mathcal{V}^{\otimes}(A, B, C_z, D_{zu})$.

(*c*) $\operatorname{im} E_r \subseteq \mathcal{V}^{\otimes}(A, B, C_z, D_{zu}) + \mathcal{S}^*(A, B, C_z, D_{zu})$.

(*d*) $\ker C_z \supseteq C_y^{-1} \operatorname{im} D_y \cap \mathcal{S}^{\otimes}(A, E_r Q, C_y, D_y Q)$.

(*e*) $\mathcal{S}^{\otimes}(A, E_r Q, C_y, D_y Q) \subseteq \mathcal{V}^{\otimes}(A, B, C_z, D_{zu})$.

(*f*) *Let F and K be such that $A + BF$ and $A + KC_y$ are Schur-stable. For any invariant zero λ on the unit circle of the system characterized by the quadruple (A, B, C_z, D_{zu}),*

$$\operatorname{im} C_z(\lambda I - A - BF)^{-1} E_r \subseteq \operatorname{im}[C_z(\lambda I - A - BF)^{-1} B + D_{zu}],$$

while for any invariant zero λ on the unit circle of the system characterized by the quadruple (A, E_r, C_y, D_y),

$$\ker R \cap \ker[C_y(\lambda I - A - KC_y)^{-1} E_r + D_y] \subseteq \ker C_z(\lambda I - A - C_y K)^{-1} E_r.$$

(*vi*) ***Measurement feedback + j-th order PD controllers of the type Σ_{C4} as given in (13.5) with j fixed:***

(*a*) (A, B) *is* $\mathbb{C}^{\ominus}$*-stabilizable and* (C_y, A) *is* $\mathbb{C}^{\ominus}$*-detectable.*

(*b*) $E_r \ker R \subseteq B \ker D_{zu} + \mathcal{V}^{\otimes}(A, B, C_z, D_{zu})$.

(*c*) $\operatorname{im} E_r \subseteq \mathcal{V}^{\otimes}(A, B, C_z, D_{zu}) + \mathcal{S}_j^*(A, B, C_z, D_{zu})$.

(*d*) $\ker C_z \supseteq C_y^{-1} \operatorname{im} D_y \cap \mathcal{S}^{\otimes}(A, E_r Q, C_y, D_y Q)$.

(*e*) $\mathcal{S}^{\otimes}(A, E_r Q, C_y, D_y Q) \subseteq \mathcal{V}^{\otimes}(A, B, C_z, D_{zu})$.

(*f*) *Let* F *and* K *be such that* $A + BF$ *and* $A + KC_y$ *are Schur-stable. For any invariant zero* λ *on the unit circle of the system characterized by the quadruple* (A, B, C_z, D_{zu}),

$$\operatorname{im} C_z(\lambda I - A - BF)^{-1} E_r \subseteq \operatorname{im}[C_z(\lambda I - A - BF)^{-1} B + D_{zu}],$$

while for any invariant zero λ *on the unit circle of the system characterized by the quadruple* (A, E_r, C_y, D_y),

$$\ker R \cap \ker[C_y(\lambda I - A - KC_y)^{-1} E_r + D_y] \subseteq \ker C_z(\lambda I - A - C_y K)^{-1} E_r.$$

Remark 13.2.7 *Theorem 13.2.3 considers the case when* $\mathbb{C}^g$ *is* $\mathbb{C}^{\ominus}$. *If* $\mathbb{C}^g$ *is not* $\mathbb{C}^{\ominus}$ *but it is any closed set within the unit circle, the results of Theorem 13.2.3 do not hold as given. To be specific, when* $\mathbb{C}^g$ *is a closed set within the unit circle, by relaxing the requirement of exact disturbance decoupling to almost disturbance decoupling the solvability conditions do not change at all. In other words, whenever* $\mathbb{C}^g$ *is a closed set within the unit circle, the solvability conditions as given in Theorem 13.2.1 with the super-scripts* $^{\ominus}$ *replaced by* ^{g} *hold irrespective of whether we require exact or almost disturbance decoupling.*

Remark 13.2.8 *For state feedback + 0-th order PD controllers with* $R = I$, *i.e. for full information feedback controllers of the type* $u = Fx + H_0 r$, *the solvability conditions given in item (iii) of the above theorem still hold, however in this case* $\mathcal{S}_0^*(A, B, C_z, D_{zu})$ *can equivalently be replaced by* $B \ker D_{zu}$.

Proof : Part (i) and (iv) are standard results which can for instance be found in [61] since due to the relationship between L_∞ induced operator norm and H_∞ norm as can be found in [8], we can use the results from H_∞ almost disturbance decoupling. Also, (ii) is a consequence of (iii) and (v) is a consequence of (vi) when we note that from Definition 1.2.2 that $\mathcal{S}_i^*$ is increasing and equal to $\mathcal{S}^*$ after a finite number of steps.

It remains to show (iii) and (vi). It is easy to see that conditions in (vi) are necessary. Conditions (vi)a), (vi)d), (vi)e), and (vi)f) are necessary to solve the almost disturbance decoupling problem with measurement feedback for $r = Q\tilde{r}$ which is obviously necessary for our problem as well since in this case $Rr = 0$. Conditions (vi)b) and (vi)c) are necessary to solve almost disturbance decoupling with state feedback + PD information and are hence obviously also necessary to solve the measurement feedback + PD information disturbance decoupling problem. Finally for (iii) it is obviously necessary that disturbance decoupling with state feedback is solvable for $r = Q\tilde{r}$ which yields conditions (iii)a) and (iii)b). The necessity of (iii)c) follows almost directly from the interpretation of $\mathcal{V}^\otimes(A, B, C_z, D_{zu})$ as an almost invariant subspace and the definition of $\mathcal{S}_i^*(A, B, C_z, D_{zu})$.

Finally, we show that the conditions in (iii) and (vi) are also sufficient. We already noted in the proof of Theorem 13.2.1 that there exists $F_0, F_1, \ldots, F_i$ such that for any $\tilde{E}$ such that $\operatorname{im}\tilde{E} \subseteq \mathcal{S}_i^*(A, B, C_z, D_{zu})$ we obtain for

$$u_r = \sum_{i=0}^{j} BF_i\tilde{E}\rho^i r,$$

and for zero initial conditions, $z = 0$ for the system

$$\begin{aligned} \rho x &= Ax + Bu_r + \tilde{E}r, \\ z &= C_z x + D_{zu}u_r. \end{aligned}$$

Similarly for all $\varepsilon > 0$ there exists a F_ε such that $A + BF_\varepsilon$ is Schur-stable and for any $\operatorname{im}\tilde{E} \subseteq \mathcal{V}^\otimes(A, B, C_z, D_{zu})$ with $u = F_\varepsilon x$ and for zero initial conditions, we get $\|z\|_\infty \leqslant \varepsilon\|\tilde{E}r\|$ for the system

$$\begin{aligned} \rho x &= Ax + Bu + \tilde{E}r, \\ z &= C_z x + D_{zu}u. \end{aligned}$$

Combining the above it is easy to see that there exists $F_\varepsilon, F_0, F_1, \ldots, F_i$ such that for any $\tilde{E}$ such that

$$\operatorname{im}\tilde{E} \subseteq \mathcal{V}^g(A, B, C_z, D_{zu}) + \mathcal{S}_i^*(A, B, C_z, D_{zu}),$$

for

$$u_r = Fx + \sum_{i=0}^{j} BF_i\tilde{E}\rho^i r,$$

and for zero initial conditions, we obtain $\|z\|_\infty \leqslant \varepsilon \|\tilde{E}r\|$ for the system

$$\begin{aligned} \rho x &= Ax + Bu_r + \tilde{E}w \\ z &= C_z x + D_{zu} u_r. \end{aligned}$$

Our claim is that the controller

$$u_r = F_\varepsilon x + \sum_{i=0}^{j} BF_i E_r R^\dagger \rho^i \bar{r},$$

will then solve our almost disturbance decoupling problem where $R^\dagger$ denotes the Moore-Penrose generalized inverse of R. We note that we can always decompose $r = r_1 + r_2 = R^\dagger Rr + (I - R^\dagger R)r$. If $r = r_1$ then $E_r R^\dagger \bar{r} = E_r r_1$ and condition (iii)c) guarantees $\|z\|_\infty \leqslant \varepsilon \|E_r r_1\|_\infty$. On the other hand if $r = r_2$ then $\bar{r} = 0$ and hence $u = Fx$. But then condition (iii)b) guarantees that $\|z\|_\infty \leqslant \varepsilon \|E_r r_2\|_\infty$. Due to linearity of the system and controller, any linear combination of r_1 and r_2 then also yields an output z with the property

$$\|z\|_\infty \leqslant \varepsilon \|E_r r_1\|_\infty + \varepsilon \|E_r r_2\|_\infty \leqslant \varepsilon M \|r\|_\infty.$$

Next we consider the measurement feedback problem. According to [58, 61] we can solve the almost disturbance decoupling problem for the following system

$$\begin{aligned} \rho x &= Ax + Bu + E_r Qr \\ y &= C_y x + D_y Qr \\ z &= C_z x + D_{zu} u \end{aligned} \tag{13.11}$$

by a controller of the form

$$\begin{aligned} \rho v &= Av + Bu + K_\varepsilon (C_y v - y) \\ u &= \tilde{F}_\varepsilon v + N_\varepsilon (C_y v - y), \end{aligned} \tag{13.12}$$

and because the F_ε we constructed before has the property that it achieves almost disturbance decoupling for any $\tilde{E}$ with $\tilde{E} \subseteq \mathcal{V}^\otimes(A, B, C_z, D_{zu})$ we know that we can choose $\tilde{F}_\varepsilon = F_\varepsilon$.

We claim that the following controller (13.13) then solves the almost disturbance decoupling problem with internal stability for the original system (13.1),

$$\begin{aligned} \rho v &= Av + Bu + K_\varepsilon (C_y v + D_y R^\dagger \bar{r} - y) + E_r R^\dagger \bar{r} \\ u &= F_\varepsilon v + N_\varepsilon (C_y v + D_y R^\dagger \bar{r} - y) + \textstyle\sum_{i=0}^{j} BF_i E_r R^\dagger \rho^i \bar{r}. \end{aligned} \tag{13.13}$$

That this controller achieves internal stability is obvious. In order to show that it achieves almost disturbance decoupling we again decompose $r = r_1 + r_2 = R^{\dagger}Rr + (I - R^{\dagger}R)r$. If $r = r_2$ then $r = Q\bar{r}$ and $\bar{r} = 0$. Then from the fact that the controller (13.12) achieves $\|z\| \leqslant \varepsilon\|r_2\|$ for the system (13.11), we immediately find that the controller (13.13) achieves $\|z\| \leqslant \varepsilon\|r_2\|$ for the original system (13.1). On the other hand if $r = r_1$ then it is immediate that, in the interconnection of (13.13) and (13.1), we have for zero initial conditions $x = v$ and $y = C_y v + D_y r$. But then

$$u = Fx + \sum_{i=0}^{j} BF_i E_r \rho^i r,$$

and we already know that in that case we have $\|z\| \leqslant \varepsilon\|E_r r_1\|$. Since $r_2 = 0$ yields $\|z\| \leqslant \varepsilon\|E_r r_1\|$ and $r_1 = 0$ yields $\|z\| \leqslant \varepsilon\|r_2\|$ and system and controller are linear we find that also for arbitrary linear combinations of r_1 and r_2 we get $\|z\| \leqslant M_2\varepsilon\|r\|$ for some M_2 and hence we can make the induced norm arbitrarily small. ■

13.3 Generalized output regulation

In this section we revisit output regulation with a generalized model for the exosystem. As discussed earlier, in classical output regulation problems, both the external disturbances that act on the given plant as well as the desired output signals that are to be tracked are modeled by an autonomous exosystem. Unlike in the classical case, we propose here a non-autonomous exosystem, i.e. a system driven by a reference signal denoted by $r(t)$. As will be seen later on such a non-autonomous model for the exosystem leads us to achieve all the objectives outlined in introduction, namely, (1) broadening of the class of signals that one can track and broadening of the class of signals that one can reject, and (2) formulating and studying a hierarchy of output regulation problems in which the requirements on steady state tracking error are gradually relaxed from being exactly zero to almost zero or in some acceptable range.

The basic configuration of the proposed output regulation scheme is depicted in Figure 13.2.

The analytical model of the plant and the exosystem as depicted in Figure

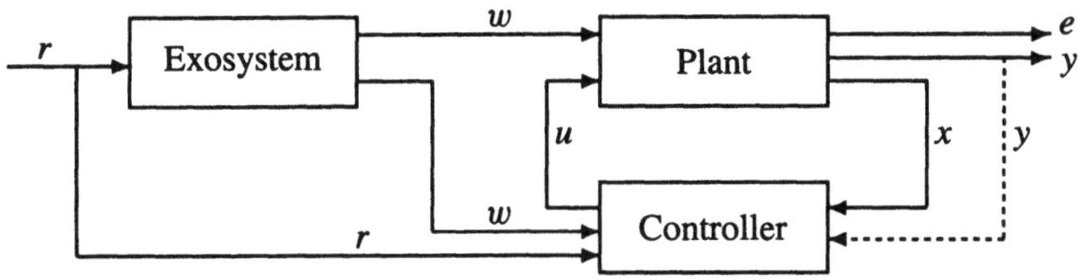

Figure 13.2: Configuration of the proposed output regulation scheme

13.2 is described by the multivariable linear system given below,

$$\begin{aligned} \rho x &= Ax + Bu + E_w w \\ \rho w &= Sw + D_{wr} r \\ y &= C_y x + D_{yw} w + D_{yr} r \\ e &= C_e x + D_{eu} u + D_{ew} w \end{aligned} \tag{13.14}$$

where, as usual, the first equation of this system describes a plant, with state $x \in \mathbb{R}^n$, and input $u \in \mathbb{R}^m$, subject to the effect of a *disturbance* represented by $E_w w$. The third equation describes the measurement signal $y \in \mathbb{R}^p$ and the last equation defines the error $e \in \mathbb{R}^q$ between the actual plant output $C_e x + D_{eu} u$ and a *reference* signal $-D_{ew} w$ which the plant output is required to track. The second equation, describes the non-autonomous exosystem, with state $w \in \mathbb{R}^s$ and an external driving signal r. The exosystem models the class of disturbance or reference signals taken into consideration.

The system represented by equation (13.14) can also be viewed as a master/slave system. In this point of view, the master system consists of the exosystem driven by the signal r, namely

$$\rho w = Sw + D_{wr} r,$$

where as the slave consists of the plant,

$$\begin{aligned} \rho x &= Ax + Bu + E_w w \\ y &= C_y x + D_{yw} w + D_{yr} r. \end{aligned}$$

The controller is to be designed so that the slave obeys the master such that the so called error signal e,

$$e = C_e x + D_{eu} u + D_{ew} w,$$

has certain desirable properties. Such a master/slave configuration of Figure 13.1 has applications not only here in connection with output regulation but also elsewhere.

A straightforward advantage in utilizing the configuration of Figure 13.2 and the analytical model (13.14) is this. In the classical case where the exosystem is autonomous, one has to know explicitly all the frequency components of the signal that needs to be tracked in order to come up with a model for the exosystem. However, it is clear that, by utilizing a non-autonomous exosystem driven by a reference signal $r(t)$ or $r(k)$, one does not necessarily need to have such a knowledge. In fact, by an appropriate selection of the driving signal r, a non-autonomous exosystem can be constructed so that any arbitrarily specified signal can be modeled as a signal that needs to be tracked. Moreover, the class of external disturbances that could act on the given plant can significantly be broadened.

The following example illustrates how a common tracking problem that is often discussed in the literature can be cast as a special case of the formulation of Figure 13.2 and system (13.14).

Example 13.3.1 *Consider a continuous-time linear system,*

$$\begin{aligned} \dot{x} &= Ax + Bu \\ y &= C_y x, \end{aligned}$$

where $x \in \mathbb{R}^n$ is the state, $u \in \mathbb{R}^m$ is the control input and $y \in \mathbb{R}^p$ is the output. Suppose we want to design a control law that would cause the output y to track any desired $y_d \in \mathcal{C}^{(s-1)}$ for some fixed $s \geqslant 1$. This problem can be recast to fit the above formulation of the output regulation problem. Indeed, the exosystem in this case is given by

$$\begin{aligned} \dot{w}_i &= w_{i+1}, \ i = 1, 2, \cdots, s-1 \\ \dot{w}_s &= y_d^{(s-1)}. \end{aligned}$$

By considering $y_d^{(s-1)}$ as the reference signal r, the above exosystem equation can be rewritten as

$$\rho w = Sw + D_{wr} r,$$

where S and D_{wr} are of dimension $s \times s$ and $s \times 1$ respectively, and

$$S = \begin{pmatrix} 0 & I \\ 0 & 0 \end{pmatrix}, \quad \text{and} \quad D_{wr} = \begin{pmatrix} 0 \\ 1 \end{pmatrix}.$$

The plant is not disturbed by w, i.e. $E_w = 0$. The error signal e is formed by

$$e = C_e x + D_{ew} w$$

where $C_e = C_y$, $D_{eu} = 0$, and $D_{ew} = \begin{pmatrix} 1 & 0 & \cdots & 0 \end{pmatrix}$.

Remark 13.3.1 *The problem dealt with in [29, 30] is a special case of the above example. In fact, by taking the dimension of the exosystem equal to* 1, *one obtains the problem dealt with in [29, 30].*

Obviously, the proposed scheme of Figure 13.2 and equation (13.14) is a generalization of the classical output regulation scheme as it introduces the driving signal r to the exosystem. Although such an introduction of the driving signal r appears to be simple and straightforward, as alluded to earlier, it opens up several avenues to generalize the classical output regulation theory. As pointed out earlier, the goal of the classical output regulation theory has been to render the steady state tracking error to zero even under the influence of persistent disturbances. In this regard, the following section studies a hierarchy of problems seeking different properties for the steady state error.

As in the previous chapters, we need in this chapter as well the following three assumptions.

A.1. The pair (A, B) is stabilizable.

A.2. The matrix S is anti-Hurwitz-stable for continuous-time systems and anti-Schur-stable for discrete-time systems.

A.3. The pair $(\begin{pmatrix} C_y & D_{yw} \end{pmatrix}, \begin{pmatrix} A & E_w \\ 0 & S \end{pmatrix})$ is detectable.

13.4 Steady state tracking performance

In this section, we formulate and study three hierarchically ordered layers of problems all dealing with steady state tracking performance. The first layer of problems concerns with the strongest requirement on the steady state tracking error. As in the case of classical output regulation problems, we seek here to render the steady state tracking error exactly zero. The second layer concerns with a weaker requirement of rendering the steady state tracking error arbitrarily close to zero. The third layer imposes the weakest requirement. Here we seek the norm of the amplitude of steady state tracking error be in a specified acceptable range.

Since our goal is output regulation, various controllers we utilize here are called as usual regulators. Six classes of regulators can be used, (1) state feedback regulators, (2) state feedback + PD regulators, (3) state feedback + j-th order PD regulators, (4) measurement feedback regulators, (5) measurement feedback + PD regulators, and (6) measurement feedback + j-th order PD regulators. Obviously we could also use the class of measurement feedback + feedforward + PD regulators like we did in (almost) disturbance decoupling. But it was already noted there that it does not weaken the solvability conditions and generally does not help us. Really, the continuous-time equivalent of feedforward shifted values of reference signals is the use of derivative information. This is most obvious in the frequency domain but in all the solvability conditions this equivalence is also clearly present.

We devote a subsection to each layer, and study the different types of output regulation problems for each particular class of regulators. Every problem is studied not only to develop the conditions under which it can be solved, but also to synthesize appropriate regulators that can solve such a problem. Our study benefits significantly by the notions of exact and almost disturbance decoupling which we reviewed in Subsection 13.2.

13.4.1 Exact steady state tracking – exact output regulation

In this subsection, as in the case of classical output regulation, our concern is to render the steady state tracking error exactly zero. We formulate and study six different problems each using a particular class of regulators.

We first have the following precise problem statement.

Problem 13.4.1 *Let a system of the form (13.14) be given. The* exact output regulation problem via state feedback regulators *is to find, if possible, a state feedback law of the form $u = Fx + Gw$ such that the following properties hold:*

(i) The system $\rho x = (A + BF)x$ is asymptotically stable.

(ii) For all initial conditions, and for any signal r (piecewise continuous in continuous-time) the solution of

$$\begin{aligned} \rho x &= (A + BF)x + (E_w + BG)w \\ \rho w &= Sw + D_{wr}r \\ e &= (C_e + D_{eu}F)x + (D_{ew} + D_{eu}G)w \end{aligned} \tag{13.15}$$

satisfies $\lim_{t\to\infty} e(t) = 0$.

We have the following solvability conditions for the exact output regulation problem via state feedback regulators.

Theorem 13.4.1 *Consider the system given in (13.14) for both continuous- and discrete-time systems. Let Assumptions A.1 and A.2 be satisfied. Then, the exact output regulation problem via state feedback regulators is solvable if and only if the following conditions are true:*

(i) *There exist matrices Π and Γ that solve the regulator equation,*

$$\begin{aligned} \Pi S &= A\Pi + B\Gamma + E_w, \\ 0 &= C_e\Pi + D_{eu}\Gamma + D_{ew}. \end{aligned} \tag{13.16}$$

(ii) *For continuous-time systems we have*

$$\operatorname{im} \Pi D_{wr} \subseteq \mathcal{V}^-(A, B, C_e, D_{eu}),$$

and for discrete-time systems we have

$$\operatorname{im} \Pi D_{wr} \subseteq \mathcal{V}^{\ominus}(A, B, C_e, D_{eu}).$$

Moreover, by knowing Π and Γ that solve the regulator equation (13.16), and by developing a feedback law $\tilde{u} = F\tilde{x}$ that solves the exact disturbance decoupling problem via state feedback for the system,

$$\begin{aligned} \rho\tilde{x} &= A\tilde{x} + B\tilde{u} - \Pi D_{wr} r \\ e &= C_e\tilde{x} + D_{eu}\tilde{u}, \end{aligned} \tag{13.17}$$

we can solve the posed exact output regulation problem by utilizing the feedback law,

$$u = Fx + (\Gamma - F\Pi)w. \tag{13.18}$$

Proof : Suppose that we have a controller of the form $u = Fx + Gw$ that achieves output regulation. First we assume $r = 0$. Then it is easy to check that $u = Fx + Gw$ yields $\lim_{t\to\infty} e(t) = 0$ for all initial conditions if and only if there exists a Π such that

$$\Pi S = (A + BF)\Pi + (E_w + BG), \quad C_e\Pi + D_{eu}(G + F\Pi) + D_{ew} = 0.$$

If we choose $\Gamma = G + F\Pi$ then it is obvious that (i) is satisfied.

Next we consider the general case where r is not necessarily 0. Then for our controller $u = Fx + Gw$, which is known to achieve exact output regulation, the closed-loop system in the new coordinate $\tilde{x} = x - \Pi w$ can be written as in equation (13.17) where

$$\tilde{u} = F\tilde{x}. \tag{13.19}$$

It is clear then that the exact disturbance decoupling problem via state feedback for the system (13.17) is solved by the feedback law (13.19), and hence by using Theorem 13.2.1 we find that (ii) is satisfied.

In order to show sufficiency we assume that (i) and (ii) are satisfied. By (ii) we know that there exists a feedback law $u = F\tilde{x}$ for the system (13.17) that achieves exact disturbance decoupling problem via state feedback. But then it is easy to check that (13.18) solves the exact output regulation problem via state feedback. ■

Sometimes we know all the derivatives of the signal $r(t)$ in continuous-time systems, or all the forward shifted values of $r(k)$ in discrete-time systems. In this case, the solvability conditions can be relaxed. This is formulated in the following problem.

Problem 13.4.2 *Let a system of the form (13.14) be given. The* exact output regulation problem via state feedback + PD regulators *is to find, if possible, a feedback law of the form*

$$u = Fx + Gw + \sum_{i=0}^{j} H_i \rho^i \bar{r} \tag{13.20}$$

for some $j \geqslant 0$ with $\bar{r}(t) = Rr(t)$, such that the following properties hold:

(i) *The system $\rho x = (A + BF)x$ is asymptotically stable.*

(ii) *For all initial conditions, and for all signals r (in continuous-time $r \in \mathcal{C}^j$) the solution of*

$$\begin{aligned} \rho x &= (A + BF)x + (E_w + BG)w + B \textstyle\sum_{i=0}^{j} H_i \rho^i \bar{r} \\ \rho w &= Sw + D_{wr} r \\ e &= (C_e + D_{eu}F)x + (D_{ew} + D_{eu}G)w + D_{eu} \textstyle\sum_{i=0}^{j} H_i \rho^i \bar{r} \end{aligned} \tag{13.21}$$

is such that $\lim_{t\to\infty} e(t) = 0$.

The solvability conditions for the above problem are given in the following theorem.

Theorem 13.4.2 *Consider the system given in (13.14) for both continuous- and discrete-time systems. Let Assumptions A.1 and A.2 be satisfied. Then, the exact output regulation problem via state feedback + PD regulators is solvable if and only if the following conditions are true:*

(*i*) *There exist matrices* Π *and* Γ *that solve the regulator equation (13.16).*

(*ii*) *For continuous-time systems we have*

$$\begin{aligned}\Pi D_{wr}\ker R &\subseteq \mathcal{V}^-(A,B,C_e,D_{eu}),\\ \operatorname{im}\Pi D_{wr} &\subseteq \mathcal{V}^-(A,B,C_e,D_{eu})+\mathcal{S}^*(A,B,C_e,D_{eu}),\end{aligned}$$

and for discrete-time systems we have

$$\begin{aligned}\Pi D_{wr}\ker R &\subseteq \mathcal{V}^{\ominus}(A,B,C_e,D_{eu}),\\ \operatorname{im}\Pi D_{wr} &\subseteq \mathcal{V}^{\ominus}(A,B,C_e,D_{eu})+\mathcal{S}^*(A,B,C_e,D_{eu}).\end{aligned}$$

Moreover, under the above conditions, for some $j \geq 0$*, there exists a feedback law of the form* $\tilde{u} = F\tilde{x} + \sum_{i=0}^{j} H_i\rho^i\bar{r}$ *that solves the exact disturbance decoupling problem for the system (13.17). In that case, utilizing such* F *and* H_i*, the exact output regulation problem via state feedback + PD regulators is solved by the feedback law,*

$$u = Fx + (\Gamma - F\Pi)w + \sum_{i=0}^{j} H_i\rho^i\bar{r}. \tag{13.22}$$

Proof : It follows from the straightforward application of Theorems 13.2.1 and 13.4.1. ∎

In the case when we know only some derivatives of the signal r in continuous-time systems, or some forward shifted values of the sequence r in discrete-time systems, we can also relax the solvability conditions to some extent. We begin with the following problem formulation.

Problem 13.4.3 *Let a system of the form (13.14) be given. The* exact output regulation problem via state feedback + j-th order PD regulators *is to find, if possible, a feedback law of the form (13.20) for a specified* $j \geqslant 0$ *such that the following properties hold:*

(*i*) *The system* $\rho x = (A + BF)x$ *is asymptotically stable.*

(*ii*) *For all initial conditions and for all signals* r *(in continuous-time* $r \in \mathcal{C}^j$*) the solution of (13.21) is such that* $\lim_{t\to\infty} e(t) = 0$.

The solvability conditions for the above problem are given as follows.

Theorem 13.4.3 *Consider the system given in (13.14) for both continuous- and discrete-time systems. Let Assumptions A.1 and A.2 be satisfied. Then, the exact output regulation problem via state feedback + j-th order PD regulators is solvable if and only if the following conditions are true:*

(*i*) *There exist matrices* Π *and* Γ *that solve the regulator equation (13.16).*

(*ii*) *For continuous-time systems we have*

$$\begin{aligned} \Pi D_{wr} \ker R &\subseteq \mathcal{V}^-(A, B, C_e, D_{eu}), \\ \operatorname{im} \Pi D_{wr} &\subseteq \mathcal{V}^-(A, B, C_e, D_{eu}) + \mathcal{S}_j^*(A, B, C_e, D_{eu}), \end{aligned}$$

and for discrete-time systems we have

$$\begin{aligned} \Pi D_{wr} \ker R &\subseteq \mathcal{V}^\ominus(A, B, C_e, D_{eu}), \\ \operatorname{im} \Pi D_{wr} &\subseteq \mathcal{V}^\ominus(A, B, C_e, D_{eu}) + \mathcal{S}_j^*(A, B, C_e, D_{eu}). \end{aligned}$$

Moreover, by knowing Π *and* Γ *that solve the regulator equation (13.16), and by developing a feedback law* $\tilde{u} = F\tilde{x} + \sum_{i=0}^{j} H_i \rho^i \bar{r}$ *that solves the exact disturbance decoupling problem via state feedback for the system (13.17). In that case, utilizing such* F *and* H_i*, we can solve the posed exact output regulation problem by utilizing the feedback law (13.22).*

Proof : It follows from the straightforward application of Theorems 13.2.1 and 13.4.1. ■

So far we have considered state variable feedback regulators. We next move on to consider the case when only the measurement variable y is available for feedback. We have the following basic problem.

Problem 13.4.4 *Let a system of the form (13.14) be given. The* exact output regulation problem via measurement feedback regulators *is to find, if possible, a measurement feedback law of the form,*

$$\begin{aligned} \rho v &= A_c v + B_c y \\ u &= C_c v + D_c y, \end{aligned} \tag{13.23}$$

such that the following properties hold:

(*i*) *The system*

$$\begin{aligned} \rho x &= (A + BD_cC_y)x + BC_cv \\ \rho v &= B_cC_yx + A_cv \end{aligned} \tag{13.24}$$

is asymptotically stable.

(*ii*) *For all initial conditions, and all signals r (piecewise continuous in continuous-time), the solution e of the interconnection of (13.23) and (13.14) satisfies* $\lim_{t\to\infty} e(t) = 0$.

We have the following result concerning the above problem.

Theorem 13.4.4 *Consider the system given in (13.14) for both continuous- and discrete-time systems. Let Assumptions A.1, A.2, and A.3 be satisfied. Let* $\mathbb{C}^g = \mathbb{C}^-$ *for continuous-time systems, and* $\mathbb{C}^g = \mathbb{C}^{\ominus}$ *for discrete-time systems. Then, the exact output regulation problem via measurement feedback regulators is solvable if and only if the following conditions are true:*

(*i*) *There exist matrices* Π *and* Γ *that solve the regulator equation (13.16).*

(*ii*) $\operatorname{im} \Pi D_{wr} \subseteq \mathcal{V}^g(A, B, C_e, D_{eu}) + \operatorname{im} B \ker D_{eu}$.

(*iii*) $\ker \begin{pmatrix} C_e & D_{eu} \end{pmatrix} \supseteq \mathcal{S}^g(A_e, D_{wr,e}, C_{y,e}, D_{yr}) \cap C_{y,e}^{-1} \operatorname{im} D_{yr}$.

(*iv*) $\mathcal{S}^g(A_e, D_{wr,e}, C_{y,e}, D_{yr}) \subseteq \mathcal{V}^g(A, B, C_e, D_{eu}) \oplus \mathbb{R}^s$.

Here A_e, $D_{wr,e}$, *and* $C_{y,e}$ *are defined by*

$$A_e = \begin{pmatrix} A & -B\Gamma \\ 0 & S \end{pmatrix}, \quad D_{wr,e} = \begin{pmatrix} -\Pi D_{wr} \\ D_{wr} \end{pmatrix}, \quad \textit{and}$$
$$C_{y,e} = \begin{pmatrix} C_y & C_y\Pi + D_{yw} \end{pmatrix}.$$

Moreover, if conditions (i)-(iv) are satisfied, then there exists a proper controller characterized by the matrix quadruple $(\bar{A}_c, \bar{B}_c, \bar{C}_c, \bar{D}_c)$ *that solves the exact disturbance decoupling problem via measurement feedback for the system (13.25),*

$$\begin{aligned} \rho\bar{x} &= \begin{pmatrix} A & -B\Gamma \\ 0 & S \end{pmatrix}\bar{x} + \begin{pmatrix} 0 & B \\ I & 0 \end{pmatrix}\bar{u} + \begin{pmatrix} -\Pi D_{wr} \\ D_{wr} \end{pmatrix} r \\ e &= \begin{pmatrix} C_e & -D_{eu}\Gamma \end{pmatrix}\bar{x} + \begin{pmatrix} 0 & D_{eu} \end{pmatrix}\bar{u} \\ \bar{y} &= \begin{pmatrix} C_y & C_y\Pi + D_{yw} \end{pmatrix}\bar{x} + D_{yr} r. \end{aligned} \tag{13.25}$$

In that case the exact output regulation problem via measurement feedback regulators is solved by the controller or regulator,

$$\Sigma_C : \begin{cases} \rho v_1 = & S v_1 + \bar{C}_{c,1} v_2 + \bar{D}_{c,1}(y + (D_{yw} + C_y \Pi) v_1) \\ \rho v_2 = & \bar{A}_c v_2 + \bar{B}_c (y + (D_{yw} + C_y \Pi) v_1) \\ u = & -\Gamma v_1 + \bar{C}_{c,2} v_2 + \bar{D}_{c,2}(y + (D_{yw} + C_y \Pi) v_1), \end{cases} \tag{13.26}$$

where $\bar{C}_{c,1}$, $\bar{C}_{c,2}$, $\bar{D}_{c,1}$, and $\bar{D}_{c,2}$ are obtained by partitioning $\bar{C}_c$ and $\bar{D}_c$ in conformity with the partitioning of $\bar{A}_e$,

$$\bar{C}_c = \begin{pmatrix} \bar{C}_{c,1} \\ \bar{C}_{c,2} \end{pmatrix}, \quad \text{and } \bar{D}_c = \begin{pmatrix} \bar{D}_{c,1} \\ \bar{D}_{c,2} \end{pmatrix}. \tag{13.27}$$

Proof : Suppose that we have a controller of the form (13.23) that achieves output regulation. Then, as in (2.17), it is easy to check that $\lim_{t \to \infty} e(t) = 0$ for all initial conditions if and only if there exists Π and Θ such that

$$\Pi S = A \Pi + E_w + B D_c (C_y \Pi + D_{yw}) + B C_c \Theta, \tag{13.28a}$$

$$\Theta S = A_c \Theta + B_c (C_y \Pi + D_{yw}), \tag{13.28b}$$

$$0 = C_e \Pi + D_{ew} + D_{eu} D_c (C_y \Pi + D_{yw}) + D_{eu} C_c \Theta. \tag{13.28c}$$

We first claim that, given our controller that achieves output regulation, there always exists a new controller that achieves output regulation and the same transfer matrix from w to r but in this case with a Θ which is injective. Given the results of Theorem 2.6.1, we know that in a suitable basis we have the following forms for various matrices such that the pair $(C_y \Pi_2 + D_{yw2},\ S_{22})$ is observable,

$$S = \begin{pmatrix} S_1 & S_{12} \\ 0 & S_{22} \end{pmatrix}, \quad \Theta = \begin{pmatrix} I & \Theta_{12} \\ 0 & \Theta_{22} \end{pmatrix}, \tag{13.29}$$

$$\Gamma = \begin{pmatrix} \Gamma_1 & \Gamma_2 \end{pmatrix}, \quad C_c = \begin{pmatrix} C_{c,1} & C_{c,2} \end{pmatrix}, \quad \Pi = \begin{pmatrix} \Pi_1 & \Pi_2 \end{pmatrix}, \tag{13.30}$$

$$A_c = \begin{pmatrix} S_1 & A_{c,12} \\ 0 & A_{c,22} \end{pmatrix}, \quad B_c = \begin{pmatrix} B_{c,1} \\ B_{c,2} \end{pmatrix}, \quad D_{yw} = \begin{pmatrix} -C_y \Pi_1 & D_{yw2} \end{pmatrix}. \tag{13.31}$$

Choose $B_{c,3}$ such that

$$A_{c,33} = S_{22} - B_{c,3}(C_y \Pi_2 + D_{yw2})$$

is Hurwitz-stable for continuous-time systems, or Schur-stable for discrete-time systems. Then the new controller described by,

$$A_c = \begin{pmatrix} S_1 & A_{c,12} & 0 \\ 0 & A_{c,22} & 0 \\ 0 & 0 & A_{c,33} \end{pmatrix}, \quad B_c = \begin{pmatrix} B_{c,1} \\ B_{c,2} \\ B_{c,3} \end{pmatrix}, \quad C_c = \begin{pmatrix} C_{c,1} & C_{c,2} & 0 \end{pmatrix},$$

and the direct feedthrough matrix equal to D_c (the direct feedthrough matrix from our original controller) yields the same closed-loop system from w to r and achieves output regulation. On the other hand for this controller, equation (13.28), is satisfied with the same Π but a new Θ given by

$$\Theta = \begin{pmatrix} I & \Theta_{12} \\ 0 & \Theta_{22} \\ 0 & I \end{pmatrix}.$$

Obviously this new Θ is injective.

If Θ is injective, then in a suitable basis we can guarantee that $\Theta = \begin{pmatrix} I & 0 \end{pmatrix}^{\mathrm{T}}$. In this new basis the controller takes a special form,

$$A_c = \begin{pmatrix} S - B_{c,1}(C_y\Pi + D_{yw}) & A_{c,12} \\ -B_{c,2}(C_y\Pi + D_{yw}) & A_{c,22} \end{pmatrix}, \qquad B_c = \begin{pmatrix} B_{c,1} \\ B_{c,2} \end{pmatrix},$$

$$C_c = \begin{pmatrix} \Gamma - D_c(C_y\Pi + D_{yw}) & C_{c,2} \end{pmatrix},$$

with the same direct feedthrough matrix D_c. It is then obvious that this controller solves the exact output regulation problem via measurement feedback regulators if and only if the controller

$$\begin{aligned} \rho v &= A_{c,22}v + B_{c,2}y \\ u &= \begin{pmatrix} C_{c,2} \\ A_{c,12} \end{pmatrix} v + \begin{pmatrix} D_c \\ B_{c,1} \end{pmatrix} y \end{aligned} \tag{13.32}$$

achieves exact disturbance decoupling via measurement feedback controllers for the system (13.25). ■

As in the case of state feedback regulators, whenever we know part of the signal r and some of its derivatives or forward shifted values we can use this additional information.

Problem 13.4.5 *Let a system of the form (13.14) be given. The* exact output regulation problem via measurement feedback + PD regulators *is to find, if possible, a measurement feedback law of the form*

$$\begin{aligned} \rho v &= A_c v + B_c y + \textstyle\sum_{i=0}^{j} G_i \rho^i \bar{r} \\ u &= C_c v + D_c y + \textstyle\sum_{i=0}^{j} H_i \rho^i \bar{r} \end{aligned} \tag{13.33}$$

for some $j \geqslant 0$, such that the following properties hold:

(i) *The system (13.24) is asymptotically stable.*

(ii) *For all initial conditions and all signals r (in continuous-time $r \in \mathcal{C}^j$), the solution of the interconnection of (13.14) and (13.33) yields $e(t) \to 0$ as $t \to \infty$.*

We have the following solvability conditions for the above problem.

Theorem 13.4.5 *Consider the system given in (13.14) for both continuous- and discrete-time systems. Let Assumptions A.1, A.2, and A.3 be satisfied. Let $\mathbb{C}^g = \mathbb{C}^-$ for continuous-time systems, and $\mathbb{C}^g = \mathbb{C}^\ominus$ for discrete-time systems. Then, the exact output regulation problem via measurement feedback + PD regulators is solvable if and only if the following conditions are true:*

(i) *There exist matrices Π and Γ that solve the regulator equation (13.16).*

(ii) $\operatorname{im} \Pi D_{wr} \subseteq \operatorname{im} B \ker D_{eu} + \mathcal{V}^g(A, B, C_e, D_{eu}) + \mathcal{S}^*(A, B, C_e, D_{eu})$.

(iii) $\Pi D_{wr} \ker R \subseteq \operatorname{im} B \ker D_{eu} + \mathcal{V}^g(A, B, C_e, D_{eu})$.

(iv) $\ker \begin{pmatrix} C_e & D_{eu} \end{pmatrix} \supseteq \mathcal{S}^g(A_e, D_{wr,e}Q, C_{y,e}, D_{yr}Q) \cap C_{y,e}^{-1} \operatorname{im} D_{yr}Q$, *where*

$$A_e = \begin{pmatrix} A & -B\Gamma \\ 0 & S \end{pmatrix}, \quad D_{wr,e} = \begin{pmatrix} -\Pi D_{wr} \\ D_{wr} \end{pmatrix}, \quad \text{and}$$
$$C_{y,e} = \begin{pmatrix} C_y & C_y\Pi + D_{yw} \end{pmatrix}.$$

(v) $\mathcal{S}^g(A_e, D_{wr,e}Q, C_{y,e}, D_{yr}Q) \subseteq \mathcal{V}^g(A, B, C_e, D_{eu}) \oplus \mathbb{R}^s$.

Moreover, if conditions (i)-(v) are satisfied, then there exists a controller of the form given below in (13.34) that solves the exact disturbance decoupling problem via measurement feedback + PD controllers for the system (13.25),

$$\bar{\Sigma}_{\mathrm{C}} : \begin{cases} \rho\bar{v} = \bar{A}_c\bar{v} + \bar{B}_c\bar{y} + \sum_{i=0}^{j} \bar{G}_i \rho^i \bar{r}, \\ \bar{u} = \bar{C}_c\bar{v} + \bar{D}_c\bar{y} + \sum_{i=0}^{j} \bar{H}_i \rho^i \bar{r}. \end{cases} \tag{13.34}$$

In that case the exact output regulation problem via measurement feedback + PD regulators is solved by the controller or regulator,

$$\Sigma_{\mathrm{C}} : \begin{cases} \rho v_1 = S v_1 + \bar{C}_{c,1} v_2 + \bar{D}_{c,1}\check{y} + \sum_{i=0}^{j} \bar{H}_{i2}\rho^i\bar{r} \\ \rho v_2 = \bar{A}_c v_2 + \bar{B}_c\check{y} + \sum_{i=0}^{j} \bar{G}_i \rho^i \bar{r} \\ \check{y} = y + (D_{yw} + C_y\Pi)v_1 \\ u = -\Gamma v_1 + \bar{C}_{c,2} v_2 + \bar{D}_{c,2}\check{y} + \sum_{i=0}^{j} \bar{H}_{i1}\rho^i\bar{r}, \end{cases} \tag{13.35}$$

where $\bar{C}_{c,1}$, $\bar{C}_{c,2}$, $\bar{D}_{c,1}$, $\bar{D}_{c,2}$, $\bar{H}_{i1}$, *and* $\bar{H}_{i2}$ *are obtained by partitioning* $\bar{C}_c$, $\bar{D}_c$, *and* $\bar{H}_i$ *in conformity with the partitioning of* $\bar{A}_e$,

$$\bar{C}_c = \begin{pmatrix} \bar{C}_{c,1} \\ \bar{C}_{c,2} \end{pmatrix}, \quad \bar{D}_c = \begin{pmatrix} \bar{D}_{c,1} \\ \bar{D}_{c,2} \end{pmatrix}, \quad \text{and} \quad \bar{H}_i = \begin{pmatrix} \bar{H}_{i1} \\ \bar{H}_{i2} \end{pmatrix}. \tag{13.36}$$

Proof : This follows from the same arguments as in the proof of Theorem 13.4.4 combined with the result of Theorem 13.2.1. ■

In the case when we know only some derivatives of the signal r in continuous-time systems, or some forward shifted values of the sequence r in discrete-time systems, we can also relax the solvability conditions to some extent. We begin with the following problem formulation.

Problem 13.4.6 *Let a system of the form (13.14) be given. The* exact output regulation problem via measurement feedback + j-th order PD regulators *is to find, if possible, a measurement feedback law of the form (13.33) for a specified* $j \geqslant 0$ *such that the following properties hold:*

(*i*) *The system (13.24) is asymptotically stable.*

(*ii*) *For all initial conditions and all signals* r *(in continuous-time* $r \in \mathcal{C}^j$*) the interconnection of (13.33) and (13.14) satisfies* $\lim_{t\to\infty} e(t) = 0$.

We have the following theorem that develops the solvability conditions for the above problem.

Theorem 13.4.6 *Consider the system given in (13.14) for both continuous- and discrete-time systems. Let Assumptions A.1, A.2, and A.3 be satisfied. Let* $\mathbb{C}^g = \mathbb{C}^-$ *for continuous-time systems, and* $\mathbb{C}^g = \mathbb{C}^\ominus$ *for discrete-time systems. Then, the exact output regulation problem via measurement feedback + j-th order PD regulators is solvable if and only if the following conditions are true:*

(*i*) *There exist matrices* Π *and* Γ *that solve the regulator equation (13.16).*

(*ii*) $\operatorname{im} \Pi D_{wr} \subseteq \operatorname{im} B \ker D_{eu} + \mathcal{V}^g(A, B, C_e, D_{eu}) + \mathcal{S}_j^*(A, B, C_e, D_{eu})$.

(*iii*) $\operatorname{im} \Pi D_{wr} Q \subseteq \operatorname{im} B \ker D_{eu} + \mathcal{V}^g(A, B, C_e, D_{eu})$.

(*iv*) $\ker\begin{pmatrix} C_e & D_{eu} \end{pmatrix} \supseteq \mathcal{S}^g(A_e, D_{wr,e}Q, C_{y,e}, D_{yr}Q) \cap C_{y,e}^{-1} \operatorname{im} D_{yr}Q$, *where*

$$A_e = \begin{pmatrix} A & -B\Gamma \\ 0 & S \end{pmatrix}, \quad D_{wr,e} = \begin{pmatrix} -\Pi D_{wr} \\ D_{wr} \end{pmatrix}, \quad \textit{and}$$

$$C_{y,e} = \begin{pmatrix} C_y & C_y\Pi + D_{yw} \end{pmatrix}.$$

(*v*) $\mathcal{S}^g(A_e, D_{wr,e}Q, C_{y,e}, D_{yr}Q) \subseteq \mathcal{V}^g(A, B, C_e, D_{eu}) \oplus \mathbb{R}^s$.

Moreover, if conditions (i)-(v) are satisfied, then there exists a controller of the form (13.34) that solves the exact disturbance decoupling problem via measurement feedback + j-th order PD controllers for the system (13.25). In that case the exact output regulation problem via measurement feedback + j-order PD regulators is solved by the controller or regulator (13.35) using the partitions as in (13.36).

Proof : This follows from the same arguments as in the proof of Theorem 13.4.4 combined with the result of Theorem 13.2.1. ■

13.4.2 Almost steady state tracking – almost output regulation

In the previous section, we had formulated and discussed several linear output regulation problems seeking to render the steady state tracking error exactly zero. Often in practice one may not need exact output regulation. What is needed in engineering applications, is perhaps an "almost" output regulation, i.e. tracking of a reference signal and/or rejection of a disturbance signal to an arbitrary degree of precision. Such an "almost" output regulation is the subject of this subsection. As in the previous subsection, we formulate and study six different problems each using a particular class of regulators. The proofs this time are basically the same as for exact steady state output regulation except that they are all based on almost disturbance decoupling instead of exact disturbance decoupling and hence the proofs are omitted.

It is worth noting that the solvability conditions for any given problem in this subsection are much weaker for continuous-time systems than for discrete-time systems since the former allows the utilization of high-gain feedback.

We consider first state feedback regulators.

Problem 13.4.7 *Let a system of the form (13.14) be given. The* almost output regulation problem via state feedback regulators *is to find, if possible, a family* *of state feedback laws of the form $u = F_\varepsilon x + G_\varepsilon w$ parameterized by $\varepsilon > 0$ such that for any given ε the following properties hold:*

(i) *The system $\rho x = (A + BF_\varepsilon)x$ is asymptotically stable.*

(ii) *For all initial conditions, and for all signals r (piecewise continuous in continuous-time), the solution of (13.15) where F and G are respectively replaced by F_ε and G_ε satisfies*

$$\limsup_{t\to\infty} \|e(t)\| \leqslant \varepsilon \|r\|_\infty.$$

Remark 13.4.1 *Note that part (ii) implies that for all initial conditions and $r \equiv 0$, the solution of*

$$\begin{aligned} \rho x &= (A + BF_\varepsilon)x + (E_w + BG_\varepsilon)w \\ \rho w &= Sw \\ e &= (C_e + D_{eu}F_\varepsilon)x + (D_{ew} + D_{eu}G_\varepsilon)w \end{aligned} \tag{13.37}$$

yields $e(t) \to 0$ as $t \to \infty$. In other words, the almost output regulation problem via state feedback regulators requires exact output regulation via state feedback in the classical sense.

The following theorem develops the solvability conditions for the above problem.

Theorem 13.4.7 *Consider the system given in (13.14) for both continuous- and discrete-time systems. Let Assumptions A.1 and A.2 be satisfied. Let $\mathbb{C}^g = \mathbb{C}^-$ for continuous-time systems, and $\mathbb{C}^g = \mathbb{C}^\ominus$ for discrete-time systems. Then, the almost output regulation problem via state feedback regulators is solvable if and only if the following conditions are true:*

(i) *There exist matrices Π and Γ that solve the regulator equation (13.16).*

(ii) *For continuous-time systems we have*

$$\operatorname{im} \Pi D_{wr} \subseteq \mathcal{V}^{-0}(A, B, C_e, D_{eu}) + \mathcal{S}^*(A, B, C_e, D_{eu}),$$

and for discrete-time systems we have

$$\operatorname{im} \Pi D_{wr} \subseteq \mathcal{V}^{\otimes}(A, B, C_e, D_{eu}).$$

(iii) *Let F be such that $A + BF$ is Hurwitz-stable for continuous-time systems and Schur-stable for discrete-time systems. For any invariant zero λ on the imaginary axis (continuous-time) or unit circle (discrete-time) of the system characterized by the quadruple (A, B, C_e, D_{eu}),*

$$\operatorname{im} C_e(\lambda I - A - BF)^{-1}\Pi D_{wr} \subseteq \operatorname{im}[C_e(\lambda I - A - BF)^{-1}B + D_{eu}].$$

Moreover, under the above conditions, there exists a family of feedback laws $\tilde{u} = F_\varepsilon \tilde{x}$ that solves the almost disturbance decoupling problem for the system (13.17). In that case, utilizing such F_ε, the almost output regulation problem via state feedback regulators is solved by the family of feedback laws

$$u = F_\varepsilon x + (\Gamma - F_\varepsilon \Pi)w.$$

Sometimes we know all the derivatives of the signal r in continuous-time systems, or all the forward shifted values of r in discrete-time systems. In this case, the solvability conditions can be relaxed. This is formulated in the following problem.

Problem 13.4.8 *Let a system of the form (13.14) be given. The* almost output regulation problem via state feedback + PD regulators *is to find, if possible, a state feedback law parameterized by $\varepsilon > 0$ and is of the form,*

$$u = F_\varepsilon x + G_\varepsilon w + \sum_{i=0}^{j} H_{i,\varepsilon}\rho^i \bar{r} \tag{13.38}$$

for some $j \geqslant 0$ such that the following properties hold:

(i) *The system $\rho x = (A + BF_\varepsilon)x$ is asymptotically stable.*

(ii) *For all initial conditions and all signals r (in continuous-time $r \in \mathcal{C}^j$), the solution of*

$$\begin{aligned} \rho x &= (A + BF_\varepsilon)x + (E_w + BG_\varepsilon)w + B\textstyle\sum_{i=0}^{j} H_{i,\varepsilon}\rho^i \bar{r} \\ \rho w &= Sw + D_{wr}r \\ e &= (C_e + D_{eu}F_\varepsilon)x + (D_{ew} + D_{eu}G_\varepsilon)w + D_{eu}\textstyle\sum_{i=0}^{j} H_{i,\varepsilon}\rho^i \bar{r} \end{aligned} \tag{13.39}$$

is such that

$$\limsup_{t\to\infty} \|e(t)\| \leqslant \varepsilon \|r\|_\infty.$$

The following theorem develops the solvability conditions for the above problem.

Theorem 13.4.8 *Consider the system given in (13.14) for both continuous- and discrete-time systems. Let Assumptions A.1 and A.2 be satisfied. Let* $\mathbb{C}^g = \mathbb{C}^-$ *for continuous-time systems, and* $\mathbb{C}^g = \mathbb{C}^\ominus$ *for discrete-time systems. Then, the almost output regulation problem via state feedback + PD regulators is solvable if and only if the following conditions are true:*

(i) *There exist matrices* Π *and* Γ *that solve the regulator equation (13.16).*

(ii) *For continuous-time systems we have*

$$\operatorname{im}\Pi D_{wr} \subseteq \mathcal{V}^{-0}(A, B, C_e, D_{eu}) + \mathcal{S}^*(A, B, C_e, D_{eu}),$$

and for discrete-time systems we have

$$\Pi D_{wr} \ker R \subseteq \mathcal{V}^{\otimes}(A, B, C_e, D_{eu}),$$
$$\operatorname{im}\Pi D_{wr} \subseteq \mathcal{V}^{\otimes}(A, B, C_e, D_{eu}) + \mathcal{S}^*(A, B, C_e, D_{eu}).$$

(iii) *Let* F *be such that* $A + BF$ *is Hurwitz-stable for continuous-time systems and Schur-stable for discrete-time systems. For any invariant zero* λ *on the imaginary axis (continuous-time) or unit circle (discrete-time) of the system characterized by the quadruple* (A, B, C_e, D_{eu}),

$$\operatorname{im} C_e(\lambda I - A - BF)^{-1}\Pi D_{wr} \subseteq \operatorname{im}[C_e(\lambda I - A - BF)^{-1}B + D_{eu}].$$

Moreover, under the above conditions, for some $j \geq 0$, *there exists a family of feedback laws of the form* $\tilde{u} = F_\varepsilon\tilde{x} + \sum_{i=0}^{j} H_{i,\varepsilon}\rho^i\bar{r}$ *that solves the almost disturbance decoupling problem for the system (13.17). In that case, utilizing such* F_ε *and* $H_{i,\varepsilon}$, *the almost output regulation problem via state feedback + PD regulators is solved by the feedback law*

$$u = F_\varepsilon x + (\Gamma - F\Pi)w + \sum_{i=0}^{j} H_{i,\varepsilon}\rho^i\bar{r}. \tag{13.40}$$

In the case when we know only some derivatives of the signal r in continuous-time systems, or some forward shifted values of the sequence r in discrete-time systems, we can also relax the solvability conditions to some extent. We begin with the following problem formulation.

Problem 13.4.9 *Let a system of the form (13.14) be given. The* almost output regulation problem via state feedback + j-th order PD regulators *is to find, if possible, a feedback law of the form (13.38) for a specified* $j \geqslant 0$ *such that the following properties hold:*

(i) *The system* $\rho x = (A + BF_\varepsilon)x$ *is asymptotically stable.*

(ii) *For all signals* r *(in continuous-time* $r \in \mathcal{C}^j$*) and for all initial conditions, the solution of (13.39) is such that*

$$\limsup_{t\to\infty} \|e(t)\| \leqslant \varepsilon \|r\|_\infty.$$

The following theorem develops the solvability conditions for the above problem.

Theorem 13.4.9 *Consider the system given in (13.14) for both continuous- and discrete-time systems. Let Assumptions A.1 and A.2 be satisfied. Let* $\mathbb{C}^g = \mathbb{C}^-$ *for continuous-time systems, and* $\mathbb{C}^g = \mathbb{C}^\ominus$ *for discrete-time systems. Then, the almost output regulation problem via state feedback +* j*-th order PD regulators is solvable if and only if the following conditions are true:*

(i) *There exist matrices* Π *and* Γ *that solve the regulator equation (13.16).*

(ii) *For continuous-time systems we have*

$$\operatorname{im} \Pi D_{wr} \subseteq \mathcal{V}^{-0}(A, B, C_e, D_{eu}) + \mathcal{S}^*(A, B, C_e, D_{eu}),$$

and for discrete-time systems we have

$$\Pi D_{wr} \ker R \subseteq \mathcal{V}^{\otimes}(A, B, C_e, D_{eu}),$$
$$\operatorname{im} \Pi D_{wr} \subseteq \mathcal{V}^{\otimes}(A, B, C_e, D_{eu}) + \mathcal{S}_j^*(A, B, C_e, D_{eu}).$$

(iii) *Let* F *be such that* $A + BF$ *is Hurwitz-stable for continuous-time systems and Schur-stable for discrete-time systems. For any invariant zero* λ *on the imaginary axis (continuous-time) or unit circle (discrete-time) of the system characterized by the quadruple* (A, B, C_e, D_{eu}),

$$\operatorname{im} C_e(\lambda I - A - BF)^{-1} \Pi D_{wr} \subseteq \operatorname{im}[C_e(\lambda I - A - BF)^{-1} B + D_{eu}].$$

Moreover, under the above conditions, there exists a family of parameterized feedback laws of the form $\tilde{u} = F_\varepsilon \tilde{x} + \sum_{i=0}^{j} H_{i,\varepsilon}\rho^i \bar{r}$ *that solves the almost disturbance decoupling problem for the system (13.17). Then, utilizing such* F_ε *and* $H_{i,\varepsilon}$*, the posed almost output regulation problem is solved by the feedback law (13.40).*

So far we have considered state variable feedback regulators. We next move on to consider the case when only the measurement variable y is available for feedback. We have the following basic problem.

Problem 13.4.10 *Let a system of the form (13.14) be given. The* almost output regulation problem via measurement feedback regulators *is to find, if possible, a family of measurement feedback laws parameterized by* $\varepsilon > 0$ *and is of the form,*

$$\begin{aligned} \rho v &= A_{c,\varepsilon} v + B_{c,\varepsilon} y \\ u &= C_{c,\varepsilon} v + D_{c,\varepsilon} y, \end{aligned} \tag{13.41}$$

such that for any $\varepsilon > 0$ *the following properties hold:*

(i) *The system*

$$\begin{aligned} \rho x &= (A + BD_{c,\varepsilon}C_y)x + BC_{c,\varepsilon} v \\ \rho v &= B_{c,\varepsilon} C_y x + A_{c,\varepsilon} v \end{aligned} \tag{13.42}$$

is asymptotically stable.

(ii) *For all initial conditions, and all signals* r *(piecewise continuous in continuous-time) the solution* e *of the interconnection of (13.41) and (13.14) is such that*

$$\limsup_{t\to\infty} \|e(t)\| \leqslant \varepsilon \|r\|_\infty.$$

Remark 13.4.2 *Note that, similar to the state feedback case, part (ii) implies that for all initial conditions, for* $r \equiv 0$*, and for any* $\varepsilon > 0$ *the solution of*

$$\begin{aligned} \rho x &= (A + BD_{c,\varepsilon}C_y)x + BC_{c,\varepsilon} v + (E_w + BD_{c,\varepsilon}D_{yw})w \\ \rho v &= B_{c,\varepsilon} C_y x + A_{c,\varepsilon} v + B_{c,\varepsilon} D_{yw} w \\ \rho w &= Sw \\ e &= (C_e + D_{eu}D_{c,\varepsilon}C_y)x + D_{eu}C_{c,\varepsilon} v + (D_{ew} + D_{eu}D_{c,\varepsilon}D_{yw})w \end{aligned} \tag{13.43}$$

is such that $e(t) \to 0$ *as* $t \to \infty$*. In other words, the almost output regulation problem via measurement feedback regulators requires the exact regulation via measurement feedback regulators in the classical sense.*

We have the following result concerning the above problem.

Theorem 13.4.10 *Consider the system given in (13.14) for both continuous- and discrete-time systems. Let Assumptions A.1, A.2, and A.3 be satisfied. Let $\mathbb{C}^g = \mathbb{C}^-$ for continuous-time systems, and $\mathbb{C}^g = \mathbb{C}^\ominus$ for discrete-time systems. Then, the almost output regulation problem via measurement feedback regulators is solvable if and only if the following conditions are true:*

(i) There exist matrices Π and Γ that solve the regulator equation (13.16).

(ii) For continuous-time systems we have

$$\operatorname{im} \Pi D_{wr} \subseteq \mathcal{V}^{-0}(A, B, C_e, D_{eu}) + \mathcal{S}^*(A, B, C_e, D_{eu}),$$

and for discrete-time systems we have

$$\operatorname{im} \Pi D_{wr} \subseteq \mathcal{V}^{\otimes}(A, B, C_e, D_{eu}) + \operatorname{im} B \ker D_{eu}.$$

(iii) For continuous-time systems we have

$$\begin{aligned}&\ker \begin{pmatrix} C_e & D_{eu} \end{pmatrix}\\ &\qquad \supseteq \mathcal{S}^{-0}(A_e, D_{wr,e}, C_{y,e}, D_{yr}) \cap \mathcal{V}^*(A_e, D_{wr,e}, C_{y,e}, D_{yr}).\end{aligned}$$

where

$$A_e = \begin{pmatrix} A & -B\Gamma \\ 0 & S \end{pmatrix}, \quad D_{wr,e} = \begin{pmatrix} -\Pi D_{wr} \\ D_{wr} \end{pmatrix}, \quad \text{and}$$
$$C_{y,e} = \begin{pmatrix} C_y & C_y\Pi + D_{yw} \end{pmatrix}.$$

Similarly, for discrete-time systems we have

$$\ker \begin{pmatrix} C_e & D_{eu} \end{pmatrix} \supseteq \mathcal{S}^{\otimes}(A_e, D_{wr,e}, C_{y,e}, D_{yr}) \cap C_{y,e}^{-1} \operatorname{im} D_{yr}.$$

(iv) For continuous-time systems we have

$$\begin{aligned}&\mathcal{S}^{-0}(A_e, D_{wr,e}, C_{y,e}, D_{yr}) \cap \mathcal{V}^*(A_e, D_{wr,e}, C_{y,e}, D_{yr})\\ &\qquad \subseteq \left(\mathcal{V}^{-0}(A, B, C_e, D_{eu}) + \mathcal{S}^*(A, B, C_e, D_{eu})\right) \oplus \mathbb{R}^s.\end{aligned}$$

Similarly, for discrete-time systems we have

$$\mathcal{S}^{\otimes}(A_e, D_{wr,e}, C_{y,e}, D_{yr}) \subseteq \mathcal{V}^{\otimes}(A, B, C_e, D_{eu}) \oplus \mathbb{R}^s.$$

(v) *Let F and K be such that $A+BF$ and $A_K = A_e + KC_{y,e}$ are Hurwitz-stable for continuous-time systems and Schur-stable for discrete-time systems. For any invariant zero λ on the imaginary axis (continuous-time) or unit circle (discrete-time) of the system characterized by the quadruple (A, B, C_e, D_{eu}), we have*

$$\operatorname{im} C_e(\lambda I - A - BF)^{-1}\Pi D_{wr} \subseteq \operatorname{im}[C_e(\lambda I - A - BF)^{-1}B + D_{eu}],$$

while for any invariant zero λ on the imaginary axis (continuous-time) or unit circle (discrete-time) of the system characterized by the quadruple $(A_e, D_{wr,e}, C_{y,e}, D_{yr})$, we have

$$\ker\left[C_{y,e}(\lambda I - A_K)^{-1}D_{wr,e} + D_{yr}\right] \subseteq \ker\begin{pmatrix} C_e & -D_{eu}\Gamma \end{pmatrix}(\lambda I - A_K)^{-1} D_{wr,e}.$$

Moreover, whenever the above conditions hold, there exists a family of proper controllers characterized by the quadruples $(\bar{A}_{c,\varepsilon}, \bar{B}_{c,\varepsilon}, \bar{C}_{c,\varepsilon}, \bar{D}_{c,\varepsilon})$ that solves the almost disturbance decoupling for the system (13.25). In that case, the almost output regulation problem via measurement feedback regulators is solved by the family of controllers,

$$\Sigma_C : \begin{cases} \rho v_1 = & S v_1 + \bar{C}_{c,1,\varepsilon} v_2 + \bar{D}_{c,1,\varepsilon}(y + (D_{yw} + C_y\Pi)v_1) \\ \rho v_2 = & \bar{A}_{c,\varepsilon} v_2 + \bar{B}_{c,\varepsilon}(y + (D_{yw} + C_y\Pi)v_1) \\ u \;\; = & -\Gamma v_1 + \bar{C}_{c,2,\varepsilon} v_2 + \bar{D}_{c,2,\varepsilon}(y + (D_{yw} + C_y\Pi)v_1), \end{cases} \quad (13.44)$$

where $\bar{C}_{c,1,\varepsilon}$, $\bar{C}_{c,2,\varepsilon}$, $\bar{D}_{c,1,\varepsilon}$, and $\bar{D}_{c,2,\varepsilon}$ are obtained by partitioning $\bar{C}_{c,\varepsilon}$ and $\bar{D}_{c,\varepsilon}$ in conformity with the partitioning of $\bar{A}_e$,

$$\bar{C}_{c,\varepsilon} = \begin{pmatrix} \bar{C}_{c,1,\varepsilon} \\ \bar{C}_{c,2,\varepsilon} \end{pmatrix}, \quad \textit{and } \bar{D}_{c,\varepsilon} = \begin{pmatrix} \bar{D}_{c,1,\varepsilon} \\ \bar{D}_{c,2,\varepsilon} \end{pmatrix}. \quad (13.45)$$

As in the case of state feedback regulators, whenever we know all the derivatives of the signal r in continuous-time systems or all the forward shifted values of r in discrete-time systems, the solvability conditions can be relaxed. We begin with the following problem formulation.

Problem 13.4.11 *Let a system of the form (13.14) be given. The* almost output regulation problem via measurement feedback + PD regulators *is to find, if* *possible, a family of measurement feedback laws parameterized by $\varepsilon > 0$ and is of the form*

$$\begin{aligned} \rho v &= A_{c,\varepsilon} v + B_{c,\varepsilon} y + \textstyle\sum_{i=0}^{j} G_{i,\varepsilon} \rho^i \bar{r} \\ u &= C_{c,\varepsilon} v + D_{c,\varepsilon} y + \textstyle\sum_{i=0}^{j} H_{i,\varepsilon} \rho^i \bar{r}, \end{aligned} \tag{13.46}$$

for some $j \geqslant 0$ such that for any $\varepsilon > 0$ the following properties hold:

(i) *The system (13.42) is asymptotically stable.*

(ii) *For all initial conditions and for all signals r (in continuous-time $r \in \mathcal{C}^j$) the interconnection of (13.46) and (13.14) is such that*

$$\limsup_{t\to\infty} \|e(t)\| \leqslant \varepsilon \|r\|_\infty .$$

We have the following theorem that develops the solvability conditions for the above problem.

Theorem 13.4.11 *Consider the system given in (13.14) for both continuous- and discrete-time systems. Let Assumptions A.1, A.2, and A.3 be satisfied. Let $\mathbb{C}^g = \mathbb{C}^-$ for continuous-time systems, and $\mathbb{C}^g = \mathbb{C}^\ominus$ for discrete-time systems. Then, the almost output regulation problem via measurement feedback + PD regulators is solvable if and only if the following conditions are true:*

(i) *There exist matrices Π and Γ that solve the regulator equation (13.16).*

(ii) *For continuous-time systems we have*

$$\operatorname{im} \Pi D_{wr} \subseteq \mathcal{V}^{-0}(A, B, C_e, D_{eu}) + \mathcal{S}^*(A, B, C_e, D_{eu}),$$

and for discrete-time systems we have

$$\begin{aligned} \Pi D_{wr} \ker R &\subseteq \mathcal{V}^{\otimes}(A, B, C_e, D_{eu}) + B \ker D_{eu}, \\ \operatorname{im} \Pi D_{wr} &\subseteq \mathcal{V}^{\otimes}(A, B, C_e, D_{eu}) + \mathcal{S}^*(A, B, C_e, D_{eu}). \end{aligned}$$

(iii) *For continuous-time systems we have*

$$\ker \begin{pmatrix} C_e & D_{eu} \end{pmatrix} \supseteq \mathcal{S}^{-0}(A_e, D_{wr,e} Q, C_{y,e}, D_{yr} Q) \cap \mathcal{V}^*(A_e, D_{wr,e} Q, C_{y,e}, D_{yr} Q),$$

where

$$A_e = \begin{pmatrix} A & -B\Gamma \\ 0 & S \end{pmatrix}, \quad D_{wr,e} = \begin{pmatrix} -\Pi D_{wr} \\ D_{wr} \end{pmatrix}, \quad \text{and}$$

$$C_{y,e} = \begin{pmatrix} C_y & C_y \Pi + D_{yw} \end{pmatrix}.$$

Similarly, for discrete-time systems we have

$$\ker\begin{pmatrix} C_e & D_{eu}\end{pmatrix} \supseteq \mathcal{S}^{\otimes}(A_e, D_{wr,e}Q, C_{y,e}, D_{yr}Q) \cap C_{y,e}^{-1}\{\operatorname{im} D_{yr}Q\}.$$

(iv) *For continuous-time systems we have*

$$\mathcal{S}^{-0}(A_e, D_{wr,e}Q, C_{y,e}, D_{yr}Q) \cap \mathcal{V}^{*}(A_e, D_{wr,e}Q, C_{y,e}, D_{yr}Q)$$
$$\subseteq \left(\mathcal{V}^{-0}(A, B, C_e, D_{eu}) + \mathcal{S}^{*}(A, B, C_e, D_{eu})\right) \oplus \mathbb{R}^s.$$

Similarly, for discrete-time systems we have

$$\mathcal{S}^{\otimes}(A_e, D_{wr,e}Q, C_{y,e}, D_{yr}Q) \subseteq \mathcal{V}^{\otimes}(A, B, C_e, D_{eu}) \oplus \mathbb{R}^s.$$

(v) *Let F and K be such that $A+BF$ and $A_K = A_e + KC_{y,e}$ are Hurwitz-stable for continuous-time systems and Schur-stable for discrete-time systems. For any invariant zero λ on the imaginary axis (continuous-time) or unit circle (discrete-time) of the system characterized by the quadruple (A, B, C_e, D_{eu}),*

$$\operatorname{im} C_e(\lambda I - A - BF)^{-1}\Pi D_{wr}$$
$$\subseteq \operatorname{im}[C_e(\lambda I - A - BF)^{-1}B + D_{eu}],$$

while for any invariant zero λon the imaginary axis (continuous-time) or unit circle (discrete-time) of the system characterized by the quadruple $(A_e, D_{wr,e}, C_{y,e}, D_{yr})$, we have

$$\ker R \cap \ker\left[C_{y,e}(\lambda I - A_K)^{-1}D_{wr,e} + D_{yr}\right]$$
$$\subseteq \ker\begin{pmatrix} C_e & -D_{eu}\Gamma\end{pmatrix}(\lambda I - A_K)^{-1}D_{wr,e}.$$

Moreover, whenever the above conditions hold, there exists a family of parameterized feedback controllers of the form given below in (13.47) that solves the almost disturbance decoupling problem via measurement feedback + PD controllers for the system (13.25),

$$\bar{\Sigma}_{c,\varepsilon} : \begin{cases} \rho\bar{v} = \bar{A}_{c,\varepsilon}\bar{v} + \bar{B}_{c,\varepsilon}\bar{y} + \sum_{i=0}^{j} \bar{G}_{i,\varepsilon}\, \rho^i\bar{r}, \\ \bar{u} \;= \bar{C}_{c,\varepsilon}\bar{v} + \bar{D}_{c,\varepsilon}\bar{y} + \sum_{i=0}^{j} \bar{H}_{i,\varepsilon}\rho^i\bar{r}. \end{cases} \tag{13.47}$$

In that case, the almost output regulation problem via measurement feedback + PD regulators is solved by the controller or regulator,

$$\Sigma_c : \begin{cases} \rho v_1 = \; Sv_1 + \bar{C}_{c,1,\varepsilon}v_2 + \bar{D}_{c,1,\varepsilon}\check{y} + \sum_{i=0}^{j} \bar{H}_{i2,\varepsilon}\rho^i\bar{r} \\ \rho v_2 = \; \bar{A}_{c,\varepsilon}v_2 + \bar{B}_{c,\varepsilon}\check{y} + \sum_{i=0}^{j} \bar{G}_{i,\varepsilon}\, \rho^i\bar{r} \\ \check{y} \;\; = \; y + (D_{yw} + C_y\Pi)v_1 \\ u \;\; = -\Gamma v_1 + \bar{C}_{c,2,\varepsilon}v_2 + \bar{D}_{c,2,\varepsilon}\check{y} + \sum_{i=0}^{j} \bar{H}_{i1,\varepsilon}\rho^i\bar{r}, \end{cases} \tag{13.48}$$

where $\bar{C}_{c,1,\varepsilon}$, $\bar{C}_{c,2,\varepsilon}$, $\bar{D}_{c,1,\varepsilon}$, $\bar{D}_{c,2,\varepsilon}$, $\bar{H}_{i1,\varepsilon}$, and $\bar{H}_{i2,\varepsilon}$ are obtained by partitioning $\bar{C}_{c,\varepsilon}$, $\bar{D}_{c,\varepsilon}$, and $H_{i,\varepsilon}$ in conformity with the partitioning of $\bar{A}_e$,

$$\bar{C}_{c,\varepsilon} = \begin{pmatrix} \bar{C}_{c,1,\varepsilon} \\ \bar{C}_{c,2,\varepsilon} \end{pmatrix}, \quad \bar{D}_{c,\varepsilon} = \begin{pmatrix} \bar{D}_{c,1,\varepsilon} \\ \bar{D}_{c,2,\varepsilon} \end{pmatrix}, \quad \text{and} \quad \bar{H}_{i,\varepsilon} = \begin{pmatrix} \bar{H}_{i1,\varepsilon} \\ \bar{H}_{i2,\varepsilon} \end{pmatrix}. \tag{13.49}$$

In the case when we know only some derivatives of the signal r in continuous-time systems, or some forward shifted values of the sequence r in discrete-time systems, we can also relax the solvability conditions to some extent. We begin with the following problem formulation.

Problem 13.4.12 *The **almost output regulation problem via measurement feedback + j-th order PD regulators** is to find, if possible, a measurement feedback law of the form (13.46) for a fixed $j \geq 0$ and parameterized by $\varepsilon > 0$ such that for any $\varepsilon > 0$ the following properties hold:*

(i) The system (13.42) is asymptotically stable.

(ii) For all initial conditions and for all signals r (in continuous-time $r \in \mathcal{C}^j$) the solution of the interconnection of (13.46) and (13.14) satisfies

$$\limsup_{t\to\infty} \|e(t)\| \leqslant \varepsilon \|r\|_\infty.$$

We have the following theorem that develops the solvability conditions for the above problem.

Theorem 13.4.12 *Consider the system given in (13.14) for both continuous- and discrete-time systems. Let Assumptions A.1, A.2, and A.3 be satisfied. Let $\mathbb{C}^g = \mathbb{C}^-$ for continuous-time systems, and $\mathbb{C}^g = \mathbb{C}^\ominus$ for discrete-time systems. Then, the almost output regulation problem via measurement feedback + j-th order PD regulators is solvable if and only if the following conditions are true:*

(i) There exist matrices Π and Γ that solve the regulator equation (13.16).

(ii) For continuous-time systems we have

$$\operatorname{im} \Pi D_{wr} \subseteq \mathcal{V}^{-0}(A, B, C_e, D_{eu}) + \mathcal{S}^*(A, B, C_e, D_{eu}),$$

and for discrete-time systems we have

$$\Pi D_{wr} \ker R \subseteq \mathcal{V}^{\otimes}(A, B, C_e, D_{eu}) + B \ker D_{eu},$$
$$\operatorname{im} \Pi D_{wr} \subseteq \mathcal{V}^{\otimes}(A, B, C_e, D_{eu}) + \mathcal{S}_j^*(A, B, C_e, D_{eu}).$$

(iii) *For continuous-time systems we have*

$$\ker\begin{pmatrix} C_e & D_{eu}\end{pmatrix} \supseteq \mathcal{S}^{-0}(A_e, D_{wr,e}, C_{y,e}, D_{yr}) \cap \mathcal{V}^*(A_e, D_{wr,e}, C_{y,e}, D_{yr}),$$

where

$$A_e = \begin{pmatrix} A & -B\Gamma \\ 0 & S \end{pmatrix}, \quad D_{wr,e} = \begin{pmatrix} -\Pi D_{wr} \\ D_{wr} \end{pmatrix}, \quad \textit{and}$$
$$C_{y,e} = \begin{pmatrix} C_y & C_y\Pi + D_{yw}\end{pmatrix}.$$

Similarly, for discrete-time systems we have

$$\ker\begin{pmatrix} C_e & D_{eu}\end{pmatrix} \supseteq \mathcal{S}^{\otimes}(A_e, D_{wr,e}, C_{y,e}, D_{yr}) \cap C_{y,e}^{-1}\operatorname{im} D_{yr}.$$

(iv) *For continuous-time systems we have*

$$\mathcal{S}^{-0}(A_e, D_{wr,e}, C_{y,e}, D_{yr}) \cap \mathcal{V}^*(A_e, D_{wr,e}, C_{y,e}, D_{yr}) \subseteq \big(\mathcal{V}^{-0}(A, B, C_e, D_{eu}) + \mathcal{S}^*(A, B, C_e, D_{eu})\big) \oplus \mathbb{R}^s.$$

Similarly, for discrete-time systems we have

$$\mathcal{S}^{\otimes}(A_e, D_{wr,e}, C_{y,e}, D_{yr}) \subseteq \mathcal{V}^{\otimes}(A, B, C_e, D_{eu}) \oplus \mathbb{R}^s.$$

(v) *Let F and K be such that $A + BF$ and $A_K = A_e + KC_{y,e}$ are Hurwitz-stable for continuous-time systems and Schur-stable for discrete-time systems. For any invariant zero λ on the imaginary axis (continuous-time) or unit circle (discrete-time) of the system characterized by the quadruple (A, B, C_e, D_{eu}), we have*

$$\operatorname{im} C_e(\lambda I - A - BF)^{-1}\Pi D_{wr} \subseteq \operatorname{im}[C_e(\lambda I - A - BF)^{-1}B + D_{eu}],$$

while for any invariant zero λ on the imaginary axis (continuous-time) or unit circle (discrete-time) of the system characterized by the quadruple $(A_e, D_{wr,e}, C_{y,e}, D_{yr})$, we have

$$\ker R \cap \ker\big[C_{y,e}(\lambda I - A_K)^{-1}D_{wr,e} + D_{yr}\big] \subseteq \ker\begin{pmatrix} C_e & -D_{eu}\Gamma\end{pmatrix}(\lambda I - A_K)^{-1}D_{wr,e}.$$

Moreover, whenever the above conditions hold, there exists a family of parameterized feedback controllers of the form given in (13.47) that solves the almost disturbance decoupling problem via measurement feedback + j-th order PD controllers for the system (13.25). In that case, the almost output regulation problem via measurement feedback + j-th order PD regulators is solved by the controller (13.48).

13.4.3 Steady state tracking with a pre-specified level of performance – γ-level output regulation

In the previous sections the problem formulations required that the disturbance (or reference) signal was rejected (tracked) either precisely or with arbitrary precision. Clearly in many cases this is too ambitious. In this section we will show that the techniques of the previous section allow us to to formulate and solve relaxed forms of output regulation. We still require asymptotic regulation in the classical sense but this time we only require that the closed-loop transfer matrix from r to e be small in some arbitrary induced operator norm.

In the proof of Theorem 13.4.1, it is clear that a state feedback controller that achieves output regulation is of the form (13.18). The interconnection of a controller of the form (13.18) and the system (13.17) and the interconnection of the controller (13.18) and the system (13.14) yield the same closed-loop transfer matrix. Therefore, also the induced operator norm is the same. We can therefore concentrate for our design on the system (13.17) without taking output regulation into account and then obtain in a straightforward manner a controller for the original system with the same transfer matrix from r to z and which achieves regulation. In other words we can transform a control problem with the output regulation constraint into an unconstrained control problem. The same is also true for more general controllers of the form (13.20).

In the measurement feedback a similar situation arises. From the proof of Theorem 13.4.4, it is clear that a measurement feedback controller that achieves output regulation (after a transformation that does not effect output regulation or the effect of r on z) is of the form (13.26). The interconnection of a controller of the form (13.26) and the system (13.14), and the interconnection of the controller characterized by the matrix quadruple $(\bar{A}_c, \bar{B}_c, \bar{C}_c, \bar{D}_c)$ and the system (13.25), both yield the same closed-loop transfer matrix. Therefore also the induced operator norm is the same. We can therefore concentrate for our design on the system (13.17) without taking output regulation into account and then obtain in a straightforward manner a controller for the original system with the same transfer matrix from r to z and which achieves output regulation. In other words, we can transform a control problem with the output regulation constraint into an unconstrained control problem. Again, this is also true for more general controllers of the form (13.5). Using these arguments we can obtain a number of results for the case that (almost) output regulation is not possible.

As usual, we consider first state feedback regulators.

Problem 13.4.13 *Let a system of the form (13.14) be given. The* output regulation problem with L_∞ (or ℓ_∞) performance level γ via state feedback regulators *is to find, if possible, a state feedback law of the form $u = Fx + Gw$ such that the following properties hold:*

(i) *The system $\rho x = (A + BF)x$ is asymptotically stable.*

(ii) *For all $\varepsilon > 0$ and for all signals r (piecewise continuous in continuous-time) there exists a $T > 0$ such that*

$$\|e\|_{\infty,[T,\infty)} \leqslant \gamma \|r\|_{\infty,[0,\infty)} + \varepsilon. \tag{13.50}$$

Remark 13.4.3 *Note that the above problem formulation is equivalent to the definition used in [79].*

We note that in the above problem, if we choose zero initial conditions in part (ii), we need that $\|e\|_{\infty,[0,\infty)} \leqslant \gamma \|r\|_{\infty,[0,\infty)}$. On the other hand by choosing $r \equiv 0$, we need that for all initial conditions the solution of (13.15) satisfies $\lim_{t\to\infty} e(t) = 0$. In other words, the controller must achieve output regulation in the classical sense. Also, we would like to note that the role of ε and T in the above conditions is to ensure that the effect of initial conditions is small enough (less than ε) if we wait long enough (T large enough).

We can also formulate the above problem with other performance criteria such as an L_p-L_q (or ℓ_p-ℓ_q) induced operator norm. In that case we require that for all initial conditions, and for all signals r (piecewise continuous in continuous-time) we have

$$\lim_{T\to\infty} \frac{\|e\|_{p,[0,T])} - \gamma \|r\|_{q,[0,T]}}{T} \leqslant 0. \tag{13.51}$$

Note that the problem formulations (13.50) is not suitable for L_p (or ℓ_p) norms with p less than ∞, and conversely (13.51) is not suitable for $p = \infty$ or $q = \infty$, but the two problem formulations are obviously closely connected. In [43] an H_∞ control problem with unstable weights was formulated. If these *weights are at the input, then this is intrinsically equivalent to minimizing the criterion in (13.51). This has been further exploited in [46–48]. However, the motivation for this problem with an H_∞ criterion is rather weak. In [43] the main argument was that it can be used to solve the problem of H_∞ optimal control with the output regulation constraint. However, for that problem we have a complete solution in Chapters 10 and 11 and we feel that the problem formulations in those chapters are more appropriate since the performance from r to z is not of interest; we only want to guarantee that we achieve output regulation.*

We have the following solvability conditions for the above problem.

Theorem 13.4.13 *Consider the system given in (13.14) for both continuous- and discrete-time systems. Let Assumptions A.1 and A.2 be satisfied. Then, the output regulation problem with L_∞ (or ℓ_∞) performance level γ via state feedback regulators is solvable if and only if the following conditions are true:*

(i) There exist matrices Π and Γ that solve the regulator equation (13.16).

(ii) There exists a stabilizing state feedback law $\tilde{u} = F\tilde{x}$ for the system (13.17) such that the $L_\infty[0,\infty)$ (or $\ell_\infty[0,\infty)$) induced operator norm from r to e for zero initial conditions is less than γ.

Moreover, if we have F, Π, and Γ satisfying (i) and (ii), then a controller solving the output regulation problem with L_∞ (or ℓ_∞) performance level γ via state feedback regulators is given by (13.18).

Again we can use derivative information of forward shifted values of part of the signal r and in this way relax the solvability conditions. This is formulated in the following problem.

Problem 13.4.14 *Let a system of the form (13.14) be given. The* output regulation problem with L_∞ (or ℓ_∞) performance level γ via state feedback + PD regulators *is to find, if possible, a feedback law of the form (13.20) for some $j \geqslant 0$, such that the following properties hold:*

(i) The system $\rho x = (A + BF)x$ is asymptotically stable.

(ii) Given any $\varepsilon > 0$, for all initial conditions and for all signals r (in continuous-time $r \in \mathcal{C}^j$) the solution of (13.21) is such that (13.50) is true.

The solvability conditions for the above problem are given in the following theorem.

Theorem 13.4.14 *Consider the system given in (13.14) for both continuous- and discrete-time systems. Let Assumptions A.1 and A.2 be satisfied. Then, the output regulation problem with L_∞ (or ℓ_∞) performance level γ via state feedback + PD regulators is solvable if and only if the following conditions are true:*

(i) *There exist matrices Π and Γ that solve the regulator equation (13.16).*

(ii) *For some $j \geq 0$, there exists a feedback law of the form $\tilde{u} = F\tilde{x} + \sum_{i=0}^{j} H_i \rho^i \bar{r}$ for the system (13.17) such that the L_∞ (or ℓ_∞) induced operator norm from r to e for zero initial conditions is less than γ.*

Moreover, in that case, utilizing such F and H_i, the output regulation problem with L_∞ (or ℓ_∞) performance level γ via state feedback + PD regulators is solvable by the feedback law (13.22).

As usual when we know only some derivatives of the signal r in continuous-time systems, we get different solvability conditions. We begin with the following problem formulation.

Problem 13.4.15 *Let a system of the form (13.14) be given. The* output regulation problem with L_∞ (or ℓ_∞) performance level γ via state feedback + j-th order PD regulators *is to find, if possible, a feedback law of the form (13.20), such that the following properties hold:*

(i) *The system $\rho x = (A + BF)x$ is asymptotically stable.*

(ii) *Given any $\varepsilon > 0$, for all initial conditions, and for all signals r (in continuous-time $r \in \mathcal{C}^j$) the solution of (13.21) is such that (13.50) is true.*

The solvability conditions for the above problem are given as follows.

Theorem 13.4.15 *Consider the system given in (13.14) for both continuous- and discrete-time systems. Let Assumptions A.1 and A.2 be satisfied. Then, the output regulation problem with L_∞ (or ℓ_∞) performance level γ via state feedback + j-th order PD regulators is solvable if and only if the following conditions are true:*

(i) *There exist matrices* Π *and* Γ *that solve the regulator equation (13.16).*

(ii) *For a specified* $j \geqslant 0$, *there exists a feedback law of the form* $\tilde{u} = F\tilde{x} + \sum_{i=0}^{j} H_i \rho^i r$ *for the system (13.17) such that the* $L_\infty[0,\infty)$ *(or* $\ell_\infty[0,\infty)$*) induced operator norm from r to e for zero initial conditions is less than* γ.

Moreover, in that case, utilizing such F and H_i, *the output regulation problem with* L_∞ *(or* ℓ_∞*) performance level* γ *via state feedback + j-th order PD regulators is solvable by the feedback law (13.22).*

So far we have considered state variable feedback regulators. We next move on to consider the case when only the measurement variable y is available for feedback. We have the following basic problem.

Problem 13.4.16 *The* output regulation problem with L_∞ (or ℓ_∞) performance level γ via measurement feedback *is to find, if possible, a measurement feedback law of the form (13.23) such that the following properties hold:*

(i) *The system (13.24) is asymptotically stable.*

(ii) *Given any* $\varepsilon > 0$, *for all initial conditions, and for all signals r (piecewise continuous in continuous-time), there exists a* $T > 0$ *such that the solution of the interconnection of (13.23) and (13.14) satisfies (13.50).*

We have the following solvability conditions for the above problem.

Theorem 13.4.16 *Consider the system given in (13.14) for both continuous- and discrete-time systems. Let Assumptions A.1, A.2, and A.3 be satisfied. Then, the output regulation problem with* L_∞ *(or* ℓ_∞*) performance level* γ *via measurement feedback regulators is solvable if and only if the following conditions are true:*

(i) *There exist matrices* Π *and* Γ *that solve the regulator equation (13.16).*

(ii) *There exists a stabilizing feedback of the form (13.23) for the system (13.25) such that the* $L_\infty[0,\infty)$ *(or* $\ell_\infty[0,\infty)$*) induced operator norm from r to e, for zero initial conditions, is less than* γ.

Moreover, if we have Π, Γ, *and a controller characterized by the quadruple* $(\bar{A}_c, \bar{B}_c, \bar{C}_c, \bar{D}_c)$ *satisfying (i) and (ii), then a controller solving the output regulation problem with* L_∞ *(or* ℓ_∞*) performance level* γ *via measurement feedback is given by (13.26).*

As in the case of state feedback regulators, whenever we know all the derivatives of the signal r in continuous-time systems or all the forward shifted values of r in discrete-time systems, the solvability conditions can be relaxed. We begin with the following problem formulation.

Problem 13.4.17 *Let a system of the form (13.14) be given. The* output regulation problem with L_∞ (or ℓ_∞) performance level γ via measurement feedback + PD regulators *is to find, if possible, a measurement feedback law of the form (13.33) for some $j \geqslant 0$ such that the following properties hold:*

(i) *The system (13.24) is asymptotically stable.*

(ii) *Given any $\varepsilon > 0$, for all initial conditions, and for all signals r (in continuous-time $r \in \mathcal{C}^j$) the interconnection of (13.33) and (13.14) is such that (13.50) is true.*

We have the following solvability conditions for the above problem.

Theorem 13.4.17 *Consider the system given in (13.14) for both continuous- and discrete-time systems. Let Assumptions A.1, A.2, and A.3 be satisfied. Then, the output regulation problem with L_∞ (or ℓ_∞) performance level γ via measurement feedback + PD regulators is solvable if and only if the following conditions are true:*

(i) *There exist matrices Π and Γ that solve the regulator equation (13.16).*

(ii) *For some $j \geq 0$, there exists a feedback law of the form (13.34) for the system (13.25) such that the $L_\infty[0, \infty)$ (or $\ell_\infty[0, \infty)$) induced operator norm from r to e for zero initial conditions is less than γ.*

Moreover, if we have Π, Γ, and a controller of the form (13.34) satisfying (i) and (ii), then a controller solving the output regulator problem with L_∞ (or ℓ_∞) performance level γ via measurement feedback + PD regulators is solvable by the feedback law (13.35).

If we have a limit on the number of derivatives or the number of forward shifted values of the signal r, then the solvability conditions are obviously stronger. We begin with the following problem formulation.

Problem 13.4.18 *Let a system of the form (13.14) be given. The* output regulation problem with L_∞ (or ℓ_∞) performance level γ via measurement feedback + j-th order PD regulators *is to find, if possible, a measurement feedback law of the form (13.33) for a specified $j \geqslant 0$ such that the following properties hold:*

(i) *The system (13.24) is asymptotically stable.*

(ii) *Given any $\varepsilon > 0$, for all initial conditions, and for all signals r (in continuous-time $r \in \mathcal{C}^j$) the interconnection of (13.33) and (13.14) is such that (13.50) is true.*

Remark 13.4.4 *In earlier problems we have seen that using controllers of the form (13.6) in continuous-time does not help us in weakening solvability conditions. Here we would like to note that it* does *help. The L_∞ induced operator norm can be reduced more by using future information of the exogenous signal $\bar{r} = Rr$. For details we refer to [12] where it is shown that optimal causal controllers have delays and from which it easy to conclude that future information on r would be useful. This result is in contrast to H_2 and H_∞ norms where future information on r does not help in reducing the norm. Although we do not give all the details for this case, we would like to note that the above theorem does apply with obvious changes also when using future information of r.*

We have the following solvability conditions for the above problem.

Theorem 13.4.18 *Consider the system given in (13.14) for both continuous- and discrete-time systems. Let Assumptions A.1, A.2, and A.3 be satisfied. Then, the output regulation problem with L_∞ (or ℓ_∞) performance level γ via measurement feedback + j-th order PD regulators is solvable if and only if the following conditions are true:*

(i) *There exist matrices Π and Γ that solve the regulator equation (13.16).*

(ii) *For a specified $j \geq 0$, there exists a feedback law of the form (13.34) for the system (13.25) such that the $L_\infty[0,\infty)$ (or $\ell_\infty[0,\infty)$) induced operator norm from r to e for zero initial conditions is less than γ.*

Moreover, if we have Π, Γ, and a controller of the form (13.34) satisfying (i) and (ii), then a controller solving the output regulation problem with L_∞ (or ℓ_∞) performance level γ via measurement feedback via measurement feedback + j-th order PD regulators is solvable by the feedback law (13.35).

If we view Theorems 13.4.13 and 13.4.16 in the L_2-induced operator norm, then these results can be reinterpreted as H_∞ control problems with unstable weighting functions. This problem has been considered in [47]. Our results provide an alternative way to solve this problem.

Note that the construction of appropriate regulators in all the above theorems involves the construction of controllers with different architectures that can achieve L_∞ (or ℓ_∞) induced norm of the operator from r to e smaller than a prescribed γ when applied to the auxiliary systems (13.17) and (13.25). For such a construction for the general case of L_p-L_q, we refer the reader to [70] for $p = q = 2$, to [11] for $p = q = \infty$, to [61] for the $p = 2, q = \infty$ (SISO case) and to [55] for $p = 2, q = \infty$ (MIMO case).

13.5 Classical output regulation with uncertain exosystem

In structural stability (see Section 2.8) we investigated the effect of perturbations of the system parameters. We wanted to arrive at a one fixed controller which achieves regulation for all possible system parameters in an open neighborhood around the nominal value. However, perturbation of the exosystem was explicitly excluded in the problem formulation. This was necessary because it follows directly from the internal model principle as discussed in Section 2.6 that the controller must contain a partial copy of the exosystem and then there obviously nearly always exist arbitrarily small perturbations such that after perturbation the controller no longer contains an internal copy and therefore does not achieve regulation.

As discussed in Section 12.1 the root of the problem might be that we do not pose the right question. Maybe we should require that after perturbation the asymptotic error is arbitrarily small instead of being exactly equal to zero. The latter problem can be very nicely formulated in terms of generalized regulation.

Assume S is weakly Hurwitz-stable (continuous-time) or weakly Schur-stable (discrete-time) with nominal value S_0. Then we have:

$$\rho w = Sw = S_0 w + (S - S_0)w = S_0 w + r$$

with $r = (S - S_0)w$. A natural question then is to ask for

$$\|e\|_{\infty,[T,\infty)} \leqslant \varepsilon \|w\|_{\infty,[0,\infty)}$$

for all perturbations of the exosystem such that $\|S - S_0\| \leqslant \delta$. But then a very natural sufficient condition amounts to requiring that

$$\|e\|_{\infty,[T,\infty)} \leqslant \frac{\varepsilon}{\delta} \|r\|_{\infty,[0,\infty)}$$

is satisfied for the nominal system. But this is precisely the problem that we have studied in this chapter.

This can then be combined with perturbations of the other system parameters, and it shows that generalized regulation yields a very nice tool to attack the problem of structural stability with perturbations of the exosystem.

Chapter 14

Generalized output regulation with saturating actuators – continuous-time systems

14.1 Introduction and problem formulation

We have formulated in Chapter 3 the semi-global linear feedback output regulation problem for linear systems subject to input saturation. Chapter 3 follows the traditional formulation of linear output regulation problems where the exosystem is autonomous. As a result, the disturbances and the references generated by the exosystem contain only the frequency components of the exosystem. In an effort to broaden the class of disturbance and reference signals, we formulate in this section the generalized semi-global linear feedback output regulation problem in which an external driving signal to the exosystem is included. Unlike in Chapter 13 where we considered different layers of output regulation problems under a broad category of controllers, we consider here only exact output regulation problems under the classical state as well as measurement feedback regulators. This chapter is based on the work of [33].

We consider a multivariable system with inputs that are subject to saturation together with an exosystem that generates disturbance and reference signals as described by the system,

$$
\begin{aligned}
\dot{x} &= Ax + B\sigma(u) + E_w w \\
\dot{w} &= Sw + r \\
y &= C_y x + D_{yw} w \\
e &= C_e x + D_{ew} w
\end{aligned}
\tag{14.1}
$$

where, as usual, $x \in \mathbb{R}^n$, $w \in \mathbb{R}^s$, $u \in \mathbb{R}^m$, $y \in \mathbb{R}^p$, and $e \in \mathbb{R}^q$, while $r \in \mathcal{C}^0$ is an external signal to the exosystem. Also, σ is a vector-valued saturation function as defined by (3.2), that is

$$\sigma(s) = [\bar{\sigma}(s_1), \bar{\sigma}(s_2), \cdots, \bar{\sigma}(s_m)]^{\mathrm{T}} \tag{14.2}$$

with

$$\bar{\sigma}i(s) = \begin{cases} s & \text{if } |s| \leq 1 \\ -1 & \text{if } s < -1 \\ 1 & \text{if } s > 1. \end{cases} \tag{14.3}$$

In what follows, we formulate two generalized output regulation problems, one utilizing the state feedback, and the other measurement feedback. The block diagram of Figure 14.1 depicts this.

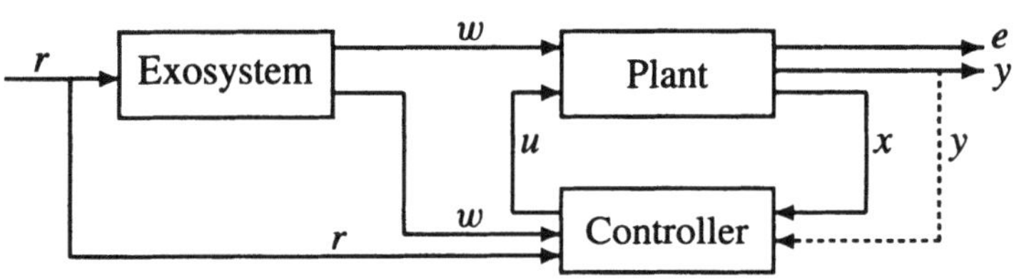

Figure 14.1: Generalized output regulation scheme

The generalized semi-global linear state feedback output regulation problem and the generalized measurement feedback output regulation problem are formulated as follows.

Problem 14.1.1 ***(Generalized semi-global linear state feedback output regulation problem)*** *Consider the system (14.1), a compact set $\mathcal{W}_0 \subset \mathbb{R}^s$, and a bounded set $\mathcal{R} \subset L_\infty$. For any a priori given (arbitrarily large) bounded set $\mathcal{X}_0 \subset \mathbb{R}^n$, find, if possible, a linear static feedback law $u = Fx + Gw + Hr$, such that the following conditions hold:*

(i) ***(Internal Stability)*** *The equilibrium point $x = 0$ of*

$$\dot{x} = Ax + B\sigma(Fx) \tag{14.4}$$

is asymptotically stable with $\mathcal{X}_0$ contained in its basin of attraction.

(*ii*) ***(Output Regulation)*** *For all* $x(0) \in \mathcal{X}_0$, $w(0) \in \mathcal{W}_0$, *and* $r \in \mathcal{R}$, *the solution of the closed-loop system satisfies*

$$\lim_{t\to\infty} e(t) = 0. \tag{14.5}$$

Problem 14.1.2 ***(Generalized semi-global linear observer based measurement feedback output regulation problem)*** *Consider the system (14.1), a compact set* $\mathcal{W}_0 \subset \mathbb{R}^s$ *and a bounded set* $\mathcal{R} \subset L_\infty$. *For any a priori given (arbitrarily large) bounded sets* $\mathcal{X}_0 \subset \mathbb{R}^n$ *and* $\mathcal{Z}_0 \subset \mathbb{R}^{n+s}$, *find, if possible, a linear observer based measurement feedback law of the form,*

$$\begin{aligned}\begin{pmatrix}\dot{\hat{x}}\\ \dot{\hat{w}}\end{pmatrix} &= \begin{pmatrix}A & E_w\\ 0 & S\end{pmatrix}\begin{pmatrix}\hat{x}\\ \hat{w}\end{pmatrix} + \begin{pmatrix}B\sigma(u)\\ r\end{pmatrix}\\ &\quad + \begin{pmatrix}K_A\\ K_S\end{pmatrix}\left(\begin{pmatrix}C_y & D_{yw}\end{pmatrix}\begin{pmatrix}\hat{x}\\ \hat{w}\end{pmatrix} - y\right)\\ u &= F\hat{x} + G\hat{w} + Hr,\end{aligned} \tag{14.6}$$

such that the following conditions hold:

(*i*) ***(Internal Stability)*** *The equilibrium point* $(x, \hat{x}, \hat{w}) = (0, 0, 0)$ *of*

$$\begin{aligned}\dot{x} &= Ax + B\sigma(F\hat{x} + G\hat{w})\\ \begin{pmatrix}\dot{\hat{x}}\\ \dot{\hat{w}}\end{pmatrix} &= \begin{pmatrix}A & E_w\\ 0 & S\end{pmatrix}\begin{pmatrix}\hat{x}\\ \hat{w}\end{pmatrix} + \begin{pmatrix}B\\ 0\end{pmatrix}\sigma(F\hat{x} + G\hat{w})\\ &\quad + \begin{pmatrix}K_A\\ K_S\end{pmatrix}(\begin{pmatrix}C_y & D_{yw}\end{pmatrix}\begin{pmatrix}\hat{x} - x\\ \hat{w}\end{pmatrix})\end{aligned} \tag{14.7}$$

is asymptotically stable with $\mathcal{X}_0 \times \mathcal{Z}_0$ *contained in its basin of attraction.*

(*ii*) ***(Output Regulation)*** *For all* $(x(0), \hat{x}(0), \hat{w}(0)) \in \mathcal{X}_0 \times \mathcal{Z}_0$, $w(0) \in \mathcal{W}_0$, *and all* $r \in \mathcal{R}$, *the solution of the closed-loop system satisfies*

$$\lim_{t\to\infty} e(t) = 0. \tag{14.8}$$

Remark 14.1.1 *As in Chapter 3, we would like to emphasize again that our definition of the generalized semi-global linear state feedback [respectively, linear observer based measurement feedback] output regulation problem does not view the set of initial conditions of the plant as given data. The set of given data consists of the models of the plant and the exosystem, the set of initial conditions for the exosystem, and the set of external inputs to the exosystem. Moreover, the generalized semi-global linear state feedback [respectively, measurement feedback] output regulation problem reduces to the* *classical semi-global linear state feedback [respectively, measurement feedback] output regulation problem as formulated in Problem 3.3.1 [respectively, Problem 3.3.2] when the external input r to the exosystem is non-existent.*

In what follows, we will give the solvability conditions for the above two problems. For clarity, we present these solvability conditions in two separate subsections, one for the state feedback case and the other for the measurement feedback case. As a special case of the generalized semi-global linear feedback output regulation problems, the solvability conditions for semi-global linear feedback tracking problems for a chain of integrators are obtained readily. The same set of solvability conditions for the global state feedback tracking problem for a chain of integrators were given earlier in [83], where nonlinear feedback laws were resorted to.

As it is clear from previous chapters, it is reasonable to formulate the following assumptions:

A.1. The pair (A, B) is stabilizable.

A.2. The matrix S is anti-Hurwitz-stable, i.e. all the eigenvalues of S are in the closed right-half plane.

A.3. The pair $\left(\begin{pmatrix} C_y & D_{yw} \end{pmatrix}, \begin{pmatrix} A & E_w \\ 0 & S \end{pmatrix}\right)$ is detectable.

14.2 State feedback output regulation problem

The solvability conditions for the generalized semi-global linear state feedback output regulation problem is given in the following theorem.

Theorem 14.2.1 *Consider the system (14.1) and given a compact set $\mathcal{W}_0 \subset \mathbb{R}^s$ and a bounded set $\mathcal{R} \subset L_\infty$. Let Assumptions A.1 and A.2 hold. Then, the generalized semi-global linear state feedback output regulation problem is solvable if the following conditions hold:*

(*i*) *A has all its eigenvalues in the closed left half plane.*

(*ii*) *There exist matrices Π and Γ such that,*

(*a*) *they solve the regulator equation (2.7), i.e.,*

$$\begin{aligned} \Pi S &= A\Pi + B\Gamma + E_w, \\ 0 &= C_e\Pi + \quad D_{ew}, \end{aligned} \tag{14.9}$$

(*b*) *for each* $r \in \mathcal{R}$, *there exists a function* $\tilde{r} \in L_\infty$ *such that* $\Pi r = B\tilde{r}$,

(*c*) *there exists a* $\delta > 0$ *and a* $T \geq 0$ *such that* $\|\Gamma w + \tilde{r}\|_{\infty,T} \leqslant 1 - \delta$ *for all* w *with* $w(0) \in \mathcal{W}_0$ *and all* $r \in \mathcal{R}$.

Remark 14.2.1 *We would like to make the following observations on the solvability conditions as given in the above theorem:*

(*i*) *As expected, the solvability conditions for the generalized semi-global linear state feedback output regulation problem as given in the above theorem reduces to those for the semi-global linear state feedback output regulation problem as formulated in Problem 3.3.1 when the external input to the exosystem is non-existent.*

(*ii*) *If* $\operatorname{im}\Pi \subseteq \operatorname{im} B$, *then Condition 2 (b) is automatically satisfied for any given set* $\mathcal{R}$.

(*iii*) *If* $\operatorname{im}\Pi \cap \operatorname{im} B = \{0\}$, *then Condition 2 (b) can never be satisfied for any given* $\mathcal{R}$ *except for* $\mathcal{R} = \{0\}$.

Proof of Theorem 14.2.1 : The proof of this theorem is similar, *mutatis mutandis*, to that of Theorem 3.3.2. As in the proof of Theorem 3.3.2, we prove this theorem by first constructing a family of linear static state feedback laws, parameterized in ε, and then showing that for each given set $\mathcal{X}_0$, there exists an $\varepsilon^* > 0$ such that for all $\varepsilon \in (0, \varepsilon^*]$, both items 1 and 2 of Problem 14.1.1 hold. The family of linear static state feedback laws we construct takes the form,

$$u = F_\varepsilon x + (\Gamma - F_\varepsilon \Pi)w + \tilde{r}, \tag{14.10}$$

where F_ε is constructed by a Riccati-based design as given by (3.10). That is,

$$F_\varepsilon = -B^{\mathrm{T}} P_\varepsilon. \tag{14.11}$$

The rest of the proof is the same as that of Theorem 3.3.2 except that (3.24) takes instead the following slightly different form,

$$\dot{\xi} = A\xi + B(\sigma(u) - \Gamma w - \tilde{r}). \tag{14.12}$$

■

Theorem 14.2.1 presents the conditions under which the Problem 14.1.1 can be solved. Also, as pointed out in the proof of Theorem 14.2.1, to solve such a problem one could construct a low-gain state feedback controller of the form (14.10) with F_ε as in (14.11). However, in order to improve the convergence of $e(t)$ to zero, one can alternatively use a low-high-gain state feedback controller. Consider the feedback control law,

$$u = -(\mu + 1)B^{\mathrm{T}} P_\varepsilon x + [(\mu + 1)B^{\mathrm{T}} P_\varepsilon \Pi + \Gamma]w + \tilde{r}, \quad \mu \geq 0 \tag{14.13}$$

where P_ε is as given in (3.9).

We have the following theorem.

Theorem 14.2.2 *Consider the system (14.1) and given a compact set $\mathcal{W}_0 \subset \mathbb{R}^s$ and a bounded set $\mathcal{R} \subset \mathcal{C}^0$. Then, under the same solvability conditions as in Theorem 14.2.1, there exists a controller, among the family of feedback control laws given by (14.13), that solves the generalized semi-global linear state feedback output regulation problem. More specifically, for any a priori given (arbitrarily large) bounded set $\mathcal{X}_0 \subset \mathbb{R}^n$, there exists an $\varepsilon^* > 0$ such that for each $\varepsilon \in (0, \varepsilon^*]$, and for each $\mu \geq 0$, the controller in the family (14.13) has the following properties:*

(i) The equilibrium point $x = 0$ of

$$\dot{x} = Ax + B\sigma(-(1 + \mu)B^T P_\varepsilon x) \tag{14.14}$$

is asymptotically stable with $\mathcal{X}_0$ contained in its basin of attraction.

(ii) For any $x(0) \in \mathcal{X}_0$, $w(0) \in \mathcal{W}$, and $r \in \mathcal{R}$, the solution of the closed-loop system satisfies

$$\lim_{t \to \infty} e(t) = 0. \tag{14.15}$$

Proof : The proof of this theorem is the same as that of Theorem 3.3.2 except that (3.24) takes instead the slightly different form given in (14.12). ∎

14.3 Dynamic measurement feedback

The solvability conditions for the generalized semi-global linear observer based measurement feedback output regulation problem is given in the following theorem.

Theorem 14.3.1 *Consider the system (14.1) and given a compact set $\mathcal{W}_0 \subset \mathbb{R}^s$ and a bounded set $\mathcal{R} \subset L_\infty$. Let Assumptions A.1, A.2, and A.3 hold. Then, the generalized semi-global linear observer based measurement feedback output regulation problem is solvable if the following conditions hold:*

(*i*) *A has all its eigenvalues in the closed left half plane.*

(*ii*) *There exist matrices Π and Γ such that,*

(*a*) *they solve the regulator equation (14.9),*

(*b*) *for each $r \in \mathcal{R}$, there exists a function $\tilde{r}(t) \in L_\infty$ such that $\Pi r(t) = B\tilde{r}(t)$ for all $t \geq 0$,*

(*c*) *there exist a $\delta > 0$ and a $T \geq 0$ such that $\|\Gamma w + \tilde{r}\|_{\infty,T} \leqslant 1 - \delta$ for all w with $w(0) \in \mathcal{W}_0$ and all $r \in \mathcal{R}$.*

Remark 14.3.1 *As expected, the solvability conditions for the generalized semi-global linear observer based measurement feedback output regulation problem as given in the above theorem reduces to those for the semi-global linear observer based measurement feedback output regulation problem as formulated in Problem 3.3.2 when the external input to the exosystem is non-existent.*

Proof of Theorem 14.3.1 : The proof of this theorem is similar, *mutatis mutandis*, to that of Theorem 3.3.4. As in the proof of Theorem 3.3.4, we prove this theorem by first constructing a family of linear observer based measurement feedback laws, parameterized in ε, and then showing that both items 1 and 2 of Problem 14.1.2 indeed hold if ε is selected appropriately. The family of observer based measurement feedback laws we construct takes the form,

$$\begin{aligned} \dot{\hat{x}} &= A\hat{x} + B\sigma(u) + E_w\hat{w} + K_A C_y(\hat{x} - x) + K_A D_{yw}(\hat{w} - w) \\ \dot{\hat{w}} &= S\hat{w} + K_S C_y(\hat{x} - x) + K_S D_{yw}(\hat{w} - w) + r \\ u &= F_\varepsilon \hat{x} + (\Gamma - F_\varepsilon \Pi)\hat{w} + \tilde{r}, \end{aligned} \tag{14.16}$$

where K_A and K_S are such that the following matrix $\bar{A}$ is Hurwitz-stable,

$$\bar{A} := \begin{pmatrix} A + K_A C_y & E_w + K_A D_{yw} \\ K_S C_y & S + K_S D_{yw} \end{pmatrix}.$$

The rest of the proof is the same as that of Theorem 3.3.4 except that (3.48) takes instead the form,

$$\begin{aligned}
\dot{\xi} &= A\xi + B\sigma\left[\Gamma(w-\tilde{w}) - F_\varepsilon(\tilde{x}-\xi-\Pi\tilde{w}) + \tilde{r}\right] \\
&\qquad +(A\Pi - \Pi S + E_w)w - \Pi r \\
\dot{\tilde{x}} &= (A + K_{\rm A}C_y)\tilde{x} + (E_w + K_{\rm A}D_{yw})\tilde{w} \\
\dot{\tilde{w}} &= K_{\rm S}C_y\tilde{x} + (S + K_{\rm S}D_{yw})\tilde{w}.
\end{aligned}$$

■

Remark 14.3.2 *From the above proof of Theorem 14.3.1, we note that the linear state feedback law (14.10) interconnected with any exponentially stable observer (where x and w are replaced by their respective estimates which converge exponentially to the real x and w as $t \to \infty$) will solve the semi-global measurement feedback output regulation problem.*

Theorem 14.3.1 presents the conditions under which the Problem 14.1.2 can be solved. Also, as pointed out in the proof of Theorem 14.3.1, to solve such a problem one could construct an observer based measurement feedback controller (14.16) with F_ε as in (3.10). However, in order to improve the convergence of $e(t)$ to zero, one can alternatively use a low-high-gain state feedback controller. Consider the family of observer based measurement feedback control laws,

$$\begin{aligned}
\dot{\hat{x}} &= A\hat{x} + B\sigma(u) + E_w\hat{w} + K_{{\rm A},\ell}C_y(\hat{x}-x) + K_{{\rm A},\ell}D_{yw}(\hat{w}-w) \\
\dot{\hat{w}} &= S\hat{w} + K_{{\rm S},\ell}C_y(\hat{x}-x) + K_{{\rm S},\ell}D_{yw}(\hat{w}-w) + r \\
u &= -(1+\mu)B^{\rm T}P_\varepsilon\hat{x} + ((\mu+1)B^{\rm T}P_\varepsilon\Pi + \Gamma)\hat{w} + \tilde{r}
\end{aligned} \tag{14.17}$$

where P_ε is as given in (3.9), and $K_{{\rm A},\ell}$ and $K_{{\rm S},\ell}$ are as given in (3.62).

Theorem 14.3.2 *Consider the system (14.1) and given a compact set $\mathcal{W}_0 \subset \mathbb{R}^s$ and a bounded set $\mathcal{R} \subset \mathcal{C}^0$. Then, under the same solvability conditions as in Theorem 14.3.1, and under the assumption that the pair*

$$\left(\begin{pmatrix} C_y & D_{yw}\end{pmatrix}, \begin{pmatrix} A & E_w \\ 0 & S\end{pmatrix}\right)$$

is observable, there exists a controller, among the family of feedback control laws given by (14.17), that solves the generalized observer based measurement feedback output regulation problem. More specifically, for any a *priori given (arbitrarily large) sets $\mathcal{X}_0 \subset \mathbb{R}^n$ and $\mathcal{Z}_0 \subset \mathbb{R}^{n+s}$, there exists an $\varepsilon^* \in (0,1]$, and for each $\varepsilon \in (0,\varepsilon^*]$, and each $\mu \geq 0$, there exists an $\ell^*(\varepsilon,\mu) > 0$ such that for each $\varepsilon \in (0,\varepsilon^*]$, each $\mu \geq 0$, and each $\ell \geq \ell^*(\varepsilon,\mu)$, the controller in the family (14.17) has the following properties:*

(*i*) *The equilibrium point $(x, \hat{x}, \hat{w}) = (0,0,0)$ of*

$$\begin{aligned}
\dot{x} &= Ax + B\sigma(u) \\
\begin{pmatrix} \dot{\hat{x}} \\ \dot{\hat{w}} \end{pmatrix} &= \begin{pmatrix} A & E_w \\ 0 & S \end{pmatrix} \begin{pmatrix} \hat{x} \\ \hat{w} \end{pmatrix} + \begin{pmatrix} B \\ 0 \end{pmatrix} \sigma(u) \\
&\quad + \begin{pmatrix} K_{\mathrm{A},\ell} \\ K_{\mathrm{S},\ell} \end{pmatrix} \left(\begin{pmatrix} C_y & D_{yw} \end{pmatrix} \begin{pmatrix} \hat{x} - x \\ \hat{w} \end{pmatrix} \right) \\
u &= -(\mu+1)B^T P_\varepsilon \hat{x} + ((\mu+1)B^T P_\varepsilon \Pi + \Gamma)\hat{w}
\end{aligned} \tag{14.18}$$

is asymptotically stable with $\mathcal{X}_0 \times \mathcal{Z}_0$ contained in its basin of attraction.

(*ii*) *For any $(x(0), \hat{x}(0), \hat{w}(0)) \in \mathcal{X}_0 \times \mathcal{Z}_0$, and $w(0) \in \mathcal{W}$, the solution of the closed-loop system satisfies*

$$\lim_{t\to\infty} e(t) = 0. \tag{14.19}$$

Proof : The proof of this theorem is similar, *mutatis mutandis*, to that of Theorem3.3.5, except that, in this case, (3.48) takes instead the form

$$\begin{aligned}
\dot{\xi} &= A\xi + B\sigma(-(1+\mu)B^{\mathrm{T}} P_\varepsilon \xi + \Gamma w - \Gamma \tilde{w} \\
&\qquad + (1+\mu)B^{\mathrm{T}} P_\varepsilon \tilde{x} - (1+\mu)B^{\mathrm{T}} P_\varepsilon \Pi \tilde{w} + \tilde{r}) \\
&\qquad + (A\Pi - \Pi S + E_w)w - \Pi r \\
\dot{\tilde{x}} &= (A + K_{\mathrm{A},\ell} C_y)\tilde{x} + (E_w + K_{\mathrm{A},\ell} D_{yw})\tilde{w} \\
\dot{\tilde{w}} &= K_{\mathrm{S},\ell} C_y \tilde{x} + (S + K_{\mathrm{S},\ell} D_{yw})\tilde{w}.
\end{aligned} \tag{14.20}$$

■

Chapter 15

Generalized regulation with saturating actuators – discrete-time systems

15.1 Introduction and problem formulation

We have formulated earlier in Chapter 4 the semi-global linear feedback output regulation problems for linear systems subject to input saturation. Chapter 4 follows the traditional formulation of linear output regulation problems where the exosystem is autonomous. As a result, the disturbances and the references generated by the exosystem contain only the frequency components of the exosystem. In an effort to broaden the class of disturbance and reference signals, we formulate in this section the generalized semi-global linear feedback output regulation problem in which an external driving signal to the exosystem is included. Unlike in Chapter 13 where we considered different layers of output regulation problems under a broad category of controllers, we consider here only exact output regulation problems under the classical state as well as measurement feedback regulators.

We consider a multivariable system with inputs that are subject to saturation together with an exosystem that generates disturbance and reference signals as described by the system,

$$\begin{aligned} x(k+1) &= Ax(k) + B\sigma(u(k)) + E_w w(k) \\ w(k+1) &= Sw(k) + r(k) \\ y(k) &= C_y x(k) + D_{yw} w(k) \\ e(k) &= C_e x(k) + D_{ew} w(k), \end{aligned} \tag{15.1}$$

where, as before, $x \in \mathbb{R}^n$, $w \in \mathbb{R}^s$, $u \in \mathbb{R}^m$, $y \in \mathbb{R}^p$, and $e \in \mathbb{R}^q$, while $r \in L_\infty$ is an external signal to the exosystem, and σ is a vector-valued saturation function as defined by (4.2), that is

$$\sigma(s) = [\bar{\sigma}(s_1), \bar{\sigma}(s_2), \cdots, \bar{\sigma}(s_m)]^{\mathrm{T}} \tag{15.2}$$

with

$$\bar{\sigma}(s) = \begin{cases} s & \text{if } |s| \leq 1 \\ -1 & \text{if } s < -1 \\ 1 & \text{if } s > 1. \end{cases} \tag{15.3}$$

In what follows, we formulate two generalized output regulation problems, one utilizing the state feedback, and the other measurement feedback. The block diagram of Figure 15.1 depicts this.

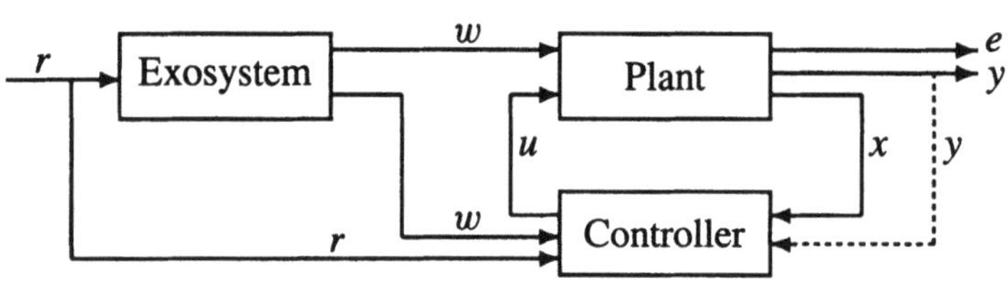

Figure 15.1: Generalized output regulation scheme

The generalized semi-global linear state feedback output regulation problem and the generalized linear observer based measurement feedback output regulation problem are formulated as follows.

Problem 15.1.1 ***(Generalized semi-global state feedback output regulation problem)*** *Consider the system (15.1), a compact set $\mathcal{W}_0 \subset \mathbb{R}^s$ and a bounded set $\mathcal{R} \subset L_\infty$. For any a priori given (arbitrarily large) bounded set $\mathcal{X}_0 \subset \mathbb{R}^n$, find, if possible, a static feedback law of the type $u = \alpha(x, w, r)$, such that the following conditions hold:*

(i) ***(Internal Stability)*** *The equilibrium point $x = 0$ of*

$$x(k+1) = Ax(k) + B\sigma(\alpha(x(k), 0, 0)) \tag{15.4}$$

is asymptotically stable with $\mathcal{X}_0$ contained in its basin of attraction.

(ii) ***(Output Regulation)*** *For all $x(0) \in \mathcal{X}_0$, $w(0) \in \mathcal{W}_0$ and $r \in \mathcal{R}$, the solution of the closed-loop system satisfies*

$$\lim_{k\to\infty} e(k) = 0. \tag{15.5}$$

Problem 15.1.2 ***(The generalized semi-global linear observer based measurement feedback output regulation problem)*** *Consider the system (15.1), a compact set $\mathcal{W}_0 \subset \mathbb{R}^s$ and a bounded set $\mathcal{R} \subset L_\infty$. For any a priori given (arbitrarily large) bounded sets $\mathcal{X}_0 \subset \mathbb{R}^n$ and $\mathcal{Z}_0 \subset \mathbb{R}^{n+s}$, find, if possible, a linear observer based measurement feedback law of the form,*

$$\begin{aligned}\begin{pmatrix}\hat{x}(k+1)\\ \hat{w}(k+1)\end{pmatrix} &= \begin{pmatrix}A & E_w\\ 0 & S\end{pmatrix}\begin{pmatrix}\hat{x}(k)\\ \hat{w}(k)\end{pmatrix} + \begin{pmatrix}B\sigma(u(k))\\ r(k)\end{pmatrix}\\ &\quad + \begin{pmatrix}K_A\\ K_S\end{pmatrix}\left(\begin{pmatrix}C_y & D_{yw}\end{pmatrix}\begin{pmatrix}\hat{x}(k)\\ \hat{w}(k)\end{pmatrix} - y(k)\right)\\ u(k) &= \alpha(\hat{x}(k), \hat{w}(k), r(k)),\end{aligned} \tag{15.6}$$

such that the following conditions hold:

(i) ***(Internal Stability)*** *The equilibrium point $(x, \hat{x}, \hat{w}) = (0, 0, 0)$ of*

$$\begin{aligned}x(k+1) &= Ax(k) + B\sigma(\alpha(\hat{x}(k), \hat{w}(k), 0))\\ \begin{pmatrix}\hat{x}(k+1)\\ \hat{w}(k+1)\end{pmatrix} &= \begin{pmatrix}A & E_w\\ 0 & S\end{pmatrix}\begin{pmatrix}\hat{x}(k)\\ \hat{w}(k)\end{pmatrix} + \begin{pmatrix}B\\ 0\end{pmatrix}\sigma(\alpha(\hat{x}(k), \hat{w}(k), 0))\\ &\quad + \begin{pmatrix}K_A\\ K_S\end{pmatrix}\begin{pmatrix}C_y & D_{yw}\end{pmatrix}\begin{pmatrix}\hat{x}(k) - x(k)\\ \hat{w}(k)\end{pmatrix}\end{aligned} \tag{15.7}$$

is asymptotically stable with $\mathcal{X}_0 \times \mathcal{Z}_0$ contained in its basin of attraction.

(ii) ***(Output Regulation)*** *For all $(x(0), \hat{x}(0), \hat{w}(0)) \in \mathcal{X}_0 \times \mathcal{Z}_0$, $w(0) \in \mathcal{W}_0$, and all $r \in \mathcal{R}$, the solution of the closed-loop system satisfies*

$$\lim_{k\to\infty} e(k) = 0. \tag{15.8}$$

Remark 15.1.1 *As in Chapter 4, we would like to emphasize again that our definition of the generalized semi-global linear state feedback [respectively, linear observer based measurement feedback] output regulation problem does not view the set of initial conditions of the plant as given data. The set of given data consists of the models of the plant and the exosystem, the set of* *initial conditions for the exosystem, and the set of external inputs to the exosystem. Moreover, the generalized semi-global linear state feedback [respectively, measurement feedback] output regulation problem reduces to the classical semi-global linear state feedback [respectively, measurement feedback] output regulation problem as formulated in Problem 4.3.1 [respectively, Problem 4.3.2] when the external input r to the exosystem is non-existent.*

Remark 15.1.2 *We would also like to emphasize that unlike the traditional output regulation problem where all interesting cases arise when the poles of the exosystem are outside or on the unit circle, for the generalized output regulation problem, there are interesting cases even when the exosystem is asymptotically stable.*

We will give the solvability conditions for the above two problems. For clarity, we present these solvability conditions in two separate subsections, one for each of the two problems.

As it is clear from previous chapters, it is reasonable to formulate the following assumptions:

A.1. The pair (A, B) is stabilizable.

A.2. The matrix S is anti-Schur-stable, i.e. all the eigenvalues of S are on or outside the unit circle.

A.3. The pair $\left(\begin{pmatrix} C_y & D_{yw} \end{pmatrix}, \begin{pmatrix} A & E_w \\ 0 & S \end{pmatrix}\right)$ is detectable.

15.2 State feedback output regulation problem

The solvability conditions for generalized semi-global state feedback output regulation problem is given in the following theorem.

Theorem 15.2.1 *Consider the system (15.1) and given a compact set $\mathcal{W}_0 \subset \mathbb{R}^s$ and a bounded set $\mathcal{R} \subset L_\infty$. Let Assumptions A.1 and A.2 hold. The generalized semi-global state feedback output regulation problem is solvable if the following conditions hold:*

(i) A has all its eigenvalues on or inside the unit circle.

(ii) There exist matrices Π and Γ such that

(*a*) *they solve the regulator equation (2.7), i.e.,*

$$\begin{aligned} \Pi S &= A\Pi + B\Gamma + E_w, \\ 0 &= C_e\Pi + \qquad D_{ew}, \end{aligned} \tag{15.9}$$

(*b*) *for each* $r \in \mathcal{R}$, *there exists a function* $\tilde{r} \in L_\infty$ *such that* $\Pi r = B\tilde{r}$,

(*c*) *there exists a* $\delta > 0$ *and a* $K \geq 0$ *such that* $\|\Gamma w + \tilde{r}\|_{\infty,K} \leqslant 1 - \delta$ *for all* w *with* $w(0) \in \mathcal{W}_0$ *and all* $r \in \mathcal{R}$.

Moreover, a linear state feedback controller of the form $u(k) = F\hat{x}(k) + G\hat{w}(k) + H\tilde{r}(k)$ *can solve the posed problem.*

Remark 15.2.1 *We would like to make the following observations on the solvability conditions as given in the above theorem:*

(*i*) *As expected, the solvability conditions for the generalized semi-global linear state feedback output regulation problem as given in the above theorem reduce to those for the classical semi-global linear state feedback output regulation problem as formulated in Problem 4.3.1 when the external input to the exosystem is non-existent.*

(*ii*) *If* $\operatorname{im}\Pi \subseteq \operatorname{im}B$, *then Condition 2 (b) is automatically satisfied for any given set* $\mathcal{R}$.

(*iii*) *If* $\operatorname{im}\Pi \cap \operatorname{im}B = \{0\}$, *then Condition 2 (b) can never be satisfied for any given* $\mathcal{R}$ *except for* $\mathcal{R} = \{0\}$.

Proof of Theorem 15.2.1: The proof of this theorem is similar, *mutatis mutandis*, to that of Theorem 4.3.1. As in the proof of Theorem 4.3.1, we prove this theorem by first constructing a family of linear static state feedback laws parameterized in ε, and then showing that for each given set $\mathcal{X}_0$, there exists an $\varepsilon^* > 0$ such that for all $\varepsilon \in (0, \varepsilon^*]$, both items 1 and 2 of Problem 15.1.1 hold. The family of linear static state feedback laws we construct takes the form,

$$u = F_\varepsilon x + (\Gamma - F_\varepsilon \Pi)w + \tilde{r} \tag{15.10}$$

where $F_\varepsilon := -(B^{\mathrm{T}} P_\varepsilon B + I)^{-1} B^{\mathrm{T}} P_\varepsilon A$ with P_ε being the solution of the Riccati equation (4.9). The rest of the proof is the same as that of Theorem 4.3.1 except that (4.20) takes the following slightly different form

$$\xi(k+1) = A\xi(k) + B(\sigma(u(k)) - \Gamma w(k) - \tilde{r}(k)).$$

■

Theorem 15.2.1 presents the conditions under which the Problem 15.1.1 can be solved. Also, as pointed out in the proof of Theorem 15.2.1, to solve such a problem one could construct a low-gain state feedback controller of the form (15.10) with F_ε constructed by a Riccati-based design as in (4.10). However, in order to improve the convergence of $e(k)$ to zero, when B is injective, one can alternatively use an improved state feedback controller. Consider the feedback control law,

$$\begin{aligned} u = {} & [F_\varepsilon + \mu\kappa(x, w, \mu)K_\varepsilon]x \\ & - [(F_\varepsilon + \mu\kappa(x, w, \mu)K_\varepsilon)\Pi - \Gamma]w + \tilde{r}, \quad \mu \in [0, 2], \end{aligned} \tag{15.11}$$

where,

$$\begin{aligned} F_\varepsilon &= -(B^{\mathrm{T}}P_\varepsilon B + I)^{-1}B^{\mathrm{T}}P_\varepsilon A, \\ K_\varepsilon &= -(B^{\mathrm{T}}P_\varepsilon B)^{-1}B^{\mathrm{T}}P_\varepsilon A, \\ A_c &= A + BF_\varepsilon, \end{aligned}$$

and P_ε being the solution of the Riccati equation (4.9) with $\lim_{\varepsilon\to 0} Q_\varepsilon = 0$. Also, the function κ is as defined by (4.30).

We have the following theorem.

Theorem 15.2.2 *Consider the system (15.1) and given a compact set $\mathcal{W}_0 \subset \mathbb{R}^s$ and a bounded set $\mathcal{R} \subset L_\infty$. Let the solvability conditions given in Theorem 15.2.1 be satisfied and let B is injective. Then, there exists a controller, among the family of controllers given in (15.11), that solves the generalized semi-global state feedback output regulation problem. More specifically, for any a priori given (arbitrarily large) bounded set $\mathcal{X} \in \mathbb{R}^n$, there exists an $\varepsilon^* \in (0, 1]$ such that for each $\varepsilon \in (0, \varepsilon^*]$ and for each $\mu \in [0, 2]$, the controller in the family (15.11) has the following properties:*

(i) The equilibrium point $x = 0$ of

$$\rho x = Ax + B\sigma(F_\varepsilon x + \mu\kappa(x, 0, \mu)K_\varepsilon x)$$

is asymptotically stable with $\mathcal{X}_0$ contained in its basin of attraction.

(ii) For any $x(0) \in \mathcal{X}_0$, $w(0) \in \mathcal{W}_0$, and $r \in \mathcal{R}$, the solution of the closed-loop system satisfies

$$\lim_{k\to\infty} e(k) = 0.$$

Proof: The proof is similar, *mutatis mutandis*, to that of Theorem 4.3.2, except that (4.41) takes the following slightly different form,

$$\rho\xi = A_c\xi + B\left[\sigma(\Gamma w + (F_\varepsilon + \mu\kappa(x, w, \mu)K_\varepsilon)\xi - \Gamma w - F_\varepsilon\xi - \tilde{r}\right].$$

15.3 Dynamic measurement feedback

The solvability conditions for the generalized semi-global linear observer based measurement feedback output regulation problem is given in the following theorem.

Theorem 15.3.1 *Consider the system (4.1) and given a compact set* $\mathcal{W}_0 \subset \mathbb{R}^s$ *and a bounded set* $\mathcal{R} \subset L_\infty$. *Let Assumptions A.1, A.2, and A.3 hold. Then, the generalized semi-global linear observer based measurement feedback output regulation problem is solvable if the following conditions hold:*

(i) A has all its eigenvalues inside or on the unit circle.

(ii) There exist matrices Π *and* Γ *such that,*

(a) they solve the linear matrix equation (15.9),

(b) for each $r \in \mathcal{R}$, *there exists a function* $\tilde{r} \in L_\infty$ *such that* $\Pi r = B\tilde{r}$ *for all* $k \geq 0$,

(c) there exist a $\delta > 0$ *and a* $K \geq 0$ *such that* $\|\Gamma w + \tilde{r}\|_{\infty,K} \leqslant 1 - \delta$ *for all* w *with* $w(0) \in \mathcal{W}_0$ *and all* $r \in \mathcal{R}$.

Moreover the function $\alpha(\hat{x}(k), \hat{w}(k), r(k))$ *in Problem 15.1.2 can be a linear function of* $x(k)$, $w(k)$, *and* $r(k)$.

Remark 15.3.1 *As expected, the solvability conditions for the generalized semi-global linear observer based measurement feedback output regulation problem as given in the above theorem reduces to those for the classical semi-global linear observer based measurement feedback output regulation problem as formulated in Problem 4.3.2 when the external input to the exosystem is non-existent.*

Proof of Theorem 15.3.1 : The proof of this theorem is similar, *mutatis mutandis*, to that of Theorem 4.3.3. As in the proof of Theorem 4.3.3, we prove this theorem by first constructing a family of linear observer based measurement feedback laws, parameterized in ε, and then showing that both items 1 and 2 of Problem 15.1.2 indeed hold. The family of linear observer based measurement feedback laws we construct takes the following form

$$\begin{aligned}
\hat{x}(k+1) &= A\hat{x}(k) + B\sigma(u(k)) + E_w\hat{w}(k) - K_A y(k) \\
&\quad + K_A(C_y\hat{x}(k) + D_{yw}\hat{w}(k)) \\
\hat{w}(k+1) &= S\hat{w}(k) - K_S y(k) + K_S(C_y\hat{x}(k) + D_{yw}\hat{w}(k)) + r(k) \\
u(k) &= F_\varepsilon\hat{x}(k) + (\Gamma - F_\varepsilon\Pi)\hat{w}(k) + \tilde{r}
\end{aligned} \tag{15.12}$$

where $F_\varepsilon := -(B^{\mathrm{T}}P_\varepsilon B + I)^{-1}B^{\mathrm{T}}P_\varepsilon A$ with P_ε being the solution of the Riccati equation (4.9), and K_A and K_S are such that the following matrix $\bar{A}$ is Schur-stable,

$$\bar{A} := \begin{pmatrix} A + K_A C_y & E_w + K_A D_{yw} \\ K_S C_y & S + K_S D_{yw} \end{pmatrix}.$$

The rest of the proof is the same as that of Theorem 4.3.3 except that (4.51) takes instead the form,

$$\begin{aligned}
\xi(k+1) &= A\xi(k) + B\sigma\,(\Gamma(w(k) - \tilde{w}(k)) \\
&\quad - F_\varepsilon(\tilde{x}(k) - \Pi\tilde{w}(k) - \xi(k)) + \tilde{r}(k)) \\
&\quad - B\Gamma w(k) - \Pi r(k) \\
\tilde{x}(k+1) &= (A + K_A C_y)\tilde{x}(k) + (E_w + K_A D_{yw})\tilde{w}(k) \\
\tilde{w}(k+1) &= K_S C_y\tilde{x}(k) + (S + K_S D_{yw})\tilde{w}(k).
\end{aligned}$$

■

Theorem 15.3.1 presents the conditions under which the Problem 15.1.2 can be solved. Also, as pointed out in the proof of Theorem 15.3.1, to solve such a problem one could construct a linear observer based controller (15.12) with F_ε as in (4.10). However, in order to improve the convergence of $e(k)$ to zero, one can alternatively use an improved measurement feedback controller. Consider the family of linear observer based measurement feedback laws,

$$\begin{aligned}
\rho\hat{x} &= A\hat{x} + B\sigma(u) + E_w\hat{w} - K_A y \\
&\quad + K_A(C_y\hat{x} + D_{yw}\hat{w}) \\
\hat{w}(k+1) &= S\hat{w} - K_S y + K_S(C_y\hat{x} + D_{yw}\hat{w}) + r \\
u &= (F_\varepsilon + \mu\kappa(\hat{x}, \hat{w}, \mu)K_\varepsilon)\hat{x} \\
&\quad - ((F_\varepsilon + \mu\kappa(\hat{x}, \hat{w}, \mu)K_\varepsilon)\Pi - \Gamma)\hat{w} + \tilde{r},
\end{aligned} \tag{15.13}$$

with P_ε being the solution of the Riccati equation (4.9), and the function κ is as defined by (4.30). The matrices K_A and K_S are chosen such that all the eigenvalues of the following matrix

$$\bar{A} = \begin{pmatrix} A + K_A C_y & E_w + K_A D_{yw} \\ K_S C_y & S + K_S D_{yw} \end{pmatrix} \tag{15.14}$$

are at the origin.

Theorem 15.3.2 *Consider the system (15.1) and given a compact set $\mathcal{W}_0 \subset \mathbb{R}^s$ and a bounded set $\mathcal{R} \subset L_\infty$. Assume that the sufficient conditions given in Theorem 15.3.1 hold. Also, assume that the pair*

$$\left(\begin{pmatrix} C_y & D_{yw} \end{pmatrix}, \begin{pmatrix} A & E_w \\ 0 & S \end{pmatrix}\right)$$

is observable and that B is injective. Then, the generalized semi-global linear observer based measurement feedback output regulation problem is solvable. More specifically, consider the family of control laws (15.13). Then, for any a priori given (arbitrarily large) bounded set $\mathcal{X}_0 \in \mathbb{R}^n$ and $\mathcal{Z}_0 \in \mathbb{R}^{n+s}$, there exists an $\varepsilon^* \in (0, 1]$ such that for each $\varepsilon \in (0, \varepsilon^*]$ and for each $\mu \in [0, 2]$, the controller in the family (15.13) has the following properties:

(i) *The equilibrium $(x, \hat{x}, \hat{w}) = (0, 0, 0)$ of*

$$\begin{aligned} \rho x(k+1) &= Ax + B\sigma(u) \\ \rho \begin{pmatrix} \hat{x} \\ \hat{w} \end{pmatrix} &= \begin{pmatrix} A & E_w \\ 0 & S \end{pmatrix} \begin{pmatrix} \hat{x} \\ \hat{w} \end{pmatrix} + \begin{pmatrix} B \\ 0 \end{pmatrix} \sigma(u) \\ &\quad + \begin{pmatrix} K_A \\ K_S \end{pmatrix} \begin{pmatrix} C_y & D_{yw} \end{pmatrix} \begin{pmatrix} \hat{x} - x \\ \hat{w} \end{pmatrix} \\ u &= (F_\varepsilon + \mu\kappa(\hat{x}, \hat{w}, \mu) K_\varepsilon)\hat{x} \\ &\quad -((F_\varepsilon + \mu\kappa(\hat{x}, \hat{w}, \mu) K_\varepsilon)\Pi - \Gamma)\hat{w} \end{aligned} \tag{15.15}$$

is asymptotically stable with $\mathcal{X}_0 \times \mathcal{Z}_0$ contained in its basin of attraction.

(ii) *For any $(x(0), \hat{x}(0) \in \mathcal{X}_0, \hat{w}(0)) \in \mathcal{Z}_0$, $w(0) \in \mathcal{W}_0$, and all $r \in \mathcal{R}$, the solution of the closed-loop system satisfies*

$$\lim_{k \to \infty} e(k) = 0. \tag{15.16}$$

Proof : The proof is similar to that of Theorem 4.3.4. ∎

Chapter 16

Generalized output regulation with actuators subject to amplitude and rate saturation

16.1 Introduction and problem formulation

We have formulated earlier in Chapter 5 the semi-global linear feedback output regulation problems for linear systems with actuators which are subject to rate and amplitude saturation. Chapter 5 follows the traditional formulation of linear output regulation problems where the exosystem is autonomous. In Chapter 13 it was argued that a more general framework has attractive features. In the last two chapters this concept was introduced for linear systems subject to amplitude saturation for continuous and discrete-time systems respectively. The objective of this chapter is to introduce this concept for systems with rate and amplitude saturation. We treat both continuous and discrete-time systems in this chapter.

We first formulate the problem of semi-global generalized output regulation via linear state feedback for linear systems with inputs subject to both amplitude and rate saturations. Consider the following continuous- or discrete-time system

$$\begin{aligned} \rho x &= Ax + B\sigma_{\alpha,\beta}(u) + E_w w \\ \rho w &= Sw + r \\ e &= C_e x + D_{ew} w \\ y &= C_y x + D_{yw} w \end{aligned} \tag{16.1}$$

where $x \in \mathbb{R}^n$, $u \in \mathbb{R}^m$, $w \in \mathbb{R}^s$, $e \in \mathbb{R}^q$ and $y \in \mathbb{R}^p$. Here $\sigma_{\alpha,\beta}$ denotes the functional differential operator which we use to enforce that the actual control input $\sigma_{\alpha,\beta}u$ will satisfy the amplitude and rate constraints. It is important to recall that $\sigma_{\alpha,\beta}$ can be viewed as a part of the controller. For details, we refer to Chapter 5. Also, as in Chapter 5, for the operator $\sigma_{\alpha,\beta}$, we denote the initial conditions by x_s, and the space of all initial signals as $\mathcal{X}_s$. Note that the exosystem is driven by an external signal r which, in continuous-time, we assume to be continuous. These non-autonomous exosystems have been defined before in Chapter 13.

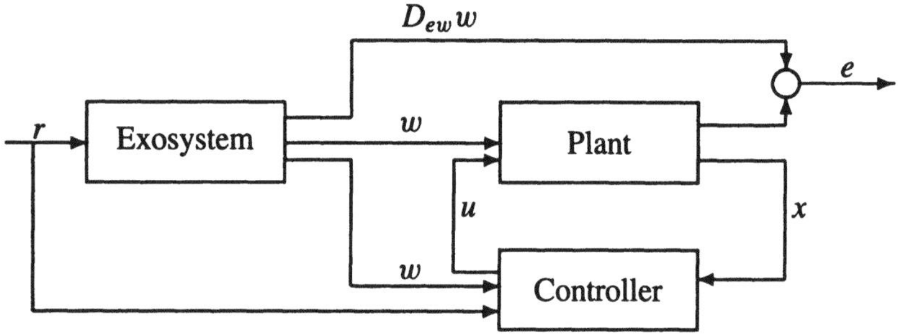

Figure 16.1: Generalized output regulation with state feedback

The problem of semi-global generalized output regulation via state feedback is formulated as follows (see Figure 16.1).

Problem 16.1.1 ***(Semi-global generalized output regulation via state feedback)*** *Consider the system (16.1), a compact set $\mathcal{W}_0 \subset \mathbb{R}^s$ and a bounded set $\mathcal{R} \subset L_\infty$. The problem of semi-global generalized output regulation via state feedback is defined as follows:*

For any a priori given (arbitrarily large) bounded set $\mathcal{X}_0 \subset \mathbb{R}^n$, find, if possible, a linear static feedback law $u = Fx + Gw + Hr$, such that the following properties hold:

(i) *The equilibrium $x = 0,\ x_s = 0$ of the system*

$$\rho x = Ax + B\sigma_{\alpha,\beta}(Fx)$$

is locally exponentially stable with $\mathcal{X}_0 \times \mathcal{X}_s$ contained in its basin of attraction.

(ii) *For all $x(0) \in \mathcal{X}_0,\ x_s \in \mathcal{X}_s,\ w(0) \in \mathcal{W}_0$ and $r \in \mathcal{R}$, the solution of the closed-loop system satisfies*

$$\lim_{t\to\infty} e(t) = 0.$$

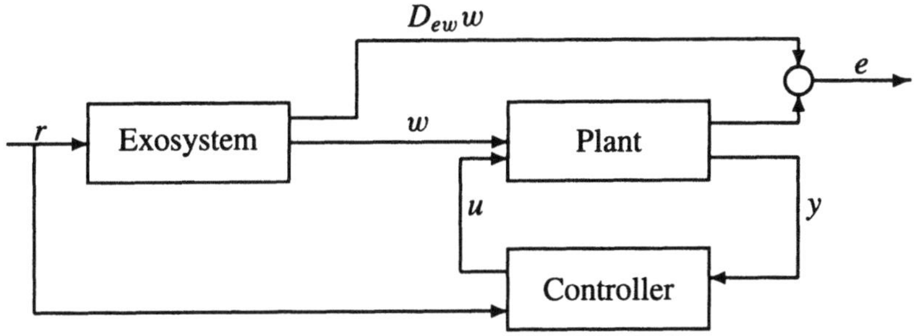

Figure 16.2: Generalized output regulation with measurement feedback

Next we consider the problem of semi-global generalized output regulation via observer based measurement feedback (see Figure 16.2).

Problem 16.1.2 ***(Semi-global generalized output regulation via measurement feedback)*** *Consider the system (16.1), a compact set $\mathcal{W}_0 \subset \mathbb{R}^s$ and a bounded set $\mathcal{R} \subset L_\infty$. The problem of semi-global generalized output regulation via observer based measurement feedback is defined as follows:*

For any a priori given (arbitrarily large) bounded sets $\mathcal{X}_0 \subset \mathbb{R}^n$ and $\mathcal{Z}_0 \subset \mathbb{R}^{n+s}$, find, if possible, a linear observer based feedback law of the form,

$$\begin{aligned}\rho\begin{pmatrix}\hat{x}\\ \hat{w}\end{pmatrix} &= \begin{pmatrix}A & E_w\\ 0 & S\end{pmatrix}\begin{pmatrix}\hat{x}\\ \hat{w}\end{pmatrix} + \begin{pmatrix}B\sigma_{\alpha,\beta}(u)\\ r\end{pmatrix}\\ &\quad + \begin{pmatrix}K_A\\ K_S\end{pmatrix}\left(\begin{pmatrix}C_y & D_{yw}\end{pmatrix}\begin{pmatrix}\hat{x}\\ \hat{w}\end{pmatrix} - y\right)\\ u &= F\hat{x} + G\hat{w} + Hr\end{aligned}$$

such that the following properties hold:

(i) *The equilibrium $(x, \hat{x}, \hat{w}) = (0, 0, 0)$, $x_s = 0$ of*

$$\begin{aligned}\rho x &= Ax + B\sigma_{\alpha,\beta}(F\hat{x} + G\hat{w}))\\ \rho\begin{pmatrix}\hat{x}\\ \hat{w}\end{pmatrix} &= \begin{pmatrix}A & E_w\\ 0 & S\end{pmatrix}\begin{pmatrix}\hat{x}\\ \hat{w}\end{pmatrix} + \begin{pmatrix}B\\ 0\end{pmatrix}\sigma_{\alpha,\beta}(u)\\ &\quad + \begin{pmatrix}K_A\\ K_S\end{pmatrix}\left(\begin{pmatrix}C_y & D_{yw}\end{pmatrix}\begin{pmatrix}\hat{x}-x\\ \hat{w}\end{pmatrix}\right)\end{aligned}$$

is locally exponentially stable with $\mathcal{X}_0 \times \mathcal{X}_s \times \mathcal{Z}_0$ contained in its basin of attraction.

(*ii*) *For all* $(x(0), \hat{x}(0), \hat{w}(0)) \in \mathcal{X}_0 \times \mathcal{Z}_0$, $x_s \in \mathcal{X}_s$, $w(0) \in \mathcal{W}_0$, *and all* $r \in \mathcal{R}$, *the solution of the closed-loop system satisfies*

$$\lim_{t \to \infty} e(t) = 0.$$

The solvability conditions for the two problems formulated above are given in the next two subsections.

16.2 Generalized output regulation via state feedback

The solvability conditions for the problem of semi-global generalized output regulation via state feedback are given in the following theorem.

Theorem 16.2.1 *Consider the system (16.1) and the given compact set* $\mathcal{W}_0 \subset \mathbb{R}^s$. *The problem of semi-global generalized output regulation via linear state feedback is solvable if the following conditions are true:*

(*i*) (A, B) *is stabilizable and* A *has all its eigenvalues in the closed left half plane (continuous-time) or in the closed unit disc (discrete-time);*

(*ii*) *There exist matrices* Π *and* Γ *such that:*

(*a*) *they solve the regulator equation,*

$$\begin{aligned} \Pi S &= A\Pi + B\Gamma + E_w, \\ 0 &= C_e\Pi + D_{ew}. \end{aligned} \tag{16.2}$$

(*b*) *for each* $r \in \mathcal{R}$, *there exists a function* $\tilde{r} \in \mathcal{C}^1$ *such that* $\Pi r = B\tilde{r}$.

(*c*) *There exist a* $\delta > 0$ *and a* $T \geqslant 0$ *such that* $\|\Gamma w + \tilde{r}\|_{\infty,T} \leqslant \alpha - \delta$ *and* $\|\Gamma S w + \Gamma r + \dot{\tilde{r}}\|_{\infty,T} \leqslant \beta - \delta$ *(continuous-time) or* $\|\Gamma(S - I)w(k) + \Gamma r(k) + \tilde{r}(k+1) - \tilde{r}(k)\|_{\infty,T} \leqslant \beta - \delta$ *(discrete-time) for all* w *with* $w(0) \in \mathcal{W}_0$ *and all* $r \in \mathcal{R}$.

Moreover, under these conditions, a family of linear static state feedback laws is given by

$$u = -B^T P_\varepsilon x + (B^T P_\varepsilon \Pi + \Gamma)w + \tilde{r} \tag{16.3}$$

in continuous-time where P_ε *is defined by the Riccati equation (5.10). For the discrete-time a family of linear static state feedback laws is given by*

$$u = -(B^T P_\varepsilon B + I)^{-1} B^T P_\varepsilon A(x - \Pi w) + \Gamma w + \tilde{r} \tag{16.4}$$

where P_ε *is defined by the Riccati equation (5.14).*

Proof : We prove this theorem only for the continuous-time system. The counterpart proof for the discrete-time system is similar. Consider the family of feedbacks given in (16.3). Let $F_\varepsilon = -B^{\mathrm{T}} P_\varepsilon$. Then as derived earlier (see (5.11)), we have

$$\|F_\varepsilon e^{(A+BF_\varepsilon)t}\|_\infty \leqslant \upsilon_\varepsilon e^{-\zeta_\varepsilon t}, \qquad \|F_\varepsilon(A+BF_\varepsilon)e^{(A+BF_\varepsilon)t}\|_\infty \leqslant \upsilon_\varepsilon e^{-\zeta_\varepsilon t} \tag{16.5}$$

where υ_ε is a positive-valued function satisfying $\lim_{\varepsilon\to 0} \upsilon_\varepsilon = 0$. By Theorem 5.3.1, there exists an ε_1^* such that for all $\varepsilon \in (0, \varepsilon_1^*]$ the closed-loop system is asymptotically stable when $w = 0$ and $r = 0$. Furthermore, $\mathcal{X}_0 \times \mathcal{X}_s$ is contained in the domain of attraction. This shows the item (i) of Problem 16.1.1.

To show that item (ii) of Problem 16.1.1 holds (i.e. the error e goes to zero asymptotically), let

$$\xi = x - \Pi w.$$

Then, using the regulator equation (16.2), we have

$$\begin{aligned}
\dot{\xi} &= \dot{x} - \Pi\dot{w} \\
&= Ax + B\sigma_{\alpha,\beta}(u) + E_w w - \Pi S w - \Pi r \\
&= A\xi + B(\sigma_{\alpha,\beta}(u) - \Gamma w - \tilde{r}).
\end{aligned}$$

By the family of state feedback laws (16.3), the closed-loop system becomes

$$\dot{\xi} = A\xi + B\left[\sigma_{\alpha,\beta}(F_\varepsilon\xi + \Gamma w + \tilde{r}) - (\Gamma w + \tilde{r})\right], \tag{16.6}$$

where $F_\varepsilon = -B^{\mathrm{T}} P_\varepsilon$. Now by condition 2(c) in the theorem, $\|\Gamma w + \tilde{r}\|_{\infty,T} < \alpha - \delta$ and $\|\Gamma\dot{w} + \dot{\tilde{r}}\|_{\infty,T} < \beta - \delta$. Also notice that $\xi(T)$ belongs to a bounded set independent of ε since $\xi(0)$ is bounded and $\xi(T)$ is determined by a linear differential equation with bounded inputs $\sigma_{\alpha,\beta}(u)$ and $\Gamma w + \tilde{r}$. If we consider the system (16.6) from time T onwards, without saturation element, we obtain

$$\dot{\xi} = (A + BF_\varepsilon)\xi. \tag{16.7}$$

Since $\xi(T)$ is bounded, (16.5) and (16.7) imply that there exists an $\varepsilon_2^* > 0$ such that, for all $\varepsilon \in (0, \varepsilon_2^*]$,

$$\|F_\varepsilon\xi\|_{\infty,T} \leqslant \delta, \qquad \|F_\varepsilon\dot{\xi}\|_{\infty,T} \leqslant \delta.$$

We can conclude then that the closed-loop system (16.6) will operate within the linear region of the saturation elements for all $t \geqslant T$ if $\varepsilon \in (0, \varepsilon_2^*]$. Hence,

$\xi(t) \to 0$ as $t \to \infty$. By the definition of $\xi(t)$ and the regulator equation (16.2), it is easy to see that

$$e(t) = C_e\xi(t) + (C_e\Pi + D_{ew})w(t) = C_e\xi(t).$$

It follows then that $e(t) \to 0$ as $t \to \infty$. We conclude the proof by taking $\varepsilon^* = \min\{\varepsilon_1^*, \varepsilon_2^*\}$. ∎

16.3 Generalized output regulation via measurement feedback

The solvability conditions for the problem of semi-global generalized output regulation via measurement feedback are given in the following theorem.

Theorem 16.3.1 *Consider the system (16.1) and the given compact set $\mathcal{W}_0 \subset \mathbb{R}^s$. The problem of semi-global generalized output regulation via measurement feedback is solvable if the following conditions hold:*

(i) (A, B) is stabilizable and A has all its eigenvalues in the closed left half plane for continuous-time systems or within or on the unit circle for discrete-time systems. Moreover, the pair

$$\left(\begin{pmatrix} C_y & D_{yw} \end{pmatrix} \begin{pmatrix} A & E_w \\ 0 & S \end{pmatrix}\right)$$

is detectable.

(ii) There exist matrices Π and Γ such that

(a) they solve the regulator equation (16.2), and

(b) there exists a $\delta > 0$ and a $T \geqslant 0$ such that $\|\Gamma w + \tilde{r}\|_{\infty,T} \leqslant \alpha - \delta$ and $\|\Gamma S w + \Gamma r + \dot{\tilde{r}}\|_{\infty,T} \leqslant \beta - \delta$ (in continuous-time) or $\|\Gamma(S - I)w(k) + \Gamma r(k) + \tilde{r}(k+1) - \tilde{r}(k)\|_{\infty,T} \leqslant \beta - \delta$ (in discrete-time) for all w with $w(0) \in \mathcal{W}_0$.

Moreover, under these conditions, in continuous-time a family of linear dynamic error feedback laws is given by,

$$\begin{aligned}\begin{pmatrix} \dot{\hat{x}} \\ \dot{\hat{w}} \end{pmatrix} &= \begin{pmatrix} A & E_w \\ 0 & S \end{pmatrix}\begin{pmatrix} \hat{x} \\ \hat{w} \end{pmatrix} + \begin{pmatrix} B\sigma_{\alpha,\beta}(u) \\ r \end{pmatrix} \\ &\quad + \begin{pmatrix} K_A \\ K_S \end{pmatrix}\left(\begin{pmatrix} C_y & D_{yw} \end{pmatrix}\begin{pmatrix} \hat{x} \\ \hat{w} \end{pmatrix} - y\right) \\ u &= F_\varepsilon \hat{x} + (\Gamma - F_\varepsilon\Pi)\hat{w} + \tilde{r},\end{aligned} \tag{16.8}$$

where $F_\varepsilon = -B^T P_\varepsilon$ and P_ε is defined by the Riccati equation (5.10), while in discrete-time a family of linear dynamic measurement feedback laws is given by,

$$\begin{aligned}\begin{pmatrix}\hat{x}(k+1)\\ \hat{w}(k+1)\end{pmatrix} &= \begin{pmatrix}A & E_w\\ 0 & S\end{pmatrix}\begin{pmatrix}\hat{x}(k)\\ \hat{w}(k)\end{pmatrix} + \begin{pmatrix}B\sigma_{\alpha,\beta}(u)(k)\\ r(k)\end{pmatrix}\\ &\quad + \begin{pmatrix}K_{\mathrm{A}}\\ K_{\mathrm{S}}\end{pmatrix}\left(\begin{pmatrix}C_y & D_{yw}\end{pmatrix}\begin{pmatrix}\hat{x}(k)\\ \hat{w}(k)\end{pmatrix} - y(k)\right)\\ u(k) &= F_\varepsilon \hat{x}(k) + (\Gamma - F_\varepsilon \Pi)\hat{w}(k) + \tilde{r}(k),\end{aligned} \tag{16.9}$$

where $F_\varepsilon = -(B^T P_\varepsilon B + I)^{-1} B^T P_\varepsilon A$ and P_ε is defined by the Riccati equation (5.14). Here K_{A} and K_{S} are chosen such that the matrix,

$$\bar{A} := \begin{pmatrix}A + K_{\mathrm{A}} C_y & E_w + K_{\mathrm{A}} D_{yw}\\ K_{\mathrm{S}} C_y & S + K_{\mathrm{S}} D_{yw}\end{pmatrix}$$

is Hurwitz-stable (continuous-time) or Schur-stable (discrete-time).

Proof : We prove this theorem only for continuous-time. The discrete-time counterpart can be proved similarly. By the given family of feedback laws, the closed-loop system consisting of the system (16.1) and the dynamic error feedback laws (16.8) can be written as,

$$\begin{aligned}\dot{x} &= Ax + B\sigma_{\alpha,\beta}(F_\varepsilon \hat{x} + (\Gamma - F_\varepsilon \Pi)\hat{w} + \tilde{r}) + E_w w\\ \dot{\hat{x}} &= A\hat{x} + B\sigma_{\alpha,\beta}(F_\varepsilon \hat{x} + (\Gamma - F_\varepsilon \Pi)\hat{w} + \tilde{r}) + E_w \hat{w}\\ &\qquad + K_{\mathrm{A}} C_y(\hat{x} - x) + K_{\mathrm{A}} D_{yw}(\hat{w} - w)\\ \dot{\hat{w}} &= S\hat{w} + K_{\mathrm{S}} C_y(\hat{x} - x) + K_{\mathrm{S}} D_{yw}(\hat{w} - w) + r.\end{aligned} \tag{16.10}$$

Introduce new state variables,

$$\begin{aligned}\xi &= x - \Pi w\\ \tilde{x} &= x - \hat{x}\\ \tilde{w} &= w - \hat{w},\end{aligned}$$

and rewrite the closed-loop system (16.10) as

$$\begin{aligned}\dot{\xi} &= A\xi + B\sigma_{\alpha,\beta}(F_\varepsilon \xi + \Gamma w - \Gamma\tilde{w} - F_\varepsilon \tilde{x} + F_\varepsilon \Pi\tilde{w} + \tilde{r})\\ &\qquad - B(\Gamma w + \tilde{r})\\ \dot{\tilde{x}} &= (A + K_{\mathrm{A}} C_y)\tilde{x} + (E_w + K_{\mathrm{A}} D_{yw})\tilde{w}\\ \dot{\tilde{w}} &= K_{\mathrm{S}} C_y \tilde{x} + (S + K_{\mathrm{S}} D_{yw})\tilde{w}.\end{aligned} \tag{16.11}$$

We first show that item (i) of Problem 16.1.2 holds. Let $w = 0$ and $r = 0$, then (16.11) reduces to

$$\begin{aligned} \dot{\xi} &= A\xi + B\sigma_{\alpha,\beta}(F_\varepsilon\xi - \Gamma\tilde{w} - F_\varepsilon\tilde{x} + F_\varepsilon\Pi\tilde{w}) \\ \begin{pmatrix} \dot{\tilde{x}} \\ \dot{\tilde{w}} \end{pmatrix} &= \begin{pmatrix} A + K_A C_y & E_w + K_A D_{yw} \\ K_S C_y & S + K_S D_{yw} \end{pmatrix} \begin{pmatrix} \tilde{x} \\ \tilde{w} \end{pmatrix}. \end{aligned} \tag{16.12}$$

Note that the matrix $\bar{A}$ defined above is Hurwitz-stable, due to the detectability, it readily follows from the second part of (16.12) that there exists a $T_1 \geqslant 0$ such that, for all possible initial conditions $(\tilde{x}(0), \tilde{w}(0))$,

$$\begin{aligned} &\|\Gamma\tilde{w}\|_{\infty,T_1} \leqslant \tfrac{\alpha}{4}, \quad \|F_\varepsilon\tilde{x}\|_{\infty,T_1} \leqslant \tfrac{\alpha}{4}, \quad \|F_\varepsilon\Pi\tilde{w}\|_{\infty,T_1} \leqslant \tfrac{\alpha}{4} \\ &\|\Gamma\dot{\tilde{w}}\|_{\infty,T_1} \leqslant \tfrac{\beta}{4}, \quad \|F_\varepsilon\dot{\tilde{x}}\|_{\infty,T_1} \leqslant \tfrac{\beta}{4}, \quad \|F_\varepsilon\Pi\dot{\tilde{w}}\|_{\infty,T_1} \leqslant \tfrac{\beta}{4} \end{aligned} \tag{16.13}$$

for all $\varepsilon \in (0, 1]$. We next consider the first equation of (16.12). $\xi(T_1)$ belongs to a bounded set independent of ε since $\xi(0)$ is bounded and since ξ is determined via a linear differential equation with bounded input $\sigma_{\alpha,\beta}(u)$. Hence there exists an M_1 such that for all possible initial conditions,

$$\|\xi(T_1)\| \leqslant M_1, \quad \text{for all } \varepsilon \in (0, 1]. \tag{16.14}$$

Let us now assume that, from time T_1 onwards, the saturation elements are nonexistent. In this case, the first equation of (16.12) can be written as

$$\dot{\xi} = (A + BF_\varepsilon)\xi - BF_\varepsilon\tilde{x} - B\Gamma\tilde{w} + BF_\varepsilon\Pi\tilde{w}. \tag{16.15}$$

Since $\tilde{x} \to 0$ and $\tilde{w} \to 0$ exponentially with a decay rate independent of ε as $t \to \infty$, it follows trivially from (16.5) that there exist an $\varepsilon_1^* > 0$ and an $M_2 > 0$ such that, for all possible initial conditions $\tilde{x}(0)$ and $\tilde{w}(0)$ and all $\varepsilon \in (0, \varepsilon_1^*]$,

$$\int_{T_1}^{\infty} \|e^{\zeta_\varepsilon\tau} B[F_\varepsilon\tilde{x}(\tau) + \Gamma\tilde{w}(\tau) - F_\varepsilon\Pi\tilde{w}(\tau)]\| d\tau \leqslant M_2.$$

This in turn shows that, for $t \geqslant T_1$,

$$\begin{aligned} \|F_\varepsilon\xi(t)\| &= \Big\| F_\varepsilon e^{(A+BF_\varepsilon)t}\xi(T_1) \\ &\quad - \int_{T_1}^{t} F_\varepsilon e^{(A+BF_\varepsilon)(t-\tau)} B\left[F_\varepsilon\tilde{x}(\tau) + \Gamma\tilde{w}(\tau) - F_\varepsilon\Pi\tilde{w}(\tau)\right] d\tau \Big\| \\ &\leqslant \upsilon_\varepsilon M_1 + \upsilon_\varepsilon \int_{T_1}^{\infty} \|e^{\zeta_\varepsilon\tau} B[F_\varepsilon\tilde{x}(\tau) + \Gamma\tilde{w}(\tau) - F_\varepsilon\Pi\tilde{w}(\tau)]\| d\tau \\ &\leqslant \upsilon_\varepsilon(M_1 + M_2). \end{aligned}$$

Choose $\varepsilon_2^* \in (0, \varepsilon_1^*]$ such that, for all $\varepsilon \in (0, \varepsilon_2^*]$,

$$\|F_\varepsilon \xi\|_{\infty,T_1} \leqslant \frac{\alpha}{4}.$$

Similarly, we can show that there exists an $\varepsilon_3^* \in (0, \varepsilon_2^*]$ such that, for all $\varepsilon \in (0, \varepsilon_3^*]$,

$$\|F_\varepsilon \dot{\xi}\|_{\infty,T_1} \leqslant \frac{\beta}{4}.$$

These two bounds, together with (16.13), shows that the system (16.12) will operate linearly after time T_1 and local exponential stability of this linear system follows from the separation principle.

In summary, we have shown that there exists an $\varepsilon_3^* > 0$ such that, for all $\varepsilon \in (0, \varepsilon_3^*]$, the equilibrium point $(\xi, \tilde{x}, \tilde{w}) = (0, 0, 0)$, $x_s = 0$ of the system (16.12) is asymptotically stable, with $\mathcal{X}_0 \times \mathcal{X}_s \times \mathcal{Z}_0$ contained in its basin of attraction.

We now proceed to show that item (ii) of Problem 16.1.2 also holds. To this end, we consider the closed-loop system (16.11). Recalling that the matrix $\bar{A}$ is Hurwitz-stable, it readily follows from the last two equations of (16.11) that there exists an $T_2 \geqslant T$ such that, for all possible initial conditions $(\tilde{x}(0), \tilde{w}(0))$,

$$\begin{array}{lll} \|\Gamma \tilde{w}\|_{\infty,T_2} \leqslant \frac{\delta}{4}, & \|F_\varepsilon \tilde{x}\|_{\infty,T_2} \leqslant \frac{\delta}{4}, & \|F_\varepsilon \Pi \tilde{w}\|_{\infty,T_2} \leqslant \frac{\delta}{4}, \\ \|\Gamma \dot{\tilde{w}}\|_{\infty,T_2} \leqslant \frac{\delta}{4}, & \|F_\varepsilon \dot{\tilde{x}}\|_{\infty,T_2} \leqslant \frac{\delta}{4}, & \|F_\varepsilon \Pi \dot{\tilde{w}}\|_{\infty,T_2} \leqslant \frac{\delta}{4} \end{array} \tag{16.16}$$

for all $\varepsilon \in (0, 1]$. We next consider the first equation of (16.11). $\xi(T_2)$ belongs to a bounded set independent of ε since $\xi(0)$ is bounded and since ξ is determined via a linear differential equation with bounded inputs $\sigma_{\alpha,\beta}(u)$ and $\Gamma w + \tilde{r}$. Hence there exists an M_3 such that for all possible initial conditions,

$$\|\xi(T_2)\| \leqslant M_3, \quad \text{for all } \varepsilon \in (0, 1].$$

Let us now assume that, from time T_2 onwards, the equation (16.11) operates without the saturation elements. In view of condition (ii)b) in the theorem, the first equation of (16.11) in the absence of the saturation elements is the same as the first equation of (16.12), and hence also reduces to (16.15) after time T_2. Hence, using a similar argument as above, we can show that there exists an $\varepsilon_3^* > 0$ such that, for all $\varepsilon \in (0, \varepsilon_3^*]$,

$$\|F_\varepsilon \xi\|_{\infty,T_2} < \frac{\delta}{4}, \qquad \|F_\varepsilon \dot{\xi}\|_{\infty,T_2} < \frac{\delta}{4}.$$

These, together with (16.16) and condition (ii)b), show that the closed-loop system (16.11) will operate linearly after time T_2, and thus the exponential stability of this linear system follows from the separation principle.

Next, by the regulator equation (16.2), it is easy to see that

$$e(t) = C_e \xi(t).$$

This implies that $e(t) \to 0$ as $t \to \infty$. Finally, taking $\varepsilon^* = \min\{\varepsilon_2^*, \varepsilon_3^*\}$, we complete our proof. ■

Chapter 17

What does one do if output regulation is not possible?

17.1 Introduction

We studied so far output regulation whether it be exact or almost type under a variety of transient, robustness and other performance constraints. Also, we studied output regulation when actuators are subject to amplitude and rate saturation. As discussed throughout the book, output regulation is not always possible. It requires certain conditions. One of the important conditions that needs to be satisfied is the solvability of a couple of algebraic equations which together are generally known as regulator equation. One of the questions that arises is this; what can a designer do if either exact or almost output regulation is not possible? To answer this question, let us first interpret the error signal e whenever output regulation is possible. Obviously, in that case e is an energy signal. On the other hand, whenever output regulation is not possible, under some mild conditions, e can be seen to be a power signal. In this case, since e does not asymptotically go to zero, one could minimize in the asymptotic sense the power of the signal e. In other words, in the classical output regulation we seek to render e asymptotically zero, where as whenever it is not possible to do so we could seek to render the power of e as small as possible.

We remark that a regulator that minimizes in the asymptotic sense the power of e has the same degree of freedom as in the case when e is an energy signal and when we seek to render it asymptotically zero. Such a freedom can be utilized to study transient performance requirements (as was done in the classical case in Chapter 6). In this regard, a couple of performance measures can be introduced and the resulting output regulation problems that achieve optimal performance can be studied. Moreover as was done in the classical case in Chapters 7 to 12, controller design problems which seek to achieve desired performance (such as H_2, H_∞, and L_1 norm based ones) with output regulation constraint in the sense of minimizing the power of error signal e can be studied as well.

17.2 Output regulation in the sense of minimizing the power of the error signal

Consider the following system,

$$\Sigma : \begin{cases} \rho x = Ax + Bu + E_w w \\ e = C_e x + D_{eu} u + D_{ew} w, \end{cases} \tag{17.1}$$

Obviously, it does not make sense to consider systems which are not stabilizable and therefore it is natural to assume that Assumption A.1 is satisfied, i.e. (A, B) is stabilizable. We know that without loss of generality we can assume that the exosystem,

$$\Sigma_E : \rho w = Sw, \tag{17.2}$$

satisfies Assumption A.2, i.e. the matrix S is anti-Hurwitz-stable for continuous-time systems and anti-Schur-stable for discrete-time systems. Then we know that the output regulation problem is solvable if and only if the regulator equations are solvable, i.e. there exist Π, Γ such that

$$\begin{aligned} \Pi S &= A\Pi + B\Gamma + E_w \\ 0 &= C_e\Pi + D_{eu}\Gamma + D_{ew}, \end{aligned}$$

As explained in introduction, the question we want to consider in this chapter is what can we do if these regulator equations are not solvable. Can we make the steady state error arbitrarily small if we cannot make it exactly equal to zero or is there a strict lower limit in what we can achieve? In the latter case, a second question of course arises as to how to characterize this lower limit. Note that if the dimension of the error signal is less than or equal to the number of inputs then the regulator equations are generically solvable and therefore we are almost sure that regulation can be achieved. On the other hand if the dimension of the error signal is larger than the number of inputs then generically the regulator equations are not solvable and therefore it is very unlikely that we can achieve output regulation.

Our first result of this chapter gives us a tool to compute the minimally achievable power of the error signal. We define the power of error signal e by

$$\|e\|^2_{\text{Power}} = \limsup_{T\to\infty} \frac{1}{T}\int_0^T \|e(t)\|^2\,dt$$

in continuous-time, while in discrete-time

$$\|e\|^2_{\text{Power}} = \limsup_{T\to\infty} \frac{1}{T}\sum_{k=0}^{T} \|e(k)\|^2\,.$$

Note that the power is not a real norm since there exist non-zero signals e which satisfy $\|e\|_{\text{Power}} = 0$. Clearly if a signal converges to zero as time tends to infinity then the power is equal to zero. It is easy to see that the power is a suitable measure for the asymptotic behavior of a signal as it is independent of the transient behavior of the signal.

The first question is whether we can always make the power of the error signal finite. The following lemma is quite obvious.

Lemma 17.2.1 *Consider the system (17.1) and the exosystem (17.2) which satisfy Assumptions A.1 and A.2. If the exosystem is weakly Hurwitz-stable in continuous-time or weakly Schur-stable in discrete-time, then there exists a stabilizing feedback such that the error signal has finite power for any initial conditions for the system and exosystem.*

Proof : If the exosystem is weakly Hurwitz-stable in continuous-time or weakly Schur-stable in discrete-time, then the state of the exosystem is bounded. A controller which stabilizes the system automatically guarantees that the error signal is bounded for any initial condition and any bounded input signal. Therefore, it has finite power. ∎

Note that if the power cannot be made finite, then there are other performance measures that can be considered. In [6] a specific criterion of "overtaking" was considered: a controller Σ_{opt} is considered optimal if for any other controller Σ_C there exists a $T_1 > 0$ such that for all initial conditions, we have

$$\int_0^T \|e_1(t)\|^2\,dt \leqslant \int_0^T \|e_2(t)\|^2\,dt$$

for all $T > T_1$ where e_1 is the error signal resulting from the controller Σ_{opt} while e_2 is the error signal resulting from the controller Σ_C. A controller which is optimal with respect to this "overtaking criterion" is also optimal in the sense of minimizing the power. However, the converse is not true because the power ignores transient effects while the "overtaking criterion" minimizes the transient effect as well. This will be briefly noted in Section 17.3.

The next question considers the problem of finding an optimal controller which minimizes the power for the error signal. For ease of exposition we assume that the the exosystem is weakly Hurwitz-stable in continuous-time or weakly Schur-stable in discrete-time in which case the above lemma guarantees that any stabilizing controller yields a finite power for the error signal.

Theorem 17.2.1 *Consider the system (17.1) and the exosystem (17.2) which satisfy Assumptions A.1 and A.2. Moreover, assume that the exosystem is weakly Hurwitz-stable in continuous-time or weakly Schur-stable in discrete-time. Consider given initial conditions x_0 and w_0 for the given system and exosystem. Then the minimum of the power of e over all stabilizing state feedback controllers of the form $u = Fx + Gw$ is given by*

$$J^*(x_0, w_0) = \inf \{ \, \|(C_e\Pi + D_{eu}\Gamma + D_{ew})w\|_{\text{Power}} \mid \Gamma, \Pi \text{ such that } \Pi S = A\Pi + B\Gamma + E_w \} \tag{17.3}$$

Moreover, the associated optimal controller is given by,

$$u = Fx + (\Gamma - F\Pi)w, \tag{17.4}$$

where F is such that $A + BF$ is Hurwitz-stable in continuous-time and Schur-stable in discrete-time while Π and Γ are the optimal solutions for the minimization problem in (17.3).

Finally, there exists Π and Γ such that for any F for which $A + BF$ is Hurwitz-stable in continuous-time and Schur-stable in discrete-time we have the controller (17.4) satisfying

$$\|e\|_{\text{Power}} = J^*(x_0, w_0)$$

for all initial conditions x_0 and w_0.

Proof : We show this theorem for continuous-time systems. The discrete-time result follows analogously. Note that we want to minimize the following criterion,

$$\limsup_{T\to\infty} \frac{1}{T} \int_0^T \|e(t)\|^2 \, dt.$$

We first minimize this criterion for a fixed T,

$$\int_0^T \|e(t)\|^2 \, dt. \tag{17.5}$$

This is a standard finite-horizon linear quadratic control problem and it is known that there exist time-varying feedbacks of the form

$$u(t) = F_T(t)x(t) + G_T(t)w(t) \tag{17.6}$$

such that the criterion (17.5) is arbitrarily close to its infimum. One particular choice for this feedback is the controller which minimizes the following perturbed criterion,

$$\int_0^T \|e(t)\|^2 + \varepsilon\|x(t)\|^2 + \varepsilon\|u(t)\|^2 \, dt. \tag{17.7}$$

For the family of controllers which minimize (17.7) it can be shown that there exist matrices F_0 and G_0 such that

$$\lim_{T\to\infty} F_T(t) = F_0, \quad \lim_{T\to\infty} G_T(t) = G_0$$

for all $t > 0$. Note that the limit is independent of t. Then a candidate near-optimal feedback for the problem of minimizing the power is given by,

$$u = F_0 x + G_0 w. \tag{17.8}$$

There are two problems that needs to be resolved before we can show that this feedback is indeed optimal. Namely stability and the interchange of a limit and an infimization. The fact that this feedback stabilizes the system for $w = 0$ is a direct consequence of linear quadratic control since it is well known that $u = F_0 x$ is a stabilizing feedback that minimizes the criterion,

$$\int_0^\infty \|e(t)\|^2 + \varepsilon\|x(t)\|^2 + \varepsilon\|u(t)\|^2 \, dt$$

when $w = 0$. The second problem is that although (17.6) is an optimal controller for the criterion

$$\int_0^T \|e(t)\|^2 + \varepsilon\|x(t)\|^2 + \varepsilon\|u(t)\|^2 \, dt, \tag{17.9}$$

it is not clear that (17.8) is an optimal controller for the criterion,

$$\limsup_{T\to\infty} \frac{1}{T} \int_0^T \|e(t)\|^2 + \varepsilon\|x(t)\|^2 + \varepsilon\|u(t)\|^2 \, dt. \tag{17.10}$$

This result was proven for linear quadratic control problems when the underlying dynamics are stabilizable. However, in our case the underlying dynamics are the dynamics of both the system and the exosystem and obviously the dynamics of the exosystem is not stabilizable. Nevertheless, in our case as well, this result can be shown, and it relies on the fact that the unstable dynamics is almost periodic (since the exosystem is weakly stable and has all its eigenvalues on the boundary of the stability domain). Since the ideas are quite simple although the details are quite technical because the exosystem is not periodic but only almost periodic, we do not give a formal proof of this claim. Note that this immediately shows that we can find a controller which is near optimal for arbitrary initial conditions for the system and exosystem.

If we apply any feedback $u = Fx + Gw$ to the system such that $A + BF$ is Hurwitz-stable in continuous-time and Schur-stable in discrete-time, then it is easily seen that there exist Π and Γ such that asymptotically we have $u = \Gamma w$ and $x = \Pi w$ where Π and Γ satisfy $\Pi S = A\Pi + B\Gamma + E_w$. But then asymptotically $e = (C_e\Pi + D_{eu}\Gamma + D_{ew})w$. Therefore it is immediate that the infimum is given by (17.3). It is obvious that the controller is then of the form (17.4).

The feedback $u = F_0 x + G_0 w$ we constructed was near optimal for all initial conditions. Therefore if we consider the associated Π and Γ to this feedback then we have a near-optimal solution of the optimization in (17.3) for any initial condition. We can then construct a family Π_i, Γ_i such that

$$J^*(x_0, w_0) = \lim_{i\to\infty} \|(C_e\Pi_i + D_{eu}\Gamma_i + D_{ew})w\|_{\text{Power}}$$

for all initial conditions. If $\Pi_i \to \Pi_0$ and $\Gamma_i \to \Gamma_0$ as $i \to \infty$, then the feedback $u = Fx + (\Gamma_0 - F\Pi_0)w$ would be optimal for all initial conditions. Actually even if the family Π_i, Γ_i is bounded then there exists a convergent subsequence which would also yield a feedback which is optimal for all initial conditions. Using the fact that we are optimizing a quadratic criterion in Π and Γ subject to a linear constraint, it is easy to show that we can always keep the family Π_i, Γ_i bounded and therefore there exists such an optimal feedback. This is done by noting that the only unbounded component of Π and Γ can occur in the kernel of $(C_e \quad D_{eu})$ and since the part in the kernel of this matrix does not effect the criterion we can keep it as small as possible subject to our requirement that Π and Γ must satisfy $\Pi S = A\Pi + B\Gamma + E_w$. In this way be obtain a bounded sequence. ■

Remark 17.2.1 *Note that the optimization in (17.3) is equal to minimizing a quadratic function of* Π *and* Γ *subject to a linear constraint and is hence* *quite feasible. In particular it is a convex optimization problem. It is however unclear to us how to conclude from the optimization in (17.3) that we can find an optimal controller which is independent from the initial conditions of the system and exosystem. Only the alternative approach of exploiting the relationship with a finite horizon linear quadratic control problem enables us to establish that fact.*

Next, we consider the measurement feedback case,

$$\Sigma : \begin{cases} \rho x = Ax + Bu + E_w w \\ y = C_y x + D_{yu} u + D_{yw} w \\ e = C_e x + D_{eu} u + D_{ew} w, \end{cases} \tag{17.11}$$

where we look for a controller of the form,

$$\Sigma_C : \begin{array}{l} \rho v = A_c v + B_c y, \\ u = C_c v + D_c y. \end{array} \tag{17.12}$$

As clearly argued in Chapter 2, we can assume essentially without loss of generality that Assumptions A.1, A.2, and A.3 stated there are satisfied.

Theorem 17.2.2 *Consider the system (17.1) and the exosystem (17.2) which satisfy Assumptions A.1, A.2 and A.3. Consider given initial conditions x_0 for the system and w_0 for the exosystem. Then the minimum of the power of e over all stabilizing measurement feedbacks of the form (17.12) is given by (17.3). Moreover, an optimal controller is given by,*

$$\begin{pmatrix} \rho\hat{x} \\ \rho\hat{w} \end{pmatrix} = \begin{pmatrix} A & E_w \\ 0 & S \end{pmatrix} \begin{pmatrix} \hat{x} \\ \hat{w} \end{pmatrix} + \begin{pmatrix} B \\ 0 \end{pmatrix} u + \begin{pmatrix} K_A \\ K_S \end{pmatrix} \left[\begin{pmatrix} C_y & D_{yw} \end{pmatrix} \begin{pmatrix} \hat{x} \\ \hat{w} \end{pmatrix} + D_{yu} u - y \right] \tag{17.13a}$$

$$u = \begin{pmatrix} F & (\Gamma - F\Pi) \end{pmatrix} \begin{pmatrix} \hat{x} \\ \hat{w} \end{pmatrix}, \tag{17.13b}$$

where F, K_A, and K_S are arbitrary matrices such that the matrices

$$A + BF \quad \textit{and} \quad \begin{pmatrix} A + K_A C_y & E_w + K_A D_{yw} \\ K_S C_y & S + K_S D_{yw} \end{pmatrix} \tag{17.14}$$

are both Hurwitz-stable for continuous-time systems and Schur-stable for discrete-time systems while Π and Γ are the optimal solutions for the minimization problem in (17.3).

Finally, there exists Π and Γ such that for any F , K_A, and K_S for which the matrices in (17.14) are Hurwitz-stable in continuous-time and Schur-stable in discrete-time, we have the controller (17.4) satisfying

$$\|e\|_{\text{Power}} = J^*(x_0, w_0)$$

for all initial conditions x_0 and w_0.

Proof : Consider the optimal state feedback controller as obtained from Theorem 17.2.1. Combine this with an arbitrary stable observer. Then we have that $\hat{x}(t) - x(t) \to 0$ and $\hat{w}(t) - w(t) \to 0$ as $t \to \infty$. But then we know that asymptotically our control input is given by (17.4). Since the transient does not affect our performance measure, we find that the controller (17.4) and the controller (17.13) achieve the same performance and hence in particular (17.13) is an optimal measurement feedback controller. ■

17.3 Transient performance

As already discussed in Chapter 6, a classical regulator problem is only concerned with asymptotic behavior and ignores the transient effect. Therefore it is of interest to use the available flexibility to minimize the transient behavior. Also in our case we have controllers of the form (17.4) in state feedback and (17.13) in measurement feedback where Π and Γ are chosen carefully to have minimal tracking error but where F, K_A, and K_S are still quite arbitrary and we can try to find suitable controllers which minimize the transient error. Obviously, we cannot minimize

$$\int_0^\infty \|e(t)\|^2 \, dt,$$

since if regulation is not possible then we cannot achieve $e(t)$ tending to zero as t tends to ∞. Here we can use the "overtaking criterion" as mentioned before which was defined in [6]. A controller Σ_{opt} is considered optimal if for any other controller Σ_C there exists a $T_1 > 0$ such that for all initial conditions:

$$\int_0^T \|e_1(t)\|^2 \, dt \leqslant \int_0^T \|e_2(t)\|^2 \, dt$$

for all $T > T_1$ where e_1 is the error signal resulting from the controller Σ_{opt} while e_2 is the error signal resulting from the controller Σ_C. We briefly sketch how this can be done. We first obtain optimal Π and Γ from theorem 17.2.2. Then we consider the following criterion:

$$\int_0^\infty \|e(t)\|^2 - \|(C_e\Pi + D_{eu}\Gamma + D_{ew})w\|^2 \, dt$$

By choosing the feedback $u = Fx + (\Gamma - F\Pi)w$ we can keep this criterion finite and hence if we minimize this criterion that it is bounded from above. Since Π and Γ are optimal we can also show that this criterion is bounded from below. Therefore the following optimization problem:

$$\min_u \int_0^\infty \|e(t)\|^2 - \|(C_e\Pi + D_{eu}\Gamma + D_{ew})w\|^2 \, dt$$

is well-posed. Note that in a sense we subtracted a known signal which is not effected by the input u so the optimal u really minimizes the error signal e. We subtracted the steady state behavior of e and therefore this criterion really minimizes the transient effect of the error signal. Note that this problem is not completely solved but it is basically a standard indefinite linear quadratic control problem and the latter is a problem which has been studied thoroughly in the literature. We refer to for instance [53].

Epilogue

As we conclude this book, we like to reflect on some aspects of this work and suggest some issues that are still open and need to be explored.

Our first concern relates to structural stability the topic of Section 2.8. This topic needs a further examination perhaps with a different perspective. The *structurally stable output regulation problem* as defined in Definition 2.8.1 concerns itself with the issue of robustness of output regulation against arbitrarily small perturbations in the given system parameters at their nominal values. To guard against arbitrarily small parameter perturbations, the problem seeks a *fixed* regulator or controller, if it exists, that solves the exact output regulation problem (while maintaining of course the internal stability of the closed-loop system) not only for the nominal parameter values of the given system but also for arbitrarily small parameter perturbations at their nominal values. Thus, even if the values of the system parameters drift but are confined to a given neighborhood, the same controller always achieves output regulation. Although the conditions needed for the existence of a controller that solves the structurally stable output regulation problem are not restrictive, there are still issues that need to be examined. As treated now, the problem does not consider any perturbations in the exosystem, i.e. the frequencies of signals that need to be tracked as well as the frequencies of the external disturbances must be known exactly. The second and more important issue is the high dimensionality of the controller that is required. As discussed in Sections 2.8 and 2.B, any controller that preserves structural stability must have in its dynamics q copies of the exosystem where q is the dimension of the error signal. This implies that the dynamic order of the required controller must be unusually large. Such a high-order controller is not practically feasible. In other words, although the structural stability output regulation problem as defined presently is mathematically interesting, it appears to be not a proper avenue to address the robustness of a regulator against arbitrarily small perturbations of the system parameters at their nominal values. As such, other avenues to deal with robustness must be explored. One promising avenue seems to be through what can be called *Practical structurally stable output regulation problem* as defined in Definition 12.1.2. This practical structurally stable output regulation problem and or any other avenues to deal with the issue of robustness of a regulator need to be examined carefully.

Our second concern relates to output regulation of linear systems with input constraints such as input amplitude and rate limitations. As presented in Chapters 3, 4, and 5, as well as in Chapters 14, 15, and 16, both classical exact output regulation as well as generalized output regulation are dealt with successfully for systems with input constraints. Moreover, the solvability conditions for the output regulation problems for such a class of systems have been obtained. Also, efficient design methodologies to synthesize appropriate regulators are developed. However, what is profoundly missing is how and when prescribed performance measures can be achieved while maintaining internal stability and asymptotic output regulation. For example, one can pose the following questions: How do we shape the transient behavior of the error signal while maintaining internal stability and asymptotic output regulation? Or more generally, how do we solve multi-objective problems in which the internal stability of the closed-loop system, asymptotic output regulation, and optimization of a given performance index are sought? We note that for linear systems without any constraints on the input amplitude and rate, we answered the first question in Chapter 6, while the second question is answered in detail in Chapters 7, 8, 9, 10, and 11. However, formulating of the proper performance criteria in order to shape transient response and considering other performance requirements and incorporating them with the output regulation problem when actuators are subject to amplitude and rate constraints still remain as very complex research tasks.

Our third concern relates to generalized output regulation. A new method of modeling for generalized output regulation as presented in Chapter 13 opens up several avenues to pose and solve different exact and almost output regulation problems. It appears to us that the full potential of new modeling introduced in Chapter 13 has not been explored completely. This and other concerns will be topics of our future work.

Bibliography

[1] J. ABEDOR, K. NAGPAL, P.P. KHARGONEKAR, AND K. POOLLA, "Robust regulation with an H_∞ constraint", in Control of uncertain dynamic systems, S.P. Bhattacharyya and L.H. Keel, eds., CRC press, San Antonio, TX, 1991, pp. 95–110.

[2] ——, "Robust regulation in the presence of norm-bounded uncertainty", IEEE Trans. Aut. Contr., 40(1), 1995, pp. 147–153.

[3] J. ABEDOR, K. NAGPAL, AND K. POOLLA, "Does robust regulation compromise H_2 performance", in Proc. 31st CDC, Tucson, Arizona, 1992, pp. 2002–2007.

[4] ——, "Robust regulation with H_2 performance", Syst. & Contr. Letters, 23, 1994, pp. 431–443.

[5] J. ACKERMANN AND T. BÜNTE, "Actuator rate limits in robust car steering control", in Proc. 36th CDC, San Diego, CA, 1997, pp. 4726–4731.

[6] Z. ARTSTEIN AND A. LEIZAROWITZ, "Tracking periodic signals with the overtaking criterion", IEEE Trans. Aut. Contr., 30(11), 1985, pp. 1123–1126.

[7] J.M. BERG, K.D. HAMMETT, C.A. SCHWARTZ, AND S.S. BANDA, "An analysis of the destabilizing effect of daisy chained rate-limited actuators", IEEE Trans. Control Systems Technology, 4(2), 1996, pp. 171–176.

[8] S.P. BOYD AND C.H. BARRATT, *Linear controller design: limits of performance*, Information and system sciences, Prentice Hall, 1991.

[9] J.J.A.M. BRANDS AND M.L.J. HAUTUS, "Asymptotic properties of matrix differential operators", J. Math. Anal. Appl., 82, 1982, pp. 199–218.

[10] C.-T. CHEN, *Linear system theory and design*, Holt, Rinehart and Winston, New York, 1984.

[11] M.A. DAHLEH AND I.J. DIAZ-BOBILLO, *Control of uncertain systems: a linear programming approach*, Prentice-Hall, 1995.

[12] M.A. Dahleh and J.B. Pearson, "L^1-optimal compensators for continuous-time systems", IEEE Trans. Aut. Contr., 32, 1987, pp. 889–895.

[13] E.J. Davison, "The output control of linear time-invariant multi-variable systems with unmeasured arbitrary disturbances", IEEE Trans. Aut. Contr., 17, 1972, pp. 621–630. Correction: Vol. 20, No. 12, 1975, pp. 824.

[14] E.J. Davison and A. Goldenberg, "The robust control of a general servomechanism problem: the servo compensator", Automatica, 11, 1975, pp. 461–471.

[15] C.A. Desoer and Y.T. Wang, "On the minimum order of a robust servocompensator", IEEE Trans. Aut. Contr., 23(1), 1978, pp. 70–73.

[16] B.A. Francis, "The linear multivariable regulator problem", SIAM J. Contr. & Opt., 15(3), 1977, pp. 486–505.

[17] ——, "The optimal linear-quadratic time invariant regulator with cheap control", IEEE Trans. Aut. Contr., 24, 1979, pp. 616–621.

[18] B. Francis, O.A. Sebakhy, and W.M. Wonham, "Synthesis of multivariable regulators: the internal model principle", Appl. Math. & Opt., 1(1), 1974, pp. 64–86.

[19] B.A. Francis and W.M. Wonham, "The internal model principle for linear multivariable regulators", Appl. Math. & Opt., 2(2), 1975, pp. 170–194.

[20] A.T. Fuller, "In-the-large stability of relay and saturating control systems with linear controller", Int. J. Contr., 10(4), 1969, pp. 457–480.

[21] F.R. Gantmacher, *The theory of matrices*, Chelsea, New York, 1959.

[22] M.L.J. Hautus, "On the solvability of linear matrix equations", COSOR memorandum 1982-07, Eindhoven University of Technology, 1982.

[23] ——, "Linear matrix equations with applications to the regulator problem", in Outils et Modèles Mathématique pour l'Automatique, I.D. Landau, ed., C.N.R.S., Paris, 1983, pp. 399–412.

[24] ——, "Operator substitution", Lin. Alg. Appl., 205-206, 1994, pp. 713–740. Third Special Issue on Systems and Control.

[25] A. Isidori and C.I. Byrnes, "Output regulation of nonlinear systems", IEEE Trans. Aut. Contr., 35(2), 1990, pp. 131–140.

[26] H. Kimura, "A new approach to the perfect regulation and the bounded peaking in linear multivariable control systems", IEEE Trans. Aut. Contr., 26(1), 1981, pp. 253–270.

[27] H. Kwakernaak and R. Sivan, *Linear Optimal control theory*, Wiley, 1972.

[28] P. Lancaster and L. Rodman, *Algebraic Riccati equations*, Clarendon Press, Oxford, 1995.

[29] D.A. Lawrence and W.J. Rugh, "Almost perfect tracking for linear systems", Syst. & Contr. Letters, 17, 1991, pp. 281–289.

[30] ——, "Correction to "Almost perfect tracking for linear systems"", Syst. & Contr. Letters, 18, 1992, pp. 323–324.

[31] Z. Lin, "Semi-global stabilization of linear systems with position and rate-limited actuators", Syst. & Contr. Letters, 30(1), 1997, pp. 1–11.

[32] ——, "Semi-global stabilization of discrete-time linear systems with position and rate limited actuators", in Proceedings of the 37th CDC, 1998, pp. 395–400.

[33] Z. Lin, R. Mantri, and A. Saberi, "Semi-global output regulation for linear systems subject to input saturation – a low-and-high gain design", Control – Th. & Adv. Techn., 4(5), 1995, pp. 2209–2232.

[34] Z. Lin, M. Pachter, S. Banda, and Y. Shamash, "Stabilizing feedback design for linear systems with rate limited actuators", in Control of uncertain systems with bounded inputs, S. Tarbouriech and G. Garcia, eds., Springer Verlag, 1997.

[35] Z. Lin and A. Saberi, "A low-and-high gain approach to semi-global stabilization and/or semi-global practical stabilization of a class of linear systems subject to input saturation via linear state and output feedback", in Proc. 32nd CDC, San Antonio, TX, 1993, pp. 1820–1821.

[36] ——, "Semi-global exponential stabilization of linear systems subject to "input saturation" via linear feedbacks", Syst. & Contr. Letters, 21(3), 1993, pp. 225–239.

[37] ——, "Semi-global exponential stabilization of linear discrete-time systems subject to 'input saturation' via linear feedbacks", Syst. & Contr. Letters, 24, 1995, pp. 125–132.

[38] ——, "A semi-global low-and-high gain design technique for linear systems with input saturation – stabilization and disturbance rejection", Int. J. Robust & Nonlinear Control, 5, 1995, pp. 381–398.

[39] Z. Lin, A. Saberi, P. Sannuti, and Y. Shamash, "A direct method of constructing H_2 suboptimal controllers – continuous-time systems", J. Optim. Th. & Appl., 99(3), 1998, pp. 585–616.

[40] ——, "A direct method of constructing H_2 suboptimal controllers – discrete-time systems", J. Optim. Th. & Appl., 99(3), 1998, pp. 617–654.

[41] Z. LIN, A. SABERI, A. STOORVOGEL, AND R. MANTRI, "An improvement to the low gain design for discrete-time linear systems subject to input saturation – solution of semi-global output regulation problems", in Proc. 13th IFAC world congress, vol. F, San Francisco, 1996, pp. 1–6.

[42] Z. LIN, A.A. STOORVOGEL, AND A. SABERI, "Output regulation for linear systems subject to input saturation", Automatica, 32(1), 1996, pp. 29–47.

[43] K.-Z. LIU AND T. MITA, "Generalized H_∞ control theory", in Proc. ACC, Chicago, IL, 1992, pp. 2245–2249.

[44] J.M. MACIEJOWSKI, *Multivariable feedback design*, Addison-Wesley, Reading, MA, 1989.

[45] R. MANTRI, A. SABERI, Z. LIN, AND A. STOORVOGEL, "Output regulation for linear discrete-time systems subject to input saturation", Int. J. Robust & Nonlinear Control, 7(11), 1997, pp. 1003–1021.

[46] G. MEINSMA, "Unstable and nonproper weights in H_∞ control", Automatica, 31, 1995, pp. 1655–1658.

[47] T. MITA, K. KURIYAMA, AND K.-Z. LIU, "H_∞ control with weighting functions having purely imaginary poles – implementing internal model to H_∞ controllers –", Int. J. Robust & Nonlinear Control, 6, 1996, pp. 537–560.

[48] T. MITA, X. XIN, AND B.D.O. ANDERSON, "Extended H_∞ control – H_∞ control with unstable weights", To appear in Automatica, 2000.

[49] H.K. OZCETIN, A. SABERI, AND Y. SHAMASH, "An H_∞ almost disturbance decoupling problem for nonstrictly proper systems: a singular perturbation approach", Control – Th. & Adv. Techn., 9(1), 1993, pp. 203–246.

[50] A.B. ÖZGÜLER, *Linear multichannel control: a system matrix approach*, Prentice Hall, Hemel Hempstead, U.K., 1994.

[51] J.B. PEARSON, R.W. SHIELDS, AND P.W. STAATS JR., "Robust solutions to linear multivariable control problems", IEEE Trans. Aut. Contr., 19, 1974, pp. 508–517.

[52] L. QIU AND E.J. DAVISON, "Performance Limitations of non-minimum phase systems in the servomechanism problem", Automatica, 29, 1993, pp. 337–349.

[53] A.C.M. RAN AND H.L. TRENTELMAN, "Linear quadratic problems with indefinite cost for discrete time systems", SIAM J. Matrix. Anal. & Appl., 14(3), 1993, pp. 776–797.

[54] H. H. ROSENBROCK, *State-space and multivariable theory*, John-Wiley, New York, 1970.

[55] M.A. ROTEA, "The generalized H_2 control problem", Automatica, 29(2), 1993, pp. 373–386.

[56] W. RUDIN, *Real and complex analysis*, McGraw-Hill, 3 Ed., 1987.

[57] A. SABERI, B. M. CHEN, AND P. SANNUTI, *Loop Transfer Recovery : Analysis and Design*, Springer Verlag, Berlin,, 1993.

[58] A. SABERI, Z. LIN, AND A.A. STOORVOGEL, "H_2 and H_∞ almost disturbance decoupling problem with internal stability", Int. J. Robust & Nonlinear Control, 6(8), 1996, pp. 789–803.

[59] A. SABERI AND P. SANNUTI, "Cheap and singular controls for linear quadratic regulators", IEEE Trans. Aut. Contr., 32, 1987, pp. 208–219.

[60] ——, "Squaring down of non-strictly proper systems", Int. J. Contr., 51(3), 1990, pp. 621–629.

[61] A. SABERI, P. SANNUTI, AND B. M. CHEN, *H_2 Optimal Control*, Prentice Hall, Englewood Cliffs, 1995.

[62] A. SABERI, A.A. STOORVOGEL, AND P. SANNUTI, "On output regulation for linear systems", Submitted for publication, 1999.

[63] ——, "On output regulation for linear systems", in Proc. 38th CDC, Phoenix, AZ, 1999. To appear.

[64] A. SABERI, A. STOORVOGEL, P. SANNUTI, AND G. SHI, "On optimal output regulation for linear systems", Submitted for publication, 1999.

[65] ALI SABERI, ANTON A. STOORVOGEL, AND ZONGLI LIN, "Generalized output regulation for linear systems", in Proc. American Control Conference, Albuquerque, NM, 1997, pp. 3909–3914.

[66] P. SANNUTI AND A. SABERI, "Special coordinate basis for multivariable linear systems – finite and infinite zero structure, squaring down and decoupling", Int. J. Contr., 45(5), 1987, pp. 1655–1704.

[67] L.M. SILVERMAN AND H.J. PAYNE, "Input-output structure of linear systems with application to the decoupling problem", SIAM J. Contr., 9(2), 1971, pp. 199–233.

[68] E.D. SONTAG AND H.J. SUSSMANN, "Nonlinear output feedback design for linear systems with saturating controls", in Proc. 29th CDC, Honolulu, 1990, pp. 3414–3416.

[69] A.G. SPARKS AND D.S. BERNSTEIN, "Asymptotic regulation with H_2 disturbance rejection", in Proc. 33rd CDC, Lake Buena Vista, FL, 1994, pp. 3614–3615.

[70] A.A. STOORVOGEL, *The H_∞ control problem: a state space approach*, Prentice-Hall, Englewood Cliffs, 1992.

[71] ——, "The singular H_2 control problem", Automatica, 28(3), 1992, pp. 627–631.

[72] ——, "Nonlinear L_1 optimal controllers for linear systems", IEEE Trans. Aut. Contr., 40(4), 1995, pp. 694–696.

[73] ——, "The H_∞ control problem with zeros on the boundary of the stability domain", Int. J. Contr., 32(1), 1996, pp. 29–47.

[74] A.A. STOORVOGEL AND A. SABERI, "The discrete algebraic Riccati equation and linear matrix inequality", Lin. Alg. Appl., 274, 1998, pp. 317–365.

[75] ——, "Output regulation of linear plants with actuators subject to amplitude and rate constraints", Int. J. Robust & Nonlinear Control, 9(10), 1999, pp. 631–657.

[76] A.A. STOORVOGEL, A. SABERI, AND B.M. CHEN, "The discrete-time H_∞ control problem with strictly proper measurement feedback", IEEE Trans. Aut. Contr., 39(9), 1994, pp. 1936–1939.

[77] A.A. STOORVOGEL, A. SABERI, AND P. SANNUTI, "Performance with regulation constraints", Available from ftp://ftp.win.tue.nl/pub/techreports/wscoas/regulation.ps, 1998.

[78] A.A. STOORVOGEL AND J.W. VAN DER WOUDE, "The disturbance decoupling problem with measurement feedback and stability including direct-feedthrough matrices", Syst. & Contr. Letters, 17(3), 1991, pp. 217–226.

[79] N. SURESHBABU AND W.J. RUGH, "Output regulation with derivative information", IEEE Trans. Aut. Contr., 40(10), 1995, pp. 1755–1766.

[80] H.J. SUSSMANN, E.D. SONTAG, AND Y. YANG, "A general result on the stabilization of linear systems using bounded controls", IEEE Trans. Aut. Contr., 39(12), 1994, pp. 2411–2425.

[81] H.J. SUSSMANN AND Y. YANG, "On the stabilizability of multiple integrators by means of bounded feedback controls", in Proc. 30th CDC, Brighton, U.K., 1991, pp. 70–72.

[82] J.J. SYLVESTER, "Sur l'équation en matrices $px = xq$", Comptes Rendus XCIX, 115/116, 1884, pp. 67–71.

[83] A.R. TEEL, *Feedback stabilization: nonlinear solutions to inherently nonlinear problems*, PhD thesis, Electronics Research Laboratory, College of Engineering, University of California, 1992.

[84] ——, "Global stabilization and restricted tracking for multiple integrators with bounded controls", Syst. & Contr. Letters, 18(3), 1992, pp. 165–171.

[85] H.L. TRENTELMAN, *Almost invariant subspaces and high gain feedback*, vol. 29 of CWI Tracts, Amsterdam, 1986.

[86] H.L. TRENTELMAN AND A.A. STOORVOGEL, "Sampled-data and discrete-time H_2 optimal control", SIAM J. Contr. & Opt., 33(3), 1995, pp. 834–862.

[87] H.L. TRENTELMAN, A.A. STOORVOGEL, AND M.L.J. HAUTUS, "Linear multivariable systems", To appear with Springer Verlag.

[88] S. WEILAND AND J.C. WILLEMS, "Almost disturbance decoupling with internal stability", IEEE Trans. Aut. Contr., 32, 1989, pp. 277–286.

[89] J.C. WILLEMS, "Almost invariant subspaces: an approach to high gain feedback design– Part I: almost controlled invariant subspaces", IEEE Trans. Aut. Contr., 26, 1981, pp. 235–252.

[90] J.C. WILLEMS AND C. COMMAULT, "Disturbance decoupling by measurement feedback with stability or pole-placement", SIAM J. Contr. & Opt., 19(4), 1981, pp. 490–504.

[91] W.M. WONHAM, *Linear multivariable control: a geometric approach*, Springer Verlag, New York, Third Ed., 1985.

[92] Q.H. WU AND M. MANSOUR, "Robust output regulation for a class of linear mulivariable systems", Syst. & Contr. Letters, 13, 1989, pp. 227–232.

[93] Y. YANG, *Global stabilization of linear systems with bounded feedback*, PhD thesis, Rutgers University, New Brunswick, 1993.

[94] ——, "Private Communications", Department of Mathematics, Rutgers University, 1993.

[95] K. ZHOU, J.C. DOYLE, AND K. GLOVER, *Robust and optimal control*, Prentice Hall, Upper Saddle River, NJ, 1996.

Index

MIX
Papier aus verantwortungsvollen Quellen
Paper from responsible sources
FSC® C105338

If you have any concerns about our products,
you can contact us on
ProductSafety@springernature.com

In case Publisher is established outside the EU,
the EU authorized representative is:
Springer Nature Customer Service Center GmbH
Europaplatz 3, 69115 Heidelberg, Germany

Printed by Libri Plureos GmbH
in Hamburg, Germany